# Methods in Enzymology

## Volume 67
## VITAMINS AND COENZYMES
### Part F

# METHODS IN ENZYMOLOGY

EDITORS-IN-CHIEF

Sidney P. Colowick     Nathan O. Kaplan

*Methods in Enzymology*

*Volume 67*

# Vitamins and Coenzymes

*Part F*

EDITED BY

## Donald B. McCormick

DEPARTMENT OF BIOCHEMISTRY
EMORY UNIVERSITY SCHOOL OF MEDICINE
ATLANTA, GEORGIA

## Lemuel D. Wright

DIVISION OF NUTRITIONAL SCIENCES
SECTION OF BIOCHEMISTRY, MOLECULAR AND CELL BIOLOGY
CORNELL UNIVERSITY
ITHACA, NEW YORK

1980

ACADEMIC PRESS

*A Subsidiary of Harcourt Brace Jovanovich, Publishers*

New York   London   Toronto   Sydney   San Francisco

ACADEMIC PRESS, INC.
111 Fifth Avenue, New York, New York 10003

*United Kingdom Edition published by*
ACADEMIC PRESS, INC. (LONDON) LTD.
24/28 Oval Road, London NW1 7DX

LIBRARY OF CONGRESS CATALOG CARD NUMBER: 54–9110

ISBN 0–12–181967–1

PRINTED IN THE UNITED STATES OF AMERICA

80 81 82 83    9 8 7 6 5 4 3 2 1

# Table of Contents

## Section I. Cobalamins and Cobamides (B$_{12}$)

# Section II.  Ubiquinone Group

# Section III.  Tocopherols

# Section IV.  Vitamin K Group

# Section V.  Vitamin A Group

## Section VI. Vitamin D Group

## Section VII. Miscellaneous Vitamins and Coenzymes

# Contributors to Volume 67

Article numbers are in parentheses following the names of contributors.
Affiliations listed are current.

J. J. AARON (19), *Department of Chemistry, Faculté des Sciences, Université de Dakar, Dakar-Fann, Senegal*

WILLIAM E. BALCH (65), *Department of Biochemistry, Stanford University School of Medicine, Stanford, California 94305*

R. F. BAYFIELD (24), *Department of Agriculture, Veterinary Research Station, Glenfield N.S.W. 2167, Australia*

WILLIAM S. BECK (7), *Harvard University, Massachusetts General Hospital, Boston, Massachusetts 02114*

JUNE E. BISHOP (48), *Department of Biochemistry, University of California, Riverside, California 92521*

INGEMAR BJÖRKHEM (45), *Department of Clinical Chemistry, Huddinge Hospital, Huddinge, Sweden*

S. L. BONTING (27, 36), *Department of Biochemistry, University of Nijmegen, 6500 HB Nijmegen, The Netherlands*

LUCILLE BREDBERG (35), *Department of Ophthalmology, University of Washington School of Medicine, Seattle, Washington 98195*

GEORGE BRITTON (31), *Department of Biochemistry, University of Liverpool, Liverpool L69 3EX, England*

DAVID J. BROWN (31), *Research Department, The Boots Company Ltd., Nottingham, England*

L. M. CANFIELD (23), *Department of Biochemistry, College of Agricultural and Life Sciences, University of Wisconsin-Madison, Madison, Wisconsin 53706*

GEORGE L. CATIGNANI (15), *Department of Food Science, North Carolina State University, Raleigh, North Carolina 27650*

JOHN S. CHANDLER (63, 64), *Department of Biochemistry, University of Arizona, Health Sciences Center, Tucson, Arizona 85724*

SYLVIA CHRISTAKOS (60), *Department of Biochemistry, University of California, Riverside, California 92521*

FRANK CHYTIL (34), *Department of Biochemistry, Vanderbilt University School of Medicine, Nashville, Tennessee 37232*

J. W. COFFEY (10), *Department of Pharmacology, Hoffmann-La Roche, Inc., Nutley, New Jersey 07110*

E. R. COLE (24), *Department of Applied Organic Chemistry, University of New South Wales, Kensington N.S.W. 2033, Australia*

BERNARD A. COOPER (9), *Division of Hematology, Royal Victoria Hospital, McGill University, Montreal H3A 1A1, Canada*

A. A. M. CRUYL (39), *Laboratoria voor Medische Biochemie en voor Klinische Analyse, Rijksuniversiteit Gent, 135 De Pintelaan, B-9000 Gent, Belgium*

VLADIMÍR DADÁK (14, 17), *Department of Biochemistry, University of J. E. Purkyně, 611 37 Brno, Czechoslovakia*

F. J. M. DAEMEN (27, 36), *Department of Biochemistry, University of Nijmegen, 6500 HB Nijmegen, The Netherlands*

W. J. DE GRIP (36), *Department of Biochemistry, University of Nijmegen, 6500 HB Nijmegen, The Netherlands*

A. P. DE LEENHEER (39), *Laboratoria voor Medische Biochemie en voor Klinische Analyse, Rijksuniversiteit Gent, 135 De Pintelaan, B-9000 Gent, Belgium*

HECTOR F. DELUCA (28, 44, 46, 50), *Department of Biochemistry, University of Wisconsin-Madison, Madison, Wisconsin 53706*

D. O. EDLUND (40), *Ames Quality Assurance, Miles Laboratories, Inc., Elkhart, Indiana 46514*

GRAHAM R. ELLIOTT (21), *British Industrial Biological Research Association, Carshalton, Surrey SM5 4DS, England*

EUGENE P. FRENKEL (6), *Department of Internal Medicine, University of Texas Health Science Center at Dallas, South-*

western Medical School, Dallas, Texas 75235

ERNEST J. FRIEDLANDER (61), Department of Chemistry, Harvard University, Cambridge, Massachusetts 02138

SABURO FUKUI (8), Department of Industrial Chemistry, Faculty of Engineering, Kyoto University, Sakyo-Ku, Kyoto 606, Japan

SIDNEY FUTTERMAN* (35), Department of Ophthalmology, University of Washington School of Medicine, Seattle, Washington 98195

J. GLOVER (33), Department of Biochemistry, University of Liverpool, Liverpool L69 3BX, England

TREVOR W. GOODWIN (31), Department of Biochemistry, University of Liverpool, Liverpool L69 3BX, England

PAMELA D. GREEN (12), Hematology Research Laboratory, Veterans Administration Medical Center, Albany, New York 12208

RALPH GREEN (13), Department of Clinical Research, Scripps Clinic and Research Foundation, La Jolla, California 92037

G. W. T. GROENENDIJK (27), Department of Biochemistry, University of Nijmegen, 6500 HB Nijmegen, The Netherlands

S. GRYS (25), Laboratory of Veterinary Hygiene, 02-156 Warsaw, Poland

JOHN G. HADDAD, JR. (53), Department of Medicine, The Jewish Hospital of St. Louis, Washington University School of Medicine, St. Louis, Missouri 63110

LAURA A. HAGAN (63, 64), Department of Biochemistry, University of Arizona, Health Sciences Center, Tucson, Arizona 85724

CHARLES A. HALL (12), Hematology Research Laboratory, Veterans Administration Medical Center, Albany, New York 12208

KIYOZO HASEGAWA (30), Department of Food Science and Nutrition, Nara Women's University, Nara 630, Japan

SAAD S. M. HASSAN (16, 37, 66), Department of Chemistry, Faculty of Science, Ain Shams University, Cairo, Egypt

MARK R. HAUSSLER (62, 63, 64), Department of Biochemistry, University of

Arizona, Health Sciences Center, Tucson, Arizona 85724

HELEN L. HENRY (52), Department of Biochemistry, University of California, Riverside, California 92521

INGER HOLMBERG (45), Department of Clinical Chemistry, Huddinge Hospital, Huddinge, Sweden

MIYOSHI IKAWA (29), Department of Biochemistry, University of New Hampshire, Durham, New Hampshire 03824

NOBUO IKEKAWA (44), Laboratory of Chemistry for Natural Products, Tokyo Institute of Technology, Nagatsuta, Midori-Ku, Yokohama, Japan

D. W. JACOBSEN (3), Department of Biochemistry, Scripps Clinic and Research Foundation, La Jolla, California 92037

P. A. A. JANSEN (27), Department of Biochemistry, University of Nijmegen, 6500 HB Nijmegen, The Netherlands

B. CONNOR JOHNSON (22), Oklahoma Medical Research Foundation, and Department of Biochemistry and Molecular Biology, University of Oklahoma Health Sciences Center, Oklahoma City, Oklahoma 73104

MAX KATZ (9), Division of Hematology, Royal Victoria Hospital, McGill University, Montreal H3A 1A1, Canada

RICHARD L. KITCHENS (6), Department of Internal Medicine, University of Texas Health Science Center at Dallas, Southwestern Medical School, Dallas, Texas 75235

TADASHI KOBAYASHI (41), Department of Hygienic Chemistry, Kobe Women's College of Pharmacy, Kobe 658, Japan

K. T. KOSHY (43), Biochemistry and Residue Analysis, The Upjohn Company, Kalamazoo, Michigan 49001

LUDMILA KŘIVÁNKOVÁ (14, 17), Department of Biochemistry, University of J. E. Purkyně, 611 37 Brno, Czechoslovakia

SUDHIR KUMAR (11), Perinatal Medicine, Christ Hospital, Oak Lawn, Illinois 60453, and Departments of Neurological Sciences and Biochemistry, Rush Medical College, Chicago, Illinois 60612

D. E. M. LAWSON (54, 57), Medical Re-

* Deceased.

search Council, Dunn Nutritional Laboratory, Cambridge CB4 1XJ, England

S. L. LEE (1), Department of Chemistry, Texas A&M University, College Station, Texas 77843

I. YA. LEVITIN (4), Institute of Organo-Element Compounds, Academy of Sciences of the USSR, Moscow B-334, USSR

ANNE M. MCCORMICK (28), Department of Biochemistry, University of Texas Health Science Center, Dallas, Texas 75235

LEO M. MEYER (11), Sickle Cell Program, Veterans Administration Hospital, St. Albans, New York 11425

YASUHIRO NAKATA (20), Eisai Kawashima Plants, Eisai Company Ltd., 1 Takehaya, Kawashima, Hajima, Gifu, Japan

JOSEPH L. NAPOLI (28), Department of Biochemistry, University of Texas Health Science Center, Dallas, Texas 75235

ANTHONY W. NORMAN (48, 49, 56, 58, 59, 60, 61), Department of Biochemistry, University of California, Riverside, California 92521

EDWARD M. ODAM (21), Pest Infestation Control Laboratory, Ministry of Agriculture, Fisheries and Food, Hook Rise South, Tolworth, Surbiton, Surrey KT6 7NF, England

LAWRENCE K. OLIVER (26), Bio-Science Laboratories, San Francisco Branch, Oakland, California 94621

DAVID ONG (34), Department of Biochemistry, Vanderbilt University School of Medicine, Nashville, Tennessee 37232

NARENDRA J. PATEL (31), Department of Biochemistry, University of Liverpool, Liverpool L69 3BX, England

PER A. PETERSON (32), Department of Cell Research, The Wallenberg Laboratories, University of Uppsala, Uppsala, Sweden

ROBERT F. PFEIFER (51), Life Science Division, Waters Associates, Milford, Massachusetts 01757

J. WESLEY PIKE (62, 63, 64), Department of Biochemistry, University of Arizona, Health Sciences Center, Tucson, Arizona 85724

RUSSELL PROUGH (6), Department of Biochemistry, University of Texas Health

Science Center at Dallas, Southwestern Medical School, Dallas, Texas 75235

L. PUUTULA-RÄSÄNEN (5), Research Department, Medix Laboratories, 00101 Helsinki 10, Finland

LARS RASK (32), Department of Cell Research, The Biomedical Center, University of Uppsala, Uppsala, Sweden

PATRICIA A. ROBERTS (56), Department of Biochemistry, University of California, Riverside, California 92521

MARTIN S. ROGINSKY (51), Department of Medicine, Nassau County Medical Center, East Meadow, New York 11554, and State University of New York at Stony Brook, Stony Brook, New York 11794

JAMES A. ROMESSER (65), DuPont Company, Central Research and Development Department, Experimental Station, Wilmington, Delaware 19898

FREDERICK P. ROSS (49), Department of Medical Biochemistry, University of the Witwatersrand Medical School, Johannesburg, South Africa 2001

JOHN C. SAARI (35), Department of Ophthalmology, University of Washington School of Medicine, Seattle, Washington 98195

YOSHIKI SEINO (55), Department of Pediatrics, Osaka University Medical School, Fukushima-Ku, Osaka 553, Japan

D. V. SHAH (23), Department of Biochemistry, College of Agricultural and Life Sciences, University of Wisconsin-Madison, Madison, Wisconsin 53706

RICHARD M. SHEPARD (46), Pfizer Incorporated, Groton, Connecticut 06340

TSUNESUKE SHIMOTSUJI (55), Department of Pediatrics, Osaka University Medical School, Fukushima-Ku, Osaka 553, Japan

D. SKLAN (42), Faculty of Agriculture, Hebrew University, Rehovot, Israel

PILL-SOON SONG (2), Department of Chemistry, Texas Tech University, Lubbock, Texas 79409

U.-H. STENMAN (5), Minerva Foundation Institute for Medical Research, 00101 Helsinki 10, Finland

GENE W. STUBBS (35), Department of Ophthalmology, University of Washington School of Medicine, Seattle, Washington 98195

SOREL SULIMOVICI (51), Department of Pathology, Nassau County Medical Center, East Meadow, New York 11554, and State University of New York at Stony Brook, Stony Brook, New York 11794

J. W. SUTTIE (23), Department of Biochemistry, College of Agricultural and Life Sciences, University of Wisconsin-Madison, Madison, Wisconsin 53706

K. TAKAMURA (18), Tokyo College of Pharmacy, 1432-1 Horinouchi, Hachioji, Tokyo 192-03, Japan

YOKO TANAKA (44), Department of Biochemistry, University of Wisconsin-Madison, Madison, Wisconsin 53706

CAROL M. TAYLOR (47), Department of Pathophysiology, University of Berne, 3010 Berne, Switzerland

RICHARD F. TAYLOR (29), BioMolecular Sciences Section, Arthur D. Little, Inc., Cambridge, Massachusetts 02140

TETSUO TORAYA (8), Department of Industrial Chemistry, Faculty of Engineering, Kyoto University, Sakyo-Ku, Kyoto 606, Japan

MICHAEL G. TOWNSEND (21), Pest Infestation Control Laboratory, Ministry of Agriculture, Fisheries and Food, Hook Rise South, Tolworth, Surbiton, Surrey KT6 7NF, England

ETSUO TSUCHIDA (20), Eisai Honjo Plants, Eisai Company Ltd, 2-3-14 Minami Honjo City, Saitama, Japan

KIYOSHI TSUKIDA (38), Kobe Women's College of Pharmacy, Kobe 658, Japan

F. WATANABE (18), Tokyo College of Pharmacy, 1432-1 Horinouchi, Hachioji, Tokyo 192-03, Japan

WAYNE R. WECKSLER (58, 59), Research Department, Bio-Science Laboratories, Van Nuys, California 91405

POKSYN S. YOON (50), Department of Biochemistry, University of Connecticut, Farmington, Connecticut 06032

# Preface

Since 1970–1971, when the earlier volumes (XVIII, A, B, and C) on "Vitamins and Coenzymes" were published as part of the *Methods in Enzymology* series, there has been a considerable expansion of techniques and methodology attendant to the assay, isolation, and characterization of the vitamins and those systems responsible for their biosynthesis, transport, and metabolism. In part, this has been generated by an increasing awareness of the diversity of such vitaminic forms as comprise essential moieties of coenzymes and also through recognition of the function of some derived metabolites as hormones, regulators, and even antioxidants.

As a consequence of this new body of information and its expected impact in the stimulation of further research on vitamins and coenzymes, we have sought to provide investigators with the more current modifications of "tried and true" methods as well as those which have only now become available. Volume 67 is the third of three volumes resulting from our efforts in soliciting contributions from numerous, active experimentalists who have published most of their findings in the usual, refereed research journals. The amount of material which appeared to warrant coverage necessitated a division into three parts, each comprising a volume: 62, Part D, which covers the vitamin and coenzyme forms of ascorbate, thiamine, lipoate, pantothenate, biotin, and pyridoxine; 66, Part E, nicotinate, flavins, and pteridines; and 67, Part F, the $B_{12}$ group and those classically considered as "fat-soluble."

We should like to express our gratitude to the contributors for their willingness to supply the information requested and, in some instances, their tolerance of editorial emendations. There has been an attempt to allow such overlap as would offer flexibility in the choice of method, such as modification of an assay procedure. Where some omissions seemingly occur, these may, in some cases, be attributed to the inadvertent oversight of the editors; however, in other cases, it was felt that the topics were adequately covered in the earlier volumes on this subject or in other volumes in the *Methods in Enzymology* series.

Finally, we again wish to thank Mrs. Patricia MacIntyre for her excellent secretarial assistance and the numerous persons at Academic Press for their efficient and kind guidance.

DONALD B. MCCORMICK
LEMUEL D. WRIGHT

# METHODS IN ENZYMOLOGY

EDITED BY

## Sidney P. Colowick and Nathan O. Kaplan

VANDERBILT UNIVERSITY
SCHOOL OF MEDICINE
NASHVILLE, TENNESSEE

DEPARTMENT OF CHEMISTRY
UNIVERSITY OF CALIFORNIA
AT SAN DIEGO
LA JOLLA, CALIFORNIA

# METHODS IN ENZYMOLOGY

EDITORS-IN-CHIEF

## Sidney P. Colowick     Nathan O. Kaplan

VOLUME XIX. Proteolytic Enzymes
*Edited by* GERTRUDE E. PERLMANN AND LASZLO LORAND

VOLUME XX. Nucleic Acids and Protein Synthesis (Part C)
*Edited by* KIVIE MOLDAVE AND LAWRENCE GROSSMAN

VOLUME XXI. Nucleic Acids (Part D)
*Edited by* LAWRENCE GROSSMAN AND KIVIE MOLDAVE

VOLUME XXII. Enzyme Purification and Related Techniques
*Edited by* WILLIAM B. JAKOBY

VOLUME XXIII. Photosynthesis (Part A)
*Edited by* ANTHONY SAN PIETRO

VOLUME XXIV. Photosynthesis and Nitrogen Fixation (Part B)
*Edited by* ANTHONY SAN PIETRO

VOLUME XXV. Enzyme Structure (Part B)
*Edited by* C. H. W. HIRS AND SERGE N. TIMASHEFF

VOLUME XXVI. Enzyme Structure (Part C)
*Edited by* C. H. W. HIRS AND SERGE N. TIMASHEFF

VOLUME XXVII. Enzyme Structure (Part D)
*Edited by* C. H. W. HIRS AND SERGE N. TIMASHEFF

VOLUME XXVIII. Complex Carbohydrates (Part B)
*Edited by* VICTOR GINSBURG

VOLUME XXIX. Nucleic Acids and Protein Synthesis (Part E)
*Edited by* LAWRENCE GROSSMAN AND KIVIE MOLDAVE

VOLUME XXX. Nucleic Acids and Protein Synthesis (Part F)
*Edited by* KIVIE MOLDAVE AND LAWRENCE GROSSMAN

VOLUME XXXI. Biomembranes (Part A)
*Edited by* SIDNEY FLEISCHER AND LESTER PACKER

VOLUME XXXII. Biomembranes (Part B)
*Edited by* SIDNEY FLEISCHER AND LESTER PACKER

# Methods in Enzymology

## Volume 67
## VITAMINS AND COENZYMES
### Part F

Section I

# Cobalamins and Cobamides ($B_{12}$)

## [1] Separation of Cobyrinic Acid and Its Biosynthetic Precursors by Ion-Exchange Paper Chromatography

### By S. L. LEE

Cobyrinic acid (IV) and its biosynthetic precursors, δ-aminolevulinic acid (I), uroporphyrinogen III (II), heptacarboxylic urophorphyrinogen III (III), and S-adenosylmethionine can be rapidly separated by ion-exchange paper chromatography. This method[1] can serve as a relatively fast assay compared to the thin-layer chromatography of cobyrinic acid heptamethyl ester[2] and phenol extraction of the free acid.[2] The last two methods also suffer from incomplete recovery of cobyrinic acid.

Principle

This method is based on the fact that all the precursors of cobyrinic acid contain nitrogen atoms that can be protonated and retained by a

[1] S. L. Lee and A. I. Scott, *Anal. Biochem.* **74**, 641 (1976).
[2] A. I. Scott, B. Yagen, and E. Lee, *J. Am. Chem. Soc.* **95**, 5761 (1973).

weak, cation-exchange paper. Cobyrinic acid, which has a positive charge on the cobalt atom, exists as the diaquo form in acidic pH. However, the positive charge is prevented from interacting with the functional groups on the paper by the water ligands, resulting in a rapid elution of cobyrinic acid.

## Method

### Reagents and Materials

δ-[5-[14]C]Aminolevulinic acid (New England Nuclear)
S-Adenosyl[methyl-[14]C]methionine (New England Nuclear)
Cobyrinic acid
Uroporphyrin[3]
Heptacarboxylic uroporphyrin[3]
Whatman CM-82 paper

The precursors and cobyrinic acid (0.1 μmol per 20 μl of each, 50,000 cpm in the case of radioactive substrates) are spotted on a 30-cm-long strip of Whatman CM-82 paper, 3 cm apart from each other. A mixture of all compounds is also spotted for comparison. Descending elution of the compounds is carried out with 2% acetic acid. Migration of 10 cm takes about 40 min. S-Adenosylmethionine and δ-aminolevulinic acid can be detected by radioscanning. The porphyrins and cobyrinic acid are identified by their brown and pink color, respectively.

## Results

All the porphyrins are bound tightly to the origin of the chromatogram. S-Adenosylmethionine and δ-aminolevulinic acid are eluted about halfway on the paper whereas cobyrinic acid migrates to the solvent front (see the table).

$R_f$ VALUES OF DIAQUOCOBYRINIC ACID AND ITS PRECURSORS

| Compound | $R_f$ |
| --- | --- |
| Uroporphyrin III | 0.0 |
| Heptacarboxylic uroporphyrin III | 0.0 |
| S-Adenosylmethionine | 0.26 |
| δ-Aminolevulinic acid | 0.57 |
| Diaquocobyrinic acid | 0.98 |

[3] Uroporphyrinogen III and heptacarboxylic uroporphyrinogen III are readily oxidized in air to the fully conjugated uroporphyrin III and heptacarboxylic uroporphyrin III, respectively. For analytical purposes, only the oxidized forms are relevant.

Note

This method can be adapted to assay of cobyrinic acid formation in cell-free systems of *Propionibacterium shermanii*.[4] Incubation products have to be adjusted to pH 3–4 before application to the paper. It might be necessary to deproteinize the incubation products to minimize saturation of the paper. That separation of δ-aminolevulinic acid is less satisfactory compared with the others may be due to the formation of lactone under this condition. Of other cation exchangers tried, phosphocellulose P-81 (Whatman) is similar but with less resolution. Strong cation-exchangers, such as Dowex 50, cannot be used because they retain cobyrinic acid. The assay is suitable only for the formation of cobalt-inserted corrin systems.

Acknowledgment

The author is grateful to Dr. A. I. Scott for his encouragement and reference samples.

[4] See Scott *et al.*[2] and for reviews see A. I. Scott, *Tetrahedron* **31**, 2639 (1975).

# [2] Spectroscopic Analysis of Vitamin B$_{12}$ Derivatives

*By* PILL-SOON SONG

This contribution is intended briefly to describe general features of the absorption spectra of vitamin B$_{12}$ derivatives. The spectroscopic properties of these compounds can then be monitored to deduce structural and environmental perturbations on the corrinoid coenzyme structure and function. Both circular dichroism (CD) and fluorescence polarization spectroscopy are useful for the assignment of various absorption bands of B$_{12}$ derivatives (e.g., descobalt B$_{12}$). A number of absorption spectra of B$_{12}$ derivatives have been recorded previously.[1–3]

Spectroscopic Methods

*Absorption Spectra.* To further characterize room temperature spectra, recording of low temperature spectra in optical dewars (one each for reference and sample compartments of a Cary 118 C spectrophotometer) is particularly useful for vitamin B$_{12}$ derivatives because of markedly

[1] J. A. Hill, J. M. Pratt, and R. J. P. Williams, *J. Chem. Soc. (London)*, p. 5149 (1964).
[2] H. A. O. Hill, J. M. Pratt, and R. J. P. Williams, *Proc. R. Soc. London Ser. A* **288**, 352 (1965).
[3] J. M. Pratt and R. G. Thorp, *J. Chem. Soc. A*, p. 187 (1966).

enhanced spectral resolution. Although rectangular cuvettes present minimum optical artifacts, such as reflection compared to cylindrical cells, the former tend to crack owing to strains from freezing at low temperatures (e.g., liquid N$_2$ temperature). It is possible to align both reference and sample cylindrical cells manually in optical dewars for satisfactory low temperature absorption measurements. It should be noted that the spectral shift and "enhanced" extinction coefficient are often due to the increased refractive index of the medium at the lower temperature. Volume contraction or expansion should also be taken into consideration in calculating molar extinction coefficients at low temperature.

*Circular Dichroism Spectra.* Optically active vitamin B$_{12}$ derivatives show relatively strong CD spectra. Because separate electronic transitions possess characteristic CD signs and rotational strengths, CD spectra are also helpful in analyzing the absorption spectra of vitamin B$_{12}$ derivatives.

*Fluorescence Polarization Spectra.* The absolute polarization directions of each absorbing transition in an absorption spectrum can be determined by polarized single-crystal absorption and reflectance spectroscopy. In the absence of single crystals of vitamin B$_{12}$, relative polarization directions of a fluorescent derivative can be deduced from the fluorescence polarization spectrum under photoselection conditions. Molecular orbital (MO) calculations are also complementary. The recommended conditions for these measurements are (*a*) high viscosity (e.g., glycerol) or low temperature to minimize Brownian rotations; (*b*) low absorbance (ideally less than 0.05, practically less than 0.3) to minimize self-absorption and depolarization due to energy transfer; (*c*) narrow bandpass for excitation and emission; and (*d*) correction for optical artifacts.[4]

### Spectroscopic Assignments

Vitamin B$_{12}$ and descobalt B$_{12}$ exhibit electronic spectra characteristic of the corrin chromophore (Fig. 1). Although the corrin structure is not a fully cyclic conjugated system (Fig. 2), its 14 $\pi$-electrons system can be qualitatively treated like a chlorin- or chlorophyll-like triangular field. The fact that the absorption spectra of corrinoid compounds are generally similar to those of porphyrins and chlorophylls with characteristic visible and "Soret" bands supports the analogy made here. In spite of the oversimplification of the $\pi$-electron framework and the implied neglect of the noncoplanarity of the corrin ring in the preceding analogy, the spectral analogy between corrins and porphyrins (particularly chlorins) is useful.

Corrins shown two visible bands designated as $\alpha$ and $\beta$ bands, along

[4] T. Azumi and S. P. McGlynn, *J. Chem. Phys.* **37**, 2413 (1962).

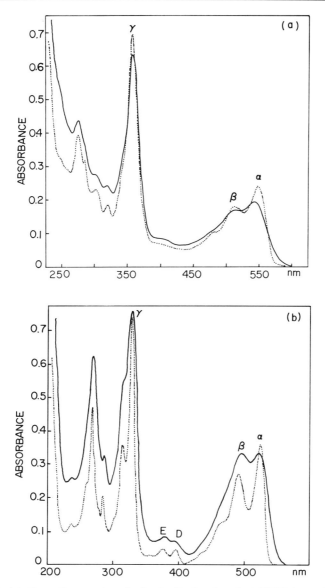

FIG. 1. (a) Absorption spectra of vitamin $B_{12}$ in ethanol at 298°K (————) and 77°K
(·········). (b) Absorption spectra of descobalt $B_{12}$ in ethanol at 298°K (————) and 77°K
(·········).

with the intense near-UV $\gamma$ band ($\simeq$ Soret) and moderately intense UV
bands. The $\alpha$ and $\beta$ bands are the 0-0 and 1-0 components of the first
allowed electronic transition. Kuhn was the first to describe the corrin

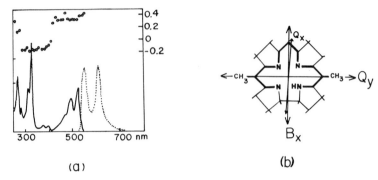

(a)                                                    (b)

FIG. 2. (a) Absorbance (———) and fluorescence excitation polarization (0; ordinate scale on the right) spectra of descobalt $B_{12}$ ($\lambda_F$ = 609 nm) in ethanol at 77°K. The uncorrected fluorescence spectrum (· · · · · ·) is also shown. (b) Predicted polarization directions [R. D. Fugate, C. A. Chin, and P.-S. Song, *Biochim. Biophys. Acta* **421**, 1 (1976)] for $\alpha(Q_y)$, $\beta(Q_x)$, and $\gamma(B_x)$ bands of descobalt $B_{12}$.

spectrum in terms of a free-electron model.[5] The four-orbital model equivalent to the Gouterman model for porphyrins[6] was proposed by Day[7] within the framework of the Hückel molecular orbital (HMO) and Pariser–Parr–Pople (PPP) MO approximations. The latter was also applied to corrins by others.[8,9] The HMO method was found to be inadequate for describing the spectral intensities of corrins. On the basis of these results, it is now well established that the visible and near-UV band intensities in corrins are mainly of the $\pi \rightarrow \pi^*$ type. This conclusion is also valid for vitamin $B_{12}$, since "free base" corrins (e.g., descobalt $B_{12}$ in Fig. 1) show the characteristic visible and near-UV spectral intensities. The $\pi \rightarrow \pi^*$ assignment for the entire UV-visible spectrum of free base corrin is clearly confirmed by the polarized phosphorescence excitation spectrum of descobalt $B_{12}$, which shows $P_0 < 0.1$ for all visible and UV peaks with respect to the out-of-plane polarized phosphorescence of $^3\pi,\pi^*$ symmetry.[10]

The low-temperature absorption spectrum (Fig. 1) of descobalt $B_{12}$ in ethanol shows well-resolved peaks at 525 ($\alpha$), 493 ($\beta$), 465 (sh), 435 (sh), 395 (band D), 375 (band E), 358 (sh), 326 ($\gamma$), 314, 300 (sh), 285, 280 (sh), 268, 260 (sh), and 238 nm with peak ratios of 1 : 0.76 : 0.36 : 0.12 : 0.16 : 0.17 : 0.12 : 2.09 : 1.04 : 0.64 : 0.48 : 0.54 : 1.39 : 0.76 : 0.43.[10]

[5] R. Eckert and H. Kuhn, *Z. Elektrochem.* **64**, 356 (1960).
[6] M. Gouterman, *J. Mol. Spectrosc.* **6**, 136 (1961).
[7] P. Day, *Theor. Chim. Acta* **7**, 328 (1967).
[8] H. Johansen and L. L. Ingraham, *J. Theor. Biol.* **23**, 191 (1969).
[9] P. O. Offenhartz, B. H. Offenhartz, and M. M. Fung, *J. Am. Chem. Soc.* **92**, 2966 (1970).
[10] R. D. Fugate, C. A. Chin, and P.-S. Song, *Biochim. Biophys. Acta* **421**, 1 (1976).

In analogy to the porphyrin notation, we assign the 525-nm peak and its vibrational satellites at 493, 465, and 435 nm to $Q_y$ 0-0, 1-0, 2-0, and 3-0, respectively, and the peaks at 395, 375, and 358 nm are designated as $Q_x$ 0-0, 1-0, and 2-0, respectively. The subscripts x and y correspond to the polarization axes (cf. Fig. 2). The 326-nm maximum may then be designated as a $B_x$ equivalent. These designations are consistent with the relative polarization data discussed below.

The lowest energy transition, $Q_y$, arises mainly from the highest occupied to lowest empty [7(HOMO) $\rightarrow$ 8(LEMO)] configuration, while $Q_x$ and $B_x$ transitions are contributed to mainly by $6 \rightarrow 8$ and $7 \rightarrow 9$ configuration mixing.[8,9] The fourth ($S_4 \leftarrow S_0$) and fifth ($S_5 \leftarrow S_0$) transitions are mainly due to $7 \rightarrow 10$ and $6 \rightarrow 9$ configurations, respectively. The latter is polarized nearly parallel to the $Q_y$ axis in descobalt $B_{12}$, whereas the $S_4$ transition is polarized between the x and y axes.[10] The predicted polarization axes for $Q_y$ and $Q_x$ bands are y and x axis, respectively (Fig. 2). The fluorescence polarization spectrum of descobalt $B_{12}$ shows that $Q_x$ and $B_x$ transitions are polarized perpendicular to the $Q_y$ band,[10,11] reminiscent of the chlorophyll $a$ spectrum.

Although the linear dichroism of vitamin $B_{12}$ in stretched polyvinyl alcohol film is too small to allow one to deduce the absolute polarization axes of the visible and near-UV bands,[5] the negative dichroism at the near-UV band is indicative of the perpendicular polarization of the $B_x$ transition with respect to the $Q_y$ polarization direction. Both CD[10,12-14] and MCD[15] spectra also suggest this polarization picture. The absolute polarizations of the absorption spectrum of a single crystal $Ni^{2+}$-corrin (nirrin) have been studied by polarized reflection spectroscopy, showing that the $Q_y$ band is polarized along the y axis and the near-UV band is oppositely polarized.[16] Thus, the polarized reflection spectroscopy result is consistent with the MO predictions.[10]

### Structure–Spectra Correlation

$Co^{3+}$-free vitamin $B_{12}$ (descobalt $B_{12}$) causes a considerable blue shift of the entire absorption spectrum (Fig. 1). The symmetric axial ligands, as

[11] A. J. Thompson, *J. Am. Chem. Soc.* **91**, 2780 (1969).

[12] M. Legrand and R. Viennet, *Bull. Soc. Chim. Fr.* **29**, 1435 (1962).

[13] R. A. Firth, H. A. O. Hill, J. M. Pratt, R. J. P. Williams, and W. R. Jackson, *Biochemistry* **6**, 2178 (1967).

[14] R. Bonnett, J. M. Godfrey, V. B. Math, D. M. Scopes, and R. N. Thomas, *J. Chem. Soc. Perkin Trans.* **1**, 252 (1973).

[15] B. Briat and C. Djerassi, *Nature (London)* **217**, 918 (1968).

[16] B. G. Anex and G. J. Eckhardt, *Abstr. Symp. Mol. Str. Spectrosc.*, Columbus, Ohio (1966).

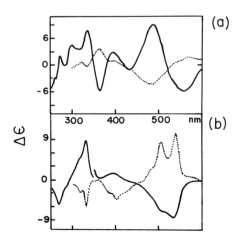

FIG. 3. (a) Circular dichroism (CD) spectra of coenzyme B$_{12}$ in methanol/ethanol (1 : 1) at 298°K (——) and 77°K (· · · ·). (b) CD spectra of descobalt B$_{12}$ in ethanol at 298°K and 77°K.

in dicyanocobalamin, show a red shift of the spectrum, compared to the absorption spectrum of vitamin B$_{12}$. The α-band maxima resolved from CD spectra at room temperature are 533 nm for descobalt B$_{12}$, 556 nm for vitamin B$_{12}$, 559 nm for coenzyme B$_{12}$, and 588 nm for dicyanocobalamin. The red shift in this series may reflect increase in the coplanarity of the corrin π-electron system and electron density on the cobalt due to the axial ligands.[7,13] From absorption and CD studies of a series of vitamin B$_{12}$ derivatives, a marked dependence of the absorption maxima and spectral shape on the electron-donating capacity of the axial ligand is found.[13] The absorption spectra of vitamin B$_{12}$ derivatives can be classified in terms of four types.[13]

The so-called "normal" spectrum is displayed by dicyanocobalamin (with electron-withdrawing ligands) and descobalt B$_{12}$, with well-resolved visible (α, β) and near-UV bands (D, E, and γ bands; see Fig. 1). The anomalous spectrum with a broad visible band (unresolved α and β bands) and structureless near-UV bands is caused by electron-donating ligands as in methylcobalamin in acid solution and coenzyme B$_{12}$ in ethanol. Two intermediate spectra resembling either of these spectra are exhibited by other derivatives. A complete documentation of the absorption spectra in this series is given elsewhere.[13]

Catalytically active coenzyme B$_{12}$ shows two opposite-sign CD bands (Fig. 3), one negative band at 559 nm assigned to α ($Q_y$, 0-0) and one positive band (β?) at 487 nm assignable to the $Q_y$, 0-1 transition. However, other derivatives, such as vitamin B$_{12}$ and dicyanocobalamin, do not show the CD splitting in the visible-band region. The unique chirality of the transition corresponding to the negative CD in coenzyme $B_{12}$ may indicate a strongly magnetic dipole-allowed transition associated with either the metal d-d transition or metal ligand charge transfer. Such a transition is then masked by the $\beta$-band in the absorption spectrum, which as a result is also broadened. CD sign inversion upon cooling to liquid $N_2$ is not related to the metal puckering or ligand inversion, since descobalt $B_{12}$ also shows the sign inversion (Fig. 3).[10]

The CD spectra of pentacoordinate $B_{12}$ $Co^{3+}$ derivatives and hexacoordinate $B_{12}$ $Co^{2+}$ derivatives are quite alike, and neither of them strongly resembles the spectra of hexacoordinate $B_{12}$ $Co^{3+}$ derivatives. Changing certain ligands, e.g., from $CN^-$ to $H_2O$, may cause a pronounced effect on the CD without significantly affecting the absorption spectrum. This is possible if the ligand substitution alters the degree of configuration interaction between $6 \rightarrow 8$ and $7 \rightarrow 9$ configurations (see above) without necessarily affecting the conformation of the corrin ring.

### Spectrophotometric Estimation of Vitamin $B_{12}$ Derivatives

Because of spectroscopic perturbations by factors such as the variable hydration of vitamin $B_{12}$ derivatives, the spectrophotometric estimation of these compounds in aqueous solutions is not achieved with high accuracy or reproducibility. Nonetheless, an extensive compilation[14] of the molar extinction coefficients for 40 vitamin $B_{12}$ derivatives should serve as a useful source of reference for the spectrophotometric analysis of these compounds.

### Acknowledgments

The work in this laboratory was supported by the Robert A. Welch Foundation (D-182) and technically assisted by Mr. Robert D. Fugate. Descobalt $B_{12}$ was supplied by Dr. John I. Toohey.

## [3] Preparation of Cryptofluorescent Analogs of Cobalamin Coenzymes[1]

### By D. W. JACOBSEN

### Preparation

*Principle.* Upper-axial ($\beta$-position) ligand analogs of methyl- and adenosylcobalamin[2] are synthesized from cob(I)alamin and alkyl halides containing fluorescent R groups [Eq. (1)]. Cob(I)alamin, which contains a cobalt atom in the $+1$ oxidation state, is prepared by the reduction of cyanocobalamin with powdered zinc in aqueous ammonium chloride.[3] Alkyl halide derivatives of fluorescent adenosine analogs (1,$N^6$-ethenoadenosine, formycin, 2,6-diaminonebularin and 2-aminonebularin) are conveniently prepared in high yield by direct halogenation at the $5'$ position with thionyl chloride in the presence of hexamethylphosphoramide.[4-8] The fluorescent R group dansyl (1-dimethylaminonaphthalene-5-sulfonyl) is coupled to cob(I)alamin after conversion to dansylamidopropyl chloride.

$$R-CH_2-X + \quad \overset{I}{Co} \quad \longrightarrow \quad \overset{\overset{\displaystyle R}{\overset{\displaystyle |}{\overset{\displaystyle CH_2}{|}}}}{Co} \quad + X^- \tag{1}$$

[1] The author acknowledges helpful discussions with his colleagues, especially Dr. F. M. Huennekens, in whose laboratory this work was conducted.

[2] A comprehensive survey of organocobaltic compounds, including the alkylcorrinoids, can be found in D. Dodd and M. D. Johnson, *Organometallic Chem. Rev.* **52**, 1 (1973).

[3] J. M. Pratt, "Inorganic Chemistry of Vitamin $B_{12}$," p. 197. Academic Press, New York, 1972.

[4] K. Kikugawa and M. Ichino, *Tetrahedron Lett.* **2**, 87 (1971).

[5] D. W. Jacobsen and R. J. Holland, *Fed. Proc., Fed. Am. Soc. Exp. Biol.* **33**, 1508 (1974).

[6] H. P. C. Hogenkamp, *Biochemistry* **13**, 2736 (1974).

[7] D. W. Jacobsen, P. M. DiGirolamo, and F. M. Huennekens, *Mol. Pharmacol.* **11**, 174 (1975).

[8] D. W. Jacobsen, R. J. Holland, Y. Montejano, and F. H. Huennekens, *J. Inorg. Biochem.* **10**, 53 (1979).

A unique feature of these cobalamin analogs is that, although the halogenated precursors are highly fluorescent, the products emit no detectable fluorescence when excited at the absorbance maxima of their respective R groups. Quenching of the $\beta$-ligand R group fluorescence occurs probably as a result of intramolecular energy transfer between the ligand and the nonfluorescent corrinoid.[8] However, cleavage of the carbon–cobalt bond and release of the $\beta$ ligand results in the reappearance of fluorescence. This cryptofluorescent property of cobalamin coenzyme analogs containing fluorescent R groups has been used to monitor their reactivity in model systems.[8]

*Reagents*

    Cyanocobalamin, crystalline (Sigma)
    1,$N^6$-Ethenoadenosine[9] (PL Biochemicals)
    Formycin (Calbiochem)
    2,6-Diaminonebularin (2,6-diaminopurine riboside; Vega Biochemicals)
    2-Aminonebularin (2-aminopurine riboside)[10]
    Dansyl chloride (Sigma)
    Phosphocellulose (Cellex P; Bio-Rad)
    Silica gel (SILICAR CC-7, 200–325 mesh; Mallinckrodt).
    SP-Sephadex (C-25-120; Sigma)
    Thionyl Chloride (J. T. Baker)
    Hexamethylphosphoramide (Aldrich)

Reagent grade solvents and other chemicals were obtained from commercial sources.

*Procedure*

*5'-Chloro-5'-deoxy-1,$N^6$-ethenoadenosine.* One gram of 1,$N^6$-ethenoadenosine (3.43 mmol) is dissolved at 4° in 10 ml of anhydrous hexamethylphosphoramide containing 1.6 ml of thionyl chloride. The solution is warmed to room temperature (23°) and stirred in a stoppered vessel for 12 hr. The reaction mixture is diluted to 250 ml with water, adjusted to pH 7.0 with 5.0 $N$ NaOH, and filtered through Whatman No. 1 paper. The filtrate is extracted 3 times with 100 ml of chloroform (which is discarded) and then concentrated to dryness on a rotary evaporator (bath temperature, 40°). The residue is extracted twice with 50 ml of warm ethanol (50°), and the combined extracts are concentrated to dryness by

[9] J. R. Barrio, J. A. Secrist, and N. J. Leonard, *Biochem. Biophys. Res. Commun.* **46**, 597 (1972).
[10] J. J. Fox, I. Wempen, A. Hampton, and I. L. Doerr, *J. Am. Chem. Soc.* **80**, 1669 (1958).

rotary evaporation at 40°. After solution of the residue in a minimum volume (approximately 25 ml) of ethanol, the product is precipitated by the addition of ether. Final purification is obtained via chromatography on silica gel: 0.25 g of precipitate is dissolved in 10 ml of $n$-butanol: water (84 : 16) and applied to a 2 × 20 cm column of silica gel. The column is eluted with the same solvent (aqueous butanol) in 10-ml fractions. Fractions 15 through 25 are combined and concentrated to dryness by rotary evaporation (40°) to produce an amorphous white powder (yield: 0.50 g, 1.48 mmol, 43%. Melting point, 191°. Thin-layer chromatography ($n$-butanol: water, 84 : 16) on cellulose, $R_f$ = 0.68 (fluorescent) and silica gel, $R_f$ = 0.62 (fluorescent). $\lambda_{max}$ (0.1 $M$ sodium phosphate, pH 7.0): 300 nm ($\epsilon_{mM}$, 3.03), 275 nm (5.86), 265 nm (5.98), and 227 nm (31.8); (0.10 $N$ HCl): 273 nm (9.51). *Anal.* Calculated for $C_{12}H_{12}N_5O_3Cl \cdot 1.5H_2O$: C, 42.80; H, 4.49; N, 20.79. Found: C, 42.65; H, 4.59; N, 18.42.

*5′-Chloro-5′-deoxyformycin.* Formycin (0.98 g, 3.66 mmol) is dissolved at 4° in 10 ml of anhydrous hexamethylphosphoramide containing 1.6 ml of thionyl chloride, warmed to 23°, and stirred for 17 hr. The solution is then diluted to 200 ml with water, adjusted to pH 3.5 with 5 $N$ NaOH, and concentrated to 50 ml by rotary evaporation (40°). The concentrate is applied to a 2.5 × 25 cm column of Dowex 50W-X4 (200–400 mesh, H$^+$ form). After washing with water (800 ml), the product is eluted with 0.5 $N$ NH$_4$OH. Fractions (15 ml) between 19 and 32 are combined and concentrated to approximately 50 ml. The product is crystallized at 0° (yield: 1.06 g, 3.48 mmol; 95%). Melting point, 210° (range 208–212° with yellowing). Thin-layer chromatography ($n$-butanol: water; 84 : 16) on cellulose, $R_f$ = 0.59 (blue fluorescence) and on silica gel, $R_f$ = 0.72 (blue fluorescence). $\lambda_{max}$ (0.1 $M$ sodium phosphate, pH 7.0) 293 nm ($\epsilon_{mM}$, 7.52); (0.1 $N$ HCl) 295 nm (7.48); (0.1 $N$ NaOH) 301 nm (5.93). *Anal.* Calculated for $C_{10}H_{12}N_5O_3Cl \cdot H_2O$: C, 39.55; H, 4.65; N, 23.06; Cl, 11.67. Found: C, 39.90; H, 4.73; N, 23.38; Cl, 12.64.

The 5′-chloro-5′-deoxy derivatives of 2-aminonebularin and 2,6-diaminonebularin are prepared as described above for 5′-chloro-5′-deoxyformycin with yields of 81 and 86%, respectively. The absorbance and fluorescence excitation and emission spectra of the 5′-chloro-5′-deoxynucleosides are similar to those reported for the parent nucleosides.[9–12]

*Dansylamidopropyl chloride.* 3-Chloropropylamine hydrochloride (1.30 g, 10.0 mmol) is dissolved at 0° in 25 ml of diethyl ether by the addition of 10 $N$ NaOH (approximately 1.2 ml). After stirring for 10 min and warming to 23°, 2.0 g of anhydrous sodium sulfate are added. The ether phase

[11] D. C. Ward, E. Reich, and L. Stryer, *J. Biol. Chem.* **244**, 1228 (1969).
[12] J. A. Secrist, J. R. Barrio, N. J. Leonard, and G. Weber, *Biochemistry* **11**, 3499 (1972).

containing 3-chloropropylamine is decanted, filtered, and concentrated to 10 ml by rotary evaporation at 25°. A suspension of dansyl chloride (1.35 g, 5.0 mmol) in 20 ml of anhydrous ether is added under argon to the 3-chloropropylamine solution over a 60-min period. Five milliliters of 0.1 $N$ HCl are added; after mixing for 10 min, the upper ether phase is decanted. The aqueous phase is extracted 3 times with 20-ml portions of ether, and the combined ether phases are concentrated by rotary evaporation to a yellow-green oil. The latter is dissolved in 28 ml of ethanol; after addition of 8 ml of water, the product is crystallized at 4°. The crystals are washed with 50% aqueous ethanol and dried *in vacuo* over $P_2O_5$ at 23° (yield: 1.30 g, 3.98 mmols, 79.5%). Melting point, 80°. Thin-layer chromatography ($n$-butanol : ethanol : $H_2O$; 250 : 80 : 175) on silica gel, $R_f = 0.90$ (yellow fluorescence). $\lambda_{max}$ (ethanol) 250 nm ($\epsilon_{mM}$, 15.4) and 334 nm (4.83). *Anal.* Calculated for $C_{15}H_{19}N_2O_2SCl$: C, 55.11; H, 5.87: N, 8.56; S, 9.82: Cl, 10.86. Found: C, 55.18; H, 5.87; N, 8.64; S, 9.84; Cl, 10.88.

*General Procedure for the Synthesis of Cobalamin Coenzyme Analogs.* Cyanocobalamin (0.50 g, 0.35 mmol)[13] is dissolved in 500 ml of water containing 50 g of ammonium chloride. After the solution is purged with argon for 20 min to remove dissolved oxygen, powdered zinc (25 g) is added, and the solution is stirred vigorously under argon for an additional 30 min at 23°. During this time, deep red cyanocob(III)alamin is initially reduced to the brown cob(II)alamin intermediate and then to the grayish-green cob(I)alamin. In order to protect the photolabile alkylcobalamin product, the next step is conducted under dim-red illumination. A 5-fold molar excess (relative to cyanocobalamin) of alkyl halide is added dropwise as an anaerobic solution (for 5'-chloronucleosides, use 50 ml of argon-purged water; for dansylamidopropyl chloride, use 50 ml of argon-purged ethanol) over a 10-min period. After mixing for an additional 50 min,[14] the reaction is terminated by filtration through Whatman No. 1 paper. The product is extracted from the filtrate 3 times with 50 ml of phenol : chloroform (1 : 1),[15] and the combined extracts are washed 5 times with an equal volume of water. The product is back-extracted into water after the addition of 3 volumes of $n$-butanol : chloroform (1 : 1) to the organic phase. The aqueous extract is then washed 3 times with an equal volume of ether to remove traces of organic solvents and concentrated to dryness by rotary evaporation (45°). Dansylamidopropylcobalamin is purified by SP-Sephadex chromatography (described in the next section),

[13] The coupling reaction can be scaled down proportionally if quantities of alkyl halide are limited.

[14] Progress of the reaction may be followed spectrophotometrically by observing a decrease in the γ-band absorbance of substrate.

[15] A. W. Johnson, L. Mervyn, N. Shaw, and E. L. Smith, *J. Chem. Soc.*, p. 4146 (1963).

TABLE I

PAPER CHROMATOGRAPHY OF COBALAMIN COENZYMES AND CRYPTOFLUORESCENT
ANALOGS[a]

| | Solvent system[b] | | |
|---|---|---|---|
| Cobalamin | A | B | C |
| 5'-Deoxyadenosyl- | 0.76 | 0.56 | 0.51 |
| Methyl- | 1.57 | 1.63 | 1.51 |
| 5'-Deoxy-1, $N^6$-ethenoadenosyl- | 0.82 | 0.58 | 0.73 |
| 5'-Deoxyformycinyl- | 0.92 | 0.51 | 0.86 |
| 5'-Deoxy-2,6-diamino-nebularinyl- | 0.50 | 0.38 | 0.46 |
| 5'-Deoxy-2-amino-nebularinyl- | 0.77 | 0.56 | 0.66 |
| Dansylamidopropyl- | 3.40 | 2.91 | 2.36 |

[a] Descending chromatography on Whatman No. 1 paper using solvent systems A, $n$-butanol : ethanol : $H_2O$ (250 : 80 : 175); B, 2-butanol : $NH_3$ : $H_2O$ (100 : 6.6 : 40); and, C, 2-butanol : $H_2O$ (95 : 40).
[b] Values represent $R_f$ cobalamin/$R_f$ cyanocobalamin. Cobalamins were visualized as quenching spots under ultraviolet light.

but the nucleoside analogs are first chromatographed on phosphocellulose as follows. The residue is dissolved in a minimum volume of 0.01 $N$ acetic acid and applied to a 2.5 × 30 cm column of phosphocellulose ($H^+$ form) that has been previously equilibrated with the same solution. The column is washed with 500 ml of 0.01 $N$ acetic acid to remove unreacted cyanocobalamin and other minor impurities. The alkylcobalamin is eluted with sodium acetate buffers as follows: 5'-deoxy-1,$N^6$-ethenoadenosylcobalamin, 25 m$M$ (pH 6.0); 5'-deoxyformycinylcobalamin, 50 m$M$ (pH 6.0); 5'-deoxy-2,6-diaminonebularinylcobalamin, 0.25 $M$ (pH 6.4); and, 5'-deoxy-2-aminonebularinylcobalamin, 0.20 $M$ (pH 6.4). Fractions containing the product are pooled and extracted with phenol : chloroform as described above. The aqueous extract is concentrated to approximately 25 ml and applied to a 4 × 45 cm column of SP-Sephadex ($Na^+$ form)[16] that has been equilibrated with water. The product is washed with 1.0 liter of water, then eluted with 50 m$M$ sodium acetate (pH 5.0). Fractions containing product are pooled, extracted with phenol:chloroform, extracted back into water, washed with ether, and concentrated. The analogs, except for dansylamidopropylcobalamin, are crystallized from acetone : $H_2O$, 85 : 15; the dansyl analog, which has re-

[16] G. Tortolani, P. Bianchini, and Y. Mantovani, *J. Chromatogr.* **53**, 577 (1970).

TABLE II

ABSORBANCE SPECTRA OF THE COBALAMIN COENZYMES AND CRYPTOFLUORESCENT ANALOGS[a]

| Cobalamin | 0.1 N HCl | | | | | 0.1 M Na phosphate, pH 7.0 | | | | |
|---|---|---|---|---|---|---|---|---|---|---|
| 5'-Deoxyadenosyl- | 265 (40.4) | 284 (22.6) | 304 (21.6) | 380 (8.0) | 458 (9.2) | 261 (33.5) | 317 (12.3) | 337 (12.2) | 376 (10.4) | 523 (8.0) |
| Methyl- | 263 (23.6) | 285 (19.0) | 304 (20.9) | 374 (8.0) | 459 (9.0) | 264 (18.2) | 314 (12.1) | 340 (12.8) | 374 (10.7) | 520 (8.6) |
| 5'-Deoxy-1,$N^6$-ethenoadenosyl- | 265 (36.7) | 285 (28.1) | 303 (22.5) | 380 (8.0) | 458 (9.1) | 264 (26.1) | 315 sh (14.0) | 340 sh (12.4) | 375 (10.2) | 523 (8.0) |
| 5'-Deoxyformycinyl- | 265 (30.5) | 287 (38.2) | 302 (28.6) | 380 (7.9) | 458 (8.4) | 266 (24.8) | — | 341 (12.7) | 375 (10.0) | 523 (8.0) |
| 5'-Deoxy-2,6-diamino-nebularinyl- | 260 (34.4) | 284 (29.2) | 295 (29.0) | 376 (8.3) | 458 (9.1) | 255 (28.3) | 316 (12.7) | 340 (12.5) | 374 (10.8) | 520 (8.9) |
| 5'-Deoxy-2-amino-nebularinyl- | 262 (27.7) | 284 (22.4) | 303 (24.1) | 375 (7.9) | 454 (9.2) | 264 (21.2) | 303 (18.5) | 340 sh (12.5) | 373 (10.3) | 519 (8.0) |
| Dansylamidopropyl- | 263 (27.0) | 283 (25.2) | 298 (25.1) | 378 (7.6) | 454 (7.7) | 240 sh (35.0) | 316 (17.7) | 340 (15.9) | 375 sh (10.6) | 512 (8.8) |

[a] Main absorbance bands in nanometers. Millimolar extinction coefficients are given in parentheses. Shoulder is denoted by sh.

sisted crystallization, is lyophilized and stored as a dry powder. Cobalamin products are obtained in 46 to 80% yield and appear to be homogeneous by paper chromatography in three solvent systems (Table I). Alternative syntheses of 1,$N^6$-ethenoadenosylcobalamin have been reported.[17]

### Characterization

*Absorbance Spectra.* The principal absorbance bands of the five cryptofluorescent analogs and the cobalamin coenzymes at pH 1 and 7 are summarized in Table II. In general, the nucleoside analogs and adenosylcobalamin have similar absorbance features in the visible region of the spectrum with a prominent but somewhat broad band appearing between 519 and 523 nm at neutral pH. However, these compounds have dissimilar absorbance patterns in the ultraviolet, where individual contributions from the $\beta$-ligand nucleosides are significant. The absorbance maximum (512 nm) of dansylamidopropylcobalamin at pH 7.0 is well below that of methylcobalamin (520 nm). Absorbance maxima of the 5 analogs and coenzymes shift to lower wavelengths (454 to 459 nm) at pH 1.0. This characteristic behavior of most alkylcobalamins is thought to derive from the protonation and uncoordination of the lower ($\alpha$) axial-ligand base, dimethylbenzimidazole.

*Fluorescence.* The analogs appear as quenching rather than fluorescent spots on paper chromatograms illuminated with ultraviolet light (Table I). Dansylamidopropylcobalamin, however, begins to fluoresce after 30 sec of continuous exposure. Analogs in solution are likewise nonfluorescent until photolyzed. Spectral characteristics of the resulting fluorescent products are listed in Table III. The relative fluorescence intensities and emission maxima of the photolytic products are similar but do not always coincide with those of the free halogenated precursors. In the case of 5'-deoxy-1,$N^6$-ethenoadenosylcobalamin, the major aerobic and anaerobic photolysis products have been identified as the 5'-aldehyde of 1,$N^6$-ethenoadenosine and 1,$N^6$-etheno-5',8-*cyclic*-5'-deoxyadenosine, respectively.[8] As shown in Table III, adenosylcobalamin itself is cryptofluorescent in that the weak fluorescence of the $\beta$ ligand, 5'-deoxyadenosine, at pH 1 can be observed only after ligand release from the corrin macrocycle. Methylcobalamin and its photolysis products are, of course, nonfluorescent.

*Reactivity.* The cryptofluorescent analogs, like the cobalamin coenzymes, are stable when stored at $-20°$ in crystalline or powdered form in the dark. Dansylamidopropylcobalamin slowly decomposes in aqueous solution at 4° to aquacobalamin and an unidentified fluorescent product.

[17] I. P. Rudakova, A. M. Yurkevich and V. A. Yakovlev, *Dokl. Akad. Nauk SSSR* **218**, 588 (1974); M. Moskophidis and W. Friedrich, *Z. Naturforsch.* **30C**, 460 (1975).

TABLE III

FLUORESCENCE CHARACTERISTICS OF PHOTOLYSIS PRODUCTS OF COBALAMIN
COENZYMES AND CRYPTOFLUORESCENT ANALOGS[a]

| Cobalamin | Solvent | Excitation (nm) | Relative fluorescence[b] | Emission (shift)[c] (nm) | |
|---|---|---|---|---|---|
| 5'-Deoxyadenosyl- | 0.10 N HClO$_4$ | 280 | 0.78 | 393 | (0) |
| Methyl- | H$_2$O | 280 | 0 | — | — |
| 5'-Deoxy-1,N$^6$-etheno-adenosyl- | H$_2$O | 300 | 0.80 | 404 | (−2) |
| 5'-Deoxyformycinyl- | H$_2$O | 295 | 1.08 | 369 | (+27) |
| 5'-Deoxy-2,6-diamino-nebularinyl- | H$_2$O | 278 | 0.69 | 373 | (+13) |
| 5'-Deoxy-2-amino-nebularinyl- | H$_2$O | 303 | 0.91 | 375 | (0) |
| Dansylamidopropyl- | Isopropanol | 338 | 0.72 | 506 | (0) |

[a] Cobalamin solutions (5 $\mu M$; all nonfluorescent) were exposed to light from a 200-W tungsten bulb; after complete photolysis (about 15 min), the emission spectra were recorded (excitation wavelengths as indicated).

[b] Relative fluorescence is defined as the emission intensity of the cobalamin photolysis product(s) divided by the emission intensity of free precursor ligand determined under identical conditions.

[c] Deviation (nm) of the emission maximum of the photolysis product(s) from that of the free precursor ligand.

The latter can be conveniently removed from stock solutions by extraction with cyclohexane.

The carbon–cobalt bond of adenosylcobalamin and its nucleoside analogs are cleaved by cyanide anion. In the case of the coenzyme, the concerted reaction products are dicyanocobalamin, whose formation can be followed spectrophotometrically at 368 nm, the cyanohydrin anomers of D-erythro-2,3-dihydroxypent-4-enol, and adenine.[18] Cyanide-mediated cleavage also occurs readily with 5'-deoxy-1,N$^6$-ethenoadenosylcobalamin, 5'-deoxy-2,6-diaminonebularinylcobalamin, and 5'-deoxy-2-aminonebularinylcobalamin. The kinetics of these reactions can be followed spectro*fluoro*metrically as the fluorescent bases are released.[8] Pseudo-first-order rate constants obtained by this method are in good agreement with those obtained by absorbance methods. The rate of 5'-deoxyformycinylcobalamin decomposition by cyanide is extremely slow, and dansylamidopropylcobalamin (like propyl- and methylcobalamin) is completely unreactive.

[18] H. P. C. Hogenkamp and T. G. Oikawa, *J. Biol. Chem.* **239**, 1911 (1964).

# [4] Polarography of Compounds of the Vitamin $B_{12}$ Series in Dimethylformamide

## By I. Ya. Levitin

It has been established that the cobalt atom in vitamin $B_{12}$ derivatives can occur in three oxidation states ranging from I to III. Naturally, redox processes are characteristic of these compounds and deeply involved in their metabolism.[1] Therefore, one should expect electrochemical methods to be a valuable tool of research in the field. This has been proved true at least as far as polarography is concerned. All but a few of the studies were carried out in natural water media, and interesting information on redox and complexation equilibria and kinetics were obtained (see articles cited in footnotes [2–4] and references cited therein). Nevertheless, at the moment polarography in an aprotic solvent, namely, dimethylformamide (DMF), seems to provide better opportunities for assay of vitamin $B_{12}$ derivatives, particularly 5'-deoxyadenosylcobalamin.[5]

Essentials of conventional (dc) polarography have already been outlined in this treatise.[6] Here we will consider the ac technique, which manifests considerable advantages in the case of electrochemically reversible processes (i.e., if charge transfers are so fast that steady redox equilibria may be essentially established), e.g. those of oxycobalamin (vitamin $B_{12a}$).

$$B_{12a} \overset{e}{\rightleftarrows} B_{12r} \overset{e}{\rightleftarrows} B_{12s} \qquad (1)$$

Further, certain features of polarography in nonaqueous media should be mentioned.

### Foundations

*Alternating Current Polarography.* There is a comprehensive treatise on the subject.[7] The simplest version of the ac techniques involves modulat-

[1] J. M. Pratt, "Inorganic Chemistry of Vitamin $B_{12}$." Academic Press, New York, 1972.

[2] H. P. C. Hogenkamp and S. Holmes, *Biochemistry* **9**, 1886 (1970).

[3] I. Ya. Levitin, I. P. Rudakova, A. M. Yurkevich, and M. E. Vol'pin, *Zh. Obshch. Khim.* **42**, 1202 (1972).

[4] D. Lexa and J. M. Saveant, *J. Am. Chem. Soc.* **98**, 2652 (1976).

[5] I. Ya. Levitin, I. P. Rudakova, A. L. Sigan, T. A. Pospelova, A. M. Yurkevich, and M. E. Vol'pin, *Zh. Obshch. Khim.* **45**, 1879 (1975).

[6] E. Knobloch, this series, Vol. 18B, p. 305.

[7] B. Breyer and H. H. Bauer, "Alternating Current Polarography and Tensammetry." Wiley (Interscience), New York, 1963.

ing of slow-scanned voltage applied to a dropping mercury electrode with a small low-frequency sinusoidal signal. Here an alternating current arises and is measured at the same frequency and plotted vs the dc potential. In the case of an electrochemical reaction, a peak usually develops. As compared with the corresponding conventional polarographic wave, it is much narrower. Thus the resolving power of ac polarographic analysis proves substantially higher. Heights of the peaks can be used for calibration in the same manner as those of conventional polarographic waves. It is pertinent to emphasize that in many cases the peak currents are limited by rates of charge transfers and hence depend heavily on reversibility of the electrode reactions. This feature of AC polarography can be exploited to enhance further selectivity of analytical procedures.

*Polarography in Nonaqueous Media.* For further information see a brief survey by Zuman.[8]

Wide use of such electrolytes has lead to a drastic extension of opportunities for polarographic studies and analysis. This is due to the following reasons:

1. In organic media, electrode reactions are usually less complicated with adsorption phenomena.
2. In the case of aprotic media such as DMF solutions, discharge of hydrogen ions cannot interfere in the process or overlap it. Furthermore, as these ions are here poorly available, protonation steps that are quite typical of polarographic reductions in aqueous media become less involved in the electrode reaction. Therefore, polarographic behavior of many compounds grows simpler or even approaches the pattern of a reversible process as soon as an aqueous electrolyte is changed for an aprotic one.

Since organic media feature poor conductivity, control of potential grows into a serious problem. Therefore, combined use of a three-electrode cell and potentiostatic polarographs is imperative.

Assay Method[5]

*Principle*

Cyanocobalamin, oxycobalamin, organocobalamins, and organocobinamides are reduced at a dropping mercury electrode in DMF solution, diffusion cathodic waves being observed (see the table). Reduction of oxycobalamin involves two reversible electron transfers [Eq. (1)].

[8] P. Zuman, *Electrochim. Acta* **21**, 687 (1976).

CATHODIC WAVES OF NATURALLY OCCURRING CORRINOIDS[a]

| Compound | $E_{\frac{1}{2}}$ (V) | æ ($\mu$A/mM) | $\dfrac{\partial E}{\partial \log [i/(I_d - i)]}$ (mV) |
|---|---|---|---|
| Cyanocobalamin | −0.69 | 1.56 | 144 |
| Oxycobalamin | −0.01[b] | 0.58 | 80 |
| | −0.69[b] | 0.59 | 54 |
| Methylcobalamin | −1.16[c] | 1.22 | — |
| 5′-Deoxyadenosylcobalamin | −1.07 | 1.32 | 63 |
| 5′-Deoxyadenosylcobamide | −1.07 | 1.31 | 65 |

[a] Solution (0.1 $M$) of $NaClO_4$ in dimethylformamide as supporting electrolyte; dropping mercury electrode with drop time 5.64 sec, and mercury flow rate 1.27 mg/sec while immersed in 0.1 $M$ aqueous solution of KCl and open circuit; calomel reference electrode with 4 $M$ aqueous solution of LiCl; 25°.

[b] Reversibility of the wave was proved with an alternating current technique (that of Kalousek's commutator).

[c] Distorted wave.

Each of the other corrinoids give one irreversible two-electron wave.[9] The cathodic waves of organocobalamins and organocobinamides as well as the well-defined second cathodic wave of oxycobalamin can be used for assay. In particular, if mixtures of oxycobalamin and an organocobalamin must be analyzed, a combination of dc and ac polarography is quite effective (Fig. 1).

*Experimental Section*

*Auxiliary Chemicals.* Commercial DMF of "pure" grade was dried over calcium hydride and distilled under reduced pressure (water pump). The purified solvent was protected from moisture and light during storage. Sodium perchlorate of the same grade was dried over $P_2O_5$ under heating *in vacuo* (100°/1 mm) for a day and stored over the same agent.

[9] *Note added in proof:* D. Lexa and J.-M. Saveant [*J. Am. Chem. Soc.* **100**, 3220 (1978)] have succeeded in obtaining evidence that organocobalamins and organocobinamides RCo(III) can in principle be reduced in a one-electron reversible step to the related RCo(II) complexes just as can model synthetic organocobalt chelates. Using cyclic voltammetry technique, these authors traced the formation and decay of extremely labile methylcob(II)alamin and slightly more stable methylcob(II)inamide as intermediates in the electrochemical reduction of the corresponding Co(III) derivatives under specific conditions (low temperatures, high voltage scanning rates, mixture of DMF and 1-propanol as solvent).

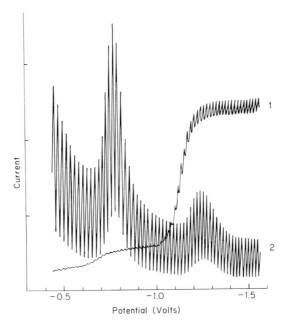

FIG. 1. Direct current (1) and alternating current (2) polarograms of a sample of 5'-deoxyadenosylcobalamin containing 7% of oxycobalamin. The ac polarogram was taken at a modulating frequency of 60 Hz and an amplitude of 10 mV and detected at the fundamental harmonic. Current scales are 1.00 and 0.45 mA/division for dc and ac polarograms, respectively. Other conditions were as given in footnote $a$ to the table.

*Cell and Electrodes.* A three-electrode cell must be used. Its design should provide application of a salt bridge (between main compartment and reference electrode) and allow protection of a solution under analysis against light. A dropping mercury electrode adjustable to a drop time of 3–5 sec and a flow rate 1.3–2.0 mg/sec can be recommended. As to potential reference, in our laboratory a separable calomel electrode with a 4 $M$ solution of LiCl is employed; this salt was selected instead of conventional KCl to avoid precipitation of the latter at the liquid junction. The reference electrode has an agar gel–filter paper plug at the tip. Either a mercury (pool) or a platinum (foil, gauze, or wire) auxiliary electrode may be used.

*Instrumentation.* Any universal (both dc and ac) potentiostatic polarograph may be used.

*Procedure.* Supporting electrolyte solutions (0.1 $M$ NaClO$_4$ in DMF) were prepared every day: they could not be preserved for long. These solutions were also used to fill the salt bridge.

All protracted operations with organocobalamins and organocobinamides were performed in the dark.

Samples of vitamin $B_{12}$ derivatives that had been dried *in vacuo* over $CaCl_2$ were dissolved in the supporting electrolyte which had been deoxygenated by bubbling pure nitrogen. Nitrogen flow above surface of the solution was maintained during measurement.

Concentrations of vitamin $B_{12}$ derivatives ranged from 0.1 to 0.5 m$M$.

Temperature should be selected in the range of 20° to 25° and kept constant within 1–2°.

Comments

1. Consulting a relevant manual, e.g., Sawyer and Roberts,[10] may be useful.

2. Sensitivity of the analysis can be improved considerably by means of certain advanced polarographic techniques. Thus, gains by two orders of magnitude may well be expected if differential pulse polarography[11] is used.

3. For an alternative spectrochemical procedure for determination of 5'-deoxyadenosylcobalamin in the presence of other $B_{12}$ compounds, see vol. 18C.[12]

[10] D. T. Sawyer and J. L. Roberts, Jr., "Experimental Electrochemistry for Chemists." Wiley, New York, 1974.
[11] H. Siegerman, this series, Vol. 43, p. 373.
[12] D. Cavallini and R. Scandurra, this series, Vol. 18C, p. 3.

[5] Comparison of Serum Vitamin $B_{12}$ (Cobalamin) Determination by Two Isotope-Dilution Methods and by *Euglena* Assay

By U.-H. STENMAN and L. PUUTULA-RÄSÄNEN

Microbiological assays for the determination of vitamin $B_{12}$ (cobalamin)[1] in serum were introduced nearly 30 years ago.[2] Their clinical value was rapidly documented in a great number of studies (extensively reviewed by Mollin *et al.*[3]). With the aid of radioactive cobalamin, Barakat and Ekins developed the first isotope-dilution assay.[4] This type of assay

[1] Vitamin $B_{12}$ is only one form of cobalamin, i.e., cyanocobalamin. Serum cobalamin comprises several vitamins $B_{12}$ analogs.
[2] G. I. M. Ross, *Nature (London)* **166**, 270 (1950).
[3] D. L. Mollin, B. B. Anderson, and J. F. Burman, *Clin. Haematol.* **5**, 521 (1976).
[4] R. M. Barakat and R. P. Ekins, *Lancet* **2**, 25 (1961).

was greatly simplified through the introduction of coated charcoal for separation of free and protein-bound vitamin.[5] In other assay methods, separation was achieved by a solid-phase technique.[6]

We have compared a microbiological assay with two isotope-dilution assays, one using coated charcoal and the other a solid-phase technique for separation of free and bound cobalamin. Emphasis was laid on samples having low and borderline values of cobalamin in an attempt to clarify which method has the highest power to differentiate between a pathological and a normal cobalamin status. When there was a discrepancy between the values obtained by different methods, the final judgment of the cobalamin status was based on other hematological and clinical data.[7]

## Methods

### Microbiological Assay

The microbiological assay was performed according to Hutner et al.[8] in which Euglena gracilis of the z strain is used as the test organism. Cyanocobalamin standards were prepared by diluting stock solutions in water. The serum samples were diluted 1 : 20 to 1 : 80 in water. Cobalamin in serum was released from its binding proteins by immersing the diluted samples in capped glass tubes for 20 min in a boiling water bath. Care was taken to maintain sterility during all procedures.

The growth of the test organism was measured turbidimetrically in a spectrophotometer at 650 nm. The reference values for this method have been studied in about 1000 apparently healthy subjects. Using 95% confidence limits, a range of 80–490 pmol/liter has been obtained.[7,9]

### Solid-Phase Assay

The solid-phase assay used in this study was a commercially available version of the radiosorbent assay of Wide and Killander[6] (Phadebas $B_{12}$ Test, Pharmacia, Uppsala, Sweden). The assay was performed as suggested by the manufacturer. Salient features of this assay are release of serum cobalamin by boiling for 15 min in glutamic acid buffer, pH 3.3, and separation of free and bound cobalamin by centrifugation of the binding

[5] K.-S. Lau, C. Gottlieb, L. R. Wasserman, and V. Herbert, *Blood* **26**, 202 (1965).
[6] L. Wide and A. Killander, *Scand. J. Clin. Lab. Invest.* **27**, 151 (1971).
[7] L. Puutula and U.-H. Stenman, *Clin. Chim. Acta* **55**, 263 (1974).
[8] S. H. Hutner, M. K. Bach, and G. I. M. Ross, *J. Protozool.* **3**, 101 (1956).
[9] R. Gräsbeck, W. Nyberg, M. Saarni, and B. von Bonsdorff, *J. Lab. Clin. Med.* **59**, 419 (1962).

protein Sephadex complex. The cyanocobalamin standards are dissolved in buffer. The reference range of this assay is 225–740 pmol/liter (300–1000 pg/ml).

### Coated-Charcoal Assay

Our isotope dilution assay is a modification of the coated-charcoal method of Lau *et al.*[5] In this assay an intrinsic factor preparation from hog pyloric mucosa is used as binding protein, and bound and free cobalamin are separated with hemoglobin-coated charcoal.

#### Reagents

Assay buffer: The assay buffer contains 68 m$M$ NaCl, 56 m$M$ HCl, 0.11 m$M$ KCN, 1.1 m$M$ NaN$_3$, and $^{57}$Co-labeled cyanocobalamin at about 17 pmol/liter (CT 2P, Radiochemical Centre, U.K.), activity 3.8 $\mu$Ci/liter. This buffer is prepared immediately before use by adding KCN and radioactive cyanocobalamin to a stock solution containing the other constituents.

Standards prepared by adding nonradioactive cyanocobalamin to a cobalamin-free serum. The concentration of the stock solution of cyanocobalamin is measured spectrophotometrically using 27,700 cm$^{-1}$ mol$^{-1}$ at 361 nm as specific absorption coefficient. The standards containing 0, 50, 100, 200, 400, and 800 pmol/liter are treated in the same way as the samples. Premixed standards can be stored at $-20°$ for at least 6 months.

Binding protein prepared from hog pyloric mucosa by depepsination at pH 10, ammonium sulfate precipitation, and DEAE-cellulose chromatography.[10] About 40% of the binding protein is intrinsic factor, and the rest is "nonintrinsic factor" (R protein). This preparation is stored as a precipitate in ammonium sulfate and is stable for at least 2 years (available from Medix Biochemica, P.B. 819, SF-00101 Helsinki, Finland). The ammonium sulfate precipitate is dissolved in 0.154 $M$ NaCl containing 3 m$M$ NaN$_3$ and 2.5 g of human serum albumin per liter (Kabi AB, Sweden) and further diluted with the same buffer to bind about 80% of the radioactive cobalamin in the final incubation volume.

Cobalamin-free serum: Serum to be used for standards was freed from endogenous cobalamin by the method of van de Wiel.[11] In this method the cobalamin is released from its binding proteins by rais-

[10] H. Aro, R. Gräsbeck, and K. Visuri, *Scand. J. Clin. Lab. Invest. Suppl.* **108**, 28, (1969).
[11] D. F. M. van de Wiel, L. J. Koster-Otte, W. T. Goedemans, and M. G. Woldring, *Clin. Chim. Acta* **56**, 131 (1974).

TABLE I
COMPARISON OF THREE METHODS FOR SERUM COBALAMIN DETERMINATION (79 SAMPLES)
ACCORDING TO THE RELATIONSHIP $y = a_0 + ax$

| $y$ | $x$ | Intercept, $a_0$ (pmol/liter) | Slope, $a$ | Correlation coefficient, $r$ |
|---|---|---|---|---|
| Charcoal method | Microbiological method | 37.0 | 0.929 | 0.954 |
| Radiosorbent method | Microbiological method | 66.8 | 1.114 | 0.903 |
| Radiosorbent method | Charcoal method | 31.6 | 1.150 | 0.907 |

ing the pH to 13.1 and heating at 100° for 60 min. Freed cobalamin is removed by dialysis.

Hemoglobin-coated charcoal[12]: Hemoglobin is prepared by lysing twice washed packed human red cells with one volume of distilled water and half a volume of toluene. After 5 min of mixing, the lysed cells are centrifuged for 5 min at 2000 $g$. The two uppermost phases containing cell debris and toluene are discarded. The hemoglobin solution is filtered, and the hemoglobin concentration is adjusted to 10 g/liter. This solution is stored at −20° in 10-ml aliquots. The working suspension of hemoglobin-coated charcoal is prepared by adding 10 g of charcoal to 10 ml of hemoglobin solution and adding water to 200 ml.

*Assay Procedure.* Duplicate samples of 0.2 ml of serum and 2.25 ml of assay buffer are added into polypropylene tubes (17 × 110 mm). Standards in cobalamin-free serum are treated the same way. The tubes are loosely stoppered and immersed in a boiling water bath for 25 min. After cooling to room temperature, 100 $\mu$l of binding protein solution are added and the tubes are incubated for 90 min. Hemoglobin-coated charcoal (0.8 ml) is added and mixed; after 10 min the tubes are centrifuged at 2000 $g$ for 15 min. The supernatant is discarded by inversion of the tubes and blotting of the edges. The radioactivity of the charcoal is counted for 50–100 sec. The results are read from a standard curve plotted on a lin-log graph.

Results

One hundred and twenty three sera were assayed by both the radiosorbent and the microbiological methods; 79 of these sera were assayed by the coated charcoal method too. The correlation between the different methods is shown in Table I.

[12] V. Herbert, C. W. Gottlieb, and K.-S. Lau, *Blood* **28**, 130 (1966).

TABLE II

DISTRIBUTION OF COBALAMIN VALUES OF 79 PATIENTS INTO LOW, BORDERLINE,
NORMAL, AND HIGH RANGES BY THE *Euglena* AND THE ISOTOPE-DILUTION ASSAYS[a,b]

| Method | Number of patients in different ranges of vitamin $B_{12}$ concentration | | | |
| | Low | Borderline | Normal | High |
|---|---|---|---|---|
| Microbiological | 41 | 26 | 7 | 5 |
| (range, pmol/liter) | (0–79) | (80–109) | (110–490) | >490 |
| Coated charcoal | 40 | 30 | 7 | 2 |
| | (0–109) | (110–139) | (140–700) | >700 |
| Radiosorbent | 39 | 28 | 10 | 2 |
| | (0–155) | (156–224) | (225–740) | >740 |

[a] Comparative ranges (low and borderline) were calculated from the regression equations in Table I, using the microbiological assay as reference method. The calculated borderline range of the radiosorbent method was extended to 225 pmol/liter as recommended by the manufacturer.
[b] Reproduced by permission of Elsevier/North-Holland Biomedical Press from Puutula and Stenman.[7]

The microbiological method gives the lowest results, and the radiosorbent method the highest ones. The correlation between all methods is good; the best correlation is obtained between the coated-charcoal and the microbiological method.

Table II shows the distribution of values obtained by the three methods. Four ranges were chosen on the basis of values earlier obtained by the microbiological method,[7,9] low (0–79 pmol/liter), borderline (80–109 pmol/liter), normal (110–490 pmol/liter), and high (>490 pmol/liter). The corresponding low and borderline ranges for the other methods (Table II) were calculated from the regression equations shown in Table I. The distribution of values into the different groups was fairly similar for the three assay methods (Table II).

In five cases where there was a discrepancy between low values obtained by different methods, we checked the clinical condition of the patient. This revealed that the microbiological assay gave one false low value. The coated-charcoal assay gave one false normal and one borderline value; and the radiosorbent assay gave two false normal and one borderline value in cases with megaloblastic anemia. Thus it seems that the isotope-dilution assays tend to miss a few cases of megaloblastic anemia (cf. Hall[13]) if the reference range shown in Table II is used. Consequently we now use 200 pmol/liter as the lower limit of the "normal" range in the coated-charcoal method.

[13] C. Hall, *Lancet* **1,** 1255 (1977).

Discussion

Comparison of the results obtained by the different methods indicates a very good correlation, especially between the *Euglena* assay and the coated-charcoal method. A cause of the lower values obtained by the *Euglena* assay may be the lack of serum in the standards. Serum tends to inhibit the growth of *Euglena gracilis*,[3] which would result in falsely low values. Serum-based standards are important in a coated-charcoal method. The serum proteins provide additional coating of the charcoal, thus reducing adsorption of the binding protein to the charcoal.[14]

The values obtained by different radioisotope dilution methods are dependent also on the binding protein used. Assays based on R proteins (nonintrinsic factor, transcobalamin I, cobalophilin) give higher values than those using pure intrinsic factor as binding protein.[15] This difference might be caused by cobalamin analogs that bind to R proteins but not to intrinsic factor. These analogs lack B$_{12}$ activity and they are therefore not measured by microbiological cobalamin asays. However, high serum levels of these analogs may obscure true cobalamin deficiency. This problem has recently been observed in some North American studies. Of patients with cobalamin deficiency, 12%[16] to 20%[15] had normal "serum B$_{12}$" measured by R protein-based radioassays. Although our two radioassays were mainly based on R protein, we observed a rate of false normal values of only 3–4%, a figure that is to be expected statistically. Thus no assay used in this study could be considered better than the other assays. Similar results have recently been obtained in a large Danish study.[17]

The coated-charcoal method employed in this study has been designed to involve a minimum of handling steps. Thus the samples are hydrolyzed, assayed, and counted in the same tube. The assay is further simplified by inclusion of the radioactive cyanocobalamin into the hydrolysis buffer and the use of a binding protein that has a reproducible binding capacity at the acid pH of the hydrolysis buffer.

There is a slight theoretical disadvantage in counting the free fraction (the charcoal). However, complete transfer of the supernatant by decanting into a counting tube requires more technical skill than does the present method.

[14] U.-H. Stenman, *Clin. Haematol.* **5**, 473 (1976).

[15] J. F. Kolhouse, H. Kondo, N. C. Allen, E. Podell, and R. H. Allen, *N. Engl. J. Med.* **299**, 785 (1978).

[16] B. A. Cooper and V. M. Whitehead, *N. Engl. J. Med.* **299**, 816 (1978).

[17] N. H. Holländer, P. Gimsing, E. Hippe, T. Hansen, H. Funch-Roseberg, and H. Olesen, *in* "Vitamin B$_{12}$. Proceedings from the Third European Symposium on Vitamin B$_{12}$ and Intrinsic Factor" (B. Zagalak and W. Friedrich, eds.) (in press). de Gruyter, Berlin, 1979.

We used a fairly pure binding-protein preparation, which had very good stability when stored precipitated in ammonium sulfate at 4°. We added human serum albumin to diluted binding-protein solutions in order to improve stability and prevent unspecific adsorption to tube walls. Bovine serum albumin was not found to be suitable because of its cobalamin-content and a variable and fairly high cobalamin binding capacity.

The isotope-dilution methods used in this study have proved to be very reliable and clinically equally useful. A change of binding protein may become advisable in the future. We have tested a pure intrinsic factor preparation (Medix Biochemica) in a modified version of our radioassay. In this assay hydrolysis is performed in 0.05 mol/liter borate buffer containing potassium cyanide 93 $\mu$mol/liter and dithiothreitol 1 g/liter.[18] This method gives somewhat lower values than our original radioassay, confirming the findings of Kolhouse et al.[15] The clinical usefulness and the reference range of this assay have yet to be determined.

### Summary

We have described a comparison of serum cobalamin determinations by a microbiological method and two isotope-dilution methods. The microbiological assay uses Euglena gracilis of the z strain as the test organism. In one isotope-dilution method, the binding protein is coupled to Sephadex, and separation is achieved by centrifugation. In the other isotope method, a soluble binding protein is used, and separation of free and bound vitamin $B_{12}$ is achieved by adsorption with hemoglobin-coated charcoal. This assay is described in detail. It incorporates serum-based standards to eliminate protein effects, and the assay procedure has been simplified to save working time and increase precision.

[18] S. Gutcho and L. Mansbach, Clin. Chem. 23, 1609 (1977).

## [6] Measurement of Tissue Vitamin $B_{12}$ by Radioisotopic Competitive Inhibition Assay and Quantitation of Tissue Cobalamin Fractions

*By* Eugene P. Frenkel, Russell Prough, and Richard L. Kitchens

### Competitive Binding Assay

#### Principle

The measurement of tissue vitamin $B_{12}$ by the competitive inhibition approach requires an initial complete separation of all endogenous $B_{12}$ from the tissue binders and the removal or inactivation of these binders. Since $B_{12}$ in tissue exists as a group of coenzyme (and storage) forms with differing affinities for tissue binding proteins and subcellular moieties, extraction of the tissue $B_{12}$ from these binders is accomplished by vigorous boiling of the homogenate of tissue in an acetate–cyanide buffer.[1] This also results in the conversion of all the $B_{12}$ forms into that of cyanocobalamin. Prior to this extraction, addition of radiolabeled vitamin $B_{12}$ ([57Co]$B_{12}$) permits the determination of the recoverable $B_{12}$ originally present. Analysis is accomplished by the subsequent incubation of these extracts (and appropriate standards of crystalline $B_{12}$) with a predetermined amount of material (e.g., serum, intrinsic factor) with unsaturated $B_{12}$ binding sites. The percentage of [57Co]$B_{12}$ that is protein bound is inversely proportional to the total $B_{12}$ concentration. The bound $B_{12}$ can be removed by batch adsorption onto diethylaminoethyl (DEAE) cellulose and subsequent centrifugation[2] or by adsorption onto coated charcoal.[3] The percentage binding of any unknown can then be determined from previously prepared standard curves.

#### Assay Method

*Preparation of Tissue.* A weighed sample of tissue (conveniently 10–100 mg) is placed in a 20-ml grinding vessel with 5 ml of 0.9% sodium chloride, pH 5.5. Using a piston-type Teflon pestle (A. Thomas Co., Philadelphia, Pennsylvania), the tissue is completely homogenized. When a motor-

---

[1] E. P. Frenkel, J. D. White, C. Galey, and C. Croy, *Am. J. Clin. Pathol.* **66**, 863 (1976).
[2] E. P. Frenkel, S. Keller, and M. S. McCall, *J. Lab. Clin. Med.* **68**, 510 (1966).
[3] K. S. Lau, C. Gottlieb, L. R. Wasserman, and V. Herbert, *Blood* **26**, 202 (1965).

METHODS IN ENZYMOLOGY, VOL. 67

driven grinder (Hollow Spindle Stirrer, Homogenizer Drive, A. Thomas Co.) is available, homogenization requires approximately 30 sec. The homogenate is transferred to a tube, and the grinding vessel and pestle are rinsed with saline solution. For liver, spleen, and brain tissue a final concentration of 5 mg of tissue per milliliter of final homogenate is used. Tissue obtained from kidney requires a 50-fold greater dilution (0.1 mg/ml). Homogenates are stable at $-20°$ for at least one year prior to assay.

*Reagents*

Radiopure high specific activity [$^{57}$Co]$B_{12}$ (specific activity greater than 100 mCi/mg), required for accurate radioassay (Amersham-Searle Corp., Des Plaines, Illinois or N. V. Phillips-Duphar Co., Pellen, Holland). Radiochemical purity of the labeled $B_{12}$ is evaluated on arrival of each new shipment, and its integrity is checked every 2 months during its use. The integrity of the labeled $B_{12}$ is checked by two methods.[4]

Method 1. Characterization of binding kinetics of the labeled $B_{12}$. To emphasize the importance of radiochemical purity, alteration of binding kinetics occurs with impure preparations, so that preparation of a standard curve can be used to assay its integrity. This is done as shown in Fig. 1, where two separate standard curves are made. The first consists of quantities of unlabeled $B_{12}$ (100–500 pg) in acetate–cyanide buffer to which 100 pg of [$^{57}$Co]$B_{12}$ are added. This is processed as usual (see below and Fig. 1, curve A), and the percentage bound is plotted against total $B_{12}$. In a second assay (curve B in Fig. 1), all the $B_{12}$ consists of labeled material. In the presence of degradation products or contamination in excess of 5–8% of the total radiolabeled material, the two curves are nonsuperimposable.

Method 2. Electrophoretic autoradiographic evaluation of the labeled $B_{12}$. Electrophoretic separation and autoradiography of the strips with subsequent counting of the engrafted areas in a routine well counter can be used to identify a radiolabeled contaminant.[5] In the absence of such contamination, a homogeneous single band is seen on the autoradiograph.

The radiolabeled $B_{12}$ for use is diluted in acetate–cyanide buffer to a concentration of 1 ng/ml; the preparation is sterilized by Millipore filtration and stored in the dark at $4°$.

Vitamin $B_{12}$, crystalline, is diluted in the same manner as the

[4] E. P. Frenkel, M. S. McCall, and J. D. White, *Am. J. Clin. Pathol.* **53**, 891 (1970).
[5] E. P. Frenkel and J. D. White, *Lab. Invest.* **29**, 614 (1973).

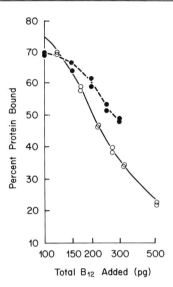

FIG. 1. Typical standard curve and evaluation of radiochemical purity. ——, A typical standard curve (A) with duplicate determinations indicated by open circles (○). The initial binding was approximately 75%. The total $B_{12}$ levels shown on the ordinate include the 100 pg of labeled $B_{12}$ used in the assay. ---, Curve (B) representing a study of radiochemical purity. The replicate samples (●) demonstrate that this assay, performed only with the radiolabeled $B_{12}$ being checked for purity, yields a curve (---) not superimposable on the classical standard curve at any concentration. This failure at superimposition identifies degradation products or contaminants.

$[^{57}Co]B_{12}$, sterilized by Millipore filtration, and frozen in small amounts sufficient for daily use.

Acetate buffer, 0.5 $M$ (pH 4.6), is prepared, using 0.5 $M$ sodium acetate and 0.5 $M$ acetic acid (glacial), and sterilized by Millipore filtration.

Sodium cyanide in deionized water is prepared in concentration of 25 $\mu$g/ml.

Acetate-cyanide reagent. This is prepared on the day of use by combining 1 volume of acetate buffer, 2 volumes of NaCN solution, and 1 volume of sterile water.

Phosphate buffer, 10 m$M$, pH 8.0, is prepared with 94.7 ml of 0.2 $M$ dibasic sodium phosphate and 5.3 ml of 0.2 $M$ monobasic sodium phosphate, and diluted to 2000 ml with deionized water.

DEAE-cellulose powder, Whatman DE-11.

Standard serum (Q-Pak-Chemistry Control Serum, unassayed, Hyland Div., Travenol Laboratories, Inc., Costa Mesa, California) Dried standard serum, 50 ml, is reconstituted and incubated with

100–400 $\mu$l of serum in acetate–cyanide reagent containing 100 pg of $[^{57}Co]B_{12}$ in a total volume of 4 ml (other "binders" are noted in Comments section). After 60 min at room temperature, phosphate buffer and cellulose are added, and the assay is carried out as outlined below. After the volume of serum required to bind 60–75% of the 100 pg $[^{57}Co]B_{12}$ is determined, the serum is stored frozen in amounts sufficient for daily use.

*Assay Procedure.* An appropriate volume of the tissue homogenate to be assayed is added to Pyrex tubes (16 × 125-mm screw-capped culture tubes) in duplicate (generally 1 ml is used, but volumes between 0.5 and 2 ml may be employed), and the total volume is brought to 2 ml by the addition of physiologic saline solution.

A standard curve is prepared for simultaneous assay by pipetting, in duplicate, 0, 50, 100, 200, and 400 pg of crystalline $B_{12}$ into 20 × 150 mm counting screw-capped culture tubes. The volume is adjusted to 3.9 ml with acetate–cyanide reagent. A counting tube for a reference standard containing 3.9 ml of the acetate–cyanide reagent is prepared. Then, 100 pg (100 $\mu$l) of $[^{57}Co]B_{12}$ are added to each tube. Thus, the total $B_{12}$ concentrations range from 100 to 500 pg in a 4-ml volume.

Simultaneously, 100 pg (100 $\mu$l) of $[^{57}Co]B_{12}$ are added to each of the 2-ml volumes of homogenates to be assayed. The tubes are mixed gently by swirling and allowed to stand for 30 min at room temperature.

After 30 min, 2 ml of acetate buffer and 4 ml of sodium cyanide solution are added to the tissue homogenates only. The tubes are capped tightly, shaken by inversion, and boiled for 30 min with occasional shaking. The tubes for the standard curve and reference standard are similarly handled, with gentle mixing but without inversion.

After boiling, the tubes are allowed to cool and the tissue homogenates are filtered through Whatman No. 40 filter paper. A 4.0-ml volume of each of the clear filtrates is then transferred to a counting tube. These tissue filtrates are counted in a well-type gamma counter. The 100-pg $[^{57}Co]B_{12}$ reference standard is simultaneously counted. The total counts recovered and the percentage of recovery for each sample are recorded. An amount of standard serum that has been previously determined to bind approximately 75% of the $[^{57}Co]B_{12}$ (usually 100–200 $\mu$l) is then added to the unknowns and the standards. The tubes are mixed by gentle swirling and allowed to incubate for 60 min at room temperature. Approximately 10 ml of phosphate buffer and approximately 0.5 g of DEAE-cellulose powder are then added, and the tubes are capped tightly and shaken vigorously for 15 min on a horizontal rotating shaker. Tubes are then centrifuged for 15 min at 1400 $g$. The supernatant fluid is aspirated and discarded, and the remaining cellulose is washed twice with 15-ml volumes of phosphate buffer. The serum-bound $B_{12}$ in the packed cellulose is then counted in a well-type gamma counter.

A standard curve is plotted on semilogarithmic paper with the total $B_{12}$ values (including the 100 pg of labeled $B_{12}$) plotted against the percentage bound being determined by dividing the counts bound to the cellulose by the counts present in the reference standard. A typical standard curve (A) is shown in Fig. 1.

The percentage bound of the unknowns, determined by dividing the counts bound to the cellulose by the counts present in the 4-ml extract, is then read from the standard curve. Corrections are made for percentage recovery, exogenous $B_{12}$ added, and the amount of tissue originally employed. The final concentration is then expressed as nanograms of $B_{12}$ per gram of tissue.

*Calculations*

$$\% \text{ Bound of unknowns} = \frac{\text{cellulose counts}}{\text{extraction filtrate counts}} \tag{1}$$

$$\text{Measured pg } B_{12} = \% \text{ bound of unknowns reading from standard curve} \tag{2}$$

$$\frac{\text{Total pg } B_{12} \text{ per amount}}{\text{of tissue assayed}} = \frac{\text{measured pg } B_{12}\text{-}[^{57}\text{Co}]\text{-}B_{12} \text{ added (100 pg)}}{\% \text{ recovery}} \tag{3}$$

$$B_{12} \text{ ng per g tissue} = \text{total pg } B_{12} \text{ per amount of tissue assayed corrected to 1 g and converted to ng} \tag{4}$$

*Composite Curve.* After a series of standard curves has been obtained with a given volume of standard serum, a composite curve can be drawn, providing the curves are relatively parallel.[2,3] Subsequent assays can then be performed with the use of two standards (at 200 and 400 pg, each with a duplicate). The average percentage bound of this standard is then corrected to the percentage bound of the 200-pg composite curve standard to determine the correction factor. The percentage bindings of the unknowns are then multiplied by this correction factor, and the corrected percentage bindings can be read from the composite curve.

Comment

Competitive inhibition methods do not have the fastidious requirements (sterility, pH, etc.) that complicate classical microbiological assay (growth curve) systems, thereby greatly simplifying the extraction procedures for the determination of $B_{12}$ in tissues. The recent identification of cobalamin analogs in human serum[6] has led to ongoing studies seeking the

[6] J. F. Kolhouse, H. Kondo, N. C. Allen, E. Podell, and R. H. Allen, *N. Engl. J. Med.* **299,** 785 (1978).

existence of such moieties in the tissues and attempts at evaluation of their potential biologic significance. These analogs have further stressed the importance of the type of binder used in the radioassay procedure. The "standard of reference" type binder is intrinsic factor (IF). Its lack of stability has limited its extended applicability. The assay described above uses transcobalamin II (serum derived) as the binder, and this has been shown to provide 88–90% the specificity of pure IF. Most of the available commercial assay methods have utilized as a binder R protein, which has only 20–30% the specificity of pure IF. Until a highly stable pure IF becomes available, the nonintrinsic factor (nonspecific R type) binding can be blocked by the addition of cobamide to the assay.[7,8] This can be done as described above by use of 10 $\mu$g of cobamide per milliliter of tissue extract added at the time of mixing with the radiolabeled B$_{12}$. If duplicate sets are assayed, the tube with the binding blocker added provides a determination of the "true" B$_{12}$ and that without its addition provides the total (true plus analogs) B$_{12}$ content.

The measurement of vitamin B$_{12}$ content in serum[2,4] or cerebrospinal fluid[9] can be performed with the same assay procedure.

## Cobalamin Separation–Quantitation

### Principle

Quantitation of the individual cobalamin forms requires an initial extraction of these forms from their tissue binding proteins.[10] IUPAC–IUB recommended abbreviations are used: hydroxycobalamins (OH-Cbl), cyanocobalamin (CN-Cbl), adenosylcobalamin (Ado-Cbl), and methylcobalamin (Me-Cbl). Extraction for cobalamin fractionation is easily achieved by heating the homogenates of the tissues in ethanol and centrifugation to remove the denatured protein.[11] Evaporation of the ethanol supernatant solution is followed by extraction of nonpolar material from the aqueous phase with a mixture of ether and acetone. The aqueous phase is then evaporated to dryness by lyophilization and dissolved in

[7] J. F. Kolhouse and R. H. Allen, *Anal. Biochem.* **84**, 486 (1978).

[8] R. H. Allen, B. Seetharam, N. C. Allen, E. R. Podell, and D. H. Alpers, *J. Clin. Invest.* **61**, 1628 (1978).

[9] E. P. Frenkel, M. S. McCall, and J. D. White, *Am. J. Clin. Pathol.* **55**, 58 (1971).

[10] E. P. Frenkel, R. L. Kitchens, and R. Prough, *J. Chromatogr.* **174**, 393 (1979).

[11] All handling of cobalamin samples should be carried out in a darkened room under a red or red-orange photographic safelight because of the photolability of the organocorrinoids (CN-Cbl, Ado-Cbl, and Me-Cbl). Exposure to dim light for brief periods of time (such as a pipetting procedure) can be tolerated without significant photolysis.

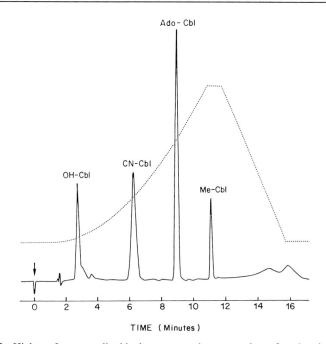

FIG. 2. High-performance liquid chromatography separation of authentic cobalamin standards. A mixture of the cobalamins [hydroxycobalamin (OH-Cbl) 7.5 μg; cyanocobalamin (CN-Cbl) 9.8 μg; deoxyadenosylcobalamin (Ado-Cbl) 6.8 μg; and methyl cobalamin (Me-Cbl) 3.4 μg] in a 15-μl volume was injected. The effluent was monitored by ultraviolet absorption at 254 nm wavelength and 0.5 a.u.f.s. deflection. ....., Methanol concentration of the eluent. This separation was done on a Waters μBondapak $C_{18}$ column.

aqueous methanol buffered with monobasic phosphate. After Millipore filtration the cobalamin fractions are separated by high-performance liquid chromatography (HPLC) (Fig. 2). The addition of radiolabeled $B_{12}$, [57Co]CN-Cbl, as an internal standard prior to the initial ethanol extraction procedure permits determination of the recoverable cobalamins in the original extract.

After HPLC separation (which requires about 10 min) quantitation of the isolated aliquots is performed by the competitive binding method described above.

### Assay Method

*Preparation of Tissue Homogenates.* A weighed sample of tissue (routinely, 2 g of liver or a convenient aliquot of other tissue) is homogenized in 9 volumes of 90% ethanol for 2 min in a Sorvall Omnimixer or Waring blender at full speed. The homogenates may be stored at $-20°$ prior to assay.

*Reagents and Materials*

[57]Co-CN-Cbl: Radiopure, high specific activity [57]Co-labeled cyano-cobalamin (Amersham-Searle Corp., Des Plaines, Illinois; or N. V. Phillips-Duphar Company, Pellen, Holland) is diluted 1 : 10 in 0.9% benzyl alcohol prior to use. Radiochemical purity of [57]Co-CN-Cbl is determined by direct HPLC separation and gamma-counting of fractions collected. Small amounts (less than 5%) of a radiochemical impurity may be seen as a peak between CN-Cbl and Ado-Cbl and as OH-Cbl. The ultraviolet-absorbing peak seen with the UV detector on the HPLC corresponds to benzyl alcohol, which is added by the vendors as a preservative.

Ethanol, 90%, prepared by mixing 900 ml of absolute ethanol with 100 ml of glass-distilled water.

Ether : ethanol (3 : 1), prepared by mixing 3 volumes of diethyl ether with 1 volume of acetone. This should be made from reagent grade solvents in a closed glass bottle containing 1 : 10 volume of 5% aqueous ferrous sulfate solution to prevent peroxide formation.

Methyl alcohol. High purity grade solvent, distilled in glass (obtained from Burdick and Jackson Laboratories, Inc., Muskegan, Michigan) is suitable for HPLC use. Millipore filter prior to use with a solvent-inert filter (0.5 $\mu$m).

Sodium phosphate, 50 m$M$, monobasic (NaH$_2$PO$_4$), pH 4.6, prepared from reagent grade material (J. T. Baker, Phillipsburg, New Jersey) and glass-distilled water. Millipore filter before use (0.45 $\mu$m).

Cobalamin (mixed) standard solution, prepared by mixing a solution of approximately 0.5 mg/ml each of crystalline hydroxocobalamin (Sigma Chemical Co., St. Louis, Missouri), cyanocobalamin (Sigma Chemical Co.) coenzyme-B$_{12}$ (Ado-Cbl, Calbiochem, La Jolla, California) and methylcobalamin (Calbiochem) made up in 23% methanol in 50 m$M$ NaH$_2$ PO$_4$.

Dilute cobalamin standard solution for column calibration: Dilute cobalamin standard solution (above) 1 : 500.

*High-Performance Liquid Chromatography Instrumentation and Conditions.* A DuPont Model 850 liquid chromatograph (Wilmington, Delaware) was used equipped with a prepacked stainless steel column (3.9 mm i.d. $\times$ 30 cm) of $\mu$Bondapak C$_{18}$, 10 $\mu$m particle size (Waters Assoc., Milford, Massachusetts). Solvent intake A is placed in a reservoir containing 50 m$M$ NaH$_2$PO$_4$, pH 4.6, and intake B is placed in methanol. Solvent degassing is achieved by continuous helium sparging at a gas flow of 20 cc/min in each reservoir. The solvent flow rate is set at 1.8 ml/min and the column oven at 40°. The solvent program beginning with the time of injection is (a) isocratic elution for 2 min at 23% B followed by (b) an upward concave gradient (exponent, +2) from 23% B to 70% B in 10 min; (c) isocratic elution at 70% B for 1 min; (d) linear reversal from 70% B to 23% in 4 min; and (e) isocratic elution at 23% B for at least 5 min. After 5 min at 23%, the system is ready for another sample injection. Column effluent is monitored with the absorbance detector at 254 nm. Retention times of the cobalamins are determined by injecting a 100-$\mu$l sample of the dilute cobalamin standards. Retention time reproducibility and separation are excellent with sample volumes of 10 $\mu$l to 200 $\mu$l.

*Assay Procedure.* The homogenate of tissue is mixed with a precisely known amount of the 1:10 dilution of [$^{57}$Co]CN-Cbl (generally 100 $\mu$l or 0.1 $\mu$Ci), loosely capped, and heated for 30 min in an 80° water bath with occasional mixing, then cooled on ice and centrifuged for 10 min at 1000 $g$. The clear supernatant fraction is transferred to a 100-ml round-bottom flask and evaporated to 2–3 ml volume in a Model R Rotary Evaporator (Buchler Instruments, Fort Lee, New Jersey), and transferred to a 50-ml screw-cap glass tube for washing with 40 ml of ether:acetone (3:1). The mixture is shaken vigorously, and phases are allowed to separate. The upper solvent phase is discarded, and the extraction is repeated. The lower aqueous phase is evaporated to dryness by lyophilization.

The residue is dissolved in 1 ml of 23% methanol in 50 m$M$ NaH$_2$PO$_4$, pH 4.6, and filtered using a 0.5 $\mu$m solvent-inert Millipore filter in a stainless steel Swinney adaptor attached to a 5-ml syringe. A 100-$\mu$l aliquot of this extract is then injected directly onto the HPLC column with a microsyringe. The HPLC separation is as above; the cobalamin fractions are collected sequentially in the effluent, and $^{57}$Co activity is counted. Appropriate aliquots of the fractions are taken for B$_{12}$ assay by the competitive binding assay described above.

*Calculations*

OH-Cbl:

$$\frac{\text{pg B}_{12} \text{ assayed in}}{\text{HPLC fraction}} \times \frac{\text{[}^{57}\text{Co]CN-Cbl activity added to original homogenate}}{^{57}\text{Co activity in HPLC injected aliquot}} = \frac{\text{pg B}_{12} \text{ in original}}{\text{homogenate}}$$

For CN-Cbl, the above formula is used, calculating from the amount of [$^{57}$Co]CN-Cbl recovered in the CN-Cbl fraction.

Comment

The extraction of total B$_{12}$ from tissue binders into the ethanol solution was at least 90% complete in liver, kidney, spleen, and brain. This was determined by direct competitive binding assay of the ethanol supernatant solution as compared to the ethanol homogenate. In the complete extraction process, the overall recovery of $^{57}$Co internal standard was 72%. Some residual activity (14%) was identified in the ethanol pellet and the ether : acetone extract (4%). The remaining activity (10%) could not be identified. Exogenously added Me-Cbl and Ado-Cbl were completely recovered in their respective fractions as determined with the use of internal standards. However, OH-Cbl recovery was decreased and variable. The OH-Cbl loss could be recognized and reconciled by the difference in total tissue cobalamin content assayed and the sum of the collected cobalamin fractions.

The recovery of the cobalamins by HPLC is >99%. Only 0.2–0.3% of the total cobalamin activity was found in the effluent of any subsequent run, a trivial amount unless an injection of tissue extract follows a cobalamin standard solution. This can be eliminated by use of a separate syringe for injection of the cobalamin standards and by washing the HPLC injection valve and column after injection of cobalamin standards. An effective washing procedure entails injection of 100 $\mu$l of 100% methanol three times after the solvent program has reached 70% methanol. Then two complete programmed runs (injecting 100 $\mu$l of 23% methanol–phosphate as sample) are performed. This permits studies of a series of tissue extracts in sequence without further washing. When very large amounts of cobalamin standards (>100 ng) are used, a separate column should be used.

Results from a representative study of normal rat liver from five female Sprague–Dawley rats are given here as percentages of total tissue cobalamins (mean and standard error).

OH-Cbl =   14% ± 0.4%
CN-Cbl = 1.4% ± 0.2%
Ado-Cbl =   74% ± 2.2%
Me-Cbl =    9% ± 1.6%

### Acknowledgments

This work was carried out in the Evelyn L. Overton Hematology–Oncology Research Laboratory and was supported by NIH Grants CA23115 and 18132; Veterans Administration Hospital MRIS 1450; the Heddens-Good Foundation; the Southwestern Medical Foundation; and the McDermott Foundation.

## [7] Ribosome-Associated Vitamin $B_{12s}$ Adenosylating Enzyme of *Lactobacillus leichmannii*[1]

*By* WILLIAM S. BECK[1]

Vitamin $B_{12s}$ [cyanocob(III)alamin] + ATP → α-(5,6-dimethylbenzimidazolyl)
deoxyadenosylcobamide + PPP

Conversion of vitamin $B_{12}$ (cyanocobalamin, CN-Cbl[2]) to the coenzyme α-(5,6-dimethylbenzimidazolyl) deoxyadenosylcobamide (adenosylcobalamin, Ado-Cbl) has been demonstrated in extracts of *Propionibacterium shermannii*[3-5] and *Clostridium tetanomorphum*.[6,7] In the reactions catalyzed by these extracts, ATP supplies the 5′-deoxyadenosyl moiety of Ado-Cbl[5,6,8] and the other reaction product, tripolyphosphate.[9] Coenzyme synthesis occurs in two steps in extracts of *C. tetanomorphum*[10]: first, a reduction of vitamin $B_{12a}$ (Co II) to vitamin $B_{12s}$ (Co I), and then the adenosylation of vitamin $B_{12s}$. The steps are catalyzed, respectively, by enzymes that are referred to here as vitamin $B_{12a}$ reductase and vitamin $B_{12s}$ adenosylating enzyme. The two activities together comprise the coenzyme synthetase system.

In *P. shermannii* and *C. tetanomorphum*, enzyme activity evidently occurs in soluble cell fractions. However, early studies of adenosylcobalamin-dependent ribonucleoside triphosphate reductase in

[1] Experiments from the author's laboratory were supported by Research Grant CA-03728 from the National Cancer Institute, National Institutes of Health, and by the John Phyffe Richardson Fund.

[2] This chapter employs two conventions of nomenclature and abbreviations. In accordance with the IUPAC–IUB Commission on Biochemical Nomenclature: The Nomenclature of Corrinoids (1973 Recommendations) *Biochemistry* **13**, 1955 (1974), derivatives of cobalamin are termed cyanocobalamin (CN-Cbl), adenosylcobalamin (AdoCbl), hydroxocobalamin (OH-Cbl), and methylcobalamin (MeCbl). Cyanocobalamin is also designated vitamin $B_{12}$, and the various oxidation stages are referred to by the older terms vitamin $B_{12}$ or $B_{12a}$ [which is cyanocob(III)alamin], vitamin $B_{12r}$ [cyanocob(II)alamin], and vitamin $B_{12s}$ [cyanocob(I)alamin]. TSM buffer is 5 m$M$ magnesium acetate–0.01 $M$ Tris-succinate, pH 7.2.

[3] R. O. Brady and H. A. Barker, *Biochem. Biophys. Res. Commun.* **4**, 464 (1961).

[4] J. Pawelkiewicz, B. Bartosinski, and W. Walerych, *Ann. N. Y. Acad. Sci.* **112**, 638 (1964).

[5] R. O. Brady, E. G. Castanera, and H. A. Barker, *J. Biol. Chem.* **237**, 2325 (1962).

[6] H. Weissbach, B. Redfield, and A. Peterkofsky, *J. Biol. Chem.* **236**, PC40 (1961).

[7] E. Vitols, G. Walker, and F. M. Huennekens, *Biochem. Biophys. Res. Commun.* **15**, 372 (1964).

[8] A. Peterkofsky and H. Weissbach, *Ann. N. Y. Acad. Sci.* **112**, 622 (1964).

[9] A. Peterkofsky and H. Weissbach, *J. Biol. Chem.* **238**, 1491 (1963).

[10] E. Vitols, G. A. Walker, and F. M. Huennekens, *J. Biol. Chem.* **241**, 1455 (1966).

crude extracts of *Lactobacillus leichmannii*[11] suggested that cyanocobalamin would support reductase activity only if ribosomes were present. Later work definitively established that in this organism ribosomes are the loci of an enzyme system that converts vitamin B$_{12a}$ to adenosylcobalamin in these same two steps.[12] The vitamin B$_{12a}$ reductase, as in other organisms, was found to be unstable. Adenosylating enzyme, however, was readily demonstrable. The following sections summarize the partial purification of ribosome-associated vitamin B$_{12s}$ adenosylating enzyme of *L. leichmannii* and the properties of the enzyme.

### Assay Methods

*Principles.* Adenosylcobalamin (AdoCbl) synthesized in an initial incubation containing adenosylating enzyme, vitamin B$_{12r}$ (obtained by treating hydroxocobalamin (vitamin B$_{12a}$) with KBH$_4$), ATP, and MnCl$_2$ is assayed in a second incubation by means of an AdoCbl-dependent enzyme reaction, the rate behavior of which is a function of AdoCbl concentration. Ado-Cbl in an aliquot of the initial incubation mixture can be assayed by two methods, A and B.

In method A, an aliquot of the initial incubation is incubated with excess purified ribonucleoside triphosphate reductase system with [2-$^{14}$C]CTP as substrate.[13] Conversion of radioactive CTP to dCTP is determined by thin-layer chromatography. When the concentration of AdoCbl in the reductase system is $10^{-7}$ to $10^{-6}$ $M$, rates of dCTP formation are proportional to the AdoCbl concentration.

In method B, AdoCbl formed in the initial incubation is assayed according to Abeles *et al.*[14] in an incubation containing excess diol dehydrase from *Aerobacter aerogenes* and 1,2-propanediol. Conversion of 1,2-propanediol to propionaldehyde is measured colorimetrically. When the concentration of AdoCbl in the diol dehydrase system is $10^{-8}$ to $10^{-7}$ $M$, the rate of propionaldehyde production is proportional to the AdoCbl concentration. In general, method B is used in experiments in which filtrates contained ingredients that might affect the ribonucleoside triphosphate reductase reaction. The two assays yield comparable results.[12] A unit of enzyme is defined as the amount producing 1 nmol of AdoCbl per hour under standard assay conditions.

*Procedures.* The initial incubation of the adenosylating enzyme assay is conducted in small Thunberg tubes initially containing the following

[11] W. S. Beck and J. Hardy, *Proc. Natl. Acad. Sci. U.S.A.* **54**, 286 (1965).
[12] H. Ohta and W. S. Beck, *Arch. Biochem. Biophys.* **174**, 713 (1976).
[13] M. Goulian and W. S. Beck, *J. Biol. Chem.* **241**, 4233 (1966).
[14] R. H. Abeles, C. Myers, and T. A. Smith, *Anal. Biochem.* **15**, 192 (1966).

ingredients: cyanocobalamin (or hydroxocobalamin), 15 nmol; ATP, 0.088 $\mu$mol; $MnCl_2$, 0.13 $\mu$mol; and Tris-HCl, pH 8.0, 20 $\mu$mol, in a total volume of 0.4 ml. Tubes are placed in ice, evacuated, and flushed with nitrogen. Potassium borohydride, 1 mg, is added to each tube under a stream of nitrogen, followed by enzyme-containing samples. Tubes are again evacuated, flushed with nitrogen, and kept chilled until reduction of vitamin $B_{12a}$ is complete. The tube is incubated for 45 min in the dark at 37°; then cooled to 0°; a 10- to 50-$\mu$l aliquot is assayed for AdoCbl.

*Reagents*

For initial incubation
  Cyanocobalamin, 0.2 m$M$
  ATP, 4.4 m$M$
  $MnCl_2$, 0.067 $M$
  Tris · HCl, 1 $M$, pH 8.0
For Method A
  Potassium borohydride
  Ribonucleotide triphosphate reductase from *L. leichmannii*,[13] (500 units/mg)
  [2-$^{14}$C]CTP, 10 m$M$ (0.2–0.5 mCi/mmol)
  Dihydrolipoic acid, 0.12 $M$, prepared according to Gunsalus and Razell[15]
  ATP, 0.1 $M$
  $MgCl_2$, 0.1 $M$
  Tris · HCl, 1$M$, pH 7.5
  Perchloric acid, 15%
  KOH, 2.8 $N$
  CMP, 50 m$M$
  dCMP, 50 m$M$
  LiCl, 0.15 $M$
For Method B
  Potassium phosphate buffer, 0.1 $M$, pH 8.0
  *dl*-1,2-Propanediol, 2%, redistilled before use
  Diol dehydrase from *Aerobacter aerogenes* (ATCC 8724),[16] (10–12 units/mg) HCl, 2 $N$
  2,4-Dinitrophenylhydrazine reagent of Böhme and Winkler[17]
  Adenosylcobalamin standard, 0.1 m$M$

*Assay of AdoCbl by Method A.* A 10-$\mu$l aliquot containing 2–20 pmol of AdoCbl is added to a standard ribonucleoside triphosphate reductase sys-

[15] I. C. Gunsalus and W. E. Razell, this series, Vol. 3, p. 941.
[16] H. A. Lee and R. H. Abeles, *J. Biol. Chem.* **238**, 2367 (1963).
[17] H. Böhme and O. Z. Winkler, *Anal. Chem.* **142**, 1 (1954).

tem containing (in nanomoles): [2-$^{14}$C]CTP, 52 (10–25 nCi); dihydrolipoic acid, 600; ATP, 160; MgCl$_2$, 400; Tris · HCl (pH 7.5), 1250; and purified reductase from *L. leichmannii*, 2 μg. Incubations are conducted for 20 min at 37° in 10-ml conical screw-cap centrifuge tubes, and reactions are terminated with 5 μl of 15% perchloric acid. Tubes are tightly capped, completely immersed in a steam bath for 20 min, and then chilled. Mixtures are adjusted to pH 7 with 5 μl of a KOH solution, the concentration of which is just sufficient to neutralize the mixture (approximately 2.8 N). A 2-μl aliquot is then added of a solution containing 2 volumes of 50 mM CMP and 1 volume of 50 mM dCMP. A 10-μl sample of the incubation supernatant solution is applied to poly(ethyleneimine) cellulose thin-layer plates prepared according to Randerath and Randerath,[18] except that vinyl sheet was used in place of glass. Plates are developed with 0.15 M LiCl that has been saturated with boric acid and neutralized with concentrated NH$_3$. The well separated dCMP and CMP spots are visualized by ultraviolet light and cut out. The vinyl back of each spot is attached to a planchet with a drop of rubber cement, and the preparations are assayed for radioactivity in a gas flow counter.

*Assay of AdoCbl by Method B.* A 50-μl aliquot of filtrate of the original incubation mixture, containing 10–100 pmol of AdoCbl is incubated in a 1-ml mixture containing potassium phosphate buffer, pH 8.0, 40 μmol; *dl*-1,2-propanediol, 0.4 mg; and diol dehydrase from *Aerobacter aerogenes* (ATCC 8724),[16] 0.3 mg. Test tubes are kept in a 37° bath until temperature equilibration is reached. The reaction is started by addition of diol dehydrase, allowed to proceed for 30 min, and terminated by addition of 0.1 ml of 2 N HCl. A 1-ml aliquot of the 2,4-dinitrophenylhydrazine reagent of Böhme and Winkler[17] is pipetted directly into the reaction mixture, and aldehyde is determined by measuring optical density at 550 nm. With each assay, two AdoCbl standards and two blanks are included. One blank contains no AdoCbl; the other contains the unknown AdoCbl solution but no enzyme. The second blank is generally negligible. All assays are carried out in the dark. Flashlights are used to provide light enough for essential operations.

## Preparation of Ribosomes

*Cultivation of Bacteria.* Stock cultures of *L. leichmannii* (ATCC 7830) are maintained by biweekly transfer in stabs of Difco vitamin B$_{12}$ agar. New series of daily-transfer working stock cultures are started from a stock culture stab every 2 or 3 weeks.[19]

---

[18] K. Randerath and E. Randerath, *J. Chromatogr.* **16**, 111 (1964).
[19] W. S. Beck, S. Hook, and B. H. Barnett, *Biochim. Biophys. Acta* **55**, 455 (1962).

The basic medium used is Difco vitamin $B_{12}$ assay medium (USP XIV), among whose ingredients are adenine, guanine, uracil, and xanthine in concentrations of 20 mg each per liter.[20] To this medium is added cyanocobalamin, 0.075 $\mu$g/liter,[21] or deoxyadenosine, 15 mg/liter. Inoculum cultures are prepared by transferring a loopful of bacteria from a fresh working stock culture to a test tube containing 10 ml of Micro-Inoculum Broth (Difco). The organisms have a generation time of 65–70 min and are in the logarithmic phase of growth from the fourth to approximately the twelfth hour of growth. After incubation at 37°, turbidity is measured in a Klett colorimeter using a No. 66 filter.[19] Large-scale cultures of L. leichmannii are grown by inoculating 15 liters of medium containing supplemental deoxyadenosine (15 mg/liter) with washed cells from a 10-ml overnight culture in Micro-Inoculum Broth (Difco). After cultivation for 12 hr at 37° in a large carboy without agitation (turbidity, about 210 Klett units, No. 66 filter), the culture is diluted with 60 liters of fresh medium identical in composition except that deoxyadenosine is omitted. The culture is incubated in carboys for 4 hr more, cooled under tap water, and harvested in a Sharples centrifuge. Cells (wet weight, about 125 g) are washed with saline, suspended in 330 ml of TSM buffer, and disrupted for 10 min by sonic oscillation in a cooled vessel. Cells harvested from smaller cultures are washed, suspended in 3 volumes of buffer, and exposed to sonic oscillation for 5 min.

*Preparation of Ribosome Suspensions.* Sonic extracts are adjusted to pH 6.5 with 5 $N$ NaOH and freed of debris and unbroken cells by slow centrifugation. The supernatant fraction is then centrifuged for 4 hr in a Spinco Model L ultracentrifuge at 144,000 $g$ (No. 50 rotor). Supernatant fractions are removed, and pellets are suspended in TSM buffer to a concentration of 20 mg/ml. Protein is determined by the method of Lowry et al.[22]

Purification Procedure

Ribosomes from large-scale cultures of L. leichmannii are washed in TSM buffer and stored at $-20°$ until use. All purification steps are performed at 0–4°. A typical purification is summarized in Table I.

*Salt Extraction of Ribosomes.* To dissociate enzyme from ribosomes, CsCl in TSM buffer is added with stirring to a suspension of ribosomes (20

---

[20] Difco Laboratories, *Difco Supplementary Literature*, Detroit (1962).
[21] This is a slightly limiting concentration of vitamin $B_{12}$.[19] A concentration that would just permit optimal growth is 0.15 $\mu$g/liter. The lower concentration is chosen because intracellular vitamin $B_{12}$ may repress synthesis of adenosylating enzyme.[12]
[22] O. H. Lowry, N. J. Rosebrough, A. L. Farr, and R. J. Randall, *J. Biol. Chem.* **193**, 265 (1951).

| Step | Total protein (mg) | Total units | Percent recovery | Specific activity (units/mg) |
|---|---|---|---|---|
| Crude sonic extract | 9500 | 20,000 | 100 | 2.1 |
| Washed ribosomes | 1633 | 12,900 | 65 | 7.9 |
| CsCl extract of ribosomes | 1200 | 20,700 | 104 | 17.2 |
| Ammonium sulfate, 32–75% | 960 | 19,700 | 99 | 20.5 |
| Hydroxyapatite chromatography | 115 | 6,500 | 33 | 56.5 |
| Calcium phosphate gel–cellulose chromatography | 20 | 2,800 | 14 | 140 |

[a] Purification scheme is described in the text. A unit is defined as the amount of enzyme catalyzing the synthesis of 1 nmol of AdoCbl per hour.

mg/ml in TSM buffer) to a final concentration of 2 $\overset{\circ}{M}$ (volume, approximately 100 ml). The mixture is allowed to stand in ice for 1 hr and then centrifuged at 144,000 $g$ for 11 hr in a Spinco Model L ultracentrifuge (No. 50 rotor). The upper 85–90% of each supernatant fraction is removed and fractions are pooled.

*Ammonium Sulfate Fractionation.* Since dialysis of the CsCl extracts against low ionic strength buffers causes precipitation of the enzyme, the CsCl-liberated protein is dialyzed for 3 hr against 3 liters of 32% ammonium sulfate in 50 m$M$ Tris · HCl (adjusted to pH 8.0 with $NH_4OH$). The resulting ammonium sulfate precipitate is removed by centrifugation. The supernatant fraction is dialyzed against 3 liters of 75% ammonium sulfate in 50 m$M$ Tris-HCl (adjusted to pH 8.0 with $NH_4OH$) for 7 hr. The precipitate is collected by centrifugation. This step affords a meager purification but yields a soluble protein.

*Hydroxyapatite Chromatography.* A hydroxyapatite column (2.5 × 8 cm) is washed under pressure for 24 hr with 1 liter of 20 m$M$ potassium phosphate, pH 6.9, 0.4 $M$ in NaCl. The 32–75% ammonium sulfate fraction is dissolved in 25 ml of the same buffer, dialyzed against the same buffer to remove traces of ammonium sulfate, and pumped through the column. A linear gradient is applied in a total volume of 800 ml with limits of 20 m$M$ to 0.2 $M$ potassium phosphate, pH 6.9, 0.4 $M$ in NaCl. Flow rate is 0.5 ml per min; 10-ml fractions are collected. Two major protein peaks appear. The transition zone between the two peaks contains most of the adenosylating enzyme activity; the activity peak emerges at a concentration of 0.11 $M$ on the gradient. Fractions having high activity are pooled.

*Calcium Phosphate Gel–Cellulose Chromatography.* Pooled active fractions from the hydroxyapatite column are dialyzed against 2 liters of 0.01 $M$ potassium phosphate buffer, pH 7.5, 0.0025 $M$ in mercaptoethanol. About 10 mg of protein are added to a calcium phosphate gel–cellulose column (3.5 × 20 cm) prepared according to Massey,[23] which is eluted with a linear gradient in a total volume of 600 ml with limits of 10 m$M$ and 0.15 $M$ potassium phosphate, pH 7.5, 2.5 m$M$ in mercaptoethanol. A major peak of enzyme activity occurs at a buffer concentration of 0.12 $M$. Enzyme is precipitated from pooled peak tubes by collecting a 40 to 70% ammonium sulfate fraction. Specific enzyme activity of about 280 can be obtained by preparatory gel electrophoresis, but this step has not been adequately reproducible. Studies of enzyme properties to be described below are obtained from incubations with fractions from the hydroxyapatite step (specific activity, 48–60) or calcium gel step (specific activity, 120–140).

*Enzyme Stability.* All fractions, especially those of later stages, are found to be unstable at −20° in the buffer in which they had been prepared, losing more than 50% activity in a week. Glycerol in concentrations of 25–50% decreases this loss to approximately 20%. Mercaptoethanol, 2.5 m$M$, is added empirically to all fractions. It is standard procedure to dialyze pooled active fractions from the calcium phosphate gel–cellulose column against 50 m$M$ Tris · HCl, pH 8.0 and store the resulting material in 50% glycerol at −22°. Adenosylating enzyme of ribosomes is stable under these conditions, but activity decreases after repeated freezing and thawing.

## Properties

*Cofactor Requirements.* Adenosylating enzyme purified from *L. leichmannii* ribosomes has absolute requirements for vitamin $B_{12s}$ (i.e., vitamin $B_{12a}$ plus $KBH_4$), ATP, and $Mn^{2+}$ or other cation (Table II). $Mn^{2+}$ (0.33 m$M$) maximally stimulates the reaction, but $Mg^{2+}$ and $Cs^+$ and, to a lesser extent, $Na^+$ and $K^+$ are also stimulatory. $Ca^{2+}$, $Cd^{2+}$, and $Zn^{2+}$ are inhibitory. $Mn^{2+}$ is inhibitory at a concentration of 10 m$M$. ATP cannot be replaced by ADP or AMP. Pyrophosphate and tripolyphosphate inhibit the reaction.

The enzyme reaction rate is linearly dependent upon enzyme concentration with both purified enzyme and intact ribosomes, and it is proportional to duration of incubation for periods up to 45 min provided that not more than two-thirds of the vitamin $B_{12s}$ had been consumed. When concentrations of ATP and vitamin $B_{12s}$ are systematically varied in otherwise standard incubations, typical hyperbolic curves were obtained at moder-

[23] V. Massey, *Biochim. Biophys. Acta* **37**, 310 (1960).

TABLE II
REQUIREMENTS OF ADENOSYLATING ENZYME[a]

| Expt. No. | Incubation mixture | AdoCbl formed (nmol) |
|---|---|---|
| 1 | Complete | 5.50 |
| | Minus hydroxocobalamin | 0.02 |
| | Minus $KBH_4$ | 0.02 |
| | Minus ATP | 0.02 |
| | Minus $Mn^{2+}$ | 2.02 |
| | Minus $Mn^{2+}$, plus EDTA | 0.11 |
| | Minus enzyme | 0.03 |
| 2 | Complete | 5.40 |
| | Minus $Mn^{2+}$ | 2.21 |
| | Minus $Mn^{2+}$, plus $Mg^{2+}$ | 3.78 |
| | Minus $Mn^{2+}$, plus $Cs^+$ | 3.45 |
| | Minus $Mn^{2+}$, plus $K^+$ | 2.98 |
| | Minus $Mn^{2+}$, plus $Na^+$ | 2.67 |
| | Minus $Mn^{2+}$, plus $Ca^{2+}$ | 0.01 |
| | Minus $Mn^{2+}$, plus $Zn^{2+}$ | 0.01 |
| | Minus $Mn^{2+}$, plus $Cd^{2+}$ | 0.00 |
| 3 | Complete | 5.50 |
| | Minus ATP | 0.01 |
| | Minus ATP, plus ADP | 0.80 |
| | Minus ATP, plus AMP | 0.02 |
| | Plus orthophosphate | 6.32 |
| | Plus pyrophosphate | 1.57 |
| | Plus tripolyphosphate | 0.64 |

[a] Standard incubation conditions were used with the indicated omissions and additions. The enzyme was 0.2 mg of hydroxyapatite fraction. Final concentration of added cations was 0.33 m$M$; of EDTA, 5 m$M$; of adenosine phosphate, 0.22 m$M$; and of inorganic phosphates (sodium salt), 10 m$M$. Mixtures were incubated for 30 min.

ate concentrations. Double-reciprocal plots gave apparent $K_m$s of $1.3 \times 10^{-5}$ $M$ for ATP and $0.75 \times 10^{-5}$ $M$ for vitamin $B_{12s}$. Apparent depression of activity observed at high vitamin $B_{12s}$ concentration can be shown to reflect inhibition of the ribonucleotide reductase system employed in the AdoCbl assay. The pH optimum is 8.0–8.6 with both assay procedures.

When studies are performed to determine whether other ribonucleotide triphosphates can replace ATP in the adenosylating reaction, it is found by means of the spectrophotometric assay of Vitols et al.[10] that UTP is 35% as active as ATP. GTP and CTP were 16% and 10% as active, respectively.

*Fate of ATP.* Studies with [8-$^{14}$C]ATP demonstrate that ATP is the source of the adenosyl moiety of AdoCbl (Table III). Fractionation of

TABLE III
ORIGIN OF ADENOSYL MOIETY OF AdoCbl[a]

| Incubation mixture | [8-$^{14}$C]ATP incorporated into AdoCbl (nmol) |
|---|---|
| Complete | 6.70 |
| Minus hydroxocobalamin | 0.01 |
| Minus enzyme | 0.02 |

[a] A reaction mixture containing hydroxocobalamin, 10 nmol; [8-$^{14}$C]ATP, 16.7 nmol, 0.5 $\mu$Ci; MnCl$_2$, 0.10 $\mu$mol; Tris · HCl buffer, pH 8.0, 15 $\mu$mol; and hydroxyapatite fraction, 0.15 mg, in a volume of 0.3 ml was prepared in a small Thunberg tube. After addition of KBH$_4$ under nitrogen, the tube was incubated at 37° for 45 min in the dark and cooled to 0°. An aliquot applied to a poly(ethyleneimine)-cellulose thin-layer plate was developed in the dark with 0.15 M LiCl that had been saturated with boric acid and neutralized with concentrated NH$_3$. The $R_f$ value of authentic AdoCbl in this system is 0.81. ATP remains at the origin. Radioactivity was assayed by scanning the plate in a Vanguard automatic scanner with intergrating circuit. The amount of [8-$^{14}$C]ATP incorporated into AdoCbl was calculated from the percentage of total radioactivity in the AdoCbl region. In a parallel incubation with nonradioactive ATP, 7.9 nmol of AdoCbl were formed.

reaction mixtures by Dowex 1 (Cl$^-$) after incubation with [$\gamma$-$^{32}$P]ATP demonstrates that inorganic tripolyphosphate is a stoichiometric product of the adenosylation reaction (Fig. 1).

*Distribution of Activity in Subcellular Fractions.* Much evidence suggests that the vitamin B$_{12s}$ adenosylating enzyme of *L. leichmannii* is ribosome-associated.[12] As shown in Table IV, when a sonic extract of cells is fractionated by differential sedimentation, washed ribosomes contain 66% of the adenosylating enzyme activity and their specific activity is 3.8 times that of whole sonic extract. It is noteworthy that 144,000 g supernatant fractions are essentially devoid of enzyme activity.

Further evidence of an association between the ribosomes and adenosylating enzyme can be obtained in sucrose density gradient analyses of intact and variously treated ribosomes. By the use of $^{14}$C-labeled *Escherichia coli* ribosomes [prepared by cultivating *E. coli* B in minimal medium containing 0.2% glucose and 500 $\mu$Ci per liter of [$^{14}$C]adenine (10 mCi/nmol), according to Cohen and Rickenberg[24] as a reference standard, the major $A_{260}$ peaks can be assigned sedimentation constants of 70 S, 50 S, and 30 S. Intact washed ribosomes, obtained as described in Table IV, yield the pattern in Fig. 2A. The main peak of enzyme activity coincides with the 70 S ultraviolet-absorbing peak. Specific enzyme activity is low in the 30 S region and relatively high in

[24] G. N. Cohen and H. V. Rickenberg, *Ann. Inst. Pasteur* **91**, 693 (1956).

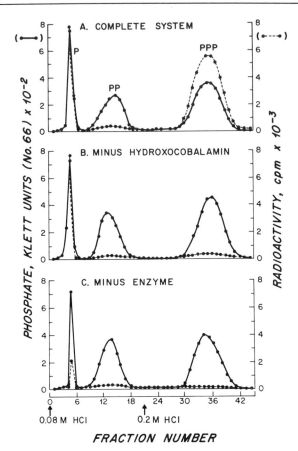

FRACTION NUMBER

FIG. 1. Identification of inorganic tripolyphosphate as a product of adenosylation reaction. Products of ATP cleavage were identified by a modification of the procedures of A. Peterkofsky and H. Weissbach [*Ann. N. Y. Acad. Sci.* **112**, 622 (1964)]. A reaction mixture containing hydroxocobalamin, 15 nmol; [$\gamma$-$^{32}$P]ATP, 46 nmol, 1 $\mu$Ci; MnCl$_2$, 0.13 $\mu$mol; Tris · HCl buffer, pH 8.0, 20 mmol; and hydroxyapatite fraction, 0.2 mg, in a volume of 0.4 mg was incubated in the dark at 37° in a Thunberg tube after reduction with KBH$_4$. The reaction was terminated by addition of 0.5 ml of a suspension of charcoal (Norit A) in 4% HClO$_4$. After filtration through glass wool, 5 $\mu$mol of potassium orthophosphate were added. The filtrate was neutralized, and KClO$_4$ was removed. The supernatant solution, diluted to 10 ml, was fractionated on a 0.5 × 2.8 cm Dowex 1 (Cl$^-$) column by elution with 60 ml of 80 m$M$ HCl and 80 ml of 0.2 $M$ HCl. Fractions (volume, 3.2 ml) were assayed for inorganic phosphate according to C. H. Fiske and Y. SubbaRow [*J. Biol. Chem.* **66**, 375 (1925)] and for radioactivity. A, Elution diagrams obtained from complete incubation mixture; B, complete mixture minus hydroxocobalamin; C, complete mixture minus enzyme.

TABLE IV
DISTRIBUTION OF ADENOSYLATING ENZYME IN SUBCELLULAR FRACTIONS

| Fraction | Total protein (%) | Total activity (%) | Specific activity (units/mg) |
|---|---|---|---|
| Whole sonic extract | 100 | 100 | 1.98 |
| Pellet from 25,000 $g$ | 27 | 29 | 2.21 |
| Washed ribosomes (144,000 $g$ pellet) | 17 | 66 | 7.50 |
| Sludge above ribosomes | 20 | 2 | 0.21 |
| Ribosome washings | 9 | 2 | 0.43 |
| Supernatant from 144,000 $g$ | 27 | 1 | 0.10 |

denser regions. Exposure of ribosomes to sodium deoxycholate causes the major $A_{260}$ peak to shift to the 50 S region (Fig. 2B). Enzyme activity continues to coincide with the $A_{260}$ peak; however, deoxycholate treatment diminishes the total amount of enzyme activity recoverable in the gradient. Exposure of ribosomes in TSM buffer to 15 m$M$ EDTA, in order to remove "nascent" proteins,[25] shifts the main $A_{260}$ peak to the 40 S region (Fig. 2C). The peak of enzyme activity remains nearly coincidental.

*Effects of Ribonuclease Treatment on Ribosomes.* When the effects of treatment of ribosomes with ribonuclease is studied by incubating washed ribosomes suspended in TSM buffer with pancreatic ribonuclease, 10 $\mu$g/ml, the time course of adenosylating enzyme activity in aliquots of the digest is that shown in Fig. 3A. After a small, regularly observed initial rise, activity slowly declines. Sucrose density gradient analyses of ribosomes digested with ribonuclease for 10 and 30 min (Figs. 3B and 3C) show accumulations of ultraviolet-absorbing material at the top and to a lesser degree in the denser regions of the gradients. Enzyme activity peaks shifts into denser regions of the gradients, where presumably polyribosomes or ribosomal aggregates are located. These results suggest that although adenosylating enzyme is only partly eliminated by ribonuclease treatment, remaining activity can be found only in association with heavy sedimenting particles.

*Removal of Enzyme from Ribosomes.* It is known that vitamin $B_{12}$ binding substances are released from *L. leichmannii* ribosomes in concentrated salt solutions (2.5 $M$ NaCl or CsCl) and, under appropriate conditions, reassociate with them.[26,27] Salt treatment concurrently dissociates ribosomes to smaller units with sedimentation constants of approximately 29 S and 38 S.

[25] K. Terao, H. Katsumi, H. Sugano, and K. Ogata, *Biochim. Biophys. Acta* 138, 369 (1967).
[26] S. Kashket, J. T. Kaufman, and W. S. Beck, *Biochim. Biophys. Acta* 64, 447 (1962).
[27] S. Kashket and W. S. Beck, *Biochem. Z.* 342, 449 (1965).

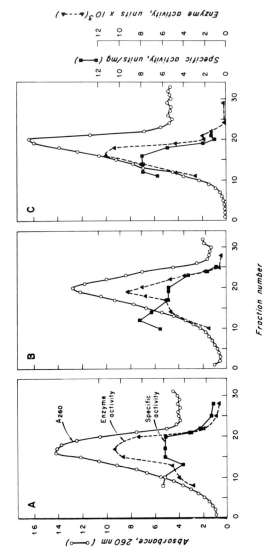

FIG. 2. Sucrose density gradient analysis of intact and variously treated ribosomes. Labels designate curves depicting absorbance at 260 nm, adenosylating enzyme activity (expressed as nanomoles of AdoCbl formed in 45 min), and specific enzyme activity (expressed as nanomoles of AdoCbl formed per $A_{260}$ unit). (A) Intact washed ribosomes, 2.7 mg, in gradient containing Tris-succinate buffer and 5 m$M$ Mg²⁺. (B) Ribosomes after exposure to 0.5% sodium deoxycholate (final concentration) for 10 min at 37° prior to centrifugation. (C) Ribosomes after exposure to 15 m$M$ EDTA (final concentration) for 10 min at 37° prior to centrifugation in a sucrose density gradient containing 15 m$M$ EDTA.

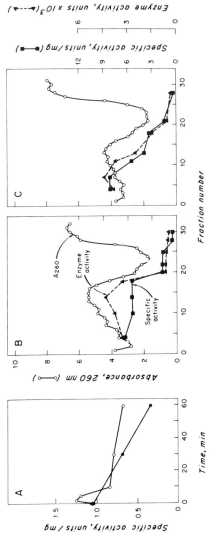

FIG. 3. Effects of ribonuclease treatment on sedimentation behavior and adenosylating enzyme activity of ribosomes. (A) Time course of effect on adenosylating enzyme activity. A buffer suspension of 0.3 mg of washed ribosomes was mixed with pancreatic ribonuclease (final concentrations, 10 μg/ml, open circles, and 20 μg/ml, closed circles), incubated at 37°, and sampled at timed intervals for enzyme assays. Abscissa, time in minutes; ordinate, specific enzyme activity in units per milligram. (B and C) Sucrose density gradient analyses of ribosomes treated with ribonuclease for 10 and 30 min, respectively. Curves are labeled as in Fig. 2.

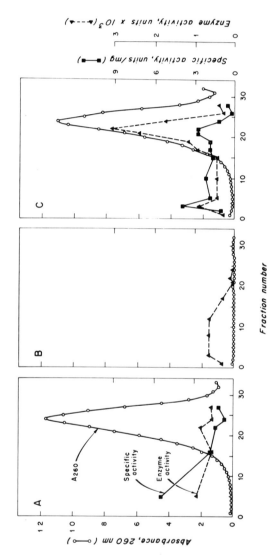

FIG. 4. Sucrose density gradient analysis studies of the reassociation of released adenosylating enzyme with "stripped" ribosomes. (A) Ribosomes (0.54 mg) previously treated with CsCl. (B) Dialyzed "strippate" (1.8 mg). (C) Pattern obtained with "stripped" ribosomes (0.54 mg), after prior addition of 1.8 mg of protein that has been released by CsCl from a separated sample of ribosomes and incubated for 20 min. Curves are labeled as in Fig. 2.

TABLE V
COMPETENCE OF WASHED AND "STRIPPED" RIBOSOMES IN PROTEIN SYNTHESIS[a]

| Expt. no. | Incubation | Radioactivity on filter (cpm) |
|---|---|---|
| 1 | Complete system (intact *L. leichmannii* ribosomes) | 3,380 |
| | Minus ribosomes | 1,200 |
| | Minus ribosomes, tRNA, supernatant | 728 |
| 2 | Complete system ("stripped" *L. leichmannii* ribosomes) | 2,900 |
| | Minus ribosomes | 880 |
| | Minus ribosomes, tRNA, supernatant | 672 |
| 3 | Complete system (intact *E. coli* ribosomes) | 17,783 |
| | Minus ribosomes | 1,183 |
| 4 | Complete system ("stripped" *E. coli* ribosomes) | 2,210 |

[a] Complete incubation mixtures contained: Tris · HCl, pH 7.9, 15 $\mu$mol; NH$_4$Cl, 20 $\mu$mol; 2-mercaptoethanol, 4 $\mu$mol; magnesium acetate, 4.5 $\mu$mol; ATP, 0.3 $\mu$mol; GTP, 0.08 $\mu$mol; phosphocreatine, 4.5 $\mu$mol; creatine kinase, 16 $\mu$g; *Escherichia coli* tRNA, 2.5 $\mu$g; washed (experiments 1 and 3) or CsCl-stripped (experiments 2 and 4) ribosomes from *Lactobacillus leichmannii* (experiments 1 and 2) or *E. coli* (experiments 3 and 4), 1.0 mg (protein); *L. leichmannii* 100,000 *g* supernatant fraction, 0.5 mg; [$^{14}$C]phenylalanine, 50 nmol (10 mCi/mol); and poly(U), 50 $\mu$g. Incubations (volume, 0.25 ml) were conducted for 30 min at 37° and terminated with 0.05 ml of 30% trichloroacetic acid. Mixtures were heated in a steam bath for 15 min and passed through a Millipore filter. Filters were assayed for radioactivity in a liquid scintillation counter.

Vitamin B$_{12s}$ adenosylating enzyme is also removed from ribosomes by salt treatment. This is shown in an experiment in which ribosomes incubated in 2.5 *M* CsCl in TSM buffer for 60 min in the cold were centrifuged at 144,000 *g* for 11 hr. The supernatant fraction is divided into 4 fractions, from the top to the bottom of the tube. Each fraction, and the resuspended pellet, is dialyzed against two changes of 2 liters of TSM buffer. Control tubes are treated similarly except for the CsCl treatment. About 70% of the ribosomal protein is recovered in the 144,000 *g* supernatant fraction in the CsCl-treated samples, whereas 90% or more of the original RNA is found in the pellet in both treated and control samples. Adenosylating enzyme is also released into the supernatant fraction by CsCl treatment. Notably, total enzyme units recovered are almost doubled by this treatment. When "stripped" ribosomes are subjected to sucrose density gradient analysis (Fig. 4A), they are virtually lacking in enzyme activity. Gradient analysis of dialyzed "strippate" (Fig. 4B) showed it to contain some enzyme activity at the bottom of the gradient. Similar results are obtained when ribosomes are treated initially with 0.5 *M* NH$_4$Cl according to Ravel.[28] Evidence that "stripped" ribosomes can reassociate with released enzyme protein is shown in Fig. 4C.

*Capacity of "Stripped" Ribosomes to Synthesize Protein.* The metabolic integrity of intact ribosomes and ribosomes stripped of adenosylating enzyme in salt solution is demonstrable in studies of the competence of such ribosomes in a standard system for the assay of amino acid incorporation into protein. In a system patterned after that of Salas *et al.*,[29] results (Table V) indicate that (*a*) intact *L. leichmannii* ribosomes are active in protein synthesis; (*b*) "stripping" them with CsCl does not greatly diminish this activity; (*c*) intact *E. coli* ribosomes are considerably more active in the system used; and (*d*) "stripping" of *E. coli* ribosomes under similar conditions does impair their activity. Although the reason for the difference between the ribosomes of the two species is not clear, it appears that adenosylating enzyme is not an integral component of *L. leichmannii* ribosomes, that the association between enzyme and ribosome is a loose and reversible one, and that removal of the enzyme does not impair ribosomal function in protein synthesis.

*Ribosome-Associated Adenosylating Enzyme in Other Species.* Studies to determine whether coenzyme synthesis is associated with ribosomes in other organisms, among them lactobacilli capable of growing on vitamin B$_{12}$, have been performed on *Lactobacillus delbrueckii* (ATCC 9649), lactobacilli indifferent to vitamin B$_{12}$, e.g., *Lactobacillus acidophilus* (ATCC 11506), and *Lactobacillus casei* (ATCC 9595), and other organisms known or presumed to contain adenosylating enzyme, e.g., *C. tetanomorphum* (ATCC 3606) and *Propionibacterium shermannii* (ATCC 9614). Such studies show that substantial adenosylating activity is associated with ribosomes only in the vitamin B$_{12}$-requiring lactobacilli *L. delbrueckii* and *L. leichmannii.* Surprisingly, *L. casei* also displays some activity. *Lactobacillus acidophilus* has no measurable activity. The bulk of the activity in *C. tetanomorphum* and *P. shermannii* is found in the 144,000 *g* supernatant fraction. Ribosomes from animal cells (rat liver, reticulocytes, and Ehrlich ascites tumor) are without detectable coenzyme synthetase or adenosylating enzyme activity.

[28] J. M. Ravel, *Proc. Natl. Acad. Sci. U.S.A.* **57**, 1181 (1967).
[29] M. Salas, M. A. Smith, W. M. Stanley, Jr., A. J. Wahba, and S. Ochoa, *J. Biol. Chem.* **240**, 3988 (1965).

## [8] Immobilized Derivatives of Vitamin B$_{12}$ Coenzyme and Its Analogs

### By TETSUO TORAYA and SABURO FUKUI

The immobilized derivatives of cobalamins are useful not only for purification of cobalamin-dependent enzymes or cobalamin-binding proteins, but also for basic studies on biospecific interactions between cobalamins and apoenzymes or cobalamin binders. Methods of preparing several immobilized cobalamins have been reported by Olesen et al.,[1] Yamada and Hogenkamp,[2] Allen et al.,[3,4] Schlecht and Müller,[5] and Lien et al.[6] Some of them proved to be effective affinity adsorbents for ribonucleotide reductase[2] and cobalamin binders.[3,4,7-10] The aspects of affinity chromatography have been described in this series[11] and are mentioned here only briefly. For application of immobilized cobalamins to the fundamental enzymological studies on the structure-binding relationship of cobalamin coenzymes, many more types of cobalamin derivatives immobilized, at a different part of the molecule, would be necessary. This chapter outlines methods for preparing several new corrinoid affinity adsorbents and also briefly describes their application for a basic study of the interaction of the cobalamin coenzyme with the apoenzyme of cobalamin-dependent diol dehydrase (1,2-propanediol hydro-lyase, EC 4.2.1.18).[12]

### Principle

Vitamin B$_{12}$ coenzyme (adenosylcobalamin) is an organocobalt compound that has a very complicated structure. The molecule is composed of the three major parts—a corrin ring (equatorial quadridentate), a lower nucleotide ligand moiety (axial fifth ligand), and an upper nucleoside

[1] H. Olesen, E. Hippe, and E. Haber, Biochim. Biophys. Acta 243, 66 (1971).
[2] R. Yamada and H. P. C. Hogenkamp, J. Biol. Chem. 247, 6266 (1972).
[3] R. H. Allen and P. W. Majerus, J. Biol. Chem. 247, 7695 (1972).
[4] R. H. Allen and C. S. Mehlman, J. Biol. Chem. 248, 3670 (1973).
[5] N. Schlecht and O. Müller, Z. Naturforsch. 28C, 351 (1973).
[6] E. L. Lien, L. Ellenbogen, P. Y. Law, and J. M. Wood, J. Biol. Chem. 249, 890 (1974).
[7] R. H. Allen and P. W. Majerus, J. Biol. Chem. 247, 7702, 7709 (1972).
[8] R. H. Allen and C. S. Mehlman, J. Biol. Chem. 248, 3660 (1973).
[9] R. L. Burger and R. H. Allen, J. Biol. Chem. 249, 7220 (1974).
[10] R. L. Burger, C. S. Mehlman, and R. H. Allen, J. Biol. Chem. 250, 7700 (1975).
[11] R. H. Allen, R. L. Burger, C. S. Mehlman, and P. W. Majerus, this series, Vol. 34, p. 305.
[12] T. Toraya and S. Fukui, J. Biol. Chem., in press.

ligand moiety (axial sixth ligand). Important information regarding the interacting sites of the coenzyme with apoenzymes would be expected to result from examining various immobilized corrinoids for their effectiveness as affinity adsorbents. For this purpose, we have synthesized several new cobalamin derivatives attached through either the corrin ring, the lower ligand, or the upper ligand.

Immobilization of corrinoids is accomplished by one of two general procedures: (a) reaction of ω-aminoalkyl-Sepharose with the carboxylic acid derivatives of corrinoids in the presence of 1-ethyl-3 (3-dimethylaminopropyl)carbodiimide; (b) reaction of cyanogen bromide-activated Sepharose with the amino derivatives of corrinoids.[12]

### Synthesis of Carboxylic Acid Derivatives of Corrinoids

*Cyanocobalamin O$^{5'}$-Succinic Acid*[13]

Cyanocobalamin (200 mg, 0.15 mmol) is dissolved in 40 ml of dimethyl sulfoxide containing 8 g of succinic anhydride and 6.4 ml of pyridine. After 12 hr at room temperature, the excess succinic anhydride is destroyed by adding 500 ml of water and keeping the pH of the reaction mixture at 6 with 10% KOH. KCN is then added to a final concentration of 10 m$M$, and the pH of the solution is readjusted to 6 with 3 $N$ HCl. After 1 hr the corrinoids are desalted by phenol extraction and applied to a column (2.5 × 20 cm) of DEAE-cellulose (acetate form). The effluent contains cyanocobalamin that has not reacted. Monosuccinyl derivatives are eluted with 0.08% acetic acid and are separated into two fractions during elution. The minor component (2% of the total cobalamin) eluted first is cyanocobalamin $O^{2'}$-succinic acid, and the main component (87% of the total cobalamin) is cyanocobalamin $O^{5'}$-succinic acid. The $O^{2'},O^{5'}$-disuccinyl derivative remains adsorbed on the column under these conditions.

*Co β-2-Carboxyethylcobalamin and Co β-3-Carboxypropylcobalamin*[12]

Cyanocobalamin (100 mg, 0.07 mmol) is dissolved in 50 ml of a 10% NH$_4$Cl solution. Two grams of zinc powder are added to the solution, and the suspension is stirred vigorously under an He atmosphere for 30 min. To the resulting brownish suspension are added 500 mg of 3-bromopropionic acid, and the mixture is allowed to stand for 15 min in the

[13] T. Toraya, K. Ohashi, H. Ueno, and S. Fukui, *Bioinorg. Chem.* **4**, 245 (1975); *Bitamin* **48**, 557 (1974).

dark. After desalting by phenol extraction, $Co\beta$-2-carboxyethylcobalamin is isolated by DEAE-cellulose (acetate form) chromatography (essentially as described for the purification of cyanocobalamin succinic acid). The yield is approximately 90%.

$Co\beta$-3-Carboxypropylcobalamin is similarly synthesized by allowing reduced cobalamin to react with 4-chlorobutyric acid. This analog can also be coupled to $\omega$-aminoalkyl-Sepharose to give a homologous alkylcobalamin agarose. Application of Sepharose-bound cobalamin of this type for purification of $N^5$-methyltetrahydrofolate-homocysteine methyltransferase has been reported recently.[14]

### 1-Carboxymethyladenosylcobalamin[12]

Adenosylcobalamin (410 mg, 0.26 mmol) is added to 2 ml of an aqueous solution of iodoacetic acid (0.2 g) that has previously been adjusted to pH 6.5 with $2 N$ LiOH. The solution is readjusted to pH 6.5 and allowed to stand in the dark at 30°. Each day, the pH is checked and is adjusted to 6.5, when necessary, with $2 N$ LiOH. After 12 days, most of the adenosylcobalamin is converted to 1-carboxymethyladenosylcobalamin, as judged from paper electrophoresis in 0.5 $N$ acetic acid. Corrinoids are desalted by phenol extraction and applied to a column (3.5 × 25 cm) of P-cellulose which has been adjusted to pH 2.7. The column is developed with distilled water. The product, 1-carboxymethyladenosylcobalamin, moves slowly as a red band and is eluted after the fast-moving yellowish brown band. 1-Carboxymethyladenosylcobalamin is purified to homogeneity by silica gel column chromatography using 1-butanol–2-propanol–water (10 : 7 : 10) as a solvent system, in a yield of approximately 35%.

### $N^6$-Carboxymethyladenosylcobalamin[12]

1-Carboxymethyladenosylcobalamin is intrinsically unstable and easily converted to $N^6$-carboxymethyladenosylcobalamin by the Dimroth rearrangement. 1-Carboxymethyladenosylcobalamin (143 mg, 0.09 mmol) is dissolved in 20 ml of 0.1 $M$ Tris. The pH of the solution is adjusted to 11 with $1 N$ NaOH, and the solution is heated at 70–72° for 1 hr to effect the rearrangement. After cooling, the solution is adjusted to pH 7.0 with $1 N$ HCl. Corrinoids are desalted and concentrated by the phenol extraction procedure and applied to a column (3.5 × 20 cm) of DEAE-cellulose (acetate form). The unrearranged 1-carboxymethyladenosylcobalamin is eluted with water. The rearranged product, $N^6$-

[14] K. Sato, E. Hiei, S. Shimizu, and R. H. Abeles, *FEBS Lett.* **85**, 73 (1978).

carboxymethyladenosylcobalamin, remains adsorbed on the top of the column and is eluted with 0.2% acetic acid, followed by desalting by phenol extraction. The overall yield is approximately 22%, based on adenosylcobalamin.

### Cyanocobalamin Monocarboxylic Acid and Polycarboxylic Acids

Since cobalamins contain seven amide side chains, cobalamin mono- and polycarboxylic acids are prepared by acid hydrolysis of cyanocobalamin.[4,15,16] Methods of preparing and immobilizing them have been described in this series[11] as well as elsewhere[2–4] and are not mentioned here. It has been reported that the three propionamide side chains are more susceptible to acid hydrolysis than the acetamide side chains.[16] Thus, preparation of the three monocarboxylic acid isomers, i.e., b-, d-, and e-carboxylic acids, and their immobilized derivatives, attached on Sepharose at different propionamide side chains, is possible.

### Synthesis of Amino Derivatives of Corrinoids

### Co β-2-Aminoethylcobalamin[12]

Cyanocobalamin (100 mg, 0.07 mmol) is dissolved in 20 ml of water. Three hundred milligrams of sodium borohydride are added to the solution, and the solution is allowed to stand for 20 min under an He atmosphere. To the resulting brownish solution is added 1 ml of ethyleneimine, and the mixture is kept in the dark for additional 20 min. The product is desalted by phenol extraction, and Co β-2-aminoethylcobalamin is obtained in an almost quantitative yield.

### Cyanocobalamin Aminododecylamide[2]

Cyanocobalamin aminododecylamide is prepared by the reaction of cyanocobalamin monocarboxylic acid with excess 1,12-diaminododecane in the presence of 1-ethyl-3(3-dimethylaminopropyl)carbodiimide (for details of experimental conditions, see the original report[2]).

[15] J. B. Armitage, J. R. Cannon, A. W. Johnson, L. F. J. Parker, E. L. Smith, W. H. Stafford, and A. R. Todd. *J. Chem. Soc., London,* 3849 (1953).
[16] K. Bernhauer, F. Wagner, H. Beisbarth, P. Rietz, and H. Vogelmann, *Biochem. Z.* **344**, 289 (1966).

General Procedures for Immobilization of Corrinoid Derivatives

*Coupling of Carboxylic Acid Derivatives of Cobalamins to*
*ω-Aminohexyl-Sepharose 4B (Procedure a)[12]*

ω-Aminohexyl-Sepharose 4B is prepared by allowing 1,6-diaminohexane to react with cyanogen bromide-activated Sepharose 4B.[17] A carboxylic acid derivative of corrinoids (approximately 50 μmol) is dissolved in 4 ml of water and added to 6 ml of packed ω-aminohexyl-Sepharose. The pH of the suspension is adjusted to 5 with 2 N HCl; with stirring continued at room temperature, 100 mg of 1-ethyl-3(3-dimethylaminopropyl)carbodiimide are added in 20-mg aliquots at 1-min intervals. After 30 and 60 min, an additional 20 mg of carbodiimide are added to the suspension, and gentle stirring is continued overnight at room temperature. The corrinoid-Sepharose is collected by vacuum filtration in a Büchner funnel, washed with copious water, 0.5 M potassium phosphate buffer (pH 7.0), and 50% aqueous dimethylformamide, and stored in 0.1 M potassium phosphate buffer (pH 7.0) at 0–5°.

*Coupling of Amino Derivatives of Corrinoids to Cyanogen*
*Bromide-Activated Sepharose 4B (Procedure b)[12]*

Sepharose 4B (20 ml of packed gel) is washed and suspended in an equal volume of water. Cyanogen bromide (250 mg per milliliter of packed Sepharose) is added to the suspension, and the mixture is stirred vigorously at room temperature. The pH of the suspension is maintained at 11 with 5 N NaOH. After the proton release has ceased, a large amount of ice is added to the suspension, and the Sepharose is washed quickly with 500 ml of cold water. The activated Sepharose is added immediately to a solution of an amino derivative of corrinoids (approximately 40 μmol) in 20 ml of 0.1 N sodium carbonate buffer (pH 10). After stirring overnight at 2–5°, the corrinoid-Sepharose is collected in a Büchner funnel, washed with copious water, 0.5 M potassium phosphate buffer (pH 7.0), and 50% aqueous dimethylformamide, and stored in 0.1 M potassium phosphate buffer (pH 7.0) at 0–5°.

Sometimes cyanogen bromide-activated groups may remain alive in corrinoid gels and react with proteins applied, resulting in their immobilization. In such cases, the excess activated groups can be blocked completely by treating cobalamin gels with 1 M glycine–1 M ethanolamine (pH 10) for either 15–16 hr at 0–5° or 2 hr at 30° (some loss of ligands occurs under these conditions).

[17] P. Cuatrecasas, *J. Biol. Chem.* **245,** 3059 (1970).

TABLE

IMMOBILIZED CORRINOID DERIVATIVES AND THEIR CORRINOID CONTENT[a]

| No. | Partial structure[b] | Site of attachment | Immobilization procedure | Amount of corrinoid bound[c] ($\mu$mol/ml packed gel) |
|---|---|---|---|---|
| (I) | | Corrin ring | a | 0.29 |
| (II) | | Corrin ring | b | 0.20 |
| (III) | | Corrin ring | b | 0.23 |
| (IV) | | Corrin ring | a | 0.50 |
| (V) | | Lower ligand | a | 0.15 |

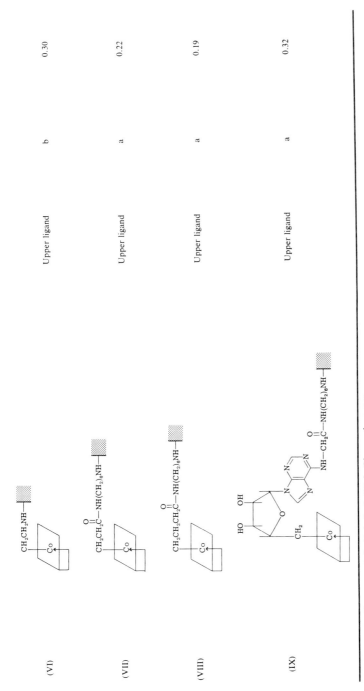

|  |  | Upper ligand | b | 0.30 |
| (VI) |  |  |  |  |
|  |  | Upper ligand | a | 0.22 |
| (VII) |  |  |  |  |
|  |  | Upper ligand | a | 0.19 |
| (VIII) |  |  |  |  |
|  |  | Upper ligand | a | 0.32 |
| (IX) |  |  |  |  |

[a] From T. Toraya and S. Fukui, J. Biol. Chem., in press.

[b] L = adenosyl, cytidyl, CH₃, CN, OH, etc.

[c] Values obtained under our experimental conditions. Immobilization of a larger amount of corrinoid would be possible by using a larger amount of the corrinoid derivatives.

Spectral Measurements of the Immobilized Corrinoids[12]

The Sepharose-bound cobalamin coenzyme or its analog is suspended in 1.2 ml of 0.05% aqueous solution of agarose. The spectra are taken using aminohexyl-Sepharose 4B in the reference cuvette. To determine the amount of corrinoid bound, the corrinoid gel is converted into the dicyano form (organo-corrinoid gels are photolyzed in the presence of 0.1 $M$ KCN with a 300 W tungsten bulb from a distance of 10–20 cm for 10 min), and the absorbance at 367 nm is measured. The molar extinction coefficient of the free dicyanocobalamin, $30.4 \times 10^3$ $M^{-1}$ cm$^{-1}$ at 367 nm,[18] is employed for estimation.

The table summarizes several examples of corrinoid derivatives immobilized on Sepharose at different sites of the molecule and the amount of corrinoid bound per 1 ml of the gels.

Change of an Upper Ligand to Cobalt of Corrinoid-Sepharose

*Conversion to the Adenosyl Coenzyme Form*[12]

A cyano form of corrinoid-Sepharose (1.5 ml) is suspended in an equal volume of 50 m$M$ potassium phosphate buffer (pH 8). To the suspension are added 50 mg of sodium borohydride under an He atmosphere. After about 10–20 minutes at room temperature, 25 mg of 5'-iodo-5'-deoxyadenosine in 3 ml of dimethylformamide are added to the suspension, and the mixture is allowed to stand for an additional 20 min with occasional shaking. The adenosyl form of corrinoid gels is collected by filtration through a glass filter, washed with copious water, 0.5 $M$ potassium phosphate buffer (pH 7.0), and 50% aqueous dimethylformamide, and stored in 0.1 $M$ potassium phosphate buffer (pH 7.0) at 0–5°.

*Conversion to Other Forms*[12]

Other alkyl forms of corrinoid gels are also synthesized by allowing the sodium borohydride-reduced cobalamin gels to react with alkyl halides. The procedure is essentially the same as that described for conversion into the adenosyl form, except that 25 mg of an appropriate alkyl halide are used instead of 5'-iodo-5'-deoxyadenosine. For instance, methyl and cytidyl forms of the gels are prepared by the use of methyl iodide and 5'-chloro-5'-deoxycytidine, respectively.

Hydroxocorrinoid-Sepharose gels are readily obtained by photolysis of adenosyl or other alkyl forms of corrinoid gels with a 300 W tungsten bulb from a distance of 10–20 cm for 10 min.

[18] H. A. Barker, R. D. Smyth, H. Weissbach, A. Munch-Petersen, J. I. Toohey, J. N. Ladd, B. E. Volcani, and R. M. Wilson, *J. Biol. Chem.* **235**, 181 (1960).

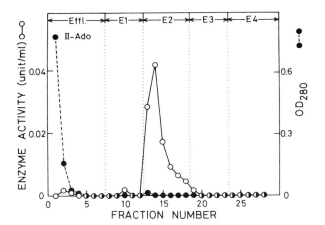

FIG. 1. Affinity chromatography of diol dehydrase on the adenosyl form of (II). About 1 unit of enzyme was applied to the adsorbent (1 ml of packed gel) in 0.1 $M$ potassium phosphate buffer (pH 8) containing 2% propanediol, in a total volume of 2 ml. Affinity chromatography was carried out as described in the text. Two-milliliter fractions were collected. Effl., effluent; E, eluate. From T. Toraya and S. Fukui, $J. Biol. Chem.$, in press.

Application to a Basic Study on Interaction between Adenosylcobalamin and Diol Dehydrase[12]

In order to get some important information on the interacting positions of cobalamins with the apoenzyme of diol dehydrase, the effectiveness as an affinity adsorbent is tested by affinity chromatography as follows: Crude extracts of $Klebsiella pneumoniae$ (formerly $Aerobacter aerogenes$) ATCC 8724 grown on 1,2-propanediol[19] or on glycerol-1,2-propanediol[20] (containing the apoenzyme of diol dehydrase) are applied to a corrinoid-Sepharose. To promote the binding, the suspension is incubated at 37° for 1–2 hr with gentle stirring. The suspension is then poured into a small column and developed successively with 0.1 $M$ potassium phosphate buffer (pH 8) containing 2% propanediol (effluent), 0.5 $M$ potassium phosphate buffer (pH 8) containing 2% propanediol (eluate 1), 0.3 $M$ Tris · HCl buffer (pH 8) containing 2% propanediol (eluate 2), 0.3 $M$ Tris · HCl buffer (pH 8) (eluate 3), and finally with 0.1 $M$ acetic acid containing 2% propanediol (eluate 4). A typical example with an adenosyl form of (II) is illustrated in Fig. 1. Diol dehydrase is completely absorbed on the gel and is not eluted by increasing the buffer concentration. The enzyme can be eluted with Tris · HCl buffer or triethanolamine · HCl buffer

[19] T. Toraya, S. Honda, S. Kuno, and S. Fukui, $J. Bacteriol.$ **135,** 726 (1978).
[20] R. H. Abeles, this series, Vol. 9, p. 686.

(pH 8) containing the substrate. The addition of potassium chloride in Tris · HCl buffer depresses the elution significantly. These findings indicate that removal of $K^+$ from the system effects dissociation of the enzyme from the affinity adsorbent, and they are consistent with the fact that $K^+$ or other monovalent cations of similar size are absolutely required for the binding of cobalamin coenzyme or its analogs to apodiol dehydrase.[21,22] Such elution, based on the dependence of the enzyme–coenzyme binding on an ionic atmosphere, would give a new and interesting example of a selective and mild elution method and might be applicable for some other systems.

Of the corrinoid-Sepharose derivatives mentioned above, the cobalamins immobilized through side chains of the corrin ring are the most effective affinity adsorbents. Although the enzyme is completely adsorbed on (II) irrespective of upper ligands, some leakage is observed with (I), and its degree is dependent on upper ligands. With its cyano form, the phenomenon is more significant than with its adenosyl form. These results suggest that a spacer of enough length is necessary for firm binding, and also that the enzyme binds to an adenosyl form of adsorbent more tightly. This indicates that the adenosyl moiety of the coenzyme is also important for specific interaction and is consistent with the finding that (IX) is not an effective affinity adsorbent. The fact that the enzyme is not adsorbed by (V) indicates that the specific interaction between enzyme and coenzyme is destroyed by attachment at the ribose of lower ligand. This is in agreement with the observation that a cyano-aqua form of nucleotide-lacking corrinoids immobilized via side chains of the corrin ring with 1,12-diaminododecane as a spacer (III–CN) does not serve as an affinity adsorbent, and it is suggestive of the important participation of the lower nucleotide ligand in the binding of cobalamins to the apoenzyme. Among the cobalamins immobilized through upper ligands, only (VIII) is a partially effective affinity adsorbent for diol dehydrase. Compound (VI) is not effective, presumably because the distance between the agarose matrix and the cobalamin moiety is too short. Compound (VII) is intrinsically unstable and is decomposed under the conditions for affinity chromatography. These aspects of the apoenzyme–coenzyme interactions, together with the results obtained by the use of free coenzyme analogs about the coenzyme's structure–function relationship,[23] provide important information for understanding the mechanism of action of vitamin $B_{12}$ coenzymes.

[21] T. Toraya, Y. Sugimoto, Y. Tamao, S. Shimizu, and S. Fukui, *Biochemistry* **10**, 3475 (1971).
[22] T. Toraya, M. Kondo, Y. Isemura, and S. Fukui, *Biochemistry* **11**, 2599 (1972).
[23] T. Toraya, K. Ushio, S. Fukui, and H. P. C. Hogenkamp, *J. Biol. Chem.* **252**, 963 (1977).

## [9] Solubilization of the Receptor for Intrinsic Factor–$B_{12}$ Complex from Guinea Pig Intestinal Mucosa

*By* MAX KATZ and BERNARD A. COOPER

The absorption of dietary vitamin $B_{12}$ from the intestine in many animal species requires the presence of intrinsic factor (IF), a glycoprotein secreted by the gastric mucosa. Normally this protein is produced in large excess over actual requirements and so is rarely a limiting factor in absorption. *In vitro* studies employing everted intestinal sacs,[1] mucosal homogenates,[2] and microvillous membranes[3] prepared from the distal, but not the proximal, small bowel show uptake of IF–$B_{12}$ complex in the presence of a physiologic pH and calcium ions. The implication of this is that a receptor for IF–$B_{12}$ complex exists at the surface of the ileal cell. Strong evidence supporting the existence of this specific receptor came from the work of Mackenzie and co-workers,[4] who showed that antisera to hamster ileal microvillous membranes, after absorption with jejunal microvilli, contained an antibody that inhibited binding of IF–$B_{12}$ by ileal mucosa. Our recent report[5] of a patient who produced an abnormal IF that was not recognized by the ileal mucosa is, by corollary, conclusive evidence that a receptor for IF–$B_{12}$ does exist on the ileal mucosa. Recent work in our laboratory has demonstrated a close relationship between the total estimated receptor available on the guinea pig intestine and the maximum amount of vitamin $B_{12}$ that can be absorbed *in vivo*.[6] It appears that the receptor for IF–$B_{12}$ is a significant limiting factor in $B_{12}$ absorption.

We have reported the solubilization of guinea pig ileal mucosal homogenates employing the detergent Triton X-100.[7] A particle-free solution was thus obtained that contained all the IF–$B_{12}$ receptors estimated to be present on the starting intestinal membranes. The techniques are herein described in the hope that better quantitation of IF–$B_{12}$ receptor will help investigators to elucidate the mechanisms of $B_{12}$ absorption and malabsorption. The method we employ for assaying and quantitating this solubilized receptor is also described.

[1] B. A. Cooper, *Medicine (Baltimore)* **43**, 689 (1964).
[2] L. W. Sullivan, V. Herbert, and W. B. Castle, *J. Clin. Invest.* **42**, 1443 (1963).
[3] R. M. Donaldson Jr., I. L. Mackenzie, and J. S. Trier, *J. Clin. Invest.* **46**, 1215 (1967).
[4] I. L. Mackenzie, R. M. Donaldson Jr., W. L. Knopp, and J. S. Trier, *J. Exp. Med.* **128**, 375 (1968).
[5] M. Katz, S. K. Lee, and B. A. Cooper, *N. Engl. J. Med.* **287**, 425 (1972).
[6] E. Stopa, M. Katz, and R. O'Brien, *Gastroenterology* **76**, 309 (1979).
[7] M. Katz and B. A. Cooper, *J. Clin. Invest.* **54**, 733 (1974).

In recent years others have reported the solubilization of IF–$B_{12}$ receptors from rat mucosa (by mechanical grinding in high pH buffers[8]) and guinea pig mucosa (by sonification[9] or by simple mechanical disruption without detergents[10]). These latter reports, however, do not quantitate the solubilized receptor, and so it is difficult to assess the efficiency of the procedures.

### Solubilization Technique

#### *Animals and Tissue Preparation*

Guinea pigs weighing approximately 250 g are maintained on Purina lab chow until use. They are fasted overnight, etherized, and weighed, and the small bowel is excised beginning from the ileocecal junction. The entire length of the small bowel is divided into distal and proximal halves. Each half is cooled and rinsed by flushing through 100 ml of 0.15 $M$ saline. Each half is then cut into manageable 15-cm long segments and placed in 100 ml of 0.15 $M$ saline kept in a beaker on melting ice. The mucosa is separated from the thin muscle layers in toothpaste fashion by pressure exerted with the edge of a glass slide on the exterior surface of each segment. Mucosal fragments so obtained are placed in 50-ml volumes of 0.15 $M$ saline in beakers that are kept on melting ice.

#### *Preparation of Cell-Free Extract*

The mucosal fragments (approximately 3 g from proximal and 2 g from distal bowel) are suspended in separate 50-ml volumes of 0.15 $M$ saline and homogenized on ice in a Virtis homogenizer run at full power for 1 min. The proximal and distal homogenates are then centrifuged for 30 min at 4° at 18,000 $g$ in a refrigerated Beckman Model L5-65 centrifuge employing an SW-27 swinging-bucket head. The sedimented homogenates are resuspended in 35-ml volumes of 0.15 $M$ saline and recentrifuged as before; the supernatant is discarded. Each pellet is weighed, resuspended in 35 ml of 3 m$M$ sodium phosphate buffer of pH 8.0, and sonified on ice employing a Branson sonifier run at maximum power for four 15-sec bursts, with an interval of 1 min between bursts to permit cooling. The volume of each suspension is now brought up to 50 ml with 3 m$M$ sodium

[8] K. Okuda and T. Fujii, *J. Lab. Clin. Med.* **89**, 172 (1977).
[9] R. P. Schneider, R. M. Donaldson Jr., and B. M. Babior, *Biochim. Biophys. Acta* **373**, 58 (1974).
[10] R. Cotter and S. P. Rothenberg, *Br. J. Haematol.* **34**, 477 (1976).

phosphate buffer, pH 8, and 2 ml are removed and frozen for later protein and DNA assays. The remaining 48 ml of each suspension are added to separate 48-ml volumes of 10% Triton X-100 in 3 m$M$ sodium phosphate buffer, pH 8.0, gently mixed for 30 min at room temperature, and left overnight at 4°. The next morning, 35 ml of each volume is centrifuged at 100,000 $g$ for 60 min, and 24 ml of the supernatants are gently removed, placed in lengths of seamless cellulose dialysis tubing (Fisher Co.), and dialyzed for 48 hr at 4° against 10 liters of doubly-distilled water.

The dialyzed samples are now concentrated to volumes less than 10 ml employing Carbowax (PEG 20,000, Union Carbide Co.). The solubilized mucosal extracts are now ready for assay. The total protein content of the starting homogenate from one guinea pig is usually some 300 mg (by the Lowry *et al.* method[11]), and the DNA content is usually some 90 mg (by the Burton method[12]).

Assay Technique

IF–B$_{12}$ alone normally filters well within the included volume of a long Sephadex G-200 column. Thus, any IF–B$_{12}$ emerging in the excluded volume of a Sephadex G-200 column, of the IF–B$_{12}$ preincubated with the solubilized mucosal extract in the presence of calcium ion, represents IF–B$_{12}$ complex that has been bound by a material of large molecular weight. Specificity of this IF–B$_{12}$ binder is proved by adding sufficient Na$_2$EDTA to chelate all available calcium, which normally eliminates uptake of IF–B$_{12}$ by receptor. An alternative means to prove specificity employs an IF–B$_{12}$ complex prepared from the gastric juice of a patient we described, whose IF will form a complex with B$_{12}$ but will not be taken up by available IF–B$_{12}$ receptor.[5] Results of a typical experiment with its control is shown in Fig. 1.[7]

Preparation of IF– B$_{12}$ Complex

*Gastric Juice.* Gastric juice is collected on ice from normal individuals by standard techniques after pentagastrin stimulation, filtered through gauze, and immediately depepsinized by raising the pH to 10 for 20 min at room temperature.[13] The pH is then brought down to 7.4, and portions are frozen for later use.

$^{57}$*Cobalt-Labeled B$_{12}$.* High specific activity [$^{57}$Co]B$_{12}$ (100 $\mu$Ci/$\mu$g),

[11] O. H. Lowry, N. J. Rosebrough, A. L. Farr, and R. J. Randall, *J. Biol. Chem.* **193**, 265 (1951).
[12] K. Burton, *Biochem. J.* **62**, 315 (1956).
[13] C. A. Flood and R. West, *Proc. Soc. Exp. Biol. Med.* **34**, 542 (1936).

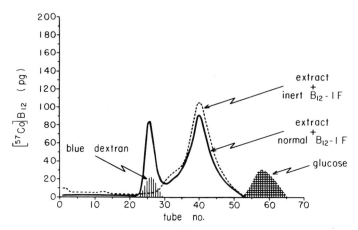

FIG. 1. Elution pattern from Sephadex G-200 column of an extract from guinea pig ileal mucosa incubated with intrinsic factor (IF)–[$^{57}$Co]B$_{12}$ (neutralized, depepsinized gastric juice incubated with [$^{57}$Co]B$_{12}$). With normal IF, some radioactivity emerged in the void volume (marked by Blue dextran 2000), whereas with biologically inert IF, all the radioactivity emerged in the included volume. The small peak that emerged at fraction number 33 probably represents the salivary vitamin B$_{12}$ binder. From M. Katz and B. A. Cooper, *J. Clin. Invest.* **54**, 733 (1974).

purchased from Amersham Searle Radiochemical Laboratories is used for all the studies.

*B$_{12}$ Binding Capacity and IF Content of Gastric Juice.* The total vitamin B$_{12}$ binding capacity and IF content of gastric juice is measured by the standard albumin–charcoal method of Gottlieb *et al.*[14] Usually the gastric juice used has a B$_{12}$ binding capacity of some 40 ng/ml, of which more than 90% is due to IF.

*Preparation of IF–[$^{57}$Co]B$_{12}$ Complex.* An incubation mixture of 0.45 ml of gastric juice, 0.4 ml of "incubation buffer" (30 m$M$ Tris-buffered 0.15 $M$ saline of pH 7.4 containing 1 m$M$ CaCl$_2$), and 0.15 ml of [$^{57}$Co]B$_{12}$ (100 ng/ml) is left at room temperature for 10 min. Residual free [$^{57}$Co]B$_{12}$ is adsorbed with 1.5 ml of albumin-coated charcoal [0.125 g of bovine serum albumin in 12.5 ml of water mixed with 0.625 g of Norit A (Fisher Co.) in 12.5 ml of water and gently stirred for 30 min at room temperature]. The charcoal is sedimented by centrifugation at 20,000 $g$ for 15 min at 0°. The supernatant is gently removed and counted with a standard in an Autowell II gamma counter (Picker Co.). An appropriate amount is then taken for the experiment of the day.

*Gel Filtration.* Sephadex G-200 (Pharmacia Inc.) is swelled in "incubation buffer" and used to pack a 2.5 × 45 cm column (Pharmacia Inc.), which is kept in a refrigerator at 4°. After packing, "incubation buffer"

[14] C. Gottlieb, K. S. Lau, L. R. Wasserman, and V. Herbert, *Blood* **25**, 875 (1965).

EFFECT OF PHYSICAL METHODS ON RELEASE OF IF–$B_{12}$ RECEPTOR FROM GUINEA PIG ILEAL
MUCOSAL HOMOGENATES[a]

| Composition of extracting fluid | Volume (ml) | Yield receptor activity (pg) |
|---|---|---|
| Sodium phosphate buffer, 3 m$M$, pH 8.0 | | |
| With 0.05% Triton X-100 | 1000 | 505 |
| Sodium phosphate buffer, 3 m$M$, pH 8.0 | | |
| Without Triton X-100 | 80 | 51 |
| With 0.05% Triton X-100 | 80 | 150 |
| With 0.05% Triton X-100, homogenate sonified[b] | 80 | 155 |
| With 5.00% Triton X-100, homogenate sonified | 80 | 440 |
| NaCl solution, with 0.05% Triton X-100, homogenate sonified | | |
| 0.15 $M$, buffered to pH 8.0 with 3 m$M$ Tris · HCl | 80 | 362 |
| 1.5 $M$, buffered to pH 8.0 with 3 m$M$ Tris · HCl | 80 | 260 |
| Tris · HCl buffer, pH 8.0 | | |
| With 0.05% Triton X-100, homogenate sonified | 80 | 205 |

[a] Ileal mucosal scrapings were first homogenized and washed three times with 20 ml of 30 m$M$ sodium phosphate-buffered 0.15 $M$ saline of pH 7.4. Equal aliquots were then subjected to the procedure listed. The resultant extracts were then incubated with 10 ng of IF–[$^{57}$Co]$B_{12}$ and assayed for receptor activity by the gel filtration method. From M. Katz and B. A. Cooper *J. Clin. Invest.* **54**, 733 (1974).

[b] In addition, as indicated, some aliquots of mucosal homogenate were further disrupted by sonification before being added to the extracting fluid.

alone is pumped in the downward direction through the column by the use of a peristaltic pump at a flow rate of 12 ml/hr. The column is calibrated by filtering a mixture of dextran blue, bovine serum albumin, and glucose. Fractions of 3.5 ml each are collected automatically and counted.

*Incubation of Mucosal Extract with* [$^{57}$Co]$B_{12}$–IF *Complex.* One-quarter of the final volume of the solubilized mucosa is incubated with 10 ng of [$^{57}$Co]$B_{12}$–IF complex in a beaker on melting ice, and the total volume of the incubation mixture is made up to 5 ml with "incubation buffer." After 30 min, 0.5 ml of a 10% glucose in "incubation buffer" solution is added; the total volume is applied to the top of the prepared Sephadex G-200 column and filtered as described above. The fractions are collected and counted. The quantity of radioactive $B_{12}$ that is shifted to the excluded volume of the column is taken as a measure of the amount of the receptor activity present in the portion of the extract assayed. Data are always presented in terms of the weight of vitamin $B_{12}$ bound, in nanograms or picograms. Filtration assay results are reproducible usually within 5% in repeated runs employing the same experimental conditions.

*Calculation of "Total Receptor."* [$^{57}$Co]B$_{12}$ present in the excluded peak is multiplied by 4 (since only one-quarter of the final volume is used in the assay), multiplied by 4 again (since only one-quarter of the original 96 ml of extraction volume was actually concentrated for assay), and multiplied by 25/24 (to correct for the 2-ml volume of the original homogenate removed for protein and DNA assays).

For convenience, the routine assay for receptor utilizes 10 ng of vitamin B$_{12}$ bound to IF (approximately 2 n$M$). Because the association constant between IF–B$_{12}$ and receptor is of the order of 10$^9$ liters per mole,[7] less than half of the added IF–B$_{12}$ is associated with receptor during assay of most receptor preparations. That the calculated "total receptor" quantity (approximately 6 ng per guinea pig) has physiologic relevance is suggested by the fact that absorption of B$_{12}$ is limited to near 6 ng even if up to 75 ng of IF–B$_{12}$ complex are fed.[6]

*Effect of Physical Methods on Release of IF–B$_{12}$ Receptor.* The effect of varying the extraction procedure is shown in the table. The procedure for extraction described above yields only 90% of the maximum IF–B$_{12}$ receptor activity that can be extracted when the sonified mucosal homogenate is suspended in 1 liter of 0.05% Triton X-100 solution. However, the procedure described here is much more convenient in that smaller volumes of fluids need to be centrifuged; it is the procedure currently in use in our laboratory.

# [10] Properties of Proteins That Bind Vitamin B$_{12}$ in Subcellular Fractions of Rat Liver

*By* J. W. COFFEY

When a trace amount of radioactive cyanocobalamin ([$^{57}$Co]CN-Cbl) is administered intravascularly to rats, the vitamin binds immediately to a serum protein and is transported to the tissues in this form. This binding protein has properties analogous to those of human transcobalamin II (TC II); therefore, the Cbl-binding protein in rat serum will be designated as rat TC II in this presentation. Studies[1] concerning the subcellular distribution of [$^{57}$Co]Cbl in rat liver at various times after the administration of [$^{57}$Co]CN-Cbl have demonstrated that (*a*) at early times (5 min) after the administration of [$^{57}$Co]CN-Cbl, the radioactive Cbl in liver homogenates is concentrated in the microsomal fraction in association with membrane

---

[1] Q. A. Pletsch and J. W. Coffey, *J. Biol. Chem.* **246,** 4619 (1971).

vesicles containing 5'-nucleotidase; (b) at intermediate times (30 min to 2 hr), lysosomal fractions isolated from liver homogenates contain significant amounts of the newly absorbed [$^{57}$Co]Cbl; and (c) at later times (24–72 hr) the [$^{57}$Co]Cbl is localized primarily in the mitochondrial and soluble fractions of liver homogenates. The [$^{57}$Co]Cbl in these various subcellular fractions is associated with binding proteins that can be distinguished by chromatographic techniques.

## Methods

*Injection of Animals.* Female rats of the Charles River strain weighing from 200–250 g are injected intracardially with 0.5 $\mu$Ci of [$^{57}$Co]CN-Cbl (approximate specific activity, 150 $\mu$Ci/$\mu$g) dissolved in 0.5 ml of 0.9% NaCl. Blood should flow freely into the syringe before and after injection of the [$^{57}$Co]CN-Cbl. Triton WR-1339 (85 mg per 100 g body weight) is administered intraperitoneally to animals whose livers are to be used for the preparation of highly purified lysosomal and mitochrondrial fractions.[2] The Triton WR-1339 is dissolved in physiological saline at a concentration of 170 mg/ml and is administered 3.5 days before the animals are sacrificed. The animals are fasted for 16 hr prior to sacrifice.

*Subcellular Fractionation of Liver.* Livers for the preparation of plasma membrane, lysosomal, soluble, and mitochondrial fractions are removed from animals sacrificed 5 min, 30 min, 24 hr, and 72 hr, respectively, after the injection of [$^{57}$Co]CN-Cbl. Animals are killed by decapitation and exsanguinated. The liver is quickly removed, immersed in a tared beaker containing cold 0.25 $M$ sucrose, weighed, blotted, and then minced finely with scalpels. The minced liver together with 3 volumes of ice-cold 0.25 $M$ sucrose is transferred to a smooth-walled glass homogenizer (Potter–Elvehjem type) fitted with a Teflon pestle. The tissue is homogenized by one down and up pass of the pestle, which is driven at about 1000 rpm by a power motor. The homogenate is separated into a nuclear fraction (N), a combined heavy plus light mitochondrial fraction (ML), a microsomal fraction (P), and a soluble fraction (S) according to the method of de Duve *et al.*[3] The N is collected by centrifugation for 10,000 $g$-min in a model PR-2 International refrigerated centrifuge equipped with a No. 269 head. The supernatant is decanted, and then the pellet (N) is homogenized for a second time in 3 volumes of 0.25 $M$ sucrose. The N is resedimented using 6000 $g$-min. The 2 supernatants are combined, and 0.25 $M$ sucrose is added to a final volume equal to 10 times the weight of the tissue. The ML

[2] A. Trouet, *Arch. Int. Physiol. Biochim.* **72**, 698 (1964).
[3] C. de Duve, B. C. Pressman, R. Gianetto, R. Wattiaux, and F. Appelmans, *Biochem. J.* **60**, 604 (1955).

is sedimented by subjecting this postnuclear supernatant to an integrated force (see below) of 250,000 $g$-min in a No. 40 rotor or of 350,000 $g$-min in a 50.1 rotor for a Beckman (Spinco) preparative ultracentrifuge. The supernatant is removed by suction, taking care to remove the fluffy pink layer which sits on top of the dense, darkly colored ML pellet. The pellet is carefully resuspended in 0.25 $M$ sucrose using a stirring rod (the pellets from 2 tubes are combined into 1 tube during this washing procedure), and the ML is resedimented under the same conditions. The supernatant and fluffy layer are again removed by suction and added to the original post-mitochondrial supernatant. The P is separated from this supernatant by centrifugation for 3,000,000 $g$-min using a No. 40 rotor. The supernatant representing the soluble fraction (S) of the cell is removed by decantation.

Centrifugation conditions are expressed in an integrated form so that forces developed during acceleration and deceleration are included. The total $g$-min generated from the time the rotor starts until it stops are related to the time integral of the squared angular velocity ($\omega$) by the expression

$$g\text{-min} = W \frac{R_{av}}{981 \times 60} \tag{1}$$

in which $R_{av}$ is the radial distance to the middle of the tube and $W$ equals the time integral of $\omega^2$ [Eq. (2)].

$$W = \int_0^t \omega^2 \, dt \tag{2}$$

Commercial integrators are available for the evaluation of the $W$ integral; however, the conditions of speed and time necessary to produce a given $W$ can be determined by a preliminary calibration of the centrifuge. The speed of the rotor is recorded at 0.5-min intervals as the rotor accelerates to and decelerates from a plateau speed. The approximate value of $W$ for each 0.5-min interval can be calculated using Eq. (3).

$$W = \frac{4\pi^2}{60} \Delta t' \left[ \frac{(rpm_a)^2 + (rpm_b)^2}{2} \right] \tag{3}$$

in which $rpm_a$ and $rpm_b$ equal the speed of the rotor at the beginning and the end of each successive 0.5-min interval and $\Delta t'$ equals the time interval in min (0.5). The $W$ values for the acceleration, plateau, and deceleration periods are summed and used to calculate the speed and time (including acceleration and deceleration periods) required to generate the desired $g$-min. If this calibration procedure is not followed, it will be difficult to reproduce exactly the fractionation procedure. Similarly, if rotors with dimensions different than those described are used, the $g$-min required to yield comparable results will be changed.

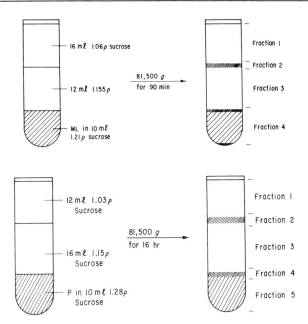

FIG. 1. Separation of subcellular fractions on discontinuous sucrose gradients. *Top:* Gradient used for the preparation of highly purified lysosomes and mitochondria. The lysosomes collect at the interface between the 1.06 and 1.155 density sucrose (fraction 2), and the purified mitochondria remain in the sample layer (fraction 4). *Bottom:* Gradient used for preparation of highly purified plasma membrane vesicles. The vesicles collect at the interface between the 1.03 and 1.15 density sucrose (fraction 2).

*Preparation of Highly Purified Lysosomal and Mitochondrial Fractions.* The administration of Triton WR-1339 to rats results in the accumulation of this detergent in the lysosomes of the liver, thereby causing a selective decrease in their equilibrium density in sucrose gradients. This change in equilibrium density permits the separation of lysosomes from the other subcellular particles found in the ML by flotation through a discontinuous sucrose gradient.[2,4] The ML pellet separated from 10 g of liver (animals pretreated with Triton WR-1339) is gently suspended in 10 ml of 45% (w/w) sucrose (density, 1.21) using a Dounce homogenizer. This suspension is layered in the bottom of a centrifuge tube for a Beckman SW-27 rotor. A 12-ml layer of 34.5% (w/w) sucrose (density, 1.155) and a 16-ml layer of 14.3% (w/w) sucrose (density, 1.06) are pipetted carefully on top, and then the tube is centrifuged at 25,000 rpm (81,500 g) for 90 min. After centrifugation, 4 fractions (Fig. 1, top) are very carefully removed from

[4] F. Leighton, B. Poole, H. Beaufay, P. Baudhuin, J. W. Coffey, S. Fowler, and C. de Duve, *J. Cell Biol.* 37, 482 (1968).

the top of the tube by suction. The suction is generated by a peristaltic pump attached to a needle that is manually lowered into the gradient as the tube empties. Fraction 2, which contains highly purified lysosomes, and fraction 4, which is greatly enriched in mitochondria, are diluted with an equal volume of 0.25 $M$ sucrose; the particles are then sedimented by centrifugation (250,000 $g$-min in No. 40 rotor).

*Separation of Plasma Membrane-Related (5′-Nucleotidase-rich) Vesicles from a Microsomal Fraction.* The method of Touster *et al.*[5] is used with slight modifications. The microsomal pellet from 5 g of liver is suspended by vigorous homogenization in 10 ml of 57.4% (w/w) sucrose (density, 1.28) and then layered in the bottom of a centrifuge tube for the SW-27 rotor. Sixteen milliliters of 34% (w/w) sucrose (density, 1.15) and 12 ml of 7.3% (w/w) sucrose (density, 1.03) are layered on top, and then the tube is centrifuged at 25,000 rpm (81,500 $g$) for 16 hr. Sucrose solutions for the preparation of this plasma membrane fraction contain 5 m$M$ Tris and are adjusted to a pH of 8.0 with HCl. After centrifugation, 5 fractions (Fig. 1, bottom) are collected as already described. Care should be taken to avoid the contamination of fraction 2 (the plasma membrane fraction) with even small amounts of fraction 3. Fraction 2 is diluted with 2 volumes of 0.25 $M$ sucrose, then the membrane vesicles are sedimented by high speed centrifugation (40,000 rpm for 1 hr in No. 40 rotor).

*Purity of Fractions.* The purity of these fractions has been assessed using marker enzymes. The acid phosphatase (EC 3.1.3.2) associated with the lysosomal fraction prepared by flotation of Triton-filled lysosomes through a discontinuous sucrose gradient is purified approximately 35-fold relative to the original homogenate with a yield (percentage of total acid phosphatase in liver) of approximately 20%.[1,4] Lysosomal proteins account for 80–90% of the protein in this fraction, and the remaining 10–20% can be attributed to proteins associated with fragments of the endoplasmic reticulum.[4] The cytochrome oxidase (EC 1.9.3.1) in fraction 4 of the gradient is purified approximately 3.5-fold relative to the homogenate with a yield varying from 80 to 90%. Mitochondrial proteins account for approximately 80% of the total protein in fraction 4; however, this mitochondrial fraction is contaminated with peroxisomes, with fragments of the endoplasmic reticulum, and perhaps with other cellular membranes.[1,4] Plasma membrane vesicles isolated from a P by flotation through a discontinuous gradient contain 5′-nucleotidase (EC 3.1.3.5) purified 30- to 40-fold relative to the homogenate with a yield of approximately 25%.[5] Under these conditions, contamination by other cellular elements is minimal.

[5] O. Touster, N. N. Aronson, J. T. Dulaney, and H. Hendrickson, *J. Cell Biol.* **47**, 604 (1970).

*Preparation of the Subcellular Fractions for Chromatography.*[6] The subcellular fractions are solubilized in 50 m$M$ sodium phosphate buffer (pH, 7.4 or 6.0—see below) containing 0.15 $M$ NaCl and 1% (w/v) Triton X-100 by stirring at 4° until the solutions become clear (6–12 hr). Pellets containing the plasma membrane fraction from 6 g of liver (~12 mg of protein), the lysosomal fraction from 15 g of liver (~30 mg of protein), and the mitochondrial fraction from 4 g of liver (~200 mg of protein) are solubilized in 3.0 ml of buffer. After solubilization, the fractions are centrifuged at 40,000 rpm in the No. 40 rotor for 1 hr to remove any insoluble material. These conditions solubilize at least 90% of the [$^{57}$Co]Cbl (50,000–100,000 cpm) associated with the pellets. Samples for chromatography on CM-cellulose are dialyzed against 1000 volumes of equilibrating buffer (see below) at 4° for 16 hr. Mouse serum or dialyzed rat serum is labeled with a slight excess of [$^{60}$Co]CN-Cbl (small amount of free [$^{60}$Co]CN-Cbl present after serum binders are saturated) and then is added to the tissue samples to provide $^{60}$Co-labeled chromatographic markers for free Cbl and for Cbl-TC II complexes. Mouse serum (0.5 ml) is used as a marker during chromatography on columns of Sephadex G-100, whereas rat serum (1.0 ml) is used during chromatography on CM-cellulose columns. A large excess of nonradioactive CN-Cbl (10 $\mu$g/ml) is added to the tissue samples prior to addition of the serum to prevent possible exchange of label between the $^{60}$Co-labeled serum binding proteins and the $^{57}$Co-labeled tissue binding proteins. All samples are adjusted to a final volume of 4 ml.

*Chromatography of Cobalamin-Binding Proteins.*[6] Sephadex G-100 is swollen in $H_2O$ for 3 days and then packed into a 2.5 × 90 cm column. The column is equilibrated with buffer (50 m$M$ sodium phosphate buffer, pH 7.4, containing 1% Triton X-100 and 0.15 $M$ NaCl) by passing the buffer through the column in an upward direction at a flow rate of 24 ml/hr using a peristaltic pump. Samples (4 ml) are pumped into the bottom of the column, and then are eluted in 4-ml fractions by upward flow. The CM-cellulose (CM-52) is suspended in 50 m$M$ sodium phosphate buffer, pH 6.0, containing 1% Triton X-100 and 0.15 $M$ NaCl and then packed into a 1.6 × 30 cm column. Buffer is pumped upward at a flow rate of 24 ml/hr until the pH and conductivity of the column eluent are the same as those of the entering buffer. The sample is pumped into the bottom of the column, and then the column is washed with approximately 200 ml of equilibrating buffer. The radioactivity bound to the CM-cellulose is eluted into 4-ml fractions using a linear gradient of NaCl in the same phosphate buffer. The concentration of NaCl ranges from 0.15 $M$ to 2.0 $M$ over a volume of 400 ml. Radioactivity in the fractions is measured using a

[6] Q. A. Pletsch, and J. W. Coffey, *Arch. Biochem. Biophys.* **151,** 157 (1972).

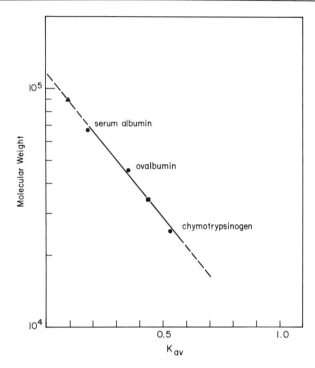

Fig. 2. Elution characteristics of rat liver cobalamin binding proteins chromatographed on Sephadex G-100. The column was calibrated with standards of known molecular weight: thyroglobulin, 680,000 (void volume); bovine serum albumin, 67,000; ovalbumin, 45,000; chymotrypsinogen A, 25,000; and cyanocobalamin, 1355 (total volume). ■, $K_{av}$ for CN-Cbl–rat TC-II, for CN-Cbl–mouse TC II, and for the complex between Cbl and the binding protein in plasma membrane and lysosomal fractions (MW approximately 35,000). ▲, Position at which the complexes between Cbl and the soluble and mitochondrial binding protein elute from the column. The mitochondrial fraction contains a second Cbl–protein complex, which elutes at the void volume ($K_{av} = 0$). No molecular weights can be assigned to these soluble and mitochondrial complexes on the basis of chromatography on Sephadex G-100, since the linear relationship of $K_{av}$ to the log of the molecular weight does not hold as the elution volume approaches the void volume.

gamma spectrometer equipped with 2 pulse-height analysis channels so that $^{57}Co$ and $^{60}Co$ can be quantitated in the same sample.

*Chromatographic Properties of Cbl-Binding Proteins in Subcellular Fractions of Rat Liver.* All the $[^{57}Co]Cbl$ solubilized from a plasma membrane or microsomal fraction (isolated 5 min after the injection of $[^{57}Co]CN-Cbl$) and 75% of the $[^{57}Co]Cbl$ solubilized from a lysosomal fraction (isolated 30 min after the injection of $[^{57}Co]CN-B_{12}$) is associated with a single binding protein that cochromatographs on Sephadex G-100 with rat or mouse $[^{60}Co]CN-Cbl–TC-II$ (Fig. 2). The remainder of the $[^{57}Co]$ in the lysosomal fraction is present in free CN-Cbl. Soluble fractions isolated 24 hr after the injection of [$^{57}$Co]CN-Cbl contain a [$^{57}$Co]Cbl–protein complex that elutes from Sephadex G-100 columns near the void volume (Fig. 2). Approximately 50% of the [$^{57}$Co]Cbl solubilized from a mitochondrial fraction isolated 72 hr after the injection of [$^{57}$Co]CN-B₁₂ is associated with binding proteins (the remainder is free), which elute as 2 peaks from columns of Sephadex G-100. One peak elutes at the void volume, and the other elutes at the same position as the binding protein in the soluble fraction (Fig. 2). However, the soluble and mitochondrial binding proteins are distinct, since the complex between [$^{57}$Co]Cbl and the soluble binding protein is stable during dialysis at pH 6.0, whereas the complex with the mitochondrial binding protein dissociates under these conditions. The [$^{57}$Co]Cbl–protein complex in the plasma membrane and lysosomal fractions also cochromatographs with [$^{60}$Co]CN-Cbl–rat TC II on CM-cellulose. Under the conditions described in this presentation, it is not possible to characterize the Cbl–binding proteins present in the soluble and mitochondrial fractions by chromatography on CM-cellulose, since these Cbl–protein complexes are either unstable or do not bind to CM-cellulose under these conditions.

## Comments

Mellman *et al.*[7] have solubilized the Cbl-binding protein from mitochondria by sonication rather than by detergent treatment. In addition, Mellman *et al.*[7] and Kolhouse and Allen[8] have characterized the high molecular weight Cbl binding proteins in soluble and mitochondrial fractions by chromatography on Sephadex G-150 and on DEAE-cellulose and have demonstrated that the binding proteins in the soluble and mitochondrial fractions are the apoenzymes for methionine synthetase (EC 2.1.1.13) and *S*-methylmalonyl-CoA mutase (EC 5.4.99.2), respectively. In agreement with these findings, the [$^{57}$Co]Cbl associated with the binding proteins in the soluble and mitochondrial fractions is present mainly as methylcobalamin and adenosylcobalamin, the coenzyme forms of vitamin B₁₂, while that associated with the binding protein (rat TC II) in the plasma membrane and lysosomal fractions is present mainly as CN-Cbl.[1,7]

[7] I. S. Mellman, P. Youngdahl-Turner, H. F. Willard, and L. E. Rosenberg, *Proc. Natl. Acad. Sci. U.S.A.* **74**, 916 (1977).
[8] J. F. Kolhouse, and R. H. Allen, *Proc. Natl. Acad. Sci. U.S.A.* **74**, 921 (1977).

# [11] Isolation and Characterization of Vitamin $B_{12}$-Binding Proteins from Human Fluids

*By* SUDHIR KUMAR and LEO M. MEYER

## Serum and Saliva $B_{12}$-Binding Proteins

Many different methods have been described for the measurement of vitamin $B_{12}$-binding proteins in human serum. The techniques of protein separation that have been used include paper and starch-gel electrophoresis,[1-3] Geon-block electrophoresis,[4] column chromatography and immunodiffusion techniques of Ouchterlony,[5] DEAE-cellulose[6-11] and CM-cellulose ion-exchange chromatography,[12,13] Sephadex G-200,[13,14] affinity chromatography,[15-17] and more recently isoelectric focusing.[18,19]

The presence of at least 3 vitamin $B_{12}$-binding proteins in human serum has been demonstrated.[6-11,14,20-22] These binders have been designated as TC I, TC II, and TC III. In earlier studies, $B_{12}$-binding proteins from normal serum were first separated into two fractions by chromatography on a DEAE-cellulose column, using a two-buffer system for elution.[14,22] These fractions were termed TC I and TC II. This was followed by gel-filtration of the TC II fraction on a Sephadex G-200 column to separate

[1] A. Miller and F. J. Sullivan, *J. Lab. Clin. Med.* **53**, 607 (1959).
[2] A. Miller and F. J. Sullivan, *J. Lab. Clin. Med.* **53**, 2135 (1959).
[3] A. Miller and F. J. Sullivan, *J. Lab. Clin. Med.* **58**, 763 (1961).
[4] C. A. Hall and A. E. Finkler, *J. Lab. Clin. Med.* **60**, 765 (1962).
[5] C. A. Hall and A. E. Finkler, *Biochim. Biophys. Acta.* **78**, 233 (1963).
[6] C. A. Hall and A. E. Finkler, *Blood* **27**, 611 (1966).
[7] P. F. Retief, C. W. Gottlieb, S. Kochwa, P. W. Pratt, and V. Herbert, *Blood* **29**, 501 (1967).
[8] J. E. Gizis, F. M. Dietrich, G. Choi, and L. M. Meyer, *J. Lab. Clin. Med.* **75**, 673 (1970).
[9] J. E. Gizis and L. M. Meyer, this series, Vol. 18C, p. 127.
[10] J. E. Gizis and L. M. Meyer, *Proc. Soc. Exp. Biol. Med.* **140**, 326 (1972).
[11] S. Kumar, L. M. Meyer, and R. A. Gams, *Proc. Soc. Exp. Biol. Med.* **147**, 377 (1974).
[12] C. A. Hall and A. E. Finkler, *J. Lab. Clin. Med.* **65**, 459 (1965).
[13] B. Hom, H. Olesen, and P. Lous, *J. Lab. Clin. Med.* **68**, 958 (1966).
[14] C. Lawrence, *Blood* **33**, 899 (1965).
[15] R. H. Allen and P. W. Majerus, *J. Biol. Chem.* **247**, 7695 (1972).
[16] R. H. Allen and P. W. Majerus, *J. Biol. Chem.* **247**, 7702 (1972).
[17] R. H. Allen and P. W. Majerus, *J. Biol. Chem.* **247**, 7709 (1972).
[18] U. H. Stenman and R. Grasbeck, *Biochim Biophys. Acta* **286**, (1972).
[19] U. H. Stenman, *Scand. J. Haematol.* **13**, 129 (1974).
[20] R. Gullberg, *Clin. Chim. Acta* **29**, 97 (1970).
[21] F. J. Bloomfield and J. M. Scott, *Br. J. Haematol* **22**, 33 (1972).
[22] I. Chanarin, J. M. England, K. L. Rowe, and J. A. Stacey, *Br. Med. J.* **2**, 441 (1972).

METHODS IN ENZYMOLOGY, VOL. 67

low molecular weight TC II from the higher molecular weight TC III, which was eluted along with TC II in the first separation on DEAE-cellulose.[14,22] The two fractions thus obtained on gel filtration were designated as TC II and TC III on the basis of their molecular weights and electrophoresis.[21] In a previous study, Gullberg[20] subjected whole serum to gel filtration with separation into two major fractions, transcobalamin large (TCL) and transcobalamin small (TCS), but no further fractionation was performed.

This chapter deals mainly with the isolation and characterization of B$_{12}$-binding proteins from human serum. However, these methods can be used also for the isolation of binding proteins from human saliva, as well as other biological fluids.

### Method

*Reagents.* Reagents are prepared in distilled water containing 0.09% methylparaben and 0.01% propylparaben (Tenneco Chemicals, New York). To dissolve the preservatives, the water is warmed to 72°, and esters are added to the water while stirring. (Pyrogen-free water should be used if the preparation is intended for *in vivo* studies.) Glassware is baked for 4 hr at 180°.

*Buffers*

   Sodium phosphate buffer, 17.5 m$M$, pH 6.3
   Sodium phosphate buffer, 30 m$M$, pH 5.9
   Sodium phosphate buffer, 60 m$M$, pH 5.85
   Sodium phosphate buffer, 100 m$M$, pH 5.8
   Sodium phosphate buffer, 250 m$M$, pH 5.4
   Sodium phosphate buffer, 5 m$M$, pH 7.4, containing 1 $M$ NaCl

[57]*Co-Labeled Vitamin B$_{12}$ Solution.* A solution is prepared containing 1000 pg/ml; specific activity about 180 $\mu$Ci/mg (Amersham).

*Preparation of Sephadex G-200.* Forty-five grams of dry Sephadex G-200 (Pharmacia) are mixed with 1000 ml of 5 m$M$ sodium phosphate buffer, pH 7.4, containing 1 $M$ NaCl and stirred to make a slurry. The suspension is allowed to remain in the cold room for at least a week to permit proper swelling of Sephadex. The suspension is then packed into a 2.6 × 33 cm column, and 1000 ml of 5 m$M$ pH 7.4, sodium phosphate buffer, containing 1 $M$ NaCl buffer is passed through the Sephadex in the cold room at a rate of 10 ml hr.

*Preparation of DEAE-Cellulose.* Forty grams of dry DEAE-cellulose (Schleicher and Schuell, No. 70, 0.94 meq/g) are allowed to sink in 2 liters of 1 $N$ NaOH. After 30 min, the suspension is filtered through a coarse fritted-glass filter funnel and washed with additional 1 $N$ NaOH until no more color is present in the filtrate. The filter cake is suspended in distilled water, filtered through the funnel, and washed with distilled water to pH 6.8. It takes approximately 30 liters of distilled water to bring the pH to that level. The cake is resuspended in 4 volumes of 17.5 m$M$ pH 6.3, sodium phosphate buffer and enough 17.5 m$M$ solution of $NaH_2PO_4 \cdot H_2O$ is added to bring the pH to 6.2. The DEAE-cellulose is washed three times in the funnel with two volumes of 17.5 m$M$, pH 6.3, sodium phosphate buffer and transferred to a beaker, where it is resuspended in sufficient buffer to make about 50 ml of suspension for each gram of dry cellulose used. The suspension is then used for packing the columns.

FOR LARGE COLUMNS. For large columns the suspension is packed into a 3 × 60-cm column, and 500 ml of 17.5 m$M$, pH 6.3, sodium phosphate buffer is passed through the cellulose in the cold room at a rate of 30 ml/hr.

FOR SMALL COLUMNS. For small columns the suspension is packed into a 0.5-cm column (5-ml plastic disposable pipette plugged with glass wool) to a height of 22 cm, and the packed column is washed with 250 ml of 17.5 m$M$, pH 6.3, sodium phosphate buffer before use. Care is taken that the columns do not dry during the washing or subsequent steps during chromatography.

### Gel Filtration on Sephadex G-200 Column

*Preparation of Serum or Saliva Samples.* Five hundred picograms of [$^{57}$Co]B$_{12}$ are added per milliliter to 20 ml of normal fresh or frozen serum or saliva, and the mixture is allowed to stand at 37° for 20 min. The reaction mixture, after counting, is dialyzed in the cold room against 10 liters of 5 m$M$, sodium phosphate buffer containing 1 $M$ NaCl, pH 7.4, for 24 hr at 4°.

*Chromatographic Procedure.* Dialyzed serum is placed on a Sephadex G-200 column (2.6 × 33 cm, void volumn 50 ml; preequilibrated with 5 m$M$ phosphate buffer, pH 7.4, containing 1 $M$ NaCl). Elution is performed with the same buffer. Two milliliter samples are collected at a flow rate of 10 ml/hr, and the fractions are counted for 2 min each in an automatic well-type scintillation counter (Nuclear-Chicago, Model 1185). Three peaks of radioactivity are obtained, although peaks 2 and 3 contain more than 98% of total radioactivity recovered. The first peak, obtained in the void volume constitutes about 1–2% of total radioactivity and is discarded. This peak seems to be similar to the one designated as TC-0.[23] Only fractions containing maximum counts corresponding to peak 2 (TCL; elution volume 65–75 ml) and peak 3 (TCS; elution volume 103–117

[23] S. Kumar, L. M. Meyer, and R. A. Gams, *Scand. J. Haematol.* **18**, 403 (1977).

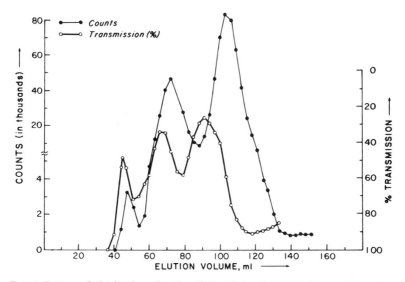

FIG. 1. Pattern of distribution of radioactivity of vitamin $B_{12}$-binding proteins of normal human serum on a Sephadex G-200 column (2.6 × 33 cm). One thousand picograms of [$^{57}Co$]$B_{12}$ were added to 2 ml of serum, and the elution was performed with 5 m$M$ sodium phosphate buffer, pH 7.4, containing 1 $M$ NaCl. Other details are as described in the text.

ml) are collected, dialyzed overnight against 4 liters of distilled water, and concentrated on a $B_{15}$ microconcentrator (Minicon™ concentrator, Amicon). More than 95% of the applied radioactivity of TCL and TCS can be recovered as TCL and TCS, respectively, when the pooled, concentrated material is resubjected to gel filtration on the same Sephadex column (Fig. 1).

### Chromatography of TCL and TCS on DEAE-Cellulose Column

*Large Columns.* Any solution remaining above the column after equilibration with 500 ml of buffer on DEAE-cellulose columns is allowed to sink into the absorbent, and the dialyzed concentrated fraction of TCL or TCS is pipetted onto the surface. This is allowed to settle into the column at a flow rate of 30 ml/hr. When all the sample has entered the absorbent, the top of the column is rinsed with three consecutive 3-ml portions of buffer. Thereupon a layer of approximately 4 cm of buffer is introduced above the adsorbent and a buffer reservoir is attached to the column by a length of Tygon tubing. The column is mounted on a fraction collector, and elution is allowed to proceed at a rate of 30 ml/hr in the cold room.

Elution of binders from the column is performed with a stepwise gra-

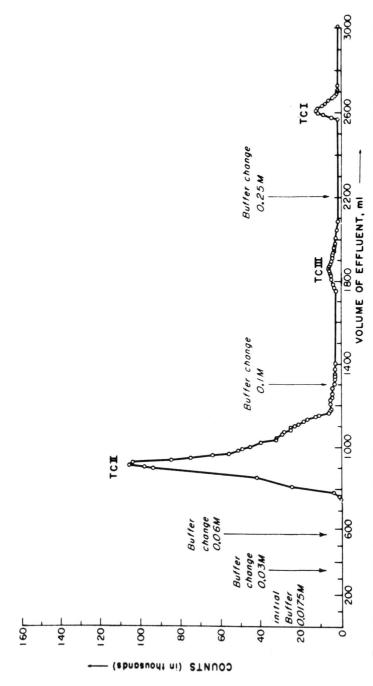

Fig. 2. Pattern of distribution of radioactivity of vitamin B$_{12}$-binding proteins of normal human serum on a large DEAE-cellulose column. The arrows indicate changes of eluting buffers.

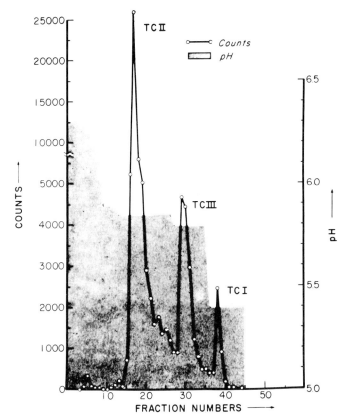

FIG. 3. Pattern of distribution of radioactivity of vitamin B$_{12}$-binding proteins of normal human serum on small DEAE-cellulose column. The shaded area indicates the pH of the eluting buffers.

dient of sodium phosphate buffers, at a flow rate of 30 ml/hr, collecting 10-ml fractions and using 350 ml of 17.5 m$M$, pH 6.3; 225 ml of 30 m$M$, pH 5.9; 725 ml of 60 m$M$, pH 5.85; 900 ml of 100 m$M$, pH 5.8; and 650 ml of 250 m$M$, pH 5.4 for elution of the binders. All the above procedures are performed in the cold room at 4°. The peaks of radioactivity are collected, dialyzed against large volumes of distilled water in the cold room for 24–48 hr, and freeze-dried (Fig. 2).

*Small Columns.* The dialyzed TCL or TCS fractions (approximately 250,000 cpm) are pipetted onto the top of the DEAE-cellulose column (5-ml pipette), and elution is performed with a stepwise gradient of sodium phosphate buffers, starting with 16 ml of 17.5 m$M$, pH 6.3; 8 ml of 30 m$M$, pH 5.9; 28 ml of 60 m$M$, pH 5.85; 16 ml of 100 m$M$, pH 5.8; and 20 ml of 250 m$M$, pH 5.4. Two-milliliter portions are collected and counted for 2 min

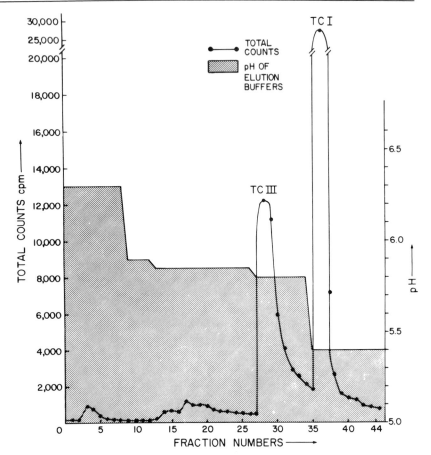

FIG. 4. Pattern of distribution of radioactivity of vitamin $B_{12}$ binders from normal human amniotic fluid on small (0.5 i.d. $\times$ 22 cm) DEAE-cellulose columns. Elution is performed with increasing concentrations of sodium phosphate buffers as described in the text.

each in an automatic well-type scintillation counter (Nuclear-Chicago, Model 1185). Relative percentages of each binder with respect to total radioactivity recovered are calculated by combining counts for each peak of radioactivity (Fig. 3).

## Properties

Two main peaks of radioactivity are obtained from Sephadex G-200 columns (TCL and TCS). Each of these peaks discloses three peaks of radioactivity when chromatographed on DEAE-cellulose columns, which are eluted with 60 m$M$ phosphate buffer, pH 5.85 (peaks A and D); 100 m$M$

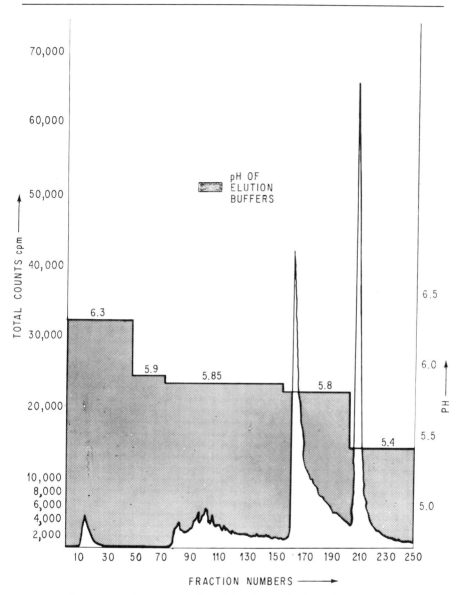

FIG. 5. Pattern of distribution of radioactivity of vitamin B$_{12}$ binders from human amniotic fluid on large (1.3 i.d. × 24 cm) DEAE-cellulose columns.

phosphate buffer, pH 5.8 (peaks B and E); and 250 m$M$ phosphate buffer, pH 5.4 (peaks C and F).[23] The fractions obtained from both TCL and TCS correspond in elution pattern to TC II, TC III, and TC I when normal human serum is chromatographed directly on DEAE-cellulose. All the binding protein peaks obtained from TCL correspond to an apparent molecular weight of about 110,000 ± 10,000, and those from TCS to 42,000 ± 1500. On electrophoresis, the major portion of fractions A and D from TCL and TCS, respectively, as determined by counting radioactivity in different areas, move as $\beta$-globulin ($R_f = 0.24$), fractions C and F between $\alpha_1$- and $\alpha_2$-globulins ($R_f = 0.67$), and fractions B and E between $\beta$- and $\gamma$-globulins ($R_f = 0.45$).

## Amniotic Fluid $B_{12}$-Binding Proteins

### Method

Procedures for the preparation of reagents, buffers, DEAE-cellulose, and the columns are as described in the preceding section on serum and saliva.

*Collection of Amniotic Fluid.* Amniotic fluid samples are obtained from pregnant women by abdominal taps, centrifuged, and stored frozen at $-20°$.

*Preparation of Amniotic Fluid Samples.* Five hundred picograms of [$^{57}$Co]$B_{12}$ are added per milliliter of fresh or frozen amniotic fluid (0.5 ml of amniotic fluid plus 250 pg of [$^{57}$Co]$B_{12}$ for small columns and 6 ml of amniotic fluid plus 3000 pg of [$^{57}$Co]$B_{12}$ for large columns), and the mixture is allowed to stand at 37° for 20 min. The reaction mixture, after counting, is dialyzed in the cold against 17.5 m$M$ sodium phosphate buffer, pH 6.3, for 24 hr at 4° and centrifuged to remove the precipitate.

### Chromatography on DEAE-Cellulose Columns

*Small Columns.* The supernatant obtained from 0.5 ml of amniotic fluid is applied to the small DEAE-cellulose column, and elution is performed with a stepwise gradient of sodium phosphate buffers, as described for serum and saliva (Fig. 4).

*Large Columns.* The supernatant obtained from 6 ml of amniotic fluid is applied on a DEAE-cellulose column (1.3 i.d. × 24 cm height; 25-ml disposable pipette plugged with glass wool) that is preequilibrated with 500 ml of 17.5 m$M$ sodium phosphate buffer, pH 6.3, before use. The elution is performed with a stepwise gradient of sodium phosphate buffers, starting with 96 ml of 17.5 m$M$, pH 6.3; 48 ml of 30 m$M$, pH 5.9; 168 ml of 60 m$M$, pH 5.85; 96 ml of 100 m$M$, pH 5.8, and 120 ml of 250 m$M$, pH 5.4. Two-milliliter fractions are collected and counted for 2 min each in an automatic well-type scintillation counter, and relative percentages of each binder are calculated as described for serum and saliva (Fig. 5).

## Properties

Two peaks of radioactivity, corresponding in elution to TC III and TC I are obtained. The distribution of TC III and TC I binding proteins in amniotic fluid from normal pregnant women is 44% and 45%, with molecular weights of 97,500 and 120,500, respectively. On electrophoresis, TC I moves as $\alpha_1$-globulin and TC III moves between $\alpha_1$- and $\alpha_2$-globulins.

## [12] Biosynthesis of Transcobalamin II

By PAMELA D. GREEN and CHARLES A. HALL

Transcobalamin II (TC II) has a vital role in cobalamin (Cbl) transport.[1-3] Previous reports indicate this Cbl-binding protein is released by the liver[4,5] and possibly several other organs.[6,7] The perfusion studies of Tan and Hansen[4] and England et al.[5] demonstrated the release of TC II into the medium of perfused mouse and rat liver, respectively. Other reports suggest extrahepatic sources of TC II synthesis.[6,7] Sonneborn et al.[6] found that hepatectomized dogs were able to restore their unsaturated serum TC II levels, and Hall and Rappazzo[7] noted release of TC II from perfused dog liver, kidney, spleen, and heart. All the above reports detected release of TC II by various organs, but none studied the de novo biosynthesis of TC II by a particular cell type.

In this chapter we describe studies of the biosynthesis de novo of TC II by two cell lines—a primary culture of rat liver parenchymal cells[8] and an established line of mouse fibroblast (L-929) cells.[9] The techniques presented are those devised or modified in our laboratory, but references and critiques of other methods are given.

[1] C. A. Hall, J. Clin. Invest. 56, 1125 (1975).
[2] N. Hakami, P. Neiman, G. P. Canellas, and J. Lazerson, N. Engl. J. Med. 285, 1163 (1971).
[3] W. H. Hitzig, U. Dohman, H. J. Pluss, and D. Vischer, J. Pediatr. 85, 622 (1974).
[4] C. H. Tan and H. J. Hansen, Proc. Soc. Exp. Biol. Med. 127, 740 (1968).
[5] J. M. England, A. S. Tavill, and I. Chanarin, Clin. Sci. Mol. Med. 45, 479 (1973).
[6] D. W. Sonneborn, G. Abouna, and Mendez-Pican, Biochim. Biophys. Acta 273, 283 (1972).
[7] C. A. Hall and M. E. Rappazzo, Proc. Soc. Exp. Biol. Med. 148, 1202 (1975).
[8] C. R. Savage, Jr. and P. D. Green, Arch. Biochem. Biophys. 173, 691 (1976).
[9] P. D. Green, C. R. Savage, Jr., and C. A. Hall, Arch. Biochem. Biophys. 176, 683 (1976).

In evaluating the de novo biosynthesis of TC II by the two cell lines, a parallel set of experiments was conducted for each cell type. To avoid repetition, the preparation and evaluation of only one cell line will be discussed in detail. The rat liver parenchymal cell study was selected because it required the preliminary steps of isolating and establishing a primary culture.

Basically the experiment was designed to detect the cellular release of a Cbl-binder, verify that the binder was TC II, and demonstrate its cellular biosynthesis. A schematic of the main steps utilized in this study is given below.

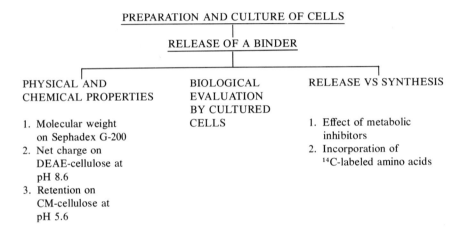

PREPARATION AND CULTURE OF CELLS

RELEASE OF A BINDER

PHYSICAL AND CHEMICAL PROPERTIES

1. Molecular weight on Sephadex G-200
2. Net charge on DEAE-cellulose at pH 8.6
3. Retention on CM-cellulose at pH 5.6

BIOLOGICAL EVALUATION BY CULTURED CELLS

RELEASE VS SYNTHESIS

1. Effect of metabolic inhibitors
2. Incorporation of $^{14}$C-labeled amino acids

### Methods

*Isolation and Culture of Rat Liver Parenchymal Cells.* The collagenase perfusion technique of Berry and Friend[10] as modified by Bonney *et al.*[11] is used to isolate the liver parenchymal cells from adult male albino rats. The procedure yields from 200 to 400 × 10$^6$ cells per rat with a viability of at least 85% as indicated by the trypan blue exclusion method. The cells are cultured in air at 37° in M-199 medium, pH 7.5, containing 25 m$M$ Hepes buffer. The initial plating medium contains 15% fetal calf serum; insulin, 10 mU/ml; penicillin, 100 units/ml; and streptomycin, 100 $\mu$g/ml. Approximately 3 to 4 × 10$^6$ cells are plated in 5 ml of medium in 25-ml Falcon flasks. The cells are allowed to settle and plate from 3–5 hr and then are maintained in serum-free medium with daily changes thereafter for about 7 days. Prior to use in subsequent experiments, the cells are exposed to dexamethasone (5 × 10$^6$ $M$) and assayed for tyrosine aminotransferase

[10] M. N. Berry and D. S. Friend, *J. Cell Biol.* **43**, 506 (1969).
[11] R. J. Bonney, J. E. Becker, P. R. Walker, and V. R. Potter, *In Vitro* **9**, 399 (1973).

(TAT) activity according to the method of Diamondstone[12] as modified by Bonney et al.[11] This is a convenient test done on primary cultures of liver parenchymal cells to confirm the existence of functioning liver cells. Our culture conditions yield functioning cells with an induction ratio of up to-7-fold on day 5 of culture.

*Tissue Culture of Mouse Fibroblast (L-929) Cells.* Mouse fibroblast (L-929) cells are grown as monolayers in either 25-ml or 250-ml Falcon flasks containing 5 ml or 15 ml of medium, respectively. The cells are initially plated in M-199 medium, pH 7.4 containing 10% newborn calf serum, 100 units of penicillin and 100 $\mu$g of streptomycin per milliliter. In output studies with the L-929 cells, the cells are changed to serum-free medium 24 hr after plating, and healthy cells maintained for 6–7 days.

When the cells are used to check the biological activity of different Cbl-binders, they are cultured in M-199 medium only and exposed to the particular Cbl-binder to be tested. The procedure for these uptake studies is described by Savage and Green.[8]

*Detection of Cbl-Binder Released by the Cultured Cells.* Since only very small amounts of TC II or unsaturated Cbl-binder are expected to be released into the culture medium by the cells, the charcoal technique of Gottlieb et al.[13] for protein-bound Cbl was modified by Savage and Green[8] for use in this study as follows.

Each tube contains 1.0 ml of sample (output medium) and 0.1 ml of bovine serum albumin (1% w/v dissolved in 0.2 $M$ sodium phosphate buffer, pH 6.2). This is followed by the addition of 0.1 ml of 0.15 $M$ NaCl containing 600 pg of $^{57}$Co-labeled cyanocobalamin (CN-[$^{57}$Co]Cbl). The tubes are mixed and allowed to incubate for 20 min at 25°. Then 0.5 ml of charcoal albumin (2.5% charcoal, w/v in 0.5% aqueous albumin) is added, and the tubes are mixed and centrifuged at 2000 $g$ for 10 min at 4°. The supernatant is removed and counted for $^{57}$Co. When a standard curve was prepared using increasing amounts of rat serum (rat TC II), the assay was found to be linear to 120 pg and sensitive to 5 pg (Fig. 1). None of the metabolic inhibitors used in the study interfered with the charcoal assay. Unlike whole human serum, rat and mouse serum contain only one Cbl-binding protein, TC II.[8,9] These sera can, therefore, be used as a source of rodent TC II without further purification.

*Preparation of Rat and Mouse TC II.* The unsaturated Cb1-binding capacity of rat and mouse serum is determined by the charcoal assay described above. A saturating amount of either CN-[$^{57}$Co]Cbl or non-labeled CN-Cbl is added to the serum and incubated for 20 min at 37°. In a few instances where the serum was examined by gel filtration, a slight

[12] T. I. Diamondstone, *Anal. Biochem.* **6,** 395 (1966).
[13] C. Gottlieb, K. S. Lau, L. R. Wasserman, and V. Herbert, *Blood* **25,** 875 (1965).

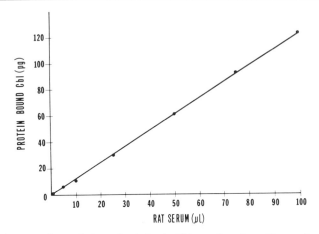

FIG. 1. Amount of protein-bound cobalamin Cbl as a function of increasing amounts of rat serum TC II. From C. R. Savage, Jr. and P. D. Green, *Arch. Biochem. Biophys.* **173**, 693 (1976).

excess of CN-[$^{57}$Co]Cbl was added as a marker for free Cbl elution on the column. After the serum is labeled, it is used immediately or stored frozen at −20°.

*Preparation of Cbl-Binder Released from Cultured Cells.* Culture medium from several preparations of each cell line (rat liver parenchymal cells and mouse fibroblast (L-929) cells) is collected from days 2–5, pooled, and centrifuged at 2000 *g* for 30 min at 4°. The supernatant is labeled with CN-[$^{57}$Co]Cbl at a 20% excess of Cbl-binding capacity, concentrated 10 to 50-fold using either dialysis tubing or PM-10 membranes. The concentrated cell output is dialyzed vs 1000 volumes of 0.15% NaCl at 4° for 18 hr to remove excess free Cbl. The cell output is then stored at −20° for use in subsequent experiments.

### Column Chromatography

*Sephadex G-200 Chromatography.* Sephadex G-200 chromatography is conducted on a 2 × 100-cm column equilibrated with 50 m$M$ sodium phosphate buffer, pH 7.4, containing 0.5 $N$ NaCl. Then 0.1–0.2 ml of rat or mouse serum and 1.0-ml aliquots of cell binder are applied to the same column and sequentially eluted at a rate of about 10 ml/hr with the equilibrating buffer.

*Ion-Exchange Chromatography.* Ion-exchange chromatography using DEAE-cellulose and CM-cellulose is carried out on 1.5 × 12-cm columns. The DEAE-cellulose column is equilibrated with 20 m$M$ Tris-Cl buffer, pH. 8.6, and developed with a salt gradient of 0.3 $M$ NaCl in the equilibrating buffer which flows into a 250-ml, constant-volume, mixing chamber containing 20 m$M$ Tris-Cl buffer, pH 8.6. The flow rate of the columns is about 12 ml/hr.

The CM-cellulose column is equilibrated with 20 m$M$ sodium acetate buffer, pH 5.6, and developed in the same manner described above using a 1.0 $M$ NaCl gradient in the equilibrating buffer.

Either 0.2 ml of whole serum, TC II prepared from ammonium sulfate precipitation as described by Begley and Hall,[14] or 1.0-ml aliquots of cell output are applied to the above ion-exchange columns and eluted as described above. The detailed preparations of each sample can be found in Savage and Green[8] and Green $et\ al.$[9]

*Affinity Chromatography.* CN-Cbl-Sepharose is prepared according to the method of Allen and Majerus.[15] A monocarboxylic acid derivative of CN-Cbl is prepared by partial acid hydrolysis. This derivative is isolated by ion-exchange chromatography and coupled to 3,3'-diaminodipropyl-amine-Sepharose by means of a carbodiimide reagent. The final preparation yields 2 $\mu$mol of CN-Cbl covalently found per milliliter of Sepharose after centrifugation at 800 $g$ for 1 min at room temperature. The 0.5 $\times$ 0.5-cm columns of CN-Cbl–Sepharose are washed according to the method of Allen and Majerus[15] as modified by Savage and Green.[8] These columns are used to analyze the [14]C-labeled TC II synthesized by the liver parenchymal cells.

Briefly, 12 flasks of 3-day-old cultures of liver parenchymal cells are exposed to 1 $\mu$Ci of [14]C-labeled amino acid hydrolyzate per flask (5 ml of medium per flask) for 48 hr. Then 6 ml of 0.2 $M$ sodium phosphate buffer, pH 5.8, containing 2% bovine serum albumin are added to the 60 ml of cell medium pooled from the 12 flasks. The pooled medium is centrifuged at 15,000 $g$ for 30 min at 4° and divided into two equal parts. One-half is applied directly to the affinity column, and the other half is saturated with nonlabeled CN-Cbl (2.5 times unsaturated Cbl-binding capacity) prior to application on the affinity column. Both columns are eluted as described in the modified procedure of Savage and Green.[8]

## Results and Discussion

The release of a Cbl-binder into the culture medium of both rat liver parenchymal cells and mouse fibroblast (L-929) cells was detected by the charcoal assay described in Methods (Table I). The amount of Cbl-binder released by the cells was a function of time in culture. Rat liver parenchymal cells released a maximum of 640 pg of Cbl-binder after 4 days of

[14] J. A. Begley and C. A. Hall, *Blood* **45**, 281 (1975).
[15] R. H. Allen and P. W. Majerus, *J. Biol. Chem.* **247**, 7709 (1972).

TABLE I

COBALAMIN BINDER RELEASED BY CULTURED CELLS

| | Cbl-binder released daily (pg CN-[$^{57}$Co]Cbl/24 hr) | |
| --- | --- | --- |
| Days in culture | Rat liver parenchymal cells[a] | Mouse fibroblast (L-929) Cells[b] |
| 2 | 134 | 242 |
| 3 | 495 | 442 |
| 4 | 630 | 279 |
| 5 | 246 | 212 |
| 6 | 210 | 390 |
| 7 | 76 | — |

[a] These data were extrapolated from C. R. Savage Jr. and P. D. Green, *Arch. Biochem. Biophys.* **173**, 695 (1976).

[b] These data were extrapolated from P. D. Green, C. R. Savage Jr., and C. A. Hall, *Arch. Biochem. Biophys.* **176**, 685 (1976).

culture, which declined to 26 pg by day 7. These results are similar to those of Bissel *et al.*,[16] who noted an elevated rate of albumin synthesis by the hepatocytes after 4 days in culture. This could represent the time needed to repair cell membranes damaged by collagenase digestion. After 5 days in culture, the cells started detaching from the flasks, which could explain the decreased production of Cbl-binder. The amount of Cbl-binder released by the mouse fibroblast (L-929) cells reached a maximum of 442 pg on day 3 and then steadily declined. There was an unexpected rise on day 6 followed by no further production of binder on day 7. The data (Table I) suggest that liver parenchymal cells are a more avid producer of Cbl-binder than mouse fibroblast cells. This hypothesis is further supported by the fact that on days 2–5 the final number of liver parenchymal cells remaining attached to the flask was about 1 to 2 × 10$^6$ cells while the mouse fibroblast cultures contained about 7 to 8 × 10$^6$ cells per flask. This large difference in final cell concentration is due to the difference in plating efficiency of both cell lines as well as the fact that mouse fibroblasts divide in culture in the presence of serum (first 24-hr incubation period).

The release of a Cbl-binder by both cell types was noted, then the physicochemical properties of liver cell binder and mouse fibroblast binder were compared with rat and mouse serum TC II, respectively. The comparative elution profiles of cell binder and rodent TC II were examined on gel filtration (Fig. 2) and ion-exchange chromatography (Fig. 3).

[16] D. M. Bissel, L. E. Hammaker, and U. A. Mayer, *J. Cell Biol.* **59**, 722 (1973).

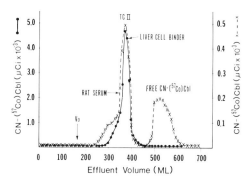

FIG. 2. Sephadex G-200 chromatography of CN-[$^{57}$Co]Cbl-labeled rat serum and liver cell binder. ---, Elution profile of rat serum; ——, liver cell binder. Rat serum and liver cell binder were run sequentially on the same column. From C. R. Savage, Jr. and P. D. Green, *Arch. Biochem. Biophys.* **173**, 697 (1976).

The results were almost identical for both cell outputs, and only the data from the liver parenchymal cell are presented. The Cbl-binder released by the cultured cells was identical to rodent TC II with respect to molecular weight and net charge by ion-exchange chromatography.

If the liver cell binder and mouse fibroblast binder were authentic TC II, they should exhibit the same biological activity as rodent TC II. Since the only *in vitro* test of the biological activity of TC II is its ability to facilitate Cbl uptake by cells, the mouse fibroblast system, being of rodent origin, was used. This later point is important because a species specificity is known to exist for the TC II's.[17] Only those TC II's (monkey and dog) known to cross-react with anti-human TC II effectively promoted Cbl uptake by HeLa cells, which are of human origin. This parallel between immunological cross-reactivity and ability to deliver Cbl to human cells suggests a specific receptor on the cell membrane capable of distinguishing TC II from different species.

The Cbl-binder released by the cells significantly enhanced Cbl uptake over that of serum TC II (Table II). One possible explanation is that rat and mouse serum TC II, being the only known Cbl-binder in the serum, carries all the native holo TC II. This could compete with the CN-[$^{57}$Co]Cbl in the serum and by dilution lower the observed Cbl uptake by the cells. To avoid the potential dilution effect of native holo TC II in a sample, we now measure its amount using the *Euglena gracilis* assay[18] and then making the necessary calculation. In assaying the liver cell binder and mouse fibroblast binder, no native holo TC II was found.[8,9]

[17] C. A. Hall and A. E. Finkler, *in* "The Cobalamins" (H. R. V. Arnstein and R. J. Wrighton, eds.) p. 49. Churchill Livingstone, Edinburgh and London, 1971.
[18] B. B. Anderson, *J. Clin. Pathol.* **17**, 14 (1964).

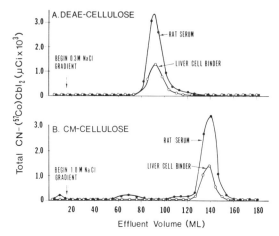

F$_{IG}$. 3. Ion-exchange chromatography of rat serum TC II and liver cell binder from ammonium sulfate precipitation on DEAE-cellulose and CM-cellulose. ●——●, Elution profile of rat serum TC II on DEAE-cellulose (A) and CM-cellulose (B). ○——○, Elution profile of liver cell binder on DEAE-cellulose (A) and CM-cellulose (B). From C. R. Savage, Jr. and P. D. Green, *Arch. Biochem. Biophys.* **173**, 697 (1976).

A similar phenomenon of a "form of TC II" being more effective than its serum counterpart was observed in the perfusion studies of Hall and Rappazzo.[7] In this study, dog "renal TC II" was more effective than dog serum TC II in promoting Cbl uptake by canine organs. There is no obvious explanation for either of these findings other than the fact that the sera (rodent and dog) used in both studies carries native holo TC II, whereas the TC II released by cells and possibly by organs do not.

The amount of free Cbl taken up by mouse fibroblasts was high when compared to that normally taken up by HeLa cells. Since mouse fibroblasts release a Cbl-binder into the culture medium, it could bind the added free Cbl and facilitate its uptake into the cell.

There was also a noticeable difference in Cbl uptake (Table II) between experiments 1 and 2. These two studies were performed independently, and, because of our limited supply of mouse fibroblast binder, the cells were exposed to only half the total amount of Cbl in the mouse fibroblast study (Table II, experiment 2) as in the liver cell study (Table II, experiment 1). The comparative results indicate, however, that Cbl uptake facilitated by cell binder was almost double that of serum TC II.

All the above data demonstrated that both rat liver parenchymal cells and mouse fibroblast (L-929) cells released a Cbl-binding protein into the culture medium which had the physicochemical and biological properties of TC II. To determine whether these cells released or actually syn-

TABLE II

UPTAKE OF COBALAMIN BY MOUSE FIBROBLAST (L-929) CELLS

| | Experiment 1[a] | Experiment 2[b] | |
|---|---|---|---|
| Form of CN-[57Co]Cbl | Cell uptake of Cbl (pg of Cbl/mg cellular protein) | Form of CN-[57Co]Cbl | Cell uptake of Cbl (pg of Cbl/mg cellular protein) |
| Rat serum TC II | 37.1 | Mouse serum TC II | 12.8 |
| Liver cell binder | 77.5 | Mouse fibroblast binder | 29.8 |
| Free CN-[57Co]Cbl | 23.2 | Free CN-[57Co]Cbl | 5.9 |

[a] Mouse fibroblasts were incubated for 2 hr at 37° with a total of 2000 pg of CN-[57Co]Cbl per flask as free Cbl or Cbl bound to the indicated protein. The values are averages of duplicate determination. From C. R. Savage, Jr. and P. D. Green, Arch. Biochem. Biophys. 173, 699 (1976).

[b] Mouse fibroblasts were incubated for 2 hr at 37° with a total of 1000 pg of CN-[57Co]Cbl per flask as free Cbl or Cbl bound to the indicated protein. The mean is an average of 4 separate experiments. From P. D. Green, C. R. Savage, Jr. and C. A. Hall, Arch. Biochem. Biophys. 176, 687 (1976).

TABLE III

INCORPORATION OF [14]C-LABELED AMINO ACIDS INTO LIVER CELL BINDER BY
HEPATOCYTES IN CULTURE[a]

| | Total amount of [14]C absorbed to CN-Cbl–Sepharose affinity column (cpm) | |
| Sample | Experiment 1 | Experiment 2 |
|---|---|---|
| Liver cell medium (unsaturated) | 714 | 737 |
| Liver cell medium saturated with unlabeled CN-Cbl | 431 | 411 |

[a] From C. R. Savage Jr. and P. D. Green, *Arch. Biochem. Biophys.* **173**, 700 (1976).

thesized TC II, they were exposed to two inhibitors of protein synthesis, cycloheximide and puromycin.

Cycloheximide (5 $\mu$g/ml) was found to be a potent inhibitor of TC II release by liver parenchymal cells. This inhibition could be partially reversed by removing the drug followed by a return of binder output to 50% of control values. Therefore at least 50% of the effect of the drug was due to inhibition of protein synthesis, not to cell death. Puromycin (5 $\mu$g/ml) produced an irreversible 35% inhibition of TC II release. By varying the concentration of this drug, different degrees of inhibition were obtained, but no condition was found that would allow significant reversal of inhibition.

Cycloheximide (5 $\mu$g/ml) and puromycin (2 $\mu$g/ml) were both found to be potent and reversible (about 50%) inhibitors of TC II release by mouse fibroblasts.

These data strongly suggest biosynthesis of TC II in that two inhibitors of protein synthesis, working at two different loci, both effectively, prevented TC II release. Since mouse fibroblast (L-929) cells were serially propagated and divided in culture, the possibility of a "storage-release" phenomenon of a performed Cbl-binder appeared remote. This was not the case with rat liver parenchymal cells, since they were a primary culture derived from organ perfusion. To eliminate this possibility, an additional study was performed with the liver parenchymal cells to determine whether they did synthesize TC II (Table III). The liver cells were exposed to [14]C-labeled amino acids for 48 hr. The medium was analyzed for [14]C-labeled TC II by affinity chromatography with an affinity column containing CN-Cbl covalently bound to the Sepharose gel. Theoretically, if a solution containing an unsaturated Cbl-binder is passed through the column, the Cbl-binder should be retained on the column. If liver parenchymal cells are capable of the de novo biosynthesis of TC II, they should be capable of utilizing the [14]C-labeled amino acids placed in the culture medium to synthesize new [14]C-labeled TC II. This new TC II should then specifically bind to the CN-Cbl–Sepharose column. Since there can be a certain degree of nonspecific absorption to the column, it would be necessary to differentiate the [14]C-labeled TC II from the other [14]C-labeled proteins synthesized by the cells. To circumvent this problem, the liver cell medium which contained several [14]C-labeled cellular proteins was divided in half. One-half was incubated with unlabeled CN-Cbl to serve as a blank for nonspecific absorption of [14]C-labeled proteins. Both halves were then run on CN-Cbl-Sepharose columns. The 50% increase in absorption of unlabeled liver cell medium over the half that had its Cbl-binders saturated was interpreted to be due to the presence of [14]C-labeled TC II synthesized de novo by the cells.

The physicochemical properties, the block by metabolic inhibitors of protein synthesis, and the biological activity taken together with the incorporation of [14]C-labeled amino acids into TC II, all indicated these cells to be capable of the de novo biosynthesis of TC II. These data support early studies of Hall and Rappazzo[7] and Sonneborn et al.,[6] which suggested that more than one cell in the body was capable of producing TC II.

### Acknowledgments

We wish to thank *Archives of Biochemistry and Biophysics* for permission to reproduce material from the Methods sections and illustrations of the following papers: C. R. Savage Jr. and Pamela D. Green, *Arch. Biochem. Biophys.* **173**, 691 (1976) and P. D. Green, C. R. Savage, Jr., and C. A. Hall, *Arch. Biochem. Biophys.* **176**, 683 (1976).

## [13] Competitive Binding Radioassays for Vitamin B₁₂ in Biological Fluids or Solid Tissues

By RALPH GREEN

### Principle

Cobalamins (Cbl) in biological fluids, solid tissues, or foods may be measured by a competitive binding radioassay. Protein-bound Cbl is released by boiling at acid pH. For solid tissue this is preceded by papain hydrolysis. Released Cbl is converted to cyanocobalamin (CN-Cbl; vitamin B₁₂) in the presence of excess cyanide. A known amount of added [57Co]CN-Cbl is allowed to compete with the extracted CN-Cbl for a limited amount of Cbl-binding protein. Bound and free portions of CN-Cbl in the mixture are separated using protein-coated charcoal. Measurement of the bound [$^{57}$Co]Cn-Cbl radioactivity allows determination of the degree of radioisotopic dilution. By reference to a curve constructed from a series of standards containing known amounts of CN-Cbl, the amount of nonradioactive CN-Cbl in a unknown specimen may be calculated.

Several radioassays for Cbl in serum and other biological fluids have been described. However, application of these techniques to assay of Cbl in solid tissues was, until recently,[1,2] neglected. The methods described here[1,3] were selected on the basis of a study of factors affecting the critical steps of a Cbl radioassay.[4] Although the methods were designed to exploit the useful Cbl-binding properties of chicken serum, the general methodology may be used with any suitable Cbl-binding protein.

## Assay Methods

### Reagents

$^{57}$Co-Labeled cyanocobalamin ([$^{57}$Co]CN-Cbl): High specific activity (100–300 $\mu$Ci/$\mu$g), [$^{57}$Co]CN-Cbl (Radiochemical Centre) diluted to a Cbl concentration of 100 pg/ml using standard CN-Cbl.

Standard CN-Cbl: Stock solution of CN-Cbl prepared (USP standards, Rockville, Maryland) to contain 1 mg/liter based on spectrophotometric determination of absorbance at 361 nm. ($\epsilon_{mM}$ = 27.7). Stock solution was diluted to give reference standards of 1000, 500, 250, and 125 pg/ml.

Sodium acetate/acetic acid (tissue-extracting buffer): 70 m$M$ sodium acetate titrated to pH 4.6 with 70 m$M$ acetic acid

Papain: Freshly made aqueous papain solution, 5 g/liter

KCN: 1 g/100 ml of water

Sodium acetate/HCl (common extracting buffer): 0.4 $M$ sodium-acetate titrated to pH 4.0 with 0.4 $N$ HCl. KCN solution is added to a cyanide concentration of 20 mg/liter.

Glycine/NaOH/NaCl (neutralizing buffer): 0.8 $M$ glycine dissolved in 0.8 $M$ NaCl and titrated to pH 10.0 with 0.8 $N$ NaOH

Saline: 9 g/liter NaCl

Cbl-binding protein: Either chicken serum (CS) or human intrinsic

[1] S. V. van Tonder, J. Metz, and R. Green, *Clin. Chim. Acta* **63**, 285 (1975).
[2] E. P. Frenkel, J. D. White, C. Galey, and C. Croy, *Am. J. Clin. Pathol.* **66**, 863 (1976).
[3] R. Green, P. A. Newmark, A. M. Musso, and D. L. Mollin, *Br. J. Haematol.* **27**, 507 (1974).
[4] P. A. Newmark, R. Green, A. M. Musso, and D. L. Mollin, *Br. J. Haematol.* **25**, 359 (1973).

factor (HIF) are prepared as follows: CS binder is crude chicken serum obtained from freshly clotted chicken blood and diluted in saline to an appropriate Cbl-binding capacity (Cbl-BC); 95% pure HIF binder is prepared from neutralized normal human gastric juice by affinity chromatography.[5] The HIF is diluted to give an appropriate Cbl-BC in 0.1 $M$ K$_2$HPO$_4$/KH$_2$PO$_4$ buffer pH 7.5 containing 1 g of bovine serum albumin per liter.

Cbl-BC of CS and HIF are determined from a binding curve by a modification of the method of Lau *et al.*[6] in which the binding of 50 pg (0.5 ml) [$^{57}$Co]CN-Cbl is measured by serial doubling dilutions of the binder (0.5 ml). Binding is carried out for 1 hr at room temperature in the presence of equal volumes (0.5 ml) of the common extracting buffer and the neutralizing buffer. Bound and free [$^{57}$Co]CN-Cbl are separated using protein-coated charcoal as described below. From the binding curve a dilution of the binder is made so that 0.25 ml binds between 40% and 60% of the 50 pg [$^{57}$Co]CN-Cbl used in the assay. Diluted CS may be stored at $-20°$ for 6 months without loss of Cbl-BC. Diluted HIF is unstable on storage and must be prepared fresh each week.

Protein-coated charcoal:[6] Charcoal is coated either with hemoglobin or albumin. A 10 g/100 ml solution of human or horse hemoglobin is stored in 5-ml aliquots at $-20°$. Before use the contents of a tube are diluted to 200 ml with water and added to an equal volume of a 5% aqueous suspension (10 g in 200 ml) of Norit A neutral activated charcoal (Amend or Nutritional Biochemicals). For albumin coating, a 1% solution of bovine serum albumin is used in place of hemoglobin. A variety of charcoals and protein or polysaccharide coating agents may be used, but it is necessary to check the performance of a particular coated-charcoal system for separating bound and free CN-Cbl.[7]

*Procedure*

*Preparation of Tissue Extract.* For assay of solid tissues, the tissue is first subjected to papain proteolytic digestion using the method of Anderson[8] modified as follows[9]: a weighed amount (5–30 mg) of tissue is placed in 1.0 ml of sodium acetate/acetic acid buffer (tissue extracting buffer)

[5] D. W. Jacobsen and Y. Montejano, "III European Symposium on Vitamin B$_{12}$ and Intrinsic Factor, Zurich." de Gruyter, Berlin, 1979.
[6] K.-S. Lau, C. Gottlieb, L. R. Wasserman, and V. Herbert, *Blood* **26**, 202 (1965).
[7] J. F. Adams and F. C. McEwan, *Br. J. Haematol.* **26**, 581 (1974).
[8] B. B. Anderson, Investigations into the *Euglena* method of assay of vitamin B$_{12}$. Ph.D. thesis, University of London (1965).
[9] I. Chanarin, "The Megaloblastic Anaemias," p. 207. Blackwell, Oxford, 1970.

ASSAY PROTOCOL (VOLUMES GIVEN IN MILLILITERS)

| Tube | $[^{57}\text{Co}]$CN-Cbl (100 pg/ml) | Unknown[a] | Saline | Standard CN-Cbl (125–1000 pg/ml)[b] | Common extracting buffer | Neutralizing buffer | Diluted B$_{12}$ binder[c] | Saline | Coated charcoal |
|---|---|---|---|---|---|---|---|---|---|
| | | | | | | 1↓[d] | | 2↓ | 3↓ |
| Unknown sample | 0.5 | 0.25 | — | — | 0.5 | 0.5 | 0.25 | — | 1.0 |
| Supernatant control (sample) | 0.5 | 0.25 | — | — | 0.5 | 0.5 | — | 0.25 | 1.0 |
| Standard curve | 0.5 | — | 0.25 or[e] | 0.25 | 0.5 | 0.5 | 0.25 | — | 1.0 |
| Supernatant control (standard) | 0.5 | — | 0.25 | — | 0.5 | 0.5 | — | 0.25 | 1.0 |
| Total radioactivity | 0.5 | — | 0.25 | — | 0.5 | 0.5 | — | 1.25 | — |

[a] Tissue extract, serum, or other biological fluid. When volume is limited, assays may be carried out on 0.1 ml.

[b] For specimens with Cbl concentration falling above this range, assay should be repeated using an appropriate dilution.

[c] Chicken serum, intrinsic factor, or other Cbl binder.

[d] Arrows: 1, Cap; boil for 15 min; 2, mix, incubate for 1 hr at room temperature; 3, mix, centrifuge, remove and count supernatants (2 ml).

[e] In addition to standard tubes containing 0.25 ml of standard CN-Cbl (125, 250, 500, 1000 pg/ml), a set of duplicate tubes containing no added (Cbl) are included to serve as zero points of the standard curve.

with 0.1 ml of papain solution and 0.1 ml of KCN solution, and the tissue is disrupted manually with a Potter–Elvehjem homogenizer. Further disruption is achieved by rapid freeze–thawing of the homogenate once using Dry-Ice–acetone. The pH is adjusted to 5.0 with the basic portion of the extracting buffer (70 m$M$ sodium acetate), and the volume is brought to 20 ml with distilled water. The diluted homogenate is placed in a shaking water bath at 60° for 1 hr. A 0.25-ml volume of the well mixed extract is used in the radioassay. The same method may be applied to solid foods.

*Assay Protocol.* The assay protocol is shown in the table. The following points should be noted: Beyond the preliminary extraction carried out on solid tissues, the procedure for tissue extracts and serum or other biological fluids is identical. All tubes in the assay are set up in duplicate in 75 × 12 mm Pyrex test tubes. Extraction at this stage is carried out with tubes in a covered boiling water bath for 15 min. To prevent evaporation during extraction, tubes are capped with tight-fitting Morton metal culture-tube closures (Scientific Products) or nonabsorbent cotton plugs. The denatured protein coagulum resulting from extraction need not be removed prior to the binding stage. Addition of the neutralizing buffer renders the pH alkaline (above 9.0) and maintains high ionic strength, conditions favoring optimal and stable binding of CN-Cbl by Cbl-binding proteins commonly used in competitive binding radioassays.[4] At each point after addition of the neutralizing buffer, the binder, and then the charcoal, tube contents are thoroughly mixed on a vortex mixer. During the 1-hr incubation for binding, it is not necessary to agitate the tube contents. Charcoal may be dispensed automatically provided the charcoal suspension is continuously agitated with a stirring magnet. After charcoal addition, tubes are centrifuged for 10 min at 1850 $g$, and 2 ml of supernatant, containing bound [⁵⁷Co]CN-Cbl are removed for radioactivity measurement.

*Charcoal Supernatant Controls.* In tubes that contain Cbl-binder, radioactivity remaining in the supernatant following treatment with coated charcoal should represent only protein-bound [⁵⁷Co]CN-Cbl. However, a small proportion of unbound [⁵⁷Co]CN-Cbl may not be removed by the coated charcoal, and it is necessary to correct the radioactivity in the supernatant accordingly. Supernatant control tubes are included for this purpose, both for the unknown samples and the standard tubes (see the table). These control tubes do not receive Cbl binder, but are treated with coated charcoal. The counts in each unknown or standard tube are corrected by subtracting counts in the appropriate supernatant control tubes. Most serum specimens have very similar supernatant control values, making it possible to use a single supernatant control prepared from pooled serum for any particular assay. However, supernatant control tubes from occasional serum specimens have considerably greater

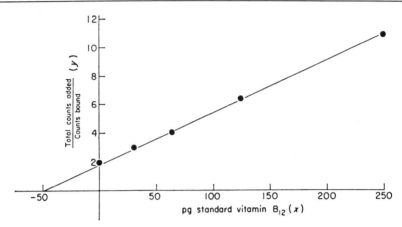

FIG. 1. A typical standard curve. Tubes containing 0, 31.25, 62.5, 125, and 250 pg of CN-Cbl standards are shown. Reproduced from R. Green, P. A. Newmark, A. M. Musso, and D. L. Mollin, *Br. J. Haematol.* **27**, 507 (1974), with permission.

counts, and it is therefore advisable to include a matching supernatant control for each specimen. Failure to do so could result in a spuriously low assay value because of spuriously high tube counts. For tissue assays, because of the wide variability between specimens, individual supernatant control tubes for each sample are essential.

*Calculation of Results.* Following the principle of a competitive binding radioassay, bound radioactivity in the supernatant is inversely related to the mass of nonradioactive Cbl present in the system. Dose-response curves for the reference standards may be represented in several ways, including some that produce convenient linearization of data. A simple method of linearization is to plot the ratio of total counts added/counts bound (after subtraction of supernatant control), on the ordinate ($y$) against the mass of standard CN-Cbl added, on the abscissa ($x$). The best-fit regression line for points on the standard curve may be readily obtained using the least-squares method, which is conveniently carried out on a small programmable calculator or computer. A typical standard curve is shown in Fig. 1.

From the linear equation $y = bx + a$, the mass of Cbl in an unknown sample $x$ may be calculated as $x = 1/b(y - a)$, where $b$ is the slope, $a$ the $y$-axis intercept, and $y$ the calculated value for total counts/counts bound in that sample. This represents the quantity of Cbl in 0.25 ml of the unknown serum or, in the case of tissue, in 0.25 ml of the 20-ml extract prepared from a weighed amount of tissue. The final calculation for Cbl concentration in a specimen of serum, other undiluted fluid, or tissue extract is

$$\text{Cbl (pg/ml)} = 4/b \left( \frac{\text{total counts}}{\text{counts bound}} \right) - a$$

For tissue Cbl concentration it is

$$\text{tissue Cbl (ng/g)} = \frac{20x}{\text{weight of tissue (mg)}}$$

where $x$ is the concentration of Cbl in the tissue extract (in pg/ml) calculated as above.

This form of data representation offers a further advantage that is useful as an internal check for assay quality control. From the parameters of the linear standard curve equation, information may be derived on the binding capacity of the binder and the mass of radioactive ligand in the system. If, as in the conditions of this assay, binding remains constant over the range of Cbl concentrations covered by the standard curve, then: Cbl-BC is given by $1/b$ (reciprocal of the slope of the linear equation); mass of radioligand [$^{57}$Co]CN-Cbl is given by $a/b$ (numerically equal to the $x$-axis intercept).[3]

### Factors Affecting the Assay[3,4]

There are three critical steps in this and other competitive binding radioassays for Cbl: step 1, extraction of endogenous Cbl; step 2, binding of the mixture of radiolabeled and endogenous CN-Cbl; step 3, separation of the mixture of bound and free [$^{57}$Co]CN-Cbl. Several factors may influence the assay during these critical steps, and these have been specifically investigated and controlled in the method described.[1,3,4]

### Step 1. Extraction

Efficacy of extraction using the common extracting buffer has been examined. After protein precipitation of serum containing added [$^{57}$Co]CN-Cbl, release of radioactivity into the supernatant ranged from 91 to 98%, and 98% or more of this radioactivity could then be removed by treatment with protein-coated charcoal, indicating that the [$^{57}$Co]CN-Cbl was not bound to any unprecipitated protein.[3]

The use of a double-extraction method for releasing Cbl from solid tissues was deemed necessary on the basis of experiments on liver obtained from an animal injected repeatedly with high specific activity [$^{57}$Co]CN-Cbl. Release of radioactivity following the double-extraction procedure adopted in the method (99.4%) was more efficient than either papain hydrolysis alone (89.4%) or single extraction with the common extracting buffer (94.7%). Moreover, double extraction consistently resulted in higher measured tissue Cbl concentrations than did papain hydrolysis alone.

### Step 2. Binding

*pH and Ionic Strength.* Cbl binders exhibit pH-dependent changes in binding capacity, attaining a maximum at alkaline pH. Cbl-BC remains unchanged over a certain pH range, which is different for each binder (for chicken serum, pH 7–12; intrinsic factor pH 5.5–9.0). Binding capacity is also affected by ionic strength, particularly below pH 5, and falls sharply as the ionic strength is decreased below 0.1. The apparent fall of Cbl-BC at low pH and ionic strength may in part be related to the phenomenon of binder–Cbl complex adsorption to tube surfaces, followed by dissociation of Cbl from the adsorbed complex. This is most liable to occur at low protein concentrations.[4,10] For these reasons, an alkaline pH and high ionic strength are used in the present assay.

*Effect of Serum or Protein.* In the present serum assay, the binding step is unknowns is carried out in the presence of the denatured serum protein coagulum. Both chicken serum and intrinsic factor binders show avid binding of CN-Cbl (association constants ($K°$) of $1.1 \times 10^{12}$ and $1.5 \times 10^{10}$ $M^{-1}$, respectively[4,10]). Specimens assayed previously,[3] using chicken serum in the presence or the absence of the protein precipitate, gave excellent overall correlation as well as good agreement for individual samples. In the author's present laboratory this has been found to hold true when purified human intrinsic factor is used as the binder.

The presence of additional or extraneous protein and its effect on the binder is also important in relation to the use of appropriate conditions in standard tubes. It is essential that binding conditions be the same in standard tubes as in unknowns. It has been established that the addition of 0.01–0.25 ml of serum has no effect on the Cbl-BC in the assay.[3,4] However, the presence or the absence of additional protein does influence the Cbl-BC of intrinsic factor, and it is necessary to add protein to the standard curve tubes in assays that use intrinsic factor as the Cbl-binding protein.[3,11-13]

### Step 3. Separation

*Stability of Binder-Cbl Complex in the Presence of Protein-Coated Charcoal.* To achieve satisfactory separation of the binder–Cbl complex from

[10] R. H. Allen, and C. S. Mehlman, *J. Biol. Chem.* **248**, 3660 (1973).

[11] P. A. Newmark and N. Patel, *Blood* **38**, 524 (1971).

[12] R. S. Hillman, M. Oakes, and C. Finholt, *Blood* **34**, 385 (1969).

[13] J. L. Raven, M. B. Robson, P. L. Walker, and P. Barkhan, *J. Clin. Pathol.* **22**, 205 (1969).

unbound Cbl, the protein-coated charcoal must not adsorb the binder–Cbl complex or cause dissociation of Cbl from the complex. Binder [$^{57}$Co]CN-Cbl complexes devoid of free [$^{57}$Co]CN-Cbl were prepared by molecular sieving and treated with protein-coated charcoals.[4] Albumin or hemoglobin-coated charcoal that adsorbed more than 99% of free [$^{57}$Co]CN-Cbl removed less than 5% of radioactivity from binder– [$^{57}$Co]CN-Cbl complexes.

### Assay Evaluation

The assays described have been evaluated[1,3] with respect to reproducibility, recovery of added CN-Cbl, assay of dilutions, and sensitivity.

*Reproducibility.* In general, the coefficient of variation for replicate assays on tissues or fluids is inversely related to the Cbl concentration. For serum the mean coefficient of variation was 5.8% within a batch and 9.9% between batches. The mean coefficient of variation in a group of liver samples was 6.7%.

*Recovery.* Standard CN-Cbl solutions (0.25 ml of solutions containing 125, 250, 500, and 1000 pg/ml) were added to serum specimens to assess recovery. The mean recovery was 102.2% (range 80 to 120%). Recovery of 500 pg of CN-Cbl added to liver extracts ranged from 95 to 106%.

*Assay of Dilutions.* Results of assays on serial dilutions of serum specimens down to 50 pg/ml did not deviate from predicted values by more than the error of the method. Assay of diluted specimens is therefore in order when the Cbl concentration exceeds the upper limit of the standard curve (1000 pg/ml).

*Sensitivity.* The lower limit of detection of Cbl is 20 pg/ml. For the tissue assay, sensitivity allows for the detection of Cbl at concentrations down to 25 ng/g. In tissues or foods with lower Cbl concentrations, it is possible to increase the lower limit of detection almost 5-fold, to 5 ng/g by reducing the final volume of the first extract from 20 ml to 4 ml.

### Comments

Results obtained with the present radioassay using chicken serum as the binder are consistently higher than with microbiological assays (*Euglena gracilis* or *Lactobacillus leichmannii*). This is true both for serum[4] and for tissue[1] specimens. Some other reports comparing radioassay with microbiological assay results for serum Cbl have found a similar discrepancy.[14,15] Mollin *et al.*[16] have recently reviewed this subject. It has not

[14] L. Wide and A. Killander, *Scand. J. Clin. Lab. Invest.* **27**, 151 (1971).
[15] J. L. Raven, M. B. Robson, J. O. Morgan, and A. V. Hoffbrand, *Br. J. Haematol.* **22**, 21 (1972).
[16] D. L. Mollin, B. B. Anderson, and J. F. Burman, *Clin. Haematol.* **5**, 521 (1976).

been clear whether this difference is the result of underestimation by the microbiological assay or overestimation by the radioassay of the actual Cbl concentration. A recent report[17] suggests that radioassays give higher results only when the relatively nonspecific R-protein binders are used for the assay. In addition to Cbl, these binders also bind Cbl analogs,[18,19] which appear to be present in serum as well as in mammalian tissues. These analogs do not have growth-promoting activity for *E. gracilis* or *L. leichmannii,* which would account for the lower values obtained using microbiological assays. When purified intrinsic factor is used as the binder, these analogs are not recognized, and radioassay results are then in closer agreement with microbiological assay values. This interesting concept requires further validation. At this time it appears that to measure true Cbl, purified intrinsic factor must be used as the binder. Alternatively, any R binder present in crude intrinsic factor preparations must be blocked by addition of cobinamides.[17] The source or significance of Cbl analogs present in biological tissues or fluid is not known. It is possible, however, to measure true Cbl using intrinsic factor binder and to measure Cbl plus Cbl-analog independently in the same specimen using an R protein as the binder.[17]

### Acknowledgments

This method was first developed in the laboratory of Dr. D. L. Mollin during a research fellowship funded by the Wellcome Trust (U.K.). Subsequent work has been supported by the South African Atomic Energy Board, and the U.S. National Institutes of Health (NS 13714).

[17] J. F. Kolhouse, H. Kondo, N. C. Allen, E. Podell, and R. H. Allen, *N. Engl. J. Med.* **299**, 785 (1978).
[18] M. B. Bunge and R. F. Schilling, *Proc. Soc. Exp. Biol. Med.* **96**, 587 (1957).
[19] C. W. Gottlieb, F. P. Retief, and V. Herbert, *Biochim. Biophys. Acta* **141**, 560 (1967).

# Section II

# Ubiquinone Group

## [14] Semimicro Extraction of Ubiquinone and Menaquinone from Bacteria

By LUDMILA KŘIVÁNKOVÁ and VLADIMÍR DADÁK

To draw correct conclusions on the function and location of quinones, it is necessary to work with a good extraction method. Many extraction procedures have been described for the isolation of quinones from various sources;[1-3] however, the recommended techniques are rather drastic and take considerable working time. The simple extraction procedure described here proved to be successful for the isolation of both ubiquinone and menaquinone from bacterial preparations. The quantitative analysis of the obtained impure extract by the difference absorption technique increases the accuracy of results and speeds up the whole method of quinone determination.

When quinones are extracted from intact bacteria, the character of the cell wall plays a significant role. In many cases, the extraction is facilitated substantially by the destruction of the wall, e.g., by ultrasonic disintegration, which makes the cell membrane more penetrable to organic solvents. The extraction procedure to be described here works with a solvent mixture that extracts quinones quantitatively both from membrane fragments and intact cells without any danger of their destruction, giving better results compared with other solvents and techniques. It also stops all the enzymic reactions by causing protein precipitation, thus being useful when the redox state of quinones is to be estimated.[4]

*Reagents*

Light petroleum (40–60°)
Methanol–light petroleum mixture, 3 : 2 (v/v)
NaOH, 2 $N$
HClO$_4$, 70%

### Extraction of Ubiquinone ($Q$) from Gram-Negative Bacteria (*Paracoccus denitrificans*)

*Basic Extraction Procedure.* Three milliliters of methanol–light petroleum mixture is added to 0.5 ml of bacterial suspension (up to 100 mg

---

[1] F. L. Crane and R. Barr, this series, Vol. 18C [220].
[2] P. J. Dunphy and A. F. Brodie, this series, Vol. 18C [233].
[3] H. Mayer and O. Isler, this series, Vol. 18C [236].
[4] A. Kröger and M. Klingenberg, *Biochem. Z.* **344**, 317 (1966).

METHODS IN ENZYMOLOGY, VOL. 67

dry weight per milliliter), and the preparation is agitated in a closed extraction tube on a laboratory mixer for 1 min. One milliliter of light petroleum is then added, the mixture is agitated again for 1 min, and the upper layer is separated by centrifugation (5 min, 2000 $g$) and removed; the residue is twice reextracted with 1-ml portions of light petroleum. In successive light petroleum fractions, there are 94, 5, and about 1% of ubiquinone. Combined fractions are evaporated *in vacuo,* and the residue is dissolved in ethanol for spectral measurement.

If the determination of ubiquinone content is performed from the difference between oxidized and reduced forms,[1] it is necessary to transform all ubiquinone in the oxidized state. The part of ubiquinone remaining in the reduced state can be autoxidized in alkaline medium during either extraction or spectral measurement. One tenth milliliter of 2 $N$ NaOH is pipetted into the mixture of methanol–light petroleum and bacterial suspension; it is agitated for 1 min before the first portion of light petroleum is added. In such an extract, quinone is fully oxidized. During the spectral determination, the reduction of ubiquinone is performed commonly by borohydride in the reference cuvette. In the sample cuvette, ubiquinone is oxidized quantitatively by adding 5 $\mu$l of 2 $N$ NaOH to 3 ml of its ethanolic solution.[5] The destruction of ubiquinone is negligible.

*Comparison of Extraction Methods.* The table compares several extraction procedures. Compared with menaquinone, as can be seen in the following part, the extraction of ubiquinone is not facilitated in the presence of perchloric acid. The extraction of bacterial fragments is easier; nevertheless, the isolated amount of ubiquinone never exceeds that obtained by the method described above.

The content of ubiquinone in *Paracoccus denitrificans* grown under anaerobic conditions was found to be 1.9 $\mu$moles of Q per gram dry weight and decreased in the presence of oxygen to 1.0 $\mu$moles of Q per gram dry weight (average of 13 batches). This is still a value about five times higher than that reported by Scholes and Smith.[6] Compared with the cytochrome *b* content in this bacterium,[7] the Q : cytochrome *b* ratio is close to the value of 6. Thus, the composition of the respiratory chain of *P. denitrificans*[8] and mammalian mitochondria is analogous also in this respect.

### Extraction of Menaquinone (MK) from Gram-Positive Bacteria (*Staphylococcus epidermidis*)

The penetration of solvents through the cell wall of *S. epidermidis* is less than with other bacteria. Membrane fragments (prepared in Braun's

[5] V. Dadák and L. Dufková, *Chem. Listy* **67,** 831 (1973).
[6] P. B. Scholes and L. Smith, *Biochim Biophys. Acta* **153,** 363 (1968).
[7] L. M. Sapshead and J. W. T. Wimpenny, *Biochim. Biophys. Acta* **267,** 388 (1972).
[8] K. Imai, A. Asano, and R. Sato, *Biochim. Biophys. Acta* **143,** 462 (1967).

COMPARISON OF EXTRACTION METHODS

| Extraction mixture | Paracoccus denitrificans[a] ($\mu$mol Q/g dry weight) | | Staphylococcus epidermidis[b] ($\mu$mol MK/g dry weight[c]) |
| --- | --- | --- | --- |
| | Intact cells | Cell fragments[d] | Intact cells |
| Methanol–light petroleum | 1.9 | 1.9 | 0.50 |
| + NaOH | 1.8 | 1.9 | 0.75 |
| + HClO$_4$ | 1.3 | 1.9 | 2.71 |
| Methanol–acetone[e] | 1.8 | 1.9 | 1.10 |
| Acetone[f] | 1.5 | 1.8 | 1.15 |

[a] *Paracoccus denitrificans* cells (1396 CCM and 8944 NCIB) were grown in the synthetic medium of J. P. Chang and J. G. Morris [*J. Gen. Microbiol.* **29**, 301 (1962)] with 50 m$M$ glucose, 20 m$M$ NO$_3^-$, and 10 $\mu M$ Fe$^{3+}$ under anaerobic conditions to the end of the logarithmic growth phase.

[b] *Staphylococcus epidermidis* 104-7 HIM was cultivated in the medium containing 12 g of Bacto-beef extract, 12 g of Bacto-peptone, and 6 g of NaCl in 1 liter of water, pH set to 7.2, to the middle of the aerobic logarithmic phase of growth. The menaquinone (MK) content at the end of growth is 3.93 $\mu$mol of MK per gram dry weight.

[c] In these experiments, 40 mg dry weight per milliliter of bacterial suspension were used.

[d] Obtained by ultrasonic disintegration.

[e] A. Kröger and V. Dadák, *Eur. J. Biochem.* **11**, 328 (1969).

[f] Q extraction: E. R. Redfearn and J. Burgos, *Nature (London)* **209**, 711 (1966); MK extraction: R. L. Lester, D. C. White, and S. L. Smith, *Biochemistry* **3**, 949 (1964).

MSK homogenizer) can be extracted easily, and pentane extraction of the lyophilized preparation is also successful. Acid hydrolysis and the prolonged time of the solvent action are the main conditions for quantitative MK isolation from intact cells.

*Modified Extraction Procedure.* One-half milliliter of the bacterial suspension (up to 100 mg dry weight per milliliter) and 4 ml of methanol–light petroleum mixture are agitated for 1 min. One-tenth milliliter of 70% HClO$_4$ is added; the suspension is mixed again for 1 min and then allowed to stand for 6–24 hr depending on the protein concentration. Up to about 40 mg dry weight per milliliter, 6 hr is sufficient; at about 80 mg/ml, the suspension must be kept in the dark for 20 hr. Afterward, MK is separated by three successive 1-ml portions of light petroleum; 90, 7, and 2% can be found in these fractions. The residue from evaporation of the combined upper layers of the solvent is dissolved in ethanol. Difference spectroscopy[9] is recommended for the quantitative determination of MK just in the crude extract.

*Note.* It should be mentioned here that neither larger amounts of HClO$_4$ nor intensive agitation during solvent action facilitate penetration. Perchloric acid must always be added to the mixture of the solvent and

[9] V. Dadák and L. Křivánková, this volume [17].

bacterial suspension, not to the bacteria alone. If higher protein concentrations than 100 mg dry weight per milliliter are to be extracted, 5 ml of the solvent mixture and 24 hr of standing ensure completeness of the extraction.

The content of MK in *S. epidermidis* depends strongly on the growth conditions. During the aerobic growth cycle, its amount increases and the maximum is achieved at the end of the logarithmic phase (3.93 $\mu$mol of MK per gram dry weight).[10] Under anaerobiosis (0.5% of glucose and 0.5% of nitrate added to the growth medium), MK content decreases to 2.2 $\mu$mol of MK per gram dry weight in accordance with changes of cytochrome content.[11]

### Extraction of Demethylmenaquinone (DMK) from *Streptococcus faecalis*

Although *S. faecalis* belongs among Gram-positive bacteria, DMK can be released easily and the "basic extraction procedure" gives satisfactory results.

[10] L. Křivánková and V. Dadák, *Scripta Fac. Sci. Nat. UJEP Brunensis, Chemia 2,* **6,** 79 (1976).
[11] N. J. Jacobs and S. F. Conti, *J. Bacteriol.* **89,** 675 (1965).

# Section III

# Tocopherols

## [15] Hepatic α-Tocopherol-Binding Protein*

### By George L. Catignani

Rat liver contains a soluble protein that binds α-tocopherol with high specificity and affinity.[1] The method of assay and some of the properties[1,2] of the binding protein are presented in this chapter.

### Animal and Tissue Preparation

Male weanling, Sprague–Dawley rats are maintained on a tocopherol-free diet[3] for a minimum of 6 weeks prior to use. Animals are killed by cervical dislocation, and the liver is immediately excised and placed on ice. The liver is perfused with ice-cold 0.9% NaCl followed by homogenization in four volumes (w/v) of cold 50 mM Tris · HCl buffer (pH 7.5) using a glass–Teflon pestle homogenizer.

*Preparation of Cytosol.* The homogenates are centrifuged at 10,000 g for 15 min at 4°; the supernatant is removed and centrifuged at 105,000 g for 1 hr at 4°. The high speed supernatant (cytosol) is collected free from the floating lipid and used immediately.

*Remarks.* Other animals besides rats may be used, assuming that they are maintained on a tocopherol-free diet appropriate for the species and for a sufficient length of time to produce biochemical tocopherol depletion as assessed by plasma tocopherol determination[4] or by the red cell hemolysis test.[5]

The use of tocopherol-depleted animals is not necessary if only the detection of the binding protein is sought; however, to demonstrate maximal binding it is essential to use depleted animals. Furthermore, quantitative comparisons between animals cannot otherwise be made owing to the presence of endogenous tissue tocopherol, which competes for binding of [³H]tocopherol.

---

* Paper No. 5913 of the Journal Series of the North Carolina Agricultural Research Service, Raleigh, N.C.
  Mention of a trademark or proprietary product does not constitute a guarantee or warranty of the product by the North Carolina Agricultural Research Service nor does it imply approval to the exclusion of other products that may be suitable.

[1] G. L. Catignani, *Biochem. Biophys. Res. Commun.* **67**, 66 (1975).
[2] G. L. Catignani and J. G. Bieri, *Biochim. Biophys. Acta* **497**, 349 (1977).
[3] R. P. Evarts and J. G. Bieri, *Lipids* **9**, 860 (1974).
[4] J. G. Bieri and R. H. Poukka Evarts, *Proc. Soc. Exp. Biol. Med.* **140**, 1162 (1972).
[5] J. G. Bieri and R. Poukka Evarts, *J. Nutr.* **100**, 557 (1970).

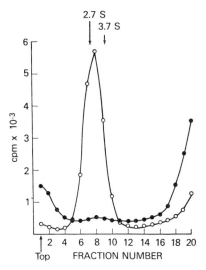

Fig. 1. Sucrose density gradient centrifugation of rat liver cytosol incubated with 100 n$M$ [³H]$\alpha$-tocopherol for 4 hr at 26° in the presence (●——●) or the absence (○——○) of a 400-fold molar excess (40 $\mu M$) of unlabeled $\alpha$-tocopherol. The sedimentation of chymotrypsinogen (2.7 S) and ovalbumin (3.7 S) is indicated by arrows.

### Binding Assay

The assay of tocopherol-binding protein (TBP) is based on the amount of radioactivity specifically bound in the 3 S region of a linear sucrose gradient. An aliquot of liver cytosol is incubated in the presence of 100 n$M$ $d$-$\alpha$-[5-methyl-³H]tocopherol (Amersham Searle, >1 Ci/mmol) added in 4–6 $\mu$l of ethanol per milliliter of cytosol, with gentle mixing. After addition of the labeled tocopherol, the sample is divided into two portions. To one portion is added a 400-fold molar excess (40 $\mu M$) of unlabeled $d$-$\alpha$-tocopherol in 2–4 $\mu$l of ethanol per milliliter of cytosol. An equal concentration of alcohol alone is added to the other portion. Both samples are incubated for 4 hr at 26°.

Linear sucrose gradients (5% to 20%) in 10 m$M$ Tris · HCl buffer (pH 7.5) containing 1 m$M$ EDTA and 10 m$M$ KCl are prepared using a Beckman Density Gradient Former. Gradients are allowed to equilibrate at 4° for 2 hr prior to use.

A 0.2-ml aliquot of cytosol is layered on top of the gradient and centrifuged in a Spinco SW 50.1 rotor at 189,000 $g$ (45,000 rpm) for 18 hr at 4°. The gradient tubes are pierced from the bottom; 0.25-ml fractions are collected directly into scintillation vials, and the radioacitivity is measured.

A typical gradient profile of liver cytosol is shown in Fig. 1. The amount of radioactivity specifically bound is calculated by subtracting the total radioactivity bound in the presence of excess unlabeled tocopherol, fractions 4–11 inclusive, from the total radioactivity bound in the absence of excess unlabeled tocopherol, fractions 4–11 inclusive.

*Remarks*

1. The radiochemical purity of the [³H]tocopherol must be greater than 85% if an accurate estimate of nonspecific binding is to be obtained. As shown in Fig. 1, when the radiochemical purity is 95%, less than 1% of the radioactivity is bound to TBP in the presence of a 400-fold excess of unlabeled tocopherol. With more impure preparations, this value increases to as much as 15% of the total bound counts, resulting in an underestimate of the specifically bound counts.

2. [³H]Tocopherol is purified by thin-layer chromatography on silica gel G containing 0.004% sodium fluorescein with benzene/ethanol (99 : 1, v/v) as the developing solvent. The tocopherol band ($R_f$ 0.6) is scraped from the plate and eluted from the silica gel with 1–2 ml of ethanol. Nine volumes of benzene are added. The purified tocopherol is divided into aliquots, flash frozen, and stored at −40°. For use, an aliquot is evaporated to dryness at room temperature by a stream of $N_2$ and redissolved in an appropriate volume of ethanol for addition to cytosol samples.

3. For assessment of specificity by competition experiments using tocopherol analogs or other compounds, each compound is added to the incubation at the desired molar excess in ethanol in the same manner as the addition of unlabeled α-tocopherol above.

## Properties of α-Tocopherol-Binding Protein (TBP)

*Tissue Localization.* TBP is detectable in the livers of both male and female rats. It is not detectable in rat heart, lung, brain, kidney, intestinal mucosa, muscle, bone marrow, testis or uterus, serum or erythrocyte hemolyzates. TBP is present in the livers of several other laboratory animals, including mouse, hamster, guinea pig, rabbit, chicken, and monkey. It is also detectable in fetal liver, but not lung or kidney. The cellular localization is apparently restricted to the liver.

*Molecular Weight.* Gel filtration on calibrated Sephadex G-100 resulted in a molecular weight estimate of 30,500.

Sucrose density gradient centrifugation yields an *s* value of 3.0 S when compared to the markers chymotrypsinogen (2.7 S) and ovalbumin (3.7 S) as shown in Fig. 1.

*Assay Conditions.* Optimal temperature and time of incubation of cytosol with [³H]α-tocopherol were established at 26° for 4 hr. Temperature (0–36°) had no effect on the extent of binding, only on the time needed to reach saturation.

Binding was maximal between pH 7.4 and pH 9.0. Below pH 7.4 binding decreased rapidly.

*Identification of Bound Tritium as [³H]α-Tocopherol.* Extraction and

TABLE I
EFFECTS OF TOCOPHEROL ANALOGS ON THE BINDING OF [³H]α-TOCOPHEROL TO ITS
HEPATIC BINDING PROTEIN[a]

| Addition | Binding (%) |
|---|---|
| None | 100 |
| α-Tocopherol | 2 |
| γ-Tocopherol | 39 |
| α-Tocopheryl acetate | 100 |
| α-Tocopherol quinone | 98 |
| α-Tocotrienol | 31 |
| Trolox | 100 |

[a] Aliquots of liver cytosol were incubated with 100 n$M$ [³H]α-tocopherol in the presence or the absence of a 400-fold molar excess of unlabeled tocopherol analog (40 $\mu M$) for 4 hr at 26°. Binding was determined by sucrose gradient assay.

analysis by thin-layer chromatography of the peak of bound radioactivity from gel filtration demonstrated that the binding protein carries α-tocopherol as its bound ligand.

*Specificity.* Specificity of TBP toward α-tocopherol was assessed in competition experiments by determining the relative affinity of the binding protein for α-tocopherol and several analogs.

Cytosols were incubated with [³H]α-tocopherol in the presence and in the absence of a 400-fold molar excess of unlabeled competitor. The data are presented in Table I. α-Tocopherol reduces the binding of [³H]α-tocopherol to 2%. γ-Tocopherol and α-tocotrienol reduce binding by 40 and 30%, respectively, whereas α-tocopheryl acetate, α-tocopherol quinone, and Trolox (6-hydroxy-2,5,7,8-tetramethylchroman-2-carboxylic acid) have no effect on binding. Similarly, retinol, retinoic acid, 1,25-dihydroxycholecalciferol, cholesterol, oleic acid, vitamin K, coenzyme Q$_9$, dexamethasone, and $N,N'$-diphenyl-$p$-phenylenediamine had no effect on binding.

Additional competition experiments directed toward the stereospecificity of the binding protein are shown in Table II. Two experiments were performed with both $d$-($RRR$)- and $dl$-(all-$rac$)-α-tocopherol. In each experiment the compound was obtained from a different source. Compounds were purified by TLC immediately prior to each experiment, and the concentrations of the $d$ and $dl$ solutions were equated by absorption at 292 nm and by colorimetric determination with FeCl$_3$-bathophenanthroline.[6]

As shown in each case, the $d$ isomer was approximately twice as

[6] J. G. Bieri, "Lipid Chromatographic Analysis" (G. U. Marinetti, ed.), Vol. 2, p. 459. Dekker, New York, 1969.

TABLE II

Effect of Varying Concentrations of Unlabeled $d$- or $dl$-α-Tocopherol on the Binding of $d$-[$^3$H]-α-Tocopherol to Its Hepatic Cytosol Binding Protein

| Expt. No. | Molar excess of unlabeled $d$- or $dl$-tocopherol | Percent binding in the presence of | |
|---|---|---|---|
| | | $d$ | $dl$ |
| 1 | 5 | 30 | 58 |
| | 10 | 18 | 37 |
| 2 | 7 | 20 | 53 |
| | 14 | 9 | 26 |

effective at reducing the amount of bound radioactivity as was the $dl$, indicating that the $l$ isomer has little or no affinity for the binding protein.

*Remarks.* The competition studies shown in Tables I and II reveal that the preferred ligand for TBP is the compound with the highest biological activity, namely $d$-α-tocopherol. Slight structural modification, as in the case of γ-tocopherol or α-tocotrienol, is sufficient to severely limit the ability to compete for binding. Further modifications, such as esterification of the chroman hydroxy group, removal of the side chain, or oxidation of the chroman ring, render these molecules totally incapable of competing with α-tocopherol for binding.

TABLE III

Subcellular Distribution of $d$-[$^3$H]-α-Tocopherol in Rat Liver *in Vitro*[a,b]

| Fraction | α-Tocopherol bound (pmol/g tissue) | Fraction of total (%) |
|---|---|---|
| Homogenate | 114.2 | (100) |
| Nuclei[c] | 35.4 | 31 |
| Mitochondria | 32.1 | 28 |
| Microsomes | 6.2 | 5.4 |
| Cytosol | 32.3 | 28 |
| Lipid[d] | 3.6 | 3.1 |

[a] Average of two rats. [$^3$H]α-Tocopherol was prepared as an emulsion in 20% ethanol–1% Tween 80. Each rat was injected intraperitoneally with 1 ml of the [$^3$H]tocopherol emulsion containing 8.4 μg of tocopherol (71.3 μCi) and was killed by cervical dislocation 4 hr later.

[b] Of the total administered dose, 8.3 and 9.1% was present in the liver after 4 hr.

[c] Crude nuclei and debris.

[d] Floating.

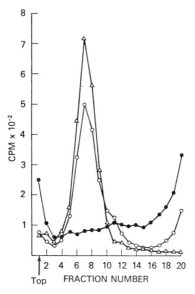

FIG. 2. Sucrose density gradient centrifugation of rat liver cytosol following an *in vivo* dose of [$^3$H]$\alpha$-tocopherol is described in Table III. Incubation for 4 hr at 0° ($\triangle$——$\triangle$), at 26° ($\bigcirc$——$\bigcirc$), or at 26° in the presence of a 400-fold excess of unlabeled $\alpha$-tocopherol ($\bullet$——$\bullet$). The quantity of excess $\alpha$-tocopherol was estimated on the basis of total radioactivity in the cytosol.

### Evidence that TBP Binds $\alpha$-Tocopherol in Vivo

Hepatic subcellular distribution of tritium following intraperitoneal administration of [$^3$H]$\alpha$-tocopherol to vitamin E-deficient rats is shown in Table III. On a per gram weight basis, about one-fourth of the radioactivity is present in the cytosol. Sucrose density gradient analysis of both supernatants revealed that over 90% of this radioactivity was specifically bound [$^3$H]$\alpha$-tocopherol.

The following experiment was carried out in an attempt to eliminate the possibility that TBP bound the [$^3$H]tocopherol only after cellular disruption. Aliquots of the *in vivo* labeled cytosol were incubated at 0° and 26° for 3 hr. Since the rate of association is temperature dependent, introduction of tocopherol into the cytoplasm by homogenization would result in more binding at 26° than at 0°. Approximately 40% more binding took place, however, at 0°, as shown in Fig. 2, suggesting that tocopherol was bound *in vivo* and that the rate of dissociation of bound tocopherol had been slowed by the lower temperature. It seems unlikely that increased proteolysis at 26° vs 0° could account for this observation, since no differential effect of temperature on extent of binding has been observed.

The function of TBP remains to be determined.

Section IV

# Vitamin K Group

## [16] Spectrophotometric Determination of the K Vitamins

*By* SAAD S. M. HASSAN

The reactivity of the quinonoid nucleus of the K vitamins toward many condensation, redox, and addition reactions permits the formation of chromogens suitable for the assay of the vitamin. The color produced depends on the nature of the reaction, type of reagent used, and the reaction conditions. As these vitamins differ solely in the nature of the substituent groups, specific methods for each have also been suggested.

A number of common reagents for the spectrophotometric and colorimetric determination of the K vitamin have been reported. Phenylhydrazine gives azo derivatives, which upon treatment with sulfuric or phosphoric acid developed a red to violet color.[1] Reduction of the vitamins, followed by treatment with excess 2,6-dichloroindophenol in butanol and measurement of the decrease in color intensity, has been employed.[2] However, the red-brown color developed by the action of alkalies and alkoxides has been employed as a simple spectrophotometric commensurable.[3]

On the other hand, specific methods for the measurement of each of the K vitamins have been developed. Vitamin $K_1$ may be determined by reaction with xanthine hydride[4] or diethyldithiocarbamate.[5] Vitamin $K_2$ can be determined by measuring the absorption difference at 244.5 nm and 271 nm, after reduction at pH 9 with potassium borohydride in propanol.[6] Vitamin $K_3$ reacts with 3-methyl-1-phenylpyrazolin-5-one[7] or 2,4-dinitrophenylhydrazine[8] with the development of characteristic colors with maximum absorption at 535 nm and 635 nm, respectively.

Spectrophotometric Determination of Vitamin $K_3$ by Reaction with Active
Methylene Compounds

Reaction of quinones having a labile $\alpha$-hydrogen atom with active methylene compounds was originally described by Kesting,[9] later by Cra-

[1] D. Greco and R. Argenziano, *Boll. Soc. Ital. Biol. Sper.* **19,** 171 (1944).
[2] J. Scudi and R. Buhs, *J. Biol. Chem.* **143,** 665 (1942).
[3] G. Carrara, L. Braidotti, and C. Guidarini, *Chim. Ind.* (*Milan*) **22,** 317 (1941).
[4] K. Schilling and H. Dam, *Acta Chem. Scand.* **12,** 347 (1958).
[5] F. Irreverre and M. Sullivan, *Science* **94,** 497 (1941).
[6] L. Křivánková and V. Dadák, *Anal. Biochem.* **75,** 305 (1976).
[7] J. Patel, R. Mehta, and M. Shastri, *Indian J. Pharm.* **37,** 141 (1975).
[8] V. Sathe, J. Dave, and C. Ramakrishnan, *Nature* (*London*) **177,** 276 (1956).
[9] W. Kesting, *Ber. Dtsch. Chem. Ges.* **62,** 1422 (1929).

METHODS IN ENZYMOLOGY, VOL. 67

ven,[10] and still later by Jeffreys.[11] Moreover, the suitability of this reaction for application to spectrophotometric determination of vitamin $K_3$ was a subject of detailed investigation by Hassan et al.[12]

It was found that vitamin $K_3$ instantaneously condensed with acetylacetone at room temperature in the presence of ammonia. An absorption maximum at 510 nm with $E_{1cm}^{1\%}$ of 17.5 is obtained. Increase in reaction temperature has no effect on the characteristics of the spectrum. Diethylmalonate under similar conditions gives a colored product with a maximum absorption at 570 nm and $E_{1cm}^{1\%}$ of 20, but hypsochromic and hyperchromic shifts are noticed when the reaction is carried out at 60°. The maximum absorption appears at 450 nm with $E_{1cm}^{1\%}$ of 32. Reaction of vitamin $K_3$ with ethylcyanoacetate is much more sensitive. At room temperature and above pH 8, the spectrum exhibits an absorption maximum at 575 nm, $E_{1cm}^{1\%}$ being 132. Higher temperatures and pH have no noticeable effect on the spectrum. A linear relationship between the absorbance and concentration in the range of 20–160 $\mu g/ml$ is obtained by the reaction of acetylacetone and diethylmalonate, and 5–100 $\mu g/ml$ with the cyanoacetate reaction.

The reaction of vitamin $K_3$ with malonitrile was recently investigated by the author.[13] A blue-colored product, with maximum absorption at 580 nm and $E_{1cm}^{1\%}$ 155, is obtained at room temperature in ammoniacal media at pH 7.5–8.5. Under these conditions, Beer's law is obeyed in the range of 3–100 $\mu g/ml$ of both menadione and menadione sodium bisulfite. Above pH 9, however, the blue color turns yellow and the sensitivity decreases. It should be noted that aqueous ammonia alone has no effect on the vitamin; it catalyzes the formation of the enolate of the condensing agents. It is possible that the reaction of malonitrile as well as other condensing agents involves a primary 1 : 4 addition of the enolate, followed by ring closure to give coumarenone derivatives. Similar reactions have been reported[14] with trimethylbenzoquinone, which is structurally and chemically close to vitamin $K_3$.

Assay Method

*Reagents*

Malonitrile, acetylacetone, diethylmalonate, and ethylcyanoacetate
Aqueous ammonia, 25% and 0.5%
Stock solution of menadione, 20 mg in 100 ml of methanol

[10] A. Craven, *J. Chem. Soc. London* p. 1605 (1931).
[11] J. Jeffreys, *J. Chem. Soc. London* p. 2153 (1959).
[12] S. S. M. Hassan, M. A. Fattah, and M. T. Zaki, *Z. Anal. Chem.* **275**, 115 (1975).
[13] S. S. M. Hassan, unpublished work, 1977.
[14] K. Finley, *in* "The Chemistry of the Quinonoid Compounds" (S. Pati, ed.), p. 1049. Wiley, New York, 1974.

*Procedure.* Transfer 0.5, 1.0, 1.5, 2.0, 2.5, 3.0, 3.5, 4.0, 4.5, and 5.0 ml aliquots of the vitamin stock solution to 10-ml measuring flasks. Add 0.5–1.0 ml of ethylcyanoacetate, acetylacetone, or diethylmalonate and about 2 ml of 25% aqueous ammonia solution to each flask. After thorough mixing of the reactants, and standing for 5 min at room temperature with the first two reagents, or at 60° with the last one, complete to the mark with methanol. Shake the solution and measure the absorbances at 570, 510, and 450 nm, respectively, using a 1.00-cm cuvette against a blank prepared under the same conditions.

With malonitrile, transfer the same aliquots of the vitamin used above, add 0.5 ml of 0.5% aqueous ammonia solution, or alternatively adjust the pH of the solution to 7.5–8.5. Add 1 ml of malonitrile, mix, and complete to 10 ml with methanol. Measure the absorbance at 580 nm using the 1.00-cm cuvette against a blank. Draw a calibration graph for each.

## Spectrophotometric Determination of Vitamins $K_1$ and $K_3$ by Reaction with Alkalies

The reaction of K vitamins with alcoholic sodium hydroxide solution has previously been described[3] with alkali solutions of concentrations not exceeding 0.2 $M$ (<0.5%). We found that the intensity of the color depends upon the nature and concentration of the alkali used.[12] With 2, 5, and 10% (w/v) sodium methoxide solutions, the $E_{1\,cm}^{1\%}$ at 450 nm for vitamin $K_3$ are 38, 49, and 72, respectively. With 5, 20, and 30% (w/v) methanolic potassium hydroxide solutions, the $E_{1\,cm}^{1\%}$ are 33, 52, and 62, respectively. The use of 10% sodium methoxide solution is therefore recommended, as the resulting color obeys Beer's law in the range of 10–100 $\mu$g of vitamin per milliliter.[12] Vitamin $K_1$ ($E_{1\,cm}^{1\%}$ 43 at 550 nm) was also satisfactorily determined by this reaction in pharmaceutical preparations.

## Assay Method

### Reagents

Methanolic potassium hydroxide, 10% w/v
Sodium methoxide, 10% w/v
Menadione stock solution, 20 mg in 100 ml methanol

*Procedure.* Transfer 1.0, 1.5, 2.0, 2.5, 3.0, 3.5, 4.0, 4.5, and 5.0 ml aliquots of the vitamin stock solution to 10-ml measuring flasks. Add 2 ml of 10% sodium methoxide or methanolic potassium hydroxide to each flask and leave for 15 min in a thermostatted water-bath at 60°. Cool to room temperature and complete to the mark with methanol. Shake the solutions and measure the absorbance at 450 nm in a 1.00-cm cuvette against a blank. Draw a calibration graph. Follow the same procedure with vitamin $K_1$, and measure the absorbance at 550 nm.

### Spectrophotometric Determination of Vitamin $K_3$ by Reaction with Piperidine

Piperidine instantaneously reacts with vitamin $K_3$ at room temperature to give a stable deep-red color.[13] The reaction is much more simple, sensitive, and selective than some of the above-mentioned reactions. Other K vitamins, however, do not respond to this reaction. Spectrophotometric studies of the reaction reveal the formation of two chromogenic species, namely, methyl piperidylhydronaphthoquinone, with maximum absorption at 510 nm, and its oxidized form, methyl piperidylnaphthoquinone, with maximum absorption at 570 nm. In the presence of excess piperidine, however, the color is solely attributed to the former chromogen whose $E_{1\,cm}^{1\%}$ is 150. The sensitivity of this reaction permits the analysis of as low as 5 $\mu$g/ml of the vitamin with an accuracy of $\pm 0.2\%$, and Beer's law is obeyed in the range of 5–200 $\mu$g/ml. Ethylenediamine and ethanolamine may be considered of comparable utility as piperidine and develop colors with maximum absorption at 440 and 480 nm, respectively.

### Assay Method

*Reagents*

Piperidine
A stock vitamin solution, 20 mg in 100 ml of methanol

*Procedure.* Transfer series of the vitamin stock solution (1–5 ml) to graduated test tubes (2 × 20 cm). Evaporate to dryness, then add 0.1 ml of piperidine and place in a boiling water-bath for 1 min; complete to 10 ml with a 1 : 1 water–alcohol mixture. Measure the absorbance at 510 nm using a 1.00-cm cuvette against a blank.

## [17] Spectroscopic Determination of Vitamin K after Reduction

### By VLADIMÍR DADÁK and LUDMILA KŘIVÁNKOVÁ

Most workers still determine the content of naphthoquinones on the basis of the absorption of the oxidized form. Arguments for the time-consuming chromatographic separation of naphthoquinones from extracts of natural material are problems connected with the quantitative transformation of naphthoquinones into the reduced state. In this assay, conditions are described that ensure a complete reduction of naphthoquinone and stability of the reduced form during the measurement.

Spectrophotometric Estimation of Menaquinone[1]

The spectra of oxidized and reduced naphthoquinones in the UV region differ considerably. In ethanol the oxidized form of menaquinone shows maximum absorbancy at 243, 248, 262, and 270 nm. Upon reduction, one absorption peak of increased intensity develops at 245 nm.[2] The isosbestic points are at 253.5 and 284.5 nm. The first data that indicate the possibility of employing the different absorptions of oxidized and reduced naphthoquinone for the determination of its concentration are given by Lester et al.[3] While working with demethylmenaquinone, they observed the unfavorable effect of the alkaline borohydride solution upon the course of reduction. By addition of 5 $\mu$l of 0.1 $M$ (5 mg/ml) aqueous solution of borohydride to an ethanolic solution of $MK_6$, a reduction of only about 25% can be achieved. The reason for the insufficient reduction of MK is obviously the alkaline reaction of the aqueous $KBH_4$ solution. The pH value of the 0.1 $M$ potassium borohydride is near 10 and, after standing for 1.5 hr, the pH moves above this value. To eliminate this problem it is recommended that one add 0.01 volume of 1 $M$ ammonium acetate buffer of pH 5.0 to the ethanolic solution of the naphthoquinone.[3] The slightly acidic medium produced by acetate buffer markedly increases the amount of reduced MK, but at the same time increases the rate of hydrolysis of borohydride. We found that, although in the aqueous solution 52% of $KBH_4$ decomposes after standing for 1.5 hr, in the presence of acetate buffer at pH 5.0, 81% decomposition takes place almost immediately. Other authors[4] working with MK in presence of acetate buffer have drawn attention to the necessity of repeated additions of borohydride to achieve complete reduction. We have found that, in some cases, addition of $KBH_4$ had to be repeated more than 10 times.[5] Even when further addition of borohydride does not increase the amount of the reduced

[1] The menaquinones are referred to as $MK_n$, where $n$ equals the number of isoprenoid units in the side chain. IUPAC–IUB Commission on Biochemical Nomenclature, *Biochim. Biophys. Acta* **107**, 1 (1965).

[2] P. J. Dunphy and A. F. Brodie, this series, Vol. 18C [233].

[3] R. L. Lester, D. C. White, and S. L. Smith, *Biochemistry* **3**, 949 (1964).

[4] M. Guerin, R. Azerad, and E. Lederer, *Bull. Soc. Chim. Biol.* **47**, 2105 (1965).

[5] L. Křivánková and V. Dadák, *Scripta Fac. Sci. Nat. UJEP Brunensis, Chemia 2*, **6**, 67 (1976).

form, this does not always signify a total transformation of MK into the reduced state. It was found that at MK concentrations of 1–30 $\mu M$, the process of reduction did not yield fully reliable results. Both at low concentrations (1–2 $\mu M$) and at concentrations greater than 10 $\mu M$, the spectrum may become stabilized even though the vitamin $K_2$ is not yet reduced completely. A large excess of borohydride supplied in fewer additions leads to irreproducible results.

### The pH Dependence of the MK Reduction

The proton concentration in the solution of MK is affected by the rise of alkaline products from borohydride hydrolysis and to some extent by the consumption of protons during quinol formation. One volume percentage of the acetate buffer was able to keep the reaction acidic even after five additions of 5 $\mu l$ of a 0.1 $M$ solution of borohydride. The addition of larger amounts of borohydride resulted in greater shifts of the pH toward the neutral and alkaline regions. Buffers of another pH were used to produce the conditions required for a complete and reliable reduction. In

DEPENDENCE OF THE REDUCTION OF $MK_6$ ON pH

|  | pH | | |
|---|---|---|---|
| Additions | Aqueous addition | Ethanolic reaction mixture[a] | Percent $MK_{red}$[b] |
| Acetic acid, 0.2 $M$ | 2.7 | 3.5 | 22 |
| Acetic acid and | 3.8 | 4.5 | 40 |
| sodium acetate, 0.2 $M$ | 4.4 | 5.0 | 65 |
|  | 5.0 | 5.7 | 91[c] |
| Malonic acid and | | | |
| lithium malonate, 0.4 $M$ | 6.2 | 6.8 | 93[d] |
| Tris and HCl, 0.2 $M$ | 6.5 | 7.0 | 10 |
|  | 7.5 | 7.3 | 45 |
|  | 8.0 | 7.6 | 63 |
|  | 8.8 | 8.5 | 89 |
|  | 9.0 | 9.0 | 100 |
| Borax and NaOH, 50 m$M$ | 9.7 | 11.0 | 0 |

[a] Aqueous ethanol, 25%; the values serve only for relative comparison of the proton concentration.

[b] An average of at least three measurements.

[c] Values varied between 80 and 100% reduction of MK. An average of six measurements (10–17 $\mu M$ solution of MK).

[d] The reduced form autoxidizes rapidly (5% of total amount in 1 min).

acetate buffers of pH 3.8 to 5.0, this requirement was not fulfilled (see the table). Acetate buffers of 0.2–1.0 $M$ at a pH higher than 5, as recommended in a previous work,[6] could not maintain a constant pH after repeated addition of borohydride doses. When a 0.2 $M$ malonate buffer of pH 6.2 was used, the MK reduction was very rapid. After the second addition of borohydride, more than 90% of reduced MK was already formed, but, at the same time, rapid reoxidation took place. The results show that buffers of a slightly acidic pH are less suitable for quantitative MK reduction. Good results can be obtained by the reduction of MK in the slightly alkaline region. In the presence of Tris · HCl buffer of pH 9.0, reduction could be completed after few additions of borohydride. When borax–NaOH buffer of a higher pH (9.7) was used, no reduction of MK by borohydride occurred.

Method

*Reagents*

Ethanol or 50% aqueous propanol, spectroscopically pure
Solid sodium or potassium borohydride or a freshly prepared solution
    of 15 mg of borohydride in 1 ml of 1 $M$
Tris · HCl buffer, pH 9.0
Lipid extracts of Gram-positive bacteria, see this volume [14].
Menaquinone samples (Hoffmann–La Roche, Basel): MK was dissolved in ethanol in 1–30 $\mu M$ concentration and kept in the dark before use. The homogenity of solid samples stored for several years in the cold and darkness was proved by thin-layer chromatography on silica gel G plates (Merck) with benzene as solvent.

*Procedure.* One-centimeter optical path cuvettes of a recording spectrophotometer are filled with 3 ml of the sample solution in ethanol, and 30 $\mu$l of 1 $M$ Tris · HCl buffer, pH 9.0, is added. Then the base line between 230 and 300 nm is recorded. Adjust the pen setting so that the drawn base line lies approximately in one-third of the chart paper. Afterward, the reduction of MK in the sample cuvette is performed by the addition of 5 $\mu$l of borohydride solution in Tris · HCl buffer or of about 0.1 mg of the solid crystals. After gentle stirring, the cuvette is allowed to stand for 5 min and the spectrum is recorded. The addition of borohydride solution or solid borohydride and 10 $\mu$l of the buffer is repeated, and spectra in the 230–300 nm region are recorded until it can be seen that the last recorded line at

[6] A. Kröger and V. Dadák, *Eur. J. Biochem.* **11**, 328 (1969).

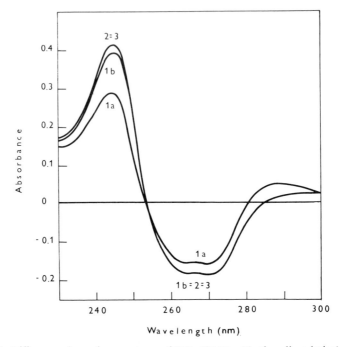

FIG. 1. Difference absorption spectrum of $MK_8$ (13.15 $\mu M$ ethanolic solution): (1a) recorded within 2 min after the addition of 5 $\mu$l of 15 mg of borohydride in 1 ml of 1 $M$ Tris · HCl buffer, pH 9.0; (1b) recorded 5 min after the first addition of borohydride solution; (2 = 3) the spectra recorded 5 min after further additions of borohydride.

245 nm falls close to the maximum seen before. As a rule, three additions are sufficient. The time span between the addition of the reductant and the recording of the spectrum is necessitated by the fact that the reduction at pH 9.0 does not proceed in only one step.[5] The shape of the differential spectrum (Fig. 1) when recorded immediately after addition of borohydride is unlike that where the disappearance of the oxidized form is associated with the corresponding change to the reduced form. The maximum of the reduced form appears within 5 min, but it may decrease slightly after longer standing (Fig. 1). The reduced MK is quite stable during the measurement. The decrease of the absorption maximum at 270 nm of the oxidized form proceeds even more rapidly if the sample is dissolved in 50% aqueous propanol.[7] Here again, the intermediate products are formed, but do not interfere with the difference spectrum in the region between 270 and 295 nm. The nature of intermediates was partially elucidated and is discussed elsewhere.[5]

[7] L. Křivánková and V. Dadák, *Anal. Biochem.* **75**, 305 (1976).

The conclusion reached regarding the easy reducibility of MK homologs was verified for other quinones of the vitamin K type. The reduction of menadione (vitamin $K_3$) in an ethanolic medium was also quantitative at pH 8.0 and could be completed also at pH values up to pH 10.0 (glycine–NaOH buffer). The procedure described here makes a complete and rapid reduction of naphthoquinones possible and was successfully tested by MK determination in crude lipid extracts of *Staphylococcus epidermidis* and by demethyl menaquinone (DMK) from *Streptococcus faecalis*. The error of estimation was always less than 3%.

### Extinction Coefficients

An exhaustive survey of the extinction coefficients of various naphthoquinones including $K_1$, MK, and DMK has been given by Dunphy and Brodie in a preceding volume of this series.[2] The difference in the molar extinction coefficients for the reduced and oxidized forms at 245 nm is mostly used for calculating the amount of naphthoquinone in the sample. The average values calculated for the molar difference extinction coefficients $\Delta\epsilon$ at 245 nm in the acidified ethanol are[2] as follows: for $K_1$ and MKs ($\epsilon_{red} - \epsilon_{ox})_{245\,nm}$ = 25,840 and for $DMK_1$ and DMKs ($\epsilon_{red} - \epsilon_{ox})_{245\,nm}$ = 19,813. The values are practically independent of the length of the isoprenic chain.[2] In our measurements, we found the millimolar extinction coefficients of the oxidized $MK_8$ in ethanol at 270 nm $\epsilon_{270\,nm}$ = 17.3 m$M^{-1}$ cm$^{-1}$ (in accordance with that reported previously[6] for $MK_7$) and $\epsilon_{248.5\,nm}$ = 20.0 m$M^{-1}$ cm$^{-1}$. The value of the millimolar extinction coefficient of reduced $MK_8$ formed in the slightly alkaline medium after the volume correction was as high as 54.3 m$M^{-1}$ cm$^{-1}$ at 245 nm; thus the ratio of the absorption at 245 nm for the reduced versus oxidized spectrum is 2.74, which is rather higher than that reported by Dunphy and Brodie for the slightly acidic medium.[2] The quantitatively reduced MK shows a minimum of absorbance in the region of 265–280 nm; any shoulder here demonstrates an incomplete reduction. The amount of MK calculated from the difference between the absorption of reduced and oxidized forms at 245 nm using $\Delta\epsilon_{245\,nm}$ = 32.2 m$M^{-1}$ cm$^{-1}$ was identical with the value received when the disappearance of the oxidized form was followed at 270 nm, $\Delta\epsilon_{270\,nm}$ = 15.0 m$M^{-1}$ cm$^{-1}$. When the recording spectrophotometer is available, it is convenient to use the difference in absorption of reduced minus oxidized forms between 245 and 270 nm. The calculated millimolar difference extinction coefficient ($\epsilon_{red} - \epsilon_{ox})_{245\,nm}$ minus ($\epsilon_{red} - \epsilon_{ox})_{270\,nm}$ was 47.2 m$M^{-1}$ cm$^{-1}$. When working in 50% propanol, the same coefficients can be used, but the maximum of the difference spectrum is shifted to 244.5 nm and the minimum to 271 nm. For a spectrophotometer equipped with a mercury lamp, the difference extinction coefficient of $MK_7$ at 265–289 nm $(\epsilon_{red} - \epsilon_{ox})_{265\,nm} - (\epsilon_{red} - \epsilon_{ox})_{289\,nm} = 14.7$ has been recommended.[6]

When a mixture of MK and ubiquinone (Q) coexists in the sample, the reduction of one quinone cannot be recorded without the interference of the other. Even here the individual amounts of Q and MK reacting in a transition into the reduced state can be calculated from the absorption differences.[8]

[8] A. Kröger, V. Dadák, M. Klingenberg, and F. Diemer, *Eur. J. Biochem.* **21**, 322 (1971).

# [18] Polarographic Determination of Vitamin $K_5$ in Aqueous Solution

*By* K. TAKAMURA and F. WATANABE

4-Amino-2-methyl-1-naphthol hydrochloride is known as a water-soluble vitamin K and is generally called vitamin $K_5$ ($VK_5$). Since aqueous media under ordinary conditions are commonly used in biological studies, a simple method for the assay of $VK_5$ under such conditions would be desirable, but none of the methods so far reported seems to be suitable for this purpose. For example, the phosphorimetric method requires a sample solution at $-196°$.[1] In determinations by gas chromatography, the conversion of $VK_5$ into its trimethylsilyl ether is required prior to measurement.[2] In spectrophotometric determinations, the use of alcoholic media is recommended because $VK_5$ is quite unstable in aqueous solutions in the presence of oxygen.[3]

Since $VK_5$ tends to be oxidized and then hydrolyzed in water,[3] the complete removal of oxygen is necessary for determinations of $VK_5$ in aqueous solutions. A polarographic technique provides a convenient procedure to fulfill this requirement. The application of direct current (dc) and alternating current (ac) polarography to the microdetermination of $VK_5$ in aqueous solution present in concentrations as low as the 0.1-ppm range will be described.

## Method and Procedure

For the measurements of dc and ac polarograms, a Yanagimoto pen-recording polarograph (Model P8) is used by us, but many other suitable

[1] J. J. Aaron and J. D. Winefordner, *Anal. Chem.* **44**, 2122 (1972).
[2] J. Cornelius and H. Y. Yang, *J. Gas Chromatogr.* **5**, 327 (1967).
[3] E. Knobloch, *Collect. Czech. Chem. Commun.* **14**, 508 (1949).

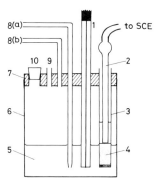

FIG. 1. Electrolysis cell setup for dc and ac polarography: 1, dropping mercury electrode; 2, agar plug containing KCl; 3, salt bridge; 4, agar plug containing $NaNO_3$; 5, sample solution; 6, beaker-style cell; 7, rubber plug; 8, nitrogen gas inlet, (a) for bubbling, (b) for flowing over solution; 9, gas outlet; 10, sample inlet. SCE, saturated calomel electrode.

instruments are commercially available. The dropping mercury electrode used has a drop time of 4.25 sec and a mercury flow rate of 1.83 mg/sec measured in an air-free base electrolyte at open circuit. A typical electrolysis cell is shown in Fig. 1. A saturated calomel electrode (SCE) is used as a reference electrode. The cell is thermostatted at $25 \pm 0.1°$ during the measurement.

4-Amino-2-methyl-1-naphthol hydrochloride ($VK_5$), which is available from Wako Pure Chemical Industries, Ltd., is purified by recrystallization from alcohol–ether in an atmosphere of nitrogen. The sample solution (aqueous solution of $VK_5$) has to be kept under anaerobic condition to prevent oxidation by dissolved oxygen (see Note: Effect of Oxygen). Pure nitrogen gas is obtained by bubbling commercial nitrogen gas through alkaline pyrogaroll solution (10%), diluted $H_2SO_4$, and distilled water.

Britton–Robinson buffer solutions of pH 3–7.5 are used as base electrolyte solutions.

The following procedure is usually used by us.

1. Pipette 20 ml of the base electrolyte solution in the electrolysis cell.

2. Remove dissolved oxygen in the base electrolyte solution by bubbling pure nitrogen gas for 10 min.

3. Add the sample[4] to the base electrolyte solution, while continuing to flush with nitrogen.

4. Stop the gas bubbling, and then flow nitrogen gas continuously over the sample solution. (Recontamination of the solution by air should be prevented during the measurement.)

[4] In the case of fluid sample, a sample volume of less than 5 ml is adapted to this procedure.

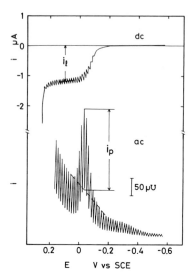

FIG. 2. Direct current (dc) and alternating current (ac) polarograms of $VK_5$ obtained at pH 5.5. $[VK_5] = 1.7 \times 10^{-4}\ M$ (dc); $7.7 \times 10^{-6}\ M$ (ac).

5. Record the dc or ac polarogram. Typical instrument settings are as follows: operating mode, dc or ac; scan rate, 0.2 V/min; scan direction, from positive to negative potentials; current range, 0.5 $\mu$A/cm (dc) or 40 $\mu\mho$/cm (ac). Initial potentials are set at the desired potentials according to the pH of the sample solution (see Note: Effect of pH; e.g., example, 0.3 V is suitable at pH 5.5).

## Results

A dc polarogram of $VK_5$ obtained in an oxygen-free base electrolyte solution at pH 5.5 is shown in Fig. 2 (upper), in which a well defined anodic wave is observed (half-wave potential $E_{1/2} = -0.08$ V). The wave is attributed to the electrode oxidation of 4-amino-2-methyl-1-naphthol (I) to form the corresponding quinoimine (II) through Eq. (1).[3]

$$ \text{(I)} \quad\longrightarrow\quad \text{(II)} \quad +\ 2e\ +\ 2H^+ \tag{1} $$

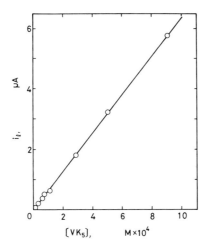

FIG. 3. Relation between $i_1$ and $VK_5$ concentration obtained at pH 5.5.

Under the present experimental condition, the limiting current ($i_1$) is dependent upon the diffusion of $VK_5$ alone, so that proportionality between $i_1$ and the concentration of $VK_5$ is to be expected.[5,6] As seen in Fig. 3, a linear relation between $i_1$ at 0.10 V and the concentration of $VK_5$ is obtained in the concentration range from $1.0 \times 10^{-5}$ to $1.0 \times 10^{-3}$ M.

The use of the ac polarographic technique[7] makes it possible to determine $VK_5$ in a much lower concentration range. The electrode oxidation of (I) gives rise to a distinct peak on an ac polarogram. A typical example obtained with $7.7 \times 10^{-6}$ M $VK_5$ at pH 5.5 is shown in Fig. 2 (lower). The peak is so high that a very low concentration of $VK_5$ can be detected by this method.

The peak height ($i_p$) is quite reproducible and the plots of $i_p$ vs the concentration of $VK_5$ are given in Fig. 4. As seen in Fig. 4, while the plot of $i_p$ vs [$VK_5$] gives a nonlinear relationship (curve a), $i_p$ is practically proportional to the $VK_5$ concentration in the limited concentration range from $5.0 \times 10^{-7}$ to $1.0 \times 10^{-5}$ M (curve b). This fact clearly indicates that the method can be applied to the microdetermination of $VK_5$ present in aqueous solution in concentrations as low as the 0.1-ppm range.

Trace amounts of $VK_5$ thus can be determined with the values of relative standard deviation of 3% (100 ppm) by dc polarography and 4% (1 ppm) by ac polarography.

[5] L. Meites, "Polarographic Techniques." Wiley (Interscience), New York, 1955.
[6] E. Knobloch, this series, Vol. 18, p. 305.
[7] D. E. Smith, *in* "Electroanalytical Chemistry" (A. J. Bard, ed.), p. 1. Dekker, New York, 1966.

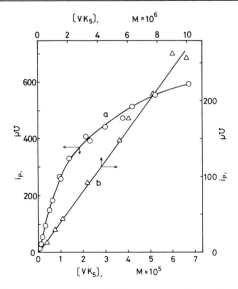

FIG. 4. Relation between $i_p$ and VK$_5$ concentration obtained at pH 5.5.

Notes

*Effect of Oxygen.* The oxidation of (I) to (II) takes place in aqueous solution in the presence of oxygen, followed by the hydrolysis of (II), according to Eqs. (2) and (3). When the sample solution of VK$_5$ is prepared without preremoval of dissolved oxygen, the reactions given by Eqs. (2) and (3) proceed with the consumption of VK$_5$. These reactions result in a lowering of the anodic wave of VK$_5$ and, at the same time, in the appearance of two cathodic waves corresponding to the electroreduction of (III) and H$_2$O$_2$ in the negative potential region. To avoid such an effect, the complete removal of oxygen from the base electrolyte solution is quite necessary prior to adding VK$_5$.

$$ \text{(I)} + O_2 \rightleftharpoons \text{(II)} + H_2O_2 \qquad (2) $$

(I)         (II)

$$+ \, H_2O \;\rightleftharpoons\; \qquad\qquad + \; NH_3 \qquad (3)$$

(II)                                           (III)

*Effect of pH.* The anodic wave of $VK_5$ is pH-dependent and the half-wave potential is shifted to the positive potential side with decreasing pH, as seen in Fig. 5. The magnitude of the changes in the half-wave potential is about 60 mV per unit pH.

At a pH below 4.6, a new anodic wave appears at a more positive potential region ($E_{1/2} = 0.26$ V, see Fig. 5). This wave is due to the anodic dissolution of mercury with chloride ion to form mercurous chloride,[8] in which chloride ion is supplied through the dissociation of $VK_5$. The wave is independent of pH and, accordingly, it disappears at higher pH levels because of the overlap with the final increase in anodic current.

Then the former wave must be employed for the determination of $VK_5$ by the present method. As known in Fig. 5, it is necessary to keep the pH of the solution higher than 3; otherwise the former wave becomes ill defined owing to overlap with the latter one. In addition, the use of alkaline media must be avoided because of instability of $VK_5$.

*Effects of Some Biological Substances.* The polarographic behavior of $VK_5$ in phosphate buffer solutions is similar to those in Britton–Robinson buffer solutions. The presence of glucose, uracil, glycine, and nicotinamide scarcely affects the polarogram of $VK_5$, but the addition of those substances in large excess results in a lowering of the ac polarographic peak in some degree, and hence the lower limit of detection varies from $10^{-7}$ to $10^{-6}$ M.

The presence of DNA and its related bases lowers the limiting current of $VK_5$, as well as the ac peak height of $VK_5$, but the determination of $VK_5$ by the standard addition method is still possible.

Cysteine and some other thiols exhibit anodic waves due to the formation of the complex with mercury on the electrode surface in the positive potential range,[9,10] and the presence of these thiols interferes significantly in the polarographic determination of $VK_5$.

[8] I. M. Kolthoff and J. J. Lingane, "Polarography," Vol. 2. Wiley (Interscience), New York, 1952.
[9] I. M. Kolthoff and C. Barnum, *J. Am. Chem. Soc.* **62**, 3061 (1940).
[10] Y. Hayakawa and K. Takamura, *J. Pharm. Soc. Jpn.* **95**, 1173 (1975).

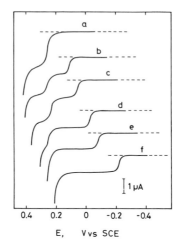

Fig. 5. Effect of pH on dc polarograms of $VK_5$. a, pH 0; b, pH 2.2; c, pH 3.3; d, pH 4.6; e, pH 5.5; f, pH 7.5. $[VK_5] = 3.5 \times 10^{-4} M$.

The effect of blood cells on $VK_5$ has been pointed out.[3,11] $VK_5$ is found to be oxidized by blood cells, probably owing to the presence of oxygen arising from oxyhemoglobin. It should be noted that $VK_5$ in any sample solution is more or less consumed through oxidation and subsequent hydrolysis unless dissolved oxygen is excluded.

[11] M. Brezina and P. Zuman, "Polarography in Medicine, Biochemistry, and Pharmacy." Wiley (Interscience), New York, 1958.

## [19] Photochemical–Fluorometric Determination of the K Vitamins

### By J. J. AARON

During the last two decades, the assay for naturally occurring and synthetic K vitamins has been carried out by numerous physical and physicochemical methods, such as colorimetry, UV spectrophotometry, polarography, coulometric titrations, chromatography, and phosphorimetry. For a survey of the existing classical methodology, the reader is referred to specialized reviews.[1–4]

[1] P. Sommer and M. Kofler, Vitam. Horm. (New York) 24, 349 (1966).
[2] R. A. Morton, Biol. Rev. Cambridge Phil. Soc. 46, 47 (1971).
[3] P. J. Dunphy and A. F. Brodie, this series, Vol. 18C, p. 407.
[4] C. A. Janicki, R. K. Gilpin, E. S. Moyer, R. H. Almond, Jr., and R. H. Erlich, Anal. Chem. 49, 110R (1977).

In this chapter we give a brief critical description of a few analytical methods used for the determination of the K vitamins. Then we report in greater detail on the use of a recently developed photochemical-fluorometric method in the assay of K vitamins.

## Classical Methods of Assay

### Colorimetric Methods

#### The Schilling–Dam Procedure

The classical colorimetric method for vitamin $K_1$ (2-methyl-3-phytyl-1,4-naphthoquinone) assay is based on the reaction between vitamin $K_1$, 5-imino-3-thione-1,2,4-dithiazolidine and potassium hydroxide in ethanol.[5] The product is a violet-blue complex, the intensity of which is measured at 410 nm. In the case of vitamin $K_3$ (menadione), a reddish complex is formed with maximum absorption at a wavelength of 440 nm. Although this method is relatively simple and does not require expensive equipment, it has several major disadvantages. The colored complex is stable for only about 24 hr. Moreover, the sensitivity of the assay is relatively low, between 25 μg/ml for vitamin $K_1$ and about 4–50 μg/ml for vitamin $K_3$ (according to the procedures), and the range of validity of the calibration curves extends only from 25 to 80 μg/ml for vitamin $K_1$ and from 10 to 200 μg/ml for vitamin $K_3$.[6]

### Spectrophotometric Methods

Spectrophotometric procedures were developed by several authors for the quantitative determination of vitamin $K_1$,[3] other naturally occurring naphthoquinones,[3] and vitamin $K_3$.[2,7,8] Since the naphthoquinones exhibit characteristic and intense absorption bands in the UV region, with maximum wavelength at about 245, 270, and 330 nm, their quantitative assay can be accomplished by extinction measurements. One of the spectrophotometric procedures is based on the observation that reduction of the naphthoquinone in an ethanolic buffer solution (aqueous pH 5.0) with sodium borohydride produces a change in extinction at the maximum wavelength of the naphthoquinone.[3] Since the change in extinction is proportional to the amount of naphthoquinone present, the vitamin K is

[5] K. Schilling and H. Dam, *Acta Chem. Scand.* **12**, 347 (1958).
[6] J. J. Aaron and J. D. Winefordner, *Anal. Chem.* **44**, 2122 (1972).
[7] C. Levorato, *Boll. Chim. Farm.* **107**, 184 (1968).
[8] V. J. Vagjand and F. S. Urodic, *Glas. Hem. Drus. Beograd* **30**, 185 (1965).

assayed by measuring either the increase in absorption at 245 nm or the decrease in absorption at 270 nm.

In the case of vitamin $K_3$, the range of validity of the spectrophotometric procedure is reported to be 0.4–15 $\mu$g/ml and the limits of detection between about 0.4 and 10 $\mu$g/ml.[7,8] The sensitivity of spectrophotometric procedures is therefore slightly better than that of colorimetric methods.

### Gas Chromatographic Methods

Vitamin $K_1$ and vitamin $K_3$ have been quantitatively determined by gas–liquid chromatography (GLC).[9,10] When one uses a hydrogen flame ionization detector, calibration plots cover a range of 0.5 to 2 $\mu$g. However, there are several inconveniences to this technique.

1. Especially designed and packed Pyrex columns have to be used with standard gas chromatographs.
2. Reproducibility is completely lost when a GLC column has been allowed to cool and is used again. Each time the instrument has to be cooled down for some reason, a new column has to be packed and preconditioned.
3. Because vitamin $K_1$ and $K_3$ are subject to heat decomposition, it is necessary to maintain the GLC column and flash heater temperatures as low as possible.

### Phosphorimetric Method

More recently, phosphorimetry was used to determine quantitatively vitamins $K_1$ and $K_3$.[6] This method is based on the direct measurement of phosphorescence intensity of vitamin $K_1$ (emission maximum at about 570 nm) and of vitamin $K_3$ (emission maximum at about 545 nm) in several rigid solvents at 77°K. Ranges of validity for a linear relationship are found to be 1–1000 $\mu$g/ml for vitamin $K_1$ and 0.07–100 $\mu$g/ml for vitamin $K_3$. Limits of detection (1 $\mu$g/ml for vitamin $K_1$ and 0.07 $\mu$g/ml for vitamin $K_3$) obtained by phosphorimetry are lower by at least one order of magnitude than are values determined by previous analytical techniques.[6]

### Photochemical–Fluorometric Method for the Determination of Vitamin $K_1$

A photochemical–fluorometric method has been recently described for the quantitative determination of vitamin $K_1$[11] and is presented in the hope that it may be helpful to other researchers.

[9] A. J. Sheppard, this series, Vol. 18C, p. 461.
[10] A. J. Sheppard and W. D. Hubbard, this series, Vol. 18C, p. 465.
[11] J. J. Aaron, J. E. Villafranca, V. R. White, and J. M. Fitzgerald, *Appl. Spectr.* **30,** 159 (1976).

*Principle*

Although vitamin $K_1$ is not naturally fluorescent, it was reported that fluorescent photolysis products are obtained upon its irradiation in ethanol or petroleum ether with the 365-nm mercury line.[12] With prolonged UV irradiation during 10–30 min, the photoinduced fluorescence intensity was shown to increase gradually to a maximum, and then decrease. Structure identification of the colored intermediates and photoproducts of vitamin $K_1$ in several polar and nonpolar solvents has also been reported, using both steady-state irradiation and flash-photolysis techniques.[13]

Digital integration over a fixed time interval or analog recording of the increasing fluorescence signal due to the photochemical product of vitamin $K_1$ can be used to determine vitamin $K_1$ initial concentration quantitatively with a low detection limit and with very good reproducibility.

*Instrumentation*

The basic system for photochemical–fluorometric measurements was designed and constructed by Lukasiewicz and Fitzgerald.[14] Figure 1 shows a simplified block diagram of the apparatus used for the photochemical–fluorometric determination of vitamin $K_1$.[11,14]

The excitation source used to photolyze vitamin $K_1$ and excite fluorescence of the photochemical product is a Hanovia Model 901B11, 200 W, combined Xe-Hg lamp, filtered by a Corning 7-54 UV transmitting-visible absorbing filter. An excitation shutter can be opened with a delay time of about 1 sec.

An American Instruments Company grating monochromator (D65-61041) with accessories (covered optical bench, tandem slit assembly, PMT housing with turret slits and shutter) is utilized. Monochromator slits are set at 2 mm. The sample cell used for photolysis–fluorescence of vitamin $K_1$ is a 25-ml, tall-form, quartz beaker (U.S. Fused Quartz, Pompton Plains, New Jersey) which is positioned rigidly inside the covered optical bench with a flat, black, aluminum box. Stirring with a commercial magnetic stirrer mounted directly under the cell holder is used for all concentrations, with the advantage of increasing maximum fluorescence emission intensity and improving reproducibility. A Corning 3-75 sharpcut (400 nm) yellow-green filter is inserted in the emission beam path between the quartz sample cell and the entrance slit of the monochromator.

[12] O. Jansson, *Acta Chim. Scand.* **24,** 2839 (1970).
[13] G. Leary and G. Porter, *J. Chem. Soc.* **A,** p. 2273 (1970).
[14] R. J. Lukasiewicz and J. M. Fitzgerald, *Anal. Chem.* **45,** 511 (1973).

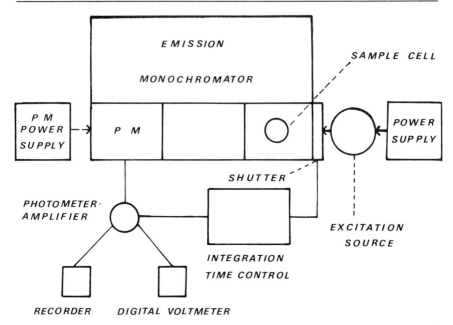

FIG. 1. Block diagram of the system for the photochemical-fluorometric measurements of vitamin $K_1$. Reprinted, with permission, from R. J. Lukasiewicz and J. M. Fitzgerald, *Anal. Chem.* **45**, 511 (1973). Copyright by the American Chemical Society.

The signal detector and amplifier include a 1P28 photomultiplier operated at 800 V and an American Instruments Model 10-261 solid state blank subtract microphotometer set at full sensitivity. The microphotometer amplifier is operated at the lowest gain.

A time-drive recorder and a fixed time interval summing digital-voltmeter (Heath Model EV-805 A; Universal Digital Instrument) are used as readout devices. For digital integration measurements, a fixed integration time of 1 or 2 min is utilized. It is obtained by means of the timing circuit of a Baird Atomic Model 540 scanning count integrator.

*Reagents and Standards*

Vitamin $K_1$ (Sigma Chemical Co), stored at 0°. The following solvents have been tested.
Ethanol (practical grade, Fisher Scientific Co.)
Isopropyl alcohol (practical grade, Fisher Scientific Co.)
Chloroform (practical grade, Fisher Scientific Co.)
Cyclohexane (practical grade, Fisher Scientific Co.)
Cyclohexene (practical grade, Matheson Co.)
Dioxane (technical and "scintillation" grade, Matheson Co.)

For the reasons given below, dioxane is the solvent of choice for the quantitative photochemical–fluorometric measurement. In order to reduce its too high and too instable fluorescence background, technical grade dioxane has to be prephotolyzed for 18 hr with 245-nm radiation. ''Scintillation'' grade dioxane is found to give a much lower and more stable fluorescence blank, which is also improved by a prephotolysis. Because of the high cost of this solvent, it is recommended that one prephotolyze the grade of dioxane that is most appropriate according to considerations of economy for quantitative determination.

Stock solutions of vitamins $K_1$ are prepared by weight. Working standards are made by volumetric dilutions of the refrigerated stock solution in dark red (low actinic) volumetric flasks, which must be kept in the refrigerator and in the dark to prevent photochemical or thermal decomposition of vitamin $K_1$. Samples should be run as rapidly as possible after preparation.

*Preliminary Procedures*

*Stability of the Photolysis-Excitation Source.* Stability of the Xe-Hg lamp used as the photolysis-excitation source is studied by observing the signal from the digital voltmeter, without the summing mode. When the digital voltmeter signal varies no more than 0.2 mV over a period of 2 min, the lamp is considered stable. Generally, warm-up time is of the order of 1–2 hr, and it decreases with lamp age.

*Uncorrected Emission Spectra of the Vitamin $K_1$ Photoproduct and Solvent Background.* A solution of vitamin $K_1$ (10 $\mu$g/ml) is irradiated during several minutes. Then the emission monochromator is scanned with the analog recorder from 300 to 600 nm; the uncorrected emission spectrum of the vitamin $K_1$ photoproduct is obtained by plotting analog recorder intensity readings as a function of wavelength. The emission spectrum of the solvent is taken by means of the same procedure, with only solvent in the sample cell. The wavelength with maximum difference between vitamin $K_1$ photoproduct and solvent background emission intensity is chosen as the optimum value for the analytical measurements of fluorescence intensities vs time.

*Choice of Solvent for Quantitative Analysis.* Fluorescence maximum wavelengths and photolysis times needed to reach the maximum emission signal value are listed for various solvents in the table. It can be seen that the wavelengths of fluorescence maxima vary from 431 to 505 according to the polarity of the solvent. Rates of formation of the fluorescent photochemical product also vary considerably according to the solvent: photo-

EMISSION WAVELENGTHS AND RECORDED EMISSION SIGNAL OF THE PHOTOCHEMICAL
PRODUCT OF VITAMIN $K_1$ IN VARIOUS SOLVENTS[a,b]

| Solvents | Emission maximum (nm) | Photolysis time[c] (min) | Maximum signal[d] (mV) |
|---|---|---|---|
| Petroleum ether | 431 | 7 | 10 |
| Dioxane | 431 | 12 | 21 |
| Isopropanol | 475 | 13 | 16 |
| Ethanol | 505 | >30 | 2 |
| Cyclohexane | 431 | 5.5 | 4 |
| Chloroform | Unstable solvent background | | |

[a] Reprinted, with permission, from J. J. Aaron, J. E. Villafranca, V. R. White, and J. M. Fitzgerald, *Appl. Spectr.* **30**, 159 (1976). Copyright by the Society for Applied Spectroscopy.

[b] All samples are 10 $\mu$g/ml solutions of vitamin $K_1$. High concentration was used to facilitate measurements.

[c] Irradiation time necessary to reach the maximum fluorescence signal value.

[d] Fluorescence signal magnitude at time listed in column 3.

lysis times range from 5.5 min for cyclohexane to more than 30 min for ethanol.

For quantitative measurements, dioxane is found to be the best solvent. It gives the largest maximum fluorescence signal in a convenient photolysis time.

### Procedures for Quantitative Assay

Two data readout procedures for the quantitative determination of vitamin $K_1$ are used; they are described below in detail.

*Digital Integration Method.* This method is based on the digital integration of the ingrowth of fluorescence at 431 nm for 1 or 2 min. With the timing circuit off and the excitation shutter closed, a solution of vitamin $K_1$ is introduced into the sample cell. The digital voltmeter is utilized with the memory and sum buttons engaged. Just before opening of the excitation shutter, the reset button on the digital voltmeter is released. The fluorescence ingrowth of the photoproduct is integrated for 1–2 min, and the integrated signal is read from the digital voltmeter. The timer and the excitation and photomultiplier shutters are turned off. The sample solution is aspirated from the cell, and the cell is rinsed between each measurement. The entire procedure takes about 2.5 min.

The procedure is repeated for several vitamin $K_1$ concentrations, each concentration being measured 3 or 4 times in order to ensure good precision. Satisfactory linear calibration curves are obtained over the range 5 to 130 ng/ml by plotting integrated fluorescence signal against initial vitamin $K_1$ concentration.

*Analog Recording Method.* In this method, the emission of the fluorescent photoproduct at 431 nm is recorded vs time on the chart recorder. The background signal is electronically blank-subtracted so that the full-scale signal displayed on the recorder is due entirely to fluorescence of the photolysis product of vitamin $K_1$ (Fig. 2). All other instrumental parameters are the same as in the digital integration method. The procedure is repeated 3 or 4 times for each concentration of vitamin $K_1$. The fluorescence signal intensity in microvolts is read from the recorder display and plotted against initial vitamin $K_1$ concentrations. Linear calibration curves are obtained over the range 5 to 200 ng/ml. A least-squares linear regression of the data indicates that the precision of the calibration plot is satisfactory.

*Comparison of the Readout Methods.* Linear vitamin $K_1$ concentration ranges obtained with the two readout methods are similar, but least-squares data treatment indicate that the precision of measurements of vitamin $K_1$ with the digital integration procedure is better than with the analog fluorescence recording method.

## Sensitivity

When compared to previously reported data, the photochemical–fluorometric method is more sensitive by at least two orders of magnitude

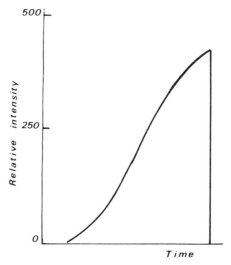

Fig. 2. Typical recorder trace of fluorescence signal of the vitamin $K_1$ photochemical product vs time. Relative intensity signal in microvolts; time of exposure about 1 min, 50 ng/ml vitamin $K_1$; background signal electronically blank subtracted.

for the quantitative determination of vitamin $K_1$. The lowest detection limit reported to date is 1000 ng/ml by phosphorimetry,[6] which is to be compared with 5 ng/ml by the photochemical–fluorometric digital or analog recording method. In terms of minimal detectable amount of vitamin $K_1$, in nanograms, a value of 500 ng has been reported for a gas chromatographic technique[9]; this compares to 125 ng (5 ng/ml, 25-ml aliquot) for the photochemical–fluorometric method. This low limit of detection could even be improved by reducing the size of the cell and by increasing slit width at the cost of precision; it would then be as low as the reported phosphorimetric minimal detectable amount of 20 ng, obtained by the means of the 20-$\mu$l volume of the sample cell.[6]

## Application of the Method

The photochemical–fluorometric method could be applied to the quantitative determination of vitamin $K_1$ extracted from bacterial, plant, or animal tissues and separated by the classical chromatographic techniques.[3] This method would be also suitable for the quantitative determination of vitamin $K_3$, which gives a highly phosphorescent photoproduct after irradiation by UV light in a polar solvent.[6,15]

## Comment

The photochemical–fluorometric method described above has proved to be very sensitive and rapid. It can be successfully used for the determination of vitamin $K_1$ in samples containing more than 5 ng of the compound per milliliter.

[15] J. J. Aaron, unpublished results, 1972.

## [20] Determination of Vitamin $K_1$ in Photodegradation Products by Gas–Liquid Chromatography

By YASUHIRO NAKATA and ETSUO TSUCHIDA[1]

Vitamin $K_1$ ($K_1$), one of the prothrombogenic vitamins, decomposes gradually on exposure to light. Ewing[2] reported initial studies of $K_1$ photodegradation, Hujisawa[3] determined the structure of a photoproduct

[1] Previous report: Y. Nakata, Y. Mita, and S. Khono, *Yakugaku Zasshi* **96**, 53 (1976).
[2] D. T. Ewing, *J. Biol. Chem.* **147**, 233 (1943).
[3] S. Hujisawa, *Yakugaku Zasshi* **87**, 1451 (1967).

TABLE I

RELATIONSHIP BETWEEN COLUMN TEMPERATURE AND RETENTION TIME OR PEAK AREA RATIO

| Column temp. (°C) | Retention time (min) | | Peak area ratio ($K_1$/Int. Std.) |
|---|---|---|---|
| | Int. Std.[a] | Vitamin $K_1$ | |
| 240 | 14.9 | 29.7 | 1.93 |
| 250 | 9.6 | 19.0 | 1.94 |
| 260 | 6.7 | 13.1 | 1.92 |
| 270 | 4.7 | 9.0 | 2.03 |

[a] Internal standard was $n$-triacontane as with conditions for GLC given in Fig. 1.

under anaerobic conditions, and Katsui et al.[4] cleared up the structures of complicated photoproducts formed under aerobic condition. Quantitative analyses of $K_1$ have been reported, e.g. column chromatography,[5] polarography,[6] spectrophotometry, spectrophotometry together with thin-layer chromatography (TLC)[7] and gas–liquid chromatography (GLC).[8] However, there are difficulties such as reproducibility in the case of column chromatography, determination during photodegradation under aerobic conditions in the case of polarography, and the rate of withdrawal from silica gel plate in case of spectrophotometry with TLC. Thus, it has been difficult to determine residual $K_1$ in many instances, and complicated photoproducts produced under aerobic or anaerobic conditions have not been reported previously in connection with this literature.

For study of photodegradation of $K_1$, the determination of $K_1$ in photoproducts was investigated by GLC, which is both a simple and efficient method, as separation and quantitative analysis has been established without interference of photoproducts. The quantitative analysis was carried out on a column packed with acid-washed and silanized Chromosorb W coated with 3% silicone OV-1 at a column temperature of 260°; $n$-triacontane was used as an internal standard.

[4] G. Katsui and M. Ohmae, Britamin 32, 308 (1965); ibid. 35, 116 (1967); ibid. 39, 181 (1969); ibid. 39, 190 (1969).
[5] G. Katsui and M. Ohmae, Bitamin 38, 182 (1968).
[6] E. B. Hersherg, J. Am. Chem. Soc. 62, 3516 (1940).
[7] G. Katsui and M. Ohmae, Bitamin 41, 173 (1970).
[8] P. P. Nair, J. Am. Oil. Chem. Soc. 40, 347 (1963); Iguchi, Yakugaku Zasshi 83, 721 (1963); K. K. Carroll, J. Am. Oil. Chem. Soc. 41, 473 (1964); W. Vetter, Helv. Chim. Acta. 50, 1866 (1967).

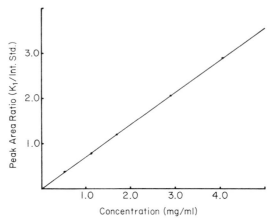

FIG. 1. Calibration curve for vitamin $K_1$ on silicone OV-1. Gas chromatograph conditions: HITACHI Model 063 type; detector, H.F.I.D.; $H_2$, 1.0 kg/cm²; air, 1.0 kg/cm²; carrier gas, $N_2$ gas; flow rate, 60 ml/min; glass tube, 3 mm × 2 m; column temperature, 260°; injection temperature, 280°; internal standard (Int. Std.), $n$-triacontane as 0.1% in isooctane.

## Results and Discussion

### Conditions for Gas–Liquid Chromatography

Preliminary gas chromatography of $K_1$ and various photoproducts under aerobic and anaerobic conditions were carried out on a column

TABLE II

REPRODUCIBILITY OF PEAK AREA RATIO OF EACH VITAMIN $K_1$

| | Peak area ratio | | |
| | | Photodegraded $K_1$[a] | |
| Run No. | Vitamin $K_1$ | Aerobic | Anaerobic |
|---|---|---|---|
| 1 | 1.92 | 1.010 | 1.10 |
| 2 | 1.93 | 0.983 | 1.11 |
| 3 | 1.95 | 0.983 | 1.09 |
| 4 | 1.91 | 0.982 | 1.08 |
| 5 | 1.93 | 0.992 | 1.11 |
| Means | 1.93 | 0.990 | 1.10 |
| SD[b] | 0.0134 | 0.0106 | 0.0117 |
| CV[c] | 0.695% | 1.07% | 1.06% |

[a] Photodegraded $K_1$: residual vitamin $K_1$ of photoproducts.
[b] SD, standard deviation.
[c] CV, coefficient of variation.

TABLE III

COMPARISON OF METHODS OF ANALYSIS

| | Aerobic conditions | | Anaerobic conditions | |
|---|---|---|---|---|
| Method | Sample A[a] (%) | Sample B[b] (%) | Sample C[a] (%) | Sample D[b] (%) |
| Gas chromatography | 78 | 58 | 83 | 60 |
| Polarography | 85 | 67 | 86 | 64 |
| TLC and spectrophotometry | 78 | 59 | — | — |

[a] Sample A, sample C: 4000–5000 lux fluorescent lamp, 20 hr.
[b] Sample B, sample D: 4000–5000 lux fluorescent lamp, 43 hr.

coated with a thin layer of silicone, for example, SE-30, OV-1, and OV-17 at high temperature (240–270°). The column packed with acid-washed and silanized Chromosorb W coated with 3% silicone OV-1 was found to be the best for retention time, column temperature, and the separation of K$_1$

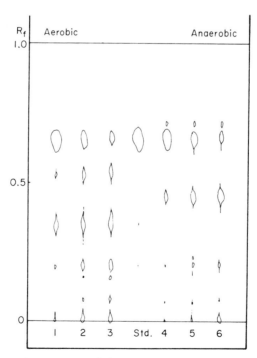

FIG. 2. Thin-layer chromatogram of photodegradation products from vitamin K$_1$ in isopropyl alcohol. Aerobic conditions were with a 4000–5000 lux fluor lamp for 20 hr (1), 43 hr (2), and 80 hr (3). Corresponding anaerobic conditions were for (4), (5), and (6). Std.: before photodegradation.

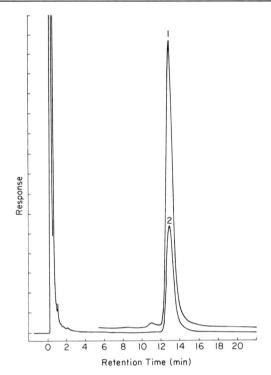

FIG. 3. Gas chromatogram of vitamin $K_1$ and photoproducts derived under aerobic conditions: 1; before photodegradation; 2; after photodegradation (80 hr).

and various photoproducts. With this column, the following studies on conditions for quantitative analysis have been done.

Table I shows the relationship between column temperature, peak area ratio, and retention time of $K_1$ and $n$-triacontane as internal standard. Since peak area ratio did not change at column temperatures of 240–260° but did at 270°, the column temperature selected was 260°, where the calibration curve for $K_1$ was drawn.

As the result, a linear relationship between concentration of $K_1$ and peak area ratio was obtained in the range as shown in Fig. 1, and the line passes through the zero point. Reproducibility of GLC was investigated with respect to peak area ratio of $K_1$ and $K_1$ in various photoproducts under aerobic or anaerobic conditions.

As shown in Table II, each coefficient of variation after photodegradation is almost the same value as before photodegradation. In order to compare GLC with another analysis of content for $K_1$ in photoproducts, samples of $K_1$ isopropanol solution were irradiated with a 4000–5000 lux fluorescent lamp for 20 and 43 hr under aerobic or anaerobic conditions, and $K_1$ in each sample was determined by GLC, polarography, and spectrophotometry with TLC.

As a result of determinations shown in Table III, each analysis was the same for residual content, except for those determined by polarography under aerobic conditions.

### Effect of Photodegradation Products of $K_1$ with GLC

To determine $K_1$ without interference of photoproducts, we investigated fully the effect of photoproducts under aerobic or anaerobic conditions and confirmed some structures of main photoproducts. Samples of $K_1$ isopropanol solution were irradiated with a 4,000–5,000 lux fluorescent lamp during 20, 43, and 80 hr under aerobic or anaerobic conditions.

Figure 2 shows a thin-layer chromatogram of photoproducts in the

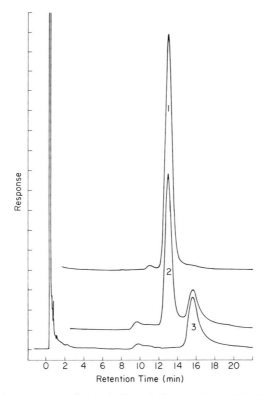

FIG. 4. Gas chromatogram of vitamin $K_1$ and photoproducts derived under anaerobic conditions: 1, before photodegradation; 2, after photodegradation (80 hr); 3, photoproducts of $R_f$ 0.40–0.50 from a thin-layer chromatogram.

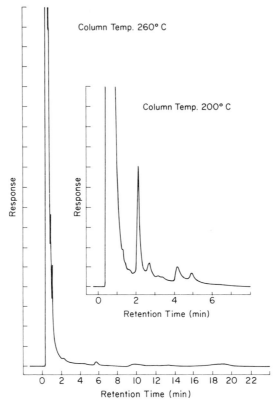

Column Temp. 260° C

Column Temp. 200° C

Response

Response

Retention Time (min)

0   2   4   6

0  2  4  6  8  10  12  14  16  18  20  22
Retention Time (min)

FIG. 5. Gas chromatogram of aerobic photoproducts from the zone of $R_f$ 0.25–0.40 on a thin-layer chromatogram.

irradiated $K_1$ isopropanol solution. In the aerobic case, the pattern of each photoproduct fits the chromatogram that Katsui *et al.*[4] reported in previous papers; the pattern of anaerobic photoproducts is the same chromatogram that Hujisawa[3] described in his papers. A gas chromatogram of $K_1$ before photodegradation was compared with $K_1$ after photodegradation for 80 hr under aerobic and anaerobic conditions. Figure 3 shows a gas chromatogram of $K_1$ and photoproducts under aerobic conditions where the peak of $K_1$ (i.e., $t_R$ is 13 min) decreases after photodegradation, and many peaks of photoproducts increase near the solvent peak. Figure 4 shows a gas chromatogram under anaerobic conditions when the peak of $K_1$ decreases and peaks of photoproducts (i.e., $t_R$ is 10 or 16 min) increase. In the anaerobic case, the main photoproduct is naphthocromenol, and the TLC spot (i.e., $R_f$ 0.45 of Fig. 2) was scraped from the silica gel plate, extracted with acetone, and injected in the GLC. As the result, peaks of

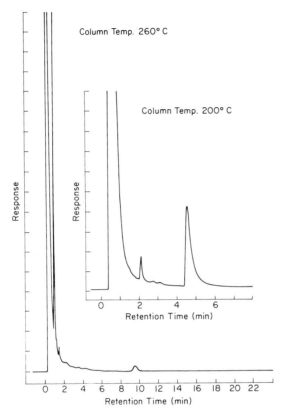

FIG. 6. Gas chromatogram of anerobic photoproducts from the zone of $R_f$ 0.05–0.25 on a thin-layer chromatogram.

the spot were observed at about 10 and 16 min of retention time, as shown in Fig. 4. On the other hand, in the case of aerobic conditions, many peaks of photoproducts were observed near the solvent peak, so it was not possible to identify directly each peak from GLC with each spot on TLC (e.g., Fig. 2). To do this, photoproducts from each zone (i.e., $R_f$ 0.05–0.25, 0.25–0.40, and 0.40–0.55 of Fig. 2) were scraped from a silica gel plate, extracted with acetone, and injected in the GLC at 200° or 260°.

Figure 5 shows a gas chromatogram of photoproducts of $R_f$ 0.25–0.40 at 200° or 260°; Fig. 6 shows a gas chromatogram of photoproducts of $R_f$ 0.05–0.25, and Fig. 7 shows a gas chromatogram of photoproducts of $R_f$ 0.40–0.55.

As shown in Fig. 5, many peaks are observed in the gas chromatogram of photoproducts ($R_f$ 0.25–0.40) run at a column temperature of 200° when decomposition was caused by the heat of injection (280–300°). However,

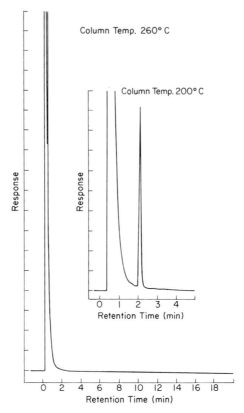

FIG. 7. Gas chromatogram of aerobic photoproducts from the zone of $R_f$ 0.40–0.50 on a thin-layer chromatogram.

peaks of photoproducts migrated near to the solvent peak at an actual column temperature of 260°, and as shown in Figs. 6 and 7, peaks of other photoproducts were also transferred near the solvent peak at 260° under aerobic conditions. A few photoproduct peaks (e.g., $t_R$ of 6, 10, and 19 min in Figs. 5 and 6) did not interfere in the peak of $K_1$ ($t_R$ was 13 min) with $n$-triacontane ($t_R$ was 6.7 min) as the internal standard. On the other hand, peaks of photoproducts formed under anaerobic conditions did not interfere with the peak of $K_1$ and $n$-triacontane, as shown in Fig. 4 (e.g., $t_R$ was 10 or 16 min). It was therefore possible to determine $K_1$ easily without interference of photoproducts by the GLC condition (e.g., Fig. 1) under aerobic and anaerobic conditions.

Katsui et al.[4] described fast decompositions as mixtures of $K_1$-hydroperoxide (I) and $K_1$-hydroxide (II) of $R_f$ 0.35, and (I) or (II) are changed to a new phthone of $R_f$ 0.53, 5-hydroxy-4-

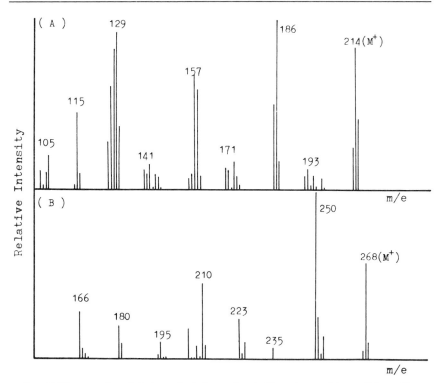

FIG. 8. Mass spectra of photoproducts from vitamin $K_1$ under aerobic conditions. Photoproducts were prepurified as: (A) spot of $R_f$ 0.20 on thin-layer chromatogram; (B) spot of $R_f$ 0.53 on thin-layer chromatogram.

methyl-6,7-benzocumaran-3-one (III) of $R_f$ 0.20, phthiocol of $R_f$ 0.10, and a polymer of $R_f$ 0.0. In the case of photoproducts formed under aerobic conditions, we confirmed the structures of some photoproducts. (I) or (II) of $R_f$ 0.35 was unstable and could not be isolated, but photoproduct A of $R_f$ 0.20 and photoproduct B of $R_f$ 0.53 were isolated.

As Figs. 8–10 show, photoproduct A of $R_f$ 0.20 was (III) and the product B of $R_f$ 0.53 was phthone when identified with mass and nuclear magnetic resonance spectrometers as Katsui *et al.* had described in previous papers.

Experimental Details

*Materials*

Vitamin $K_1$, Eisai Co. Ltd.
*n*-Triacontane (99%), Applied Science Lab. Ltd.

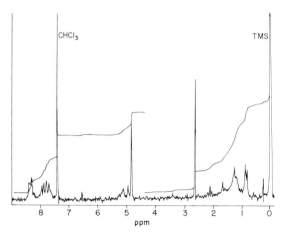

FIG. 9. Nuclear magnetic resonance spectrum of photoproduct A from vitamin $K_1$ under aerobic conditions. Photoproduct A: Spot of $R_f$ 0.20 on thin-layer chromatogram.

Silicone OV-1, 3%, on Chromosorb W(AW) DMCS 60/80 mesh, Nihon Chromato Works Ltd.

All the other chemicals are of highest grade.

*Apparatus*

Gas chromatograph equipped with a hydrogen flame ionization detector such as Hitachi Model 063

Mass spectrometer, Japan Electron Optics Lab., Model JMS-01-SG-2

NMR spectrometer, Japan Electron Optics Lab., Model JNM-PS-100 (tetramethylsilane as internal standard)

Polarograph, Yanagimoto Co. Ltd. Model P-8

Spectrophotometer, Hitachi Model 139 type.

*Preparation of Vitamin $K_1$ Solution.* Weigh 1.00 g of $K_1$, add isopropyl alcohol, dissolve, and prepare a 10 mg/ml concentration. The solution (5 ml) is injected into 10-ml normal ampoules under $N_2$ gas or $O_2$ gas, closed, and stored. These samples are protected from light.

*Exposure Procedure.* The samples of $K_1$ solution in ampoules are irradiated with a 4000–5000 lux fluorescent lamp for 20, 43, and 80 hr.

*Thin-Layer Chromatography.* Silica gel plates (Kiesel gel GF$_{254}$, Merck Co. Ltd., 250 $\mu$m, activated for 1 hr at 105°) are spotted with 0.10 mg of $K_1$, dried, and developed for a distance of 10 cm with chloroform. Compounds are detected with a UV lamp after treatment with conc. $H_2SO_4$ for 15 min at 105°.

Silica gel plates (Kiesel gel G, Merck Co. Ltd., 300 $\mu$m, activated for 1

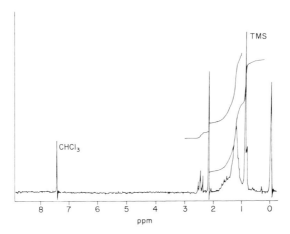

FIG. 10. Nuclear magnetic resonance spectrum of photoproduct B from vitamin $K_1$ under aerobic conditions. Photoproduct B: Spot of $R_f$ 0.53 on thin-layer chromatogram.

hr at 105°) are used for separation and isolation of photoproducts. For this, streak 0.5 ml of each irradiated $K_1$ isopropanol solution in bands 7 cm wide, dry, develop for 10 cm with $CHCl_3$. Scrape each zone ($R_f$ 0.05–0.25, 0.25–0.40, and 0.40–0.55), extract with acetone, filter, dry, dissolve in 0.5 ml of acetone, and inject into the GLC.

*Preparation of Calibration Curve.* Weigh accurately 50 mg of $K_1$, add isopropyl alcohol, dissolve, and prepare a 1 mg/ml concentration. Pipette 1.0, 2.0, 3.0, 5.0, and 7.0 ml, dry, add accurately to each 2 ml of 0.1% $n$-triacontane in isooctane, dissolve, and inject 3 $\mu$l. Measurement of peak area is by the half-width method.

*GLC of Irradiated Vitamin $K_1$ and Reproducibility of GLC.* Pipette 1.0 ml of each $K_1$ isopropanol solution, dry, add 4.0 ml of isooctane or isooctane containing 0.1% $n$-triacontane, and inject 3 $\mu$l in GLC. Measurement of peak area ratio is by the half-width value method.

*Determination of Vitamin $K_1$ in Photodegradation Products*

a. GAS CHROMATOGRAPHY. Samples of $K_1$ isopropanol solution (10 mg/ml) irradiated for 20 and 43 hr are used. Pipette 1.0 ml of each sample, dry, add 4.0 ml of isooctane containing 0.1% $n$-triacontane, dissolve, and inject 3 $\mu$l in GLC. Measurement of peak area ratio is by the half-width value method. Content of residual $K_1$ was calculated on the basis of peak area ratio before photodegradation.

Content of residual $K_1$(%)

$$= \frac{\text{peak area ratio after photodegradation}}{\text{peak area ratio before photodegradation}} \times 100$$

b. POLAROGRAPHY. Pipette 1.0 ml of irradiated sample (a) into a 50-ml volumetric flask, add accurately isopropyl alcohol, pipette 2.0 ml, and add 3.0 ml of Britton-Robinson buffer (pH 8.0) in a brown electrolyzing bottle. Sweep through $N_2$ gas for 15 min. Polarograph conditions are as follows: span volt, $-2.0$; initial volt, $-0.1$; temperature, $25 \pm 1°$; sensitivity 5 nA/mm. Calculation of content of residual $K_1$ was the same as for (a).

c. SPECTROPHOTOMETRY TOGETHER WITH THIN-LAYER CHROMATOG- RAPHY. Pipette 2.0 ml of irradiated sample (a), dry, add accurately 5.0 ml of acetone, dissolve, and spot accurately 0.5 ml in a 7-cm band on a silica gel plate. Dry, develop for 10 cm with $CHCl_3$, and scrape the zone of $K_1$ ($R_f$ 0.55–0.70). Transfer to a 50-ml volumetric flask, add accurately iso- propyl alcohol, filter, pipette 15.0 ml of $K_1$ isopropanol solution, transfer to a 50-ml volumetric flask, and add accurately isopropyl alcohol. Measure absorbance at 270 nm. The content of residual $K_1$ is calculated on the basis of absorbance before photodegradation.

$$\text{Content of residual } K_1(\%) = \frac{\text{absorbance after photodegradation}}{\text{absorbance before photodegradation}} \times 100$$

## Acknowledgment

The authors wish to thank Mr. Yuuji Naito, the president of Eisai Co. Ltd., for permis- sion to publish this study.

## [21] Assay Procedure for the Vitamin $K_1$ 2,3-Epoxide-Reducing System

By GRAHAM R. ELLIOTT, MICHAEL G. TOWNSEND, and EDWARD M. ODAM

It was first shown by Matschiner et al.[1] that a major metabolite of radioactive vitamin $K_1$ that accumulated in the liver of warfarin-treated, anticoagulant-susceptible rats was vitamin $K_1$ 2,3-epoxide (I). Evidence was obtained for the presence of low levels of this epoxide in the tissues of animals not treated with warfarin, and it was subsequently proposed that a reaction cycle exists in which the vitamin $K_1$ 2,3-epoxide produced by oxidation was reduced back to vitamin $K_1$. The hypothesis that the hypoprothrombinemia produced by warfarin was due to competition be- tween vitamin $K_1$ 2,3-epoxide and vitamin $K_1$ for an active site[2] was not

[1] J. T. Matschiner, R. G. Bell, J. M. Amelotti, and T. E. Knauer, Biochim. Biophys. Acta 201, 309 (1970).
[2] R. G. Bell and J. T. Matschiner, Nature (London) 237, 32 (1972).

(I)

substantiated.[3,4] Although vitamin $K_1$ 2,3-epoxide has a similar activity to vitamin $K_1$ in vitamin K-deficient rats, it was ineffective in warfarin-treated animals,[5] and it was suggested that warfarin inhibited the *in vivo* conversion of vitamin $K_1$ 2,3-epoxide to vitamin $K_1$ and led to the observed increase in the ratio of vitamin $K_1$ 2,3-epoxide to vitamin $K_1$.[2,6] Studies with Sprague–Dawley derived, warfarin-resistant rats showed that warfarin treatment resulted in a lower tissue ratio[7] and suggested that this strain had an altered enzyme system that was both less sensitive to inhibition by warfarin and less efficient at catalyzing the reduction of vitamin $K_1$ 2,3-epoxide to vitamin $K_1$. This was later confirmed *in vitro* using 6,7-[³H]vitamin $K_1$ 2,3-epoxide as substrate.[8,9] These findings offered an explanation of the increased dietary requirement for vitamin $K_1$ by this strain.[10] The availability of an assay for the vitamin $K_1$ 2,3-epoxide-reducing system is of importance to those (*a*) studying the site of action of warfarin, (*b*) accounting for species differences in the toxicity of anticoagulants, and (*c*) identifying compounds that, although they act at the same site as warfarin, inhibit the "resistant" enzyme and therefore have potential commercial importance as rodenticides. As the method first described[8] requires radioactive vitamin $K_1$ 2,3-epoxide, which is not readily available commercially, we developed an assay system in which high-performance liquid chromatography is used to separate and quantify the reaction product.[11]

[3] S. R. Goodman, R. M. Houser, and R. E. Olson, *Biochem. Biophys. Res. Commun.* **61**, 250 (1974).

[4] J. A. Sadowski and J. W. Suttie, *Biochemistry* **13**, 3696 (1974).

[5] R. G. Bell and J. T. Matschiner, *Arch. Biochem. Biophys.* **141**, 473 (1970).

[6] R. G. Bell, J. A. Sadowski, and J. T. Matschiner, *Biochemistry* **11**, 1959 (1972).

[7] R. G. Bell and P. T. Caldwell, *Biochemistry* **12**, 1759 (1973).

[8] J. T. Matschiner, A. Zimmermann, and R. G. Bell, *Thromb. Diath. Haemorrh. Suppl.* **57**, 43 (1974).

[9] A. Zimmermann and J. T. Matschiner, *Biochem. Pharmacol.* **23**, 1033 (1974).

[10] M. A. Hermodson, J. W. Suttie, and K. P. Link, *Am. J. Physiol.* **217**, 1316 (1969).

[11] G. R. Elliott, E. M. Odam, and M. G. Townsend, *Biochem. Soc. Trans.* **4**, 615 (1976).

Method of Assay

The method used is a modification of that described by Matschiner *et al.*[8] The vitamin $K_1$ is purchased from Sigma (London) Ltd., and the 2,3-epoxide prepared from it by the method of Tishler *et al.*[12] is chromatographically pure. Male rats (190–210 g) from a Wistar-derived colony are killed by cervical dislocation and bled by cutting the blood vessels of the neck. The liver is removed and washed in ice-cold 0.9% saline and weighed. It is macerated in 0.25 $M$ sucrose–25 m$M$ potassium phosphate buffer, pH 7.4–5 m$M$ dithiothreitol with a Potter–Elvehjem homogenizer fitted with a Teflon pestle and surrounded by iced water. The homogenate is adjusted to 10% (w/v), and a measured volume is used to produce subcellular fractions by centrifugation.[13] The incubations and all subsequent steps are carried out using amberized glassware (Aimer Ltd, London NW1, U.K.). The enzyme reaction is carried out in a 25-ml conical flask open to the air and shaken at 100 oscillations per minute in a water bath at 37°. Each flask contains a 2-ml aliquot of the postmitochondrial supernatant, 1 ml of 25 m$M$ potassium phosphate buffer, pH 7.4–5 m$M$ dithiothreitol (with or without additions such as warfarin), and a glass marble to facilitate mixing. After a 5-min preincubation, the reaction is started by the addition of vitamin $K_1$ 2,3-epoxide dissolved in 50 $\mu$l of absolute ethanol and terminated after 20 min by pouring the contents of each flask into a stoppered centrifuge tube containing 1 g of sodium chloride and 7 ml of propan-2-ol–hexane (3 : 2, v/v) kept on ice. The solvents are mixed with a vortex mixer, and the two phases are separated by centrifugation at 2000 $g$ for 15 min. The upper phase is removed, dried by passing down a 25-cm glass column containing 6 g of anhydrous sodium sulfate, and collected in a sample tube. The lower phase is then extracted twice with 5 ml of hexane in a similar way. The combined extract is maintained at 40° on a water bath, and the solvent is evaporated with a slow stream of nitrogen. The concentrated extract is transferred with diethyl ether to a stoppered, tapered 2-ml tube. The organic solvent is evaporated just to dryness, redissolved in 100 $\mu$l of chloroform containing vitamin $K_{2(30)}$[14] as an internal standard, and sealed until analyzed.

Vitamin $K_1$ and its epoxide are separated by high-performance liquid chromatography (LC 750, ARL, Luton, U.K.) using a 50 cm × 4 mm column dry-packed with Permaphase ODS and methanol–methyl cyanide–water (2 : 2 : 1, v/v) as the eluting solvent at constant pressure (2 MPa). A high-pressure syringe is used to apply up to 30-$\mu$l samples to the

[12] M. Tishler, L. F. Fieser, and N. L. Wendler, *J. Am. Chem. Soc.* **62**, 2866 (1940).
[13] P. J. Bunyan, M. G. Townsend, and A. Taylor, *Chem. Biol. Interact.* **5**, 13 (1972).
[14] Vitamin $K_{2(30)}$ was a gift from F. Hoffmann-La Roche & Co. Ltd, Basel.

column through a 7-mm silicon rubber/Teflon septum [SGE (UK) Ltd, London N.W. 2] previously punctured by a 21-gauge hypodermic syringe needle before use, as this was found to reduce the incidence of needle blockage and increase the life of the septum. The constant-wavelength detector (254 nm) gives a linear response over the range of 50–500 ng of vitamin $K_1$ and 2–16 $\mu$g of vitamin $K_1$ 2,3-epoxide when standard solutions are analyzed separately or as mixtures. Because of the large difference between the amount of vitamin $K_1$ 2,3-epoxide added as substrate and the small quantity of product formed, a change in the attenuation is necessary if all the peaks are to be kept on scale.

For rat postmitochondrial supernatant, the assay was found to be linear over the range of 50–200 mg fresh weight of liver, and 5–40 min, the maximum activity occurring with 300 $\mu M$ vitamin $K_1$ 2,3-epoxide.[11] The endogenous reducing agent *in vivo* is not known, but it has been reported that in the absence of dithiothreitol a crude liver homogenate shows a low level of activity,[8] although we were unable to confirm this. The activity of the postmitochondrial supernatant is dependent upon the dithiothreitol concentration up to at least 5 m$M$, but as the degree of inhibition of the reaction by warfarin is inversely dependent on the concentration of sulfhydryl present,[8] a compromise is necessary. The effect of several reducing agents has been studied, and dithiothreitol is reported to be the most active.[15] We have found that $N$-acetylcysteine at an equimolar sulfhydryl concentration is less effective than dithiothreitol. In the absence of dithiothreitol no postmitochondrial supernatant activity was observed with either 1 m$M$ NADPH or NADH. In the presence of dithiothreitol and either of these cofactors, there was no increase in activity above that due to dithiothreitol alone. The *cis*- and *trans*-isomers of vitamin $K_1$, separated by column chromatography[16] and oxidized to the epoxide,[12] were reduced equally at a substrate concentration of 300 $\mu M$ by the postmitochondrial reductase system. The subcellular distribution of the reduction activity is shown in Table I. From these data it is difficult to associate the system with any one specific fraction, and the data would seem to confirm the previous findings that the crude nuclear fraction, microsomes, and cytosol are required for activity.[17]

The recovery of vitamin $K_1$ and the epoxide from the assay mixture when the enzyme preparation was added after the addition of the extraction solvent was reproducible and not less than 80%. The reverse-phase

[15] R. G. Bell, personal communication, 1976.

[16] T. E. Knauer, C. Siegfried, A. K. Willingham, and J. T. Matschiner, *J. Nutr.* **105**, 1519 (1975).

[17] M. L.-L. Wai, A. K. Willingham, and J. T. Matschiner, *Fed. Proc., Fed. Am. Soc. Exp. Biol.* **34**, 683 (1975).

TABLE I
SUBCELLULAR DISTRIBUTION OF REDUCTASE ACTIVITY

| Fraction | Nanomoles of vitamin $K_1$ produced per gram fresh weight of liver equivalent per hour[a] |
|---|---|
| Whole homogenate | 832 |
| Postmitochondrial supernatant | 269 |
| Cytosol | 13 |
| Microsomes | 84 |

[a] All values were obtained by the method described for postmitochondrial supernatant.

ODS-treated silica columns do vary in their efficiency, even when material from the same manufacturer is used, and the conditions for the optimum separation must be determined for each. There is now evidence to show that the analysis of our standard solutions on a 5-$\mu$m microparticulate, reverse-phase column gives better resolution, narrower peak widths, and increased sensitivity over the columns used in our studies.[18] We have investigated a number of readily available compounds, including 2-tert-butylanthraquinone, piperonyl butoxide, linolenic acid, naphthyl esters, and $n$-alkylbenzene, for their suitability as an internal standard. It would appear that the chromatographic system is fairly specific for vitamin K-type compounds and that most others elute with the solvent front or are retained. After the use of the $p$-nitrobenzyl moiety as a chromophore by Keenan et al.[19] in their assay of polyprenols, we synthesized the corresponding phytol ester, but it is eluted from our column too close to vitamin $K_1$ 2,3-epoxide to be a suitable standard. At present, vitamin $K_{2(30)}^{14}$ is the compound of choice.

The absolute sensitivity of the assay based on [$^3$H]vitamin $K_1$ 2,3-epoxide[8] is greater than that for the method described here, but the practical difficulties in obtaining reproducible separations with a reverse-phase, thin-layer chromatographic system are a disadvantage. The enzymic assay conditions described by Matschiner et al.[8] resulted in a high conversion to vitamin $K_1$. The level of vitamin $K_1$ 2,3-epoxide used by these workers was 0.7 $\mu M$, which is well below that found to be optimal in our experiments[11] and may have been rate limiting. We have applied our assay system to several species and strains of rats (Table II). A comparison of the warfarin-susceptible and -resistant rat strains confirmed the previous findings that the activity is highest for susceptible animals.[8] A study of the vitamin $K_1$ requirements of these strains has shown that the Scottish

[18] M. R. Dilloway, personal communication, 1978.
[19] R. W. Keenan, N. Rice, and R. Quock, Biochem. J. 165, 405 (1977).

TABLE II
REDUCTASE ACTIVITIES FOR VARIOUS SPECIES AND STRAINS

| Species or strains | Nanomoles of vitamin $K_1$ produced per gram fresh weight of liver equivalent per hour[a] |
|---|---|
| Rat, laboratory | |
| Susceptible (TAS) | 170 |
| Resistant (HW) | 106 |
| Resistant (HS) | 127 |
| Mouse, laboratory, susceptible (LAC/gray) | 1270 |
| Red squirrel, wild, trapped | 74 |
| Japanese quail, laboratory strain | 53 |

[a] All values were obtained by the method described for postmitochondrial supernatant.

resistant (HS) strain is intermediate between the Welsh resistant (HW) and the susceptible (TAS) strain.[20] The reductase activity is in agreement with these findings. Although the activity of the HW strain showed less percentage inhibition by 1 $\mu M$ warfarin[21] than that from the TAS strain, the measured activities were very similar, as found previously.[11] The activity of the HW strain was not inhibited by 1 $\mu M$ difenacoum,[21] whereas that of the TAS strain was reduced by 25%. This result was unexpected, as this new rodenticide has been found to be almost equally toxic to TAS and HW rats in the laboratory[22] and to control warfarin-resistant rats under field conditions.[23] The Japanese quail showed the lowest reductase activity, and none was measurable below 200 $\mu M$ vitamin $K_1$ 2,3-epoxide. The blood levels of the vitamin K-dependent blood clotting factors of birds are much lower than those of the rat,[24] and this may be due in part to the low reductase activity.

[20] A. D. Martin, *Biochem. Soc. Trans.* **1**, 1206 (1973).
[21] Warfarin and difenacoum were given by M. R. Hadler, Sorex (London) Ltd.
[22] M. R. Hadler, R. Redfern, and F. P. Rowe, *J. Hyg.* **74**, 441 (1975).
[23] B. D. Rennison and M. R. Hadler, *J. Hyg.* **74**, 449 (1975).
[24] D. A. Walz, R. K. Kipfer, and R. E. Olson, *J. Nutr.* **105**, 972 (1975).

# [22] Vitamin K-Dependent Carboxylase

## *By* B. CONNOR JOHNSON

The vitamin K group was last covered in this series in Volume 18C (pp. 407–564) in 1971. These 155 pages covered structure, determination, isolation, and synthesis. However, until 1974 the function of vitamin K could not be covered.

The demonstration of the posttranslational site of function of vitamin K in prothrombin formation[1-3] was followed by the demonstration, in the spring of 1974, that the vitamin K-dependent step in prothrombin finalization (and other vitamin K-dependent proteins) was carboxylation.[4] This carboxylation was proved to form $\gamma$-carboxyglutamyl residues by isolation and characterization of the new amino acid.[5,6] Only the first 10 (of the total 71) glutamyl residues of prothrombin (positions 7, 8, 15, 17, 20, 21, 26, 27, 30, and 33) were carboxylated.[7-9] This vitamin K-dependent carboxylation reaction was found to proceed *in vitro* using microsomes from vitamin K-deficient rat livers,[10,11] and the system could be solubilized by various detergents.[11,12] In addition to the crude enzyme–substrate system, the reaction requires vitamin K hydroquinone, carbon dioxide, and molecular oxygen, but not ATP or biotin.[12-14] The vitamin K hydroquinone can be replaced by vitamin K quinone plus NADH.[10,11] The carboxylation is markedly stimulated by dithiothreitol, and, if particulate microsomes are used, dithiothreitol (or dithioerythritol) completely replaces NADH.

The only metabolic product of vitamin K that has been identified in the overall reaction is vitamin K epoxide.[15] The finding that the carboxylation requirement for molecular oxygen cannot be replaced by oxidizing agents such as FAD[16] suggests a coupled oxidative carboxylation, in which the

[1] P. Goswami and H. N. Munro, *Biochim. Biophys. Acta* **55**, 410 (1962).

[2] B. M. Babior, *Biochim. Biophys. Acta* **123**, 606 (1966).

[3] R. B. Hill, S. Gaetani, A. M. Paolucci, P. B. Rama Rao, R. Alden, G. S. Ranhotra, D. V. Shah, V. K. Shah, and B. C. Johnson, *J. Biol. Chem.* **243**, 3930 (1968).

[4] J.-M. Girardot, R. Delaney, and B. C. Johnson, *Biochem. Biophys. Res. Commun.* **59**, 1197 (1974).

[5] J. Stenflo, P. Fernlund, W. Egan, and P. Reopstorff, *Proc. Natl. Acad. Sci. U.S.A.* **71**, 2730 (1974).

[6] G. L. Nelsestuen, T. H. Zytokovicz, and B. Howard, *J. Biol. Chem.* **249**, 6347 (1974).

[7] S. Magnusson, L. Sottrup-Jensen, and T. E. Petersen, *FEBS Lett.* **44**, 189 (1974).

[8] S. Magnusson, T. E. Petersen, L. Sottrup-Jensen, and H. Claeys, in "Proteases and Biological Control" (E. Reich, D. B. Rifkin, and E. Shaw, eds.); Cold Spring Harbor Conferences on Cell Proliferation **2**, 123 (1975).

[9] H. R. Morris, A. Dell, T. E. Peterson, L. Sottrup-Jensen, and S. Magnusson, *Biochem. J.* **153**, 663 (1976).

[10] C. T. Esmon, J. A. Sadowski, and J. W. Suttie, *J. Biol. Chem.* **250**, 4744 (1975).

[11] D. O. Mack, E. T. Suen, J. M. Girardot, J. Miller, R. Delaney, and B. C. Johnson, *J. Biol. Chem.* **251**, 3269 (1976).

[12] C. T. Esmon and J. W. Suttie, *J. Biol. Chem.* **251**, 6238 (1976).

[13] J. M. Girardot, D. O. Mack, R. A. Floyd, and B. C. Johnson, *Biochem. Biophys. Res. Commun.* **70**, 655 (1976).

[14] P. A. Friedman and M. A. Shia, *Biochem. Biophys. Res. Commun.* **70**, 647 (1976).

[15] J. A. Sadowski, H. K. Schnoes, and J. W. Suttie, *Biochemistry* **16**, 3856 (1977).

[16] B. C. Johnson, J.-M. Girardot, E. T. Suen, D. O. Mack, R. A. Floyd, and R. Delaney, *World Rev. Nutr. Diet* **31**, 202 (1978).

energy for the reaction (replacing a proton by a carboxyl) comes from the energy change involved in oxidizing vitamin K hydroquinone to vitamin K epoxide.

It has been proposed that the vitamin K hydroquinone plus oxygen forms an "active oxygen" capable of withdrawing the $\gamma$-proton from the specific glutamyl residues designated by the carboxylase enzyme and addition of carbon dioxide to the carbanion or radical so formed, followed by epoxidation of vitamin K. It is suggested that, in order for recycling of the enzyme to occur, the epoxide must be released and reduced to the hydroquinone.

Thus, for the overall carboxylation (prothrombin finalization) reaction to occur in the living system, a series of reactions occur; these include reduction of vitamin K to its hydroquinone (vitamin K reductase[17,18]), carboxylation, epoxidation, release and reduction of the epoxide. This chapter will be primarily limited insofar as possible to the present situation with regard to the carboxylation reaction. It will include the following topics.

1. Methods for carrying out *in vitro* solubilized (or particulate) liver microsomal vitamin K-dependent carboxylation of specific glutamyl residues at the $\gamma$-position (i.e., prothrombin formation) from protein precursor
2. Nature of the carboxylase: (*a*) requirements for the reaction; (*b*) extraction of the enzyme by detergents and its sulfhydryl nature; (*c*) reductase and carboxylase; (*d*) comparison of particulate and detergent-solubilized system and enzyme losses during solubilization
3. The polypeptide substrate: (*a*) endogenous; (*b*) exogenous
4. Reaction rate and temperature dependency
5. Carboxylation inhibitors: (*a*) non-vitamin K analogs; (*b*) vitamin K analogs reversible by dithiothreitol; (*c*) vitamin K analogs not reversible by dithiothreitol
6. Hydroquinones that can replace the naturally occurring vitamin K hydroquinones
7. pH optimum
8. Monovalent ion stimulation
9. Subcellular localization of enzyme and of precursor
10. Recent studies on the carboxylation reaction

The vitamin K-dependent carboxylation reaction occurs in the liver and is necessary for the formation of activatable factors, II (prothrombin), VII, IX, and X. In addition, it is presumably required, whenever

[17] C. Martius, and R. Strufe, *Biochem. Z.* **326**, 24 (1954).
[18] R. Wallin, O. Gebhardt, and H. Prydz, *Biochem. J.* **169**, 95 (1978).

γ-carboxyglutamyl (gla) residues are found in proteins—for example, proteins C[19] and S,[20] a bone matrix protein, osteocalcin,[21-24] and, in renal calculi,[25] a kidney protein(s).[26] In addition, gla residues have been reported in certain tumors[27,28] in atheromatous plaques,[29] in certain ribosomal proteins,[30] and in chorioallantoic membrane.[31] The carboxylation reaction has been carried out *in vitro* with kidney microsomes[26] and with bone microsomes[32] as well as liver microsomes and their solubilized preparations, and recently with human placenta.[33] Two-stage prothrombin is produced by cell-free systems,[16,34,35] although the yield is poor, compared to the amount of protein carboxylation.

## Methods

*Vitamin K-Deficient Animals.* Deficiency is produced in white, male Sprague–Dawley rats by feeding them a vitamin K-deficient diet[36] while they are housed in tubular, coprophagy-preventing cages.[37] The diet is a typical synthetic diet in which the "vitamin-free" casein is replaced by soy protein plus methionine. The "deficiency" can be produced also by feeding warfarin at the level of 1.0–5.0 mg/kg body weight, 18 hr before

[19] J. Stenflo, *J. Biol. Chem.* **251**, 355 (1976).
[20] R. G. Discipio, M. A. Hermondson, S. G. Yates, and E. W. Davie, *Biochem. J.* **16**, 698 (1977).
[21] P. V. Hauschka, J. B. Lian, and P. M. Gallop, *Proc. Natl. Acad. Sci. U.S.A.* **72**, 3925 (1975).
[22] J. B. Lian, P. V. Hauschka and P. M. Gallop, *Fed. Proc., Fed. Am. Soc. Exp. Biol.* **37**, 2615 (1978).
[23] P. A. Price, J. W. Poser, and N. Raman, *Proc. Natl. Acad. Sci. U.S.A.* **73**, 3374 (1976).
[24] K. King, *Biochim. Biophys. Acta* **542**, 542 (1978).
[25] J. B. Lian, E. L. Prien, M. J. Glimcher, and P. M. Gallop, *J. Clin. Invest.* **59**, 1151 (1977).
[26] P. V. Hauschka, P. A. Friedman, H. P. Traverso, and P. M. Gallop, *Biochem. Biophys. Res. Commun.* **71**, 1207 (1976).
[27] R. Delaney, P. N. Gray, and B. C. Johnson, *Proc. South-West Oncology Group,* San Antonio, Texas, Nov. 7 (1978).
[28] R. Delaney and P. N. Gray, *Fed. Proc., Fed. Am. Soc. Exp. Biol.* **38**, 710 (1979).
[29] J. B. Lian, M. Skinner, M. J. Glimcher, and P. M. Gallop, *Biochem. Biophys. Res. Commun.* **73**, 349 (1976).
[30] J. J. Van Buskirk and W. M. Kirsch, *Biochem. Biophys. Res. Commun.* **80**, 1033 (1978).
[31] R. Tuan, W. A. Scott, and Z. A. Cohn, *J. Cell Biol.* **77**, 752 (1978).
[32] J. B. Lian and P. A. Friedman, *J. Biol. Chem.* **253**, 6623 (1978).
[33] P. A. Friedman, P. V. Hauschka, M. A. Shia, and J. K. Wallace, *Biochim. Biophys. Acta* **583**, 261 (1979).
[34] C. Vermeer, B. A. M. Soute, J. Govers-Riemslag, and H. C. Hemker, *Biochim. Biophys. Acta.* **444**, 926 (1976).
[35] J. Lowenthal and V. Jaeger, *Biochem. Biophys. Res. Commun.* **74**, 25 (1977).
[36] M. S. Mameesh, and B. C. Johnson, *Proc. Soc. Exp. Biol. Med.* **101**, 467 (1959).
[37] Y. C. Metta, L. Nash, and B. C. Johnson, *J. Nutr.* **74**, 473 (1961).

killing. Livers from warfarin-treated rats, while not as satisfactory as those from deficient rats, can be used for carboxylation studies.

*Preparation of Vitamin K-Deficient Rat Liver Microsomes.* Microsomal suspensions are prepared from the livers of vitamin K-deficient rats (prothrombin time greater than 60 sec by the one-stage assay of Quick[38]). The livers are excised and homogenized in twice their volume of 0.2 $M$ sucrose, containing 1 m$M$ benzamidine-HCl in 25 m$M$ imidazole buffer, pH 7.2. The homogenates are centrifuged for 10 min at 10,000 $g$ at 3–4° to remove nuclei, mitochondria, and cell debris. The supernatants are then centrifuged for 1 hr at 105,000 $g$ at 3–4°. The microsome pellet obtained is resuspended in half times the original volume of the above buffer, and centrifuged for 1 hr at 105,000 $g$ at 3–4°. The pellet is resuspended in one-fifth the original volume of the above buffer.[39]

*Preparation of the Soluble System.* The soluble system is prepared by diluting 1 ml of the vitamin K-deficient microsomal suspension with 0.5 ml of a solution containing 7.0% Triton X-100, 800 m$M$ NaCl, and 8 m$M$ benzamidine-HCl, in 25 m$M$ imidazole buffer, pH 7.2 (50 m$M$ potassium phosphate buffer, pH 7.4, is equally satisfactory). After centrifugation at 105,000 $g$ for 1 hr, the supernatant is collected; it contains the proteins necessary for the carboxylation reaction.[39]

*Carboxylation Assay Procedure.* The carboxylation assay is done in capped tubes with 0.15 ml of solubilized microsomes derived from 0.25 $g$ of liver per tube. The final assay volume is 0.2 ml (the method can be scaled down to 0.1 ml satisfactorily) and contains 5 mCi NaH$^{14}$CO$_3$ (specific activity 40–60 mCi/mmol), and 0.11 m$M$ vitamin $K_1$ hydroquinone. When the quinone form of vitamin $K_1$ is used, NADH is added at 1 m$M$ and dithiothreitol at 2 m$M$ (final concentrations). All additions are made in 25 m$M$ imidazole buffer, pH 7.2. Where solubility demands, the vitamin K compound may be added in 1–5 $\mu$l of pyridine or ethanol, which do not affect the carboxylation reaction. When menadione or other 2-methylnaphthoquinones, unsubstituted in the 3-position, are used, then the dithiothreitol (or other sulfhydryl compound) must be omitted from the incubation mixture, since such sulfhydryl compounds will form active compounds by addition at the open 3-position of the menadione. The *in vitro* carboxylation plateaus in about 2 min at 37° due to inactivation at this temperature. Maximum carboxylation is obtained at 20–25° and is complete in less than 5 min. At 0°, carboxylation continues linearly for about 120 min.

After incubation, the reaction is stopped by the addition of an equal volume of 10% trichloroacetic acid. The trichloroacetic acid pellet is

[38] A. L. Quick, *Am. J. Clin. Pathol.* **15**, 560 (1945).
[39] D. O. Mack, M. Wolfensberger, J. M. Girardot, J. Miller, and B. C. Johnson, *J. Biol. Chem.* **254**, 2656 (1979).

twice dissolved in 2% sodium carbonate and reprecipitated with 10% trichloroacetic acid, before a final 5% trichloroacetic acid wash. The pellet is dissolved in 1 ml of Hyamine hydroxide solution, then transferred with 14 ml of Aquasol or Handifluor to a liquid scintillation vial and counted. All incubations and assays are done in duplicate.

*Preparation of Hydroquinones.* The hydroquinone of vitamin $K_1$ and related quinones made by treatment with sodium dithionite is carried out as described by Fieser.[40] Hydroquinones are made also by sodium borohydride reduction as follows. The 1,4-naphthoquinone (20–50 mmol) is dissolved in 500 ml of ethanol and freed of oxygen (by alternately evacuating and adding nitrogen) in 1-ml hypovials. A solution of sodium borohydride in ethanol (20–50 ml) is added to the quinone solution. The mixture is kept at room temperature in the dark for 30 min and then heated in a boiling water bath until it becomes colorless. The hydroquinone solutions are stored under nitrogen at $-20°$.

*Assay of Prothrombin Formation during the in Vitro Carboxylation Incubation.* The amount of activatable prothrombin which has been formed in this *in vitro* incubation can be determined. The products of the carboxylation are adsorbed on $BaSO_4$. After washing of the $BaSO_4$, the prothrombin is eluted with 0.17 $N$ sodium citrate and assayed by the two-stage assay method[41] and counted by liquid scintillation as before. The prothrombin formed can also be precipitated by antibody to rat prothrombin and counted.[16,35] Vermeer *et al.*[34] indicated that liver microsomes contain prothrombin inhibitor (thrombin inhibitor),[42] which can be removed by fractionation of microsomes on DEAE-cellulose, then optimum prothrombin formation (by clotting assay) can be obtained in a 60-min incubation.

### Nature of the Carboxylase

*Requirements for the in Vitro Carboxylation Reaction.* The protein requirements of the system are supplied by the microsomes of the vitamin K-deficient rat. These consist of at least a carboxylase enzyme and an endogenous substrate, since the reaction is catalyzed by vitamin K hydroquinone, and a vitamin K reductase, since the reaction is also catalyzed by vitamin K plus NADH. However, the major part of the reductase actually occurs in the microsomal supernatant. In either case, the reaction requires molecular oxygen[12,16] and carbon dioxide.[43] There is no supernatant frac-

[40] L. F. Fieser, *J. Biol. Chem.* **133**, 391 (1940).
[41] A. G. Ware and W. H. Seegers, *Am. J. Clin. Pathol.* **19**, 471 (1949).
[42] B. C. Johnson, *in* "The Fat-Soluble Vitamins" (H. F. DeLuca and J. W. Suttie, eds.), p. 491. Univ. of Wisconsin Press, Madison, 1969.
[43] J. P. Jones, E. J. Gardner, T. C. Cooper, and R. E. Olson, *J. Biol. Chem.* **252**, 7738 (1977).

tion, ATP, or biotin requirement for the carboxylation in either this crude microsomal system or the solubilized microsomal system. The oxygen requirement is not replaceable by the oxidizing agent, FAD, and appears to be for molecular oxygen.

Microsomal supernatant may be necessary for the production of a good yield of two-stage prothrombin activity.[34,35,44]

*Extraction of the Enzyme from Microsomes and Its Sulfhydryl Nature.* The enzyme–endogenous substrate "complex" can be extracted from vitamin K-deficient rat liver microsomes by Triton X-100, Lubrol PX, sodium deoxycholate, etc. All these detergents somewhat inhibit the reaction when added to the microsomes, but nonetheless, highly active carboxylating micellar solutions that are highly responsive to vitamin K hydroquinone can be made. For example, treatment of the microsomal suspension described above with Triton X-100 at an extraction concentration of 2.0% yields an active system (lower Triton concentrations markedly lower the yield). After centrifugation for 1 hr at 100,000 $g$, the pellet is without carboxylating activity. Owing to the additions, the carboxylation is carried out at 1.75% Triton. The homogenization and all treatments are carried out in the presence of the serine protease blocking agent benzamidine-HCl to prevent loss of carboxylating activity (enzyme and/or substrate).

The carboxylation, by either intact microsomes or soluble system, catalyzed by vitamin K hydroquinone is rapidly and completely blocked by 0.5–2.0 m$M$ $p$-chloromercuribenzoate. This block is reversible by dithiothreitol (2–10 m$M$) (or dithioerythritol), indicating possible SH enzyme involvement in the reaction. Furthermore, the reaction is blocked completely by 2 m$M$ menadione, even in the presence of high (0.55 m$M$) levels of vitamin $K_1$ hydroquinone. Menadione is a reactive SH reagent by nucleophilic addition at the open 3-position.[45] This block is also reversible by dithiothreitol, which forms an adduct with menadione and restores the SH of the enzyme.

*Reductase and Carboxylase.* Before further discussion of the sulfhydryl nature of the enzyme, it is necessary to consider the two reactions (at least) involved in the catalysis of carboxylation by vitamin K.

The vitamin K quinone reductase and the vitamin K hydroquinone-dependent carboxylase are required to give carboxylation when vitamin K quinone is the initiator.

Vitamin K epoxide does not need to be considered here, since vitamin $K_1$ epoxide plus NADH does not promote carboxylation in the soluble system.

[44] D. V. Shah and J. W. Suttie, *Biochem. Biophys. Res. Commun.* **60**, 1397 (1974).
[45] W. J. Nickerson, G. Falcone, and G. Strauss, *Biochemistry* **2**, 537 (1963).

TABLE I
SULFHYDRYL INHIBITORS OF VITAMIN K-DEPENDENT CARBOXYLATION

| Sulfhydryl reagent | Level used (mM) | % Carboxylation inhibition | | |
| | | Inhibition when initiated by vitamin K[a] + NADH (%) | Inhibition when initiated by vitamin K hydroquinone[a] (%) | Inhibition reversed by dithiothreitol (%) |
| --- | --- | --- | --- | --- |
| None | — | 0 | 0 | — |
| N-Ethylmaleimide[b] | 10 | 92 | 50 | 0 |
| Iodoacetamide[b] | 20 | 42 | 0 | 0 |
| Pcmb[c,d] | 1 | 100 | 100 | 90 |
| Menadione[d] | 0.29 | 83 | 99 | 80 |

[a] Vitamin K and KH$_2$ used at 0.11 mM for experiments 3 and 4; for experiments 1 and 2, 0.55 mM was used.
[b] Preincubation at 0° for 60 min.
[c] p-Chloromercuribenzoate.
[d] No preincubation.

TABLE II

COMPARISON OF PARTICULATE MICROSOMAL AND TRITON-SOLUBILIZED MICROSOME
CARBOXYLATION SYSTEMS

| Reaction initiated by | Level (mM) | Percent carboxylation obtained | |
|---|---|---|---|
| | | Microsomes | Soluble systems |
| Vitamin K + NADH[a] + dithiothreitol[a] | 0.022 | 100 | 100 |
| Vitamin K + NADH[a] | 0.022 | 80 | 80 |
| Vitamin K + dithiothreitol | 0.022 | 140 | 0[a] |
| Vitamin K hydroquinone | 0.022 | 150 | 190 |
| Vitamin K epoxide + NADH[a] + dithiothreitol[a] | 0.100 | 80 | 0 |

| Reaction[b] blocked by | | Percent inhibition | |
|---|---|---|---|
| p-Chloromercuribenzoate | 1.000 | 100 | 100 |
| Tetrachloropyridinol | 0.230 | 94 | 99 |
| Lapachol[c] | 5.500 | — | 90[d] |
| Warfarin | 0.160 | 94 | 0 |

[a] NADH, 1 mM; dithiothreitol, 2 mM.

[b] Reaction initiated by vitamin K quinone (0.11 mM) plus NADH.

[c] 2-Hydroxy-3(3-methyl-2-butenyl)-1,4-naphthoquinone.

[d] Reaction initiated by 0.55 mM vitamin $K_1$ hydroquinone.

Both the vitamin K reductase and the vitamin K hydroquinone-dependent carboxylase are apparently sulfhydryl enzymes, and the relative effect of several sulfhydryl reagents on these two reactions is shown in Table I.

The inhibition observed when vitamin $K_1$ hydroquinone is used to initiate carboxylation demonstrates the presence of a required sulfhydryl group on the vitamin K hydroquinone-dependent carboxylase. The existence of conditions whereby the vitamin K plus NADH-initiated carboxylation is inhibited to a greater extent by SH reagent than the vitamin $K_1$ hydroquinone-initiated carboxylation, further substantiates the already known sulfhydryl nature of the vitamin K reductase.

Of a large number of reducing compounds tried in the reductase system of the microsome, only NADH, NADPH, dithiothreitol, dithioerythritol, and sodium borohydride have been found to be active. Reduced glutathione and $\beta$-mercaptoethanol have 10% of the activity of 2 mM dithiothreitol, when used at 4 mM. Ascorbic acid, hydroquinone, sodium bisulfite, sodium dithionite, and reduced coenzyme A do not replace NADH in the reduction of vitamin K quinone.

*Comparison of Particulate Microsomal and Detergent-Solubilized Carboxylation Systems.* While both systems carry out very satisfactory carboxylation, it is apparent from the data in Table II, that some enzymic components of the overall system have been lost by solubilization. For example, vitamin K epoxide functions satisfactorily in the particulate system, but it appears that the enzyme required to release and/or recycle it has been lost with the detergent treatment. It also appears that the step in the carboxylation blocked by warfarin has been largely lost by the solubilization. These results suggest that the step lost in the preparation of the soluble system is vitamin K epoxide reduction. Matschiner *et al.*[46] and Ren *et al.*[47] have shown the blocking of this step by warfarin *in vivo* and *in vitro*. The inability to recycle the small amount of vitamin K normally in the liver from vitamin K epoxide may be the reason for the anticoagulant action of warfarin.

Another characteristic of the particulate microsome system is the ability of dithiothreitol to completely replace NADH. In this system, NADH is in fact somewhat inhibitory, when added on top of dithiothreitol.[11] In the soluble system, on the other hand, both NADH and dithiothreitol are required for optimum carboxylation if the reaction is initiated by vitamin K quinone.

Table II indicates that the reductase step is the limiting step when carboxylation is initiated by vitamin K quinone plus NADH.

### The Polypeptide Substrate

*The Endogenous Polypeptide Substrate.* Preprothrombin (PIVKA) was reported as early as 1963 by Hemker *et al.*[48,49] Nilehn and Ganrot[50] found a protein in the circulating plasma of the warfarin-treated bovine which, while not activatable to prothrombin by the two-stage assay, cross-reacted with bovine prothrombin antibody. Grant and Suttie[51] have shown that in rat liver there are two such preprothrombins with different isoelectric points pI 7.2 and pI 5.8. Preprothrombin has been isolated and shown to be noncarboxylated prothrombin. However, these isolated preprothrombins do not serve as substrates for the described *in vitro* vitamin K-dependent carboxylation. Similarly, isolated rat prothrombin, decarboxylated by acid-heat, does not serve as a substrate. It appears plausible that the carboxylation reaction takes place in the lipid bilayer, as the

[46] J. T. Matschiner, A. Zimmerman, and R. G. Bell, *Thromb. Diath. Haemorrh. Suppl.* **52,** 45 (1974).
[47] P. Ren, P. Y. Stark, R. L. Johnson, and R. G. Bell, *J. Pharmacol. Exp. Ther.* **201,** 541 (1977).
[48] H. C. Hemker, J. J. Veltkamp, A. Hensen, and E. A. Loeliger, *Nature (London)* **200,** 589 (1963).
[49] M. J. Lindhout, B. H. M. Kop-Klaassen, P. P. M. Reekers, and H. C. Hemker, *J. Mol. Med.* **1,** 223 (1976).
[50] J. E. Nilehn and P. O. Ganrot, *Scand. J. Clin. Lab. Invest.* **22,** 17 (1968).
[51] G. A. Grant and J. W. Suttie, *Biochemistry* **15,** 5387 (1976).

preprothrombin comes off the polysome with the signal peptide still attached, and while still in a highly hydrophobic environment.

*Exogenous Substrates.* Suttie *et al.*[52] have synthesized pentapeptides of the sequences Phe-Leu-Glu-Glu-Val, Phe-Leu-Glu-Glu-Leu, and Phe-Leu-Glu-Glu-Ile, which will serve as exogenous substrates for carboxylation by the soluble system. These three pentapeptides possess the sequences from amino acid 5 to 9 that occur in bovine, human, and rat prothrombin, respectively. These pentapeptide substrates give initial rates of carboxylation about 25 times slower than carboxylation of endogenous protein. The $K_m$ value for pentapeptide is many times higher than any possible $K_m$ for protein carboxylation.

## Reaction Rate and Temperature Dependency

When $^{14}CO_2$ incorporation into protein (i.e., carboxylation) is carried out *in vitro* as described under Methods, at 37°, the extent of carboxylation reaches a plateau in about 5 min if vitamin K quinone plus NADH plus dithiothreitol are used to initiate the reaction in the soluble system. If the hydroquinone initiates the reaction, the plateau is reached in not over 2 min at 37°. At this temperature, the carboxylation system is essentially dead after these first few minutes. On the other hand, the reaction will proceed, albeit at a slower rate, even at 0°.[16,53] If the reaction is carried out at 0° for 2 hr, the plateau reached is higher than that obtainable at 37°, because the system remains alive and functioning throughout the incubation. Maximum carboxylation, over twice that at 37°, occurs at 20–25°. It can be concluded that there is a highly temperature-sensitive component(s) of the carboxylase reaction.

## Carboxylation Inhibitors

In the discussion of inhibitors and vitamin K active compounds, only carboxylation (i.e., the reaction initiated by hydroquinone in the soluble system) will be considered, although there are data on the vitamin K reductase and on the vitamin K epoxide reduction-recycling.

*Non-Vitamin K Analogs.* The carboxylation is not inhibited by cyanide, azide, the cytochrome P-450 blocking agents (e.g., CO, SKF 525A, aminoglutethimide, 7,8-benzoflavone, metyrapone), $H_2O_2$ ($H_2O_2$ + azide to block catalase), dithionite (to 10 m$M$), bisulfite, ATPase, Atabrine, Avidin, EDTA, in either the particulate or soluble microsomal system,

[52] J. W. Suttie, J. M. Hageman, S. R. Lehrman, and D. H. Rich, *J. Biol. Chem.* **251**, 5827 (1976).

[53] J. P. Jones, A. Fausto, R. M. Houser, E. J. Gardner, and R. E. Olson, *Biochem. Biophys. Res. Commun.* **72**, 589 (1976).

initiated by either vitamin K quinone plus NADH plus dithiothreitol or by vitamin K hydroquinone.

The carboxylase (vitamin K hydroquinone-initiated reaction) is inhibited by a high level of dithionite (93% at 100 m$M$), by tetrachloropyridinol (86% at 0.11 m$M$), and by Chloro-K. It is also inhibited by the spin-trapping agent 5,5-dimethyl-1-pyrroline $N$-oxide at high concentrations (77% at 72 m$M$).

*Vitamin K Analogs.* The soluble system carboxylation is not inhibited by phytol, warfarin, and menadiol. However, a number of vitamin K analogs do inhibit the carboxylation. There appear to be two types of such blocking: (*a*) those compounds that react with enzyme SH; and (*b*) those compounds that appear to be competitive for the vitamin K binding site on the carboxylase.

*SH Enzyme Inhibitors.* These inhibitors are all reversed by the addition of dithiothreitol to the reaction mixture. They include the vitamin K analogs, menadione (99% inhibition at 0.29 m$M$), 2,3-dichloro-1,4-naphthoquinone (100% at 0.3 m$M$), 5-ethoxymenadione (96.5% at 0.3 m$M$), and 1,4-naphthoquinone (98% at 0.3 m$M$), vitamin K-SII, and other 3-thioether derivatives containing charged side chains (e.g., menadione-3-thioethylamine, and menadione-3-cysteine). SH-inhibitors also include *p*-chloromercuribenzoate (100% at 0.75 m$M$), 5,5'-dithiobis(2-nitrobenzoate) (80% at 10 m$M$), $N$-ethyl maleimide (50% at 10 m$M$), and to a much smaller extent iodoacetate and iodoacetamide. Several nitroso compounds inhibit carboxylation over 90% at 10 m$M$.

*Structural Analogs Not Reversible by Dithriothreitol.* Lapachol [2-hydroxy-3-(3-methyl-2-butenyl)-1,4-naphthoquinone] (90% at 5.5 m$M$), plumbagin (5-hydroxymenadione) (100% at 0.3 m$M$), and duroquinone (2,3,4,6-tetramethylbenzoquinone) block vitamin K hydroquinone-initiated carboxylation in the soluble system, presumably by oxidation of the hydroquinone. No hydroquinone tried has completely blocked carboxylation initiated by vitamin K hydroquinone at 0.55 m$M$. The most active hydroquinone "competitive" inhibitor found so far is 2-methoxy-1,4-naphthohydroquinone (61% inhibition at 1.5 m$M$), followed by lapachol hydroquinone (48% inhibition at 1.5 m$M$). These results may indicate competition for an active site.

### Hydroquinones That Can Replace Vitamin K$_1$ Hydroquinone in the Carboxylation Reaction in the Soluble System

A number of well-known K vitamins have been compared,[14,16,53] and the "naturally" occurring K vitamins, with a methyl in the 2-position and a polyisoprene in the 3-position, have all been shown to have varying degrees of *in vitro* activity. The activity comparisons were carried out differently, and only in the case of Friedman and Shia[14] are the relative

TABLE III

VITAMIN K ANALOGS WITH NONISOPRENOID SIDE CHAINS IN THE 3-POSITION THAT HAVE
SIGNIFICANT ACTIVITY IN THE *in Vitro* CELL-FREE SOLUBILIZED-MICROSOME
CARBOXYLATION SYSTEMS[a]

| Side chain at the 3-position of 2-methyl-1,4-naphthohydroquinone | Relative carboxylation activity compared to vitamin $K_1$ hydroquinone set at 100% |
|---|---|
| Dithiothreitol | 150 |
| Dithioerythritol | 220 |
| 1,4-Butanedithiol | 146 |
| Thiomethyl | 92 |
| Thioethyl | 100 |
| 1-Thiobutyl | 64 |
| 3-Thiopropane-1,2-diol | 64 |
| Thiobenzyl | 48 |
| Thiophenyl | 15 |
| Thiopropionic acid methyl ester | 32 |
| Thioacetic acid ethyl ester | 25 |
| Methoxy | 77 |
| Ethoxy | 60 |
| *n*-Propoxy | 32 |
| Pentadecyl | 25 |

[a] The concentration for vitamin $K_1$ hydroquinone and the 3-thioethers is 0.11 m$M$, for the 3-oxygen ethers and the 3-pentadecyl derivative, it is 0.55 m$M$. In the case of the thioether compounds, an additional amount of the thiol listed on each line was added to a level of 5 m$M$ in order to provide the reducing capacity usually provided by dithiothreitol.

activities reported in terms of amount required to give the same amount of carboxylation as vitamin $K_1$. Thus they find that MK-3 is 80 times as active as vitamin $K_1$ (i.e., it required 1/80th as much MK-3 as $K_1$ for the same amount of carboxylation). On the other hand, the relative activities reported by Jones *et al.*[53] (and also those reported in Table III) are based on amounts of carboxylation obtained when the active compound is added at a plateau level for vitamin $K_1$. Thus, Jones *et al.* found essentially equal activities at 37° for vitamin $K_1$ and MK-2 through MK-6. While these compounds were studied as the quinones, the fact that vitamin K-dependent carboxylation occurred, indicates that they were reduced to their respective functional hydroquinones.

Conversion of the $\beta$-$\gamma$ double bond of the phytyl side chain from trans to cis,[14] or hydrogenation of this bond,[54] very markedly reduced the carboxylating activity of vitamin $K_1$ quinone plus NADH and dithiothreitol. However, a considerable number of nonisoprenoid derivatives of

[54] J. M. Girardot, R. Delaney, and B. C. Johnson, *SW-ACS Meet., 30th Proc., Abst.* **130** (1976).

2-methyl-1,4-naphthohydroquinone have activity.[39] Those found to function in carboxylation are given in Table III.

Not all menadione adducts, however, function in the carboxylation reaction. The inactive compounds include, in addition to menadiol and duroquinol, phthiocol, 2-ethoxy-3-monoisoprene-1,4-naphthohydroquinone (lapachol hydroquinone). Also inactive are the 3-position adducts of menadione, with cysteine, with reduced glutathione, with reduced coenzyme A, with 3-thiopropionic acid, with aminoethanethiol, with butyl aminoethanethiol, and with dimethyl aminoethanethiol.

From these data, it appears that a side chain is necessary in the 3-position [perhaps to prevent reaction of the oxidized (quinone) form with the enzyme SH] and that, while it is not necessary that this side chain be isoprenoid or even hydrophobic, it cannot contain charged groups (carboxyl and/or amine). It also appears that the 2-methyl position is required for significant function, even though the reaction apparently does not go via a chromanol methine phosphate, the chromanols being inactive[16] and the methyl group not exchanging a proton with the medium.[15,55]

While all the compounds of Table III are active in carboxylation, their activities toward the reductase (i.e., when used as the quinones) are not in the same ratio. In particular, the 2,3-pentadecyl derivative of menadione was much less active for the reductase.

## pH Optimum

The pH optimum of the carboxylase is between 7.0 and 7.6.[16]

## Salt Effect, Monovalent Ion Stimulation

The optimum monovalent ion (Na or K) level is approximately 150 mM and can be supplied as chloride or phosphate.

## Subcellular Localization of Enzyme and Precursor

The localization of the carboxylation complex within the microsome has been studied by Helgeland.[56] They concluded that the vitamin K-dependent carboxylase is localized in the rough endoplasmic reticulum, mainly on the luminal side, whereas the precursor substrate appears to be in both the luminal and membrane fraction.

## Incomplete Carboxylation

Is the endogenous precursor protein carboxylated in any sequence? Can partial carboxylation occur? These are two of many questions that

[55] J. M. Girardot, D. O. Mack, J. Price, E. Suen, and B. C. Johnson, *Fed. Proc., Fed. Am. Soc. Exp. Biol.* **36**, Abst. 190 (1977).
[56] L. Helgeland, *Biochim. Biophys. Acta* **499**, 181 (1977).

remain to be answered [in addition to the problems of isolation of functional carboxylase as a pure enzyme(s) and of the endogenous precursor in carboxylatable form]. A partial answer to the first question may be contained in the recent paper of Esnouf and Prowse,[57] who have isolated forms of prothrombin induced by warfarin therapy, which have reduced biological activity. One of these prothrombins contains on the average seven out of the possible ten γ-carboxyglutamyl residues and the other, only four. This indicates that the carboxylation of the precursor polypeptide is a sequential series of carboxylations.

### Effect of Pyridoxal Phosphate on Carboxylation

Added pyridoxal phosphate increases the initial rate of carboxylation of pentapeptide carboxylation about 4-fold,[58,59] although not affecting endogenous carboxylation. The effect is due to reaction of the pyridoxal phosphate with a protein of the carboxylation system, not to reaction of the cofactor with the pentapeptide.[59]

### Effect of Decarboxylated Protein on the Carboxylation Reaction

Endogenous substrate protein and partially decarboxylated prothrombin have been found to increase the activity of the pentapeptide carboxylation activity system although having little effect on the much more rapid endogenous protein carboxylation.[60] The data indicate that one of the reasons for the low carboxylation ability of vitamin K-treated animals is their lack of liver endogenous protein substrate.

### Carboxylation Enzyme Purification

While little real success has been had to date in isolating the pure component proteins of the system, the solubilized microsomes from the deficient rat have been fractionated into two inactive parts which when recombined restore activity.[61] Only one of these components has been shown to contain essential sulfhydryl.

[57] M. P. Esnouf and C. V. Prowse, *Biochim. Biophys. Acta* **490**, 47 (1977).
[58] J. W. Suttie, S. R. Lehrman, L. D. Geweke, J. M. Hageman, and D. H. Rich, *Biochem. Biophys. Res. Commun.* **86**, 500 (1979).
[59] A. Dubin, E. T. Suen, R. Delaney, A. Chiu, and B. C. Johnson, *Biochem. Biophys. Res. Commun.* **88**, 1024 (1979).
[60] A. Dubin, E. T. Suen, A. Chiu, R. Delaney, and B. C. Johnson, *in* "Vitamin K Metabolism and Vitamin K-Dependent Proteins" (J. W. Suttie, ed.), p. 484. Univ. Park Press, Baltimore, 1979.
[61] J. A. Price and B. C. Johnson, *in* "Vitamin K Metabolism and Vitamin K-Dependent Proteins" (J. W. Suttie, ed.), p. 500. Univ. Park Press, Baltimore, 1979.

Recent Reviews

A number of reviews of the vitamin K-dependent carboxylation system have been published.[62-70]

Acknowlegment

The original work reported here was supported, in part, by NIH Grant No. HL-17619. The work was carried out by Dr. D. O. Mack, Dr. J. M. Girardot, Dr. Max Wolfensberger, Dr. Adam Dubin, Dr. Eric Suen, Dr. Joy A. Price, Ms. Julie Miller, Ms. Vicki Bartels, Ms. Betty Wiseman, and Ms. Julia Watson with the constant advice and consultation of Dr. Robert Delaney and frequently also of Dr. Charles Esmon.

[62] S. Magnusson, T. E. Peterson, L. Sottrup-Jensen, and H. Claey, *Cold Spring Harbor Symp. Quant. Biol.* **39**, 123 (1975).
[63] E. W. Davie and K. Fujikawa, *Annu. Rev. Biochem.* **44**, 799 (1975).
[64] J. Stenflo and J. W. Suttie, *Annu. Rev. Biochem.* **46**, 157 (1977).
[65] H. Prydz, *Sem. Thromb.* **4**, 1 (1977).
[66] C. M. Jackson and J. W. Suttie, *Prog. Haematol.* **10**, 333 (1977).
[67] R. E. Olson and J. W. Suttie, *Vitam. Horm.* (*New York*) **35**, 59 (1977).
[68] B. C. Johnson, CRC Handbook Series in Nutrition and Food. In press.
[69] J. Stenflo, *Adv. Enzymol.* **46**, 1 (1978).
[70] "Vitamin K Metabolism and Vitamin K-Dependent Proteins (J. W. Suttie, ed.), (Proc. of 8th Steenbock Symp., Madison, Wis. 1979). Univ. Park Press, Baltimore, 1979.

# [23] Microsomal Vitamin K-Dependent Carboxylase

*By* J. W. Suttie, L. M. Canfield, and D. V. Shah

$$\text{Phe-Leu-Glu-Glu-Leu} + CO_2 \xrightarrow{\text{vitamin } KH_2 \text{ and } O_2} \text{Phe-Leu-Gla-Glu-Leu}$$

The physiological substrates for this enzyme are the liver microsomal precursor(s) of each of the vitamin K-dependent plasma proteins. Incorporation of $CO_2$ into Glu residues of these endogenous substrates to form γ-carboxyGlu (Gla) residues can be followed as a vitamin K-dependent fixation of $^{14}CO_2$ into trichloroacetic acid-precipitable protein. Various low-molecular-weight peptides containing Glu-Glu sequences are also substrates,[1-3] and enzyme activity assayed with peptides as substrates

[1] J. W. Suttie, J. M. Hageman, S. R. Lehrman, and D. H. Rich, *J. Biol. Chem.* **251**, 5827 (1976).
[2] R. M. Houser, D. J. Carey, K. M. Dus, G. R. Marshall, and R. E. Olson, *FEBS Lett.* **75**, 226 (1977).
[3] J. W. Suttie, S. R. Lehrman, L. O. Geweke, J. M. Hageman, and D. H. Rich, *Biochim. Biophys. Acta* **86**, 500 (1979).

can be followed by measuring the vitamin K-dependent incorporation of $^{14}CO_2$ into nonvolatile, non-trichloroacetic acid-precipitable radioactivity.

## Assay Method

The standard assay is run at 17° in 13 × 100 mm glass tubes containing 0.4 ml of solubilized microsomal preparation (see below); 50 $\mu$l of 5.0 m$M$ Phe-Leu-Glu-Glu-Leu[4] in buffer B (see below); 10 $\mu$l of 1.0 mCi/ml NaH$^{14}CO_3$; and 10 $\mu$l of 5 mg/ml vitamin $K_1$ hydroquinone in ethanol.

The reaction is started by the addition of vitamin K; the tops of the tubes are covered with parafilm, and they are incubated for 30 min in a rotary shaker. To stop the reaction, a 0.2-ml aliquot of the incubation mixture is precipitated with 1 ml of 10% trichloroacetic acid and let sit on ice for 10 min. The precipitate is centrifuged down, and the supernatant is decanted into a second tube and gassed with $CO_2$ for 4 min at room temperature. A 0.4-ml aliquot of the gassed supernatant is added to a minivial, mixed with 4.3 ml of Aquasol, and radioactivity is determined in a liquid scintillation spectrometer. It has been shown that the vitamin K-dependent nonvolatile radioactivity measured is predominately present as Gla,[1] and that most of the product is carboxylated in only the first of the two Glu residues of the substrate.[5]

## Preparation of Microsomal Enzyme

Male, 200–300-g vitamin K-deficient rats which have been fasted overnight are decapitated; the livers are quickly removed, minced, and homogenized in two parts (w/v) of ice-cold 0.25 $M$ sucrose–25 m$M$ imidazole pH 7.2 buffer (buffer A). A postmitochondrial supernatant is obtained by centrifugation of the homogenate at 10,000 $g$ for 10 min. Microsomes are prepared from the postmitochondrial supernatant by centrifugation at 105,000 $g$ for 60 min in a Beckman Model L-2 ultracentrifuge. The microsomal pellet is surface-washed twice with equivalent volumes of buffer B (0.25 $M$ sucrose–25 m$M$ imidazole–0.5 $M$ KCl–1.0 m$M$ dithiothreitol) and then resuspended with 8 strokes of a loose-fitting Dounce homogenizer (Kontes, type A pestle) in sufficient buffer B containing 1.5% Triton X-100 to give a final volume equal to that of the original postmitochondrial supernatant. This solution is recentrifuged at 105,000 $g$ for 60 min to remove a small amount of insoluble material and used in the assay.

[4] Available from Vega Biochemicals, Tucson, Arizona.
[5] J. L. Finnan and J. W. Suttie, *Fed. Proc., Fed. Am. Soc. Exp. Biol.* **38**, 376 (1979).

TABLE I

EFFECT OF HYPOPROTHROMBINEMIA AND SPECIES ON LIVER MICROSOMAL VITAMIN
K-DEPENDENT CARBOXYLASE ACTIVITY[a]

| | | Hypoprothrombinemic | |
| Species | Normal | Vitamin K deficient | Anticoagulant treated |
| --- | --- | --- | --- |
| Rat | 100 | 350 | 330 |
| Mouse | 65 | — | 165 |
| Rabbit | 55 | — | 170 |
| Hamster | 365 | 860 | — |
| Chick | 8 | — | 13 |

[a] All values are expressed relative to the activity in normal rat liver as assayed at 27° with 0.5 m$M$ Phe-Leu-Glu-Glu-Leu as a substrate. The dose and duration of anticoagulant treatment depend on the species, but in all cases were sufficient to reduce plasma prothrombin concentrations to <20% of normal.

## Variations in Enzyme Activity

Although liver from normal rats contains easily detectable carboxylase activity, this activity is greatly enhanced[6] when livers are obtained from rats made hypoprothrombinemic by the production[7,8] of a vitamin K deficiency (7 days) or by the intraperitoneal administration of Warfarin, 5 mg/kg, for 18 hr (Table I). The activity of the enzyme, as measured with Phe-Leu-Glu-Glu-Leu under the same conditions as are optimal for the rat liver enzyme, also varies significantly (Table I) when livers are obtained from different species.[9]

## Properties of the Enzyme

The enzyme is sensitive to high temperatures (Fig. 1) and is rapidly inactivated above 20°. Meaningful rates of carboxylation can therefore be measured only at 17° or below. Half-maximal activity of the enzyme is achieved at about 20 $\mu$g of vitamin KH$_2$ per milliliter of incubation; at the standard incubation conditions of nearly 100 $\mu$g/ml, the enzyme is saturated with respect to this substrate.[3] The O$_2$ dependence has not been studied in detail, but activity is not increased by raising the O$_2$ concentra-

[6] D. V. Shah and J. W. Suttie, Arch. Biochem. Biophys. 191, 571 (1978).
[7] V. C. Metta, L. Nash, and B. C. Johnson, J. Nutr. 74, 473 (1961).
[8] M. S. Mameesh and B. C. Johnson, Proc. Soc. Exp. Biol. Med. 101, 467 (1959).
[9] D. V. Shah and J. W. Suttie, Proc. Soc. Exp. Biol. Med. 161, 498 (1979).

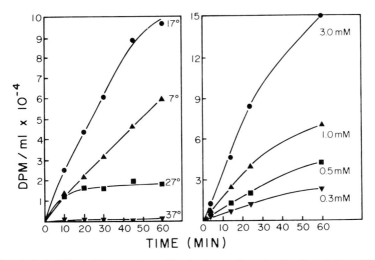

FIG. 1. *Left:* Effect of temperature of incubation on the rate of carboxylation of 0.5 mM Phe-Leu-Glu-Glu-Leu. *Right:* Effect of varying concentration of Phe-Leu-Glu-Glu-Leu on the rate of carboxylation at 17°. The apparent $K_m$ for this substrate is about 4 mM.

tion above air saturation. Under the standard assay conditions, the system is not saturated with respect to the peptide substrate (Fig. 1). However, at the concentration of substrate used (0.5 mM), an insignificant amount (<1%) of the peptide is carboxylated during the assay, and the reaction is linear over the 30-min incubation.

The system is sensitive to much the same inhibitors and conditions of stimulation as is the carboxylation of endogenous protein substrates.[3] The activity of the system is enhanced by the presence of dithiothreitol in the incubation and is stimulated 2- to 3-fold by pyridoxal phosphate in a manner not understood. The coumarin anticoagulant, Warfarin, is a significant inhibitor at relatively high concentrations (>1 mM), whereas the direct vitamin K antagonist, chloro-K (2-chloro-3-phytyl-1,4-naphthoquinone), is a significant inhibitor when present at a ratio of inhibitor to vitamin as low as 1:100. The system is inhibited by sulfhydryl poisons, such as *p*-hydroxymercuribenzoate.

The peptide Phe-Leu-Glu-Glu-Leu is homologous to a region of the bovine prothrombin precursor, and other peptides with Glu-Glu sequences are also active. Phe-Leu-Glu-Glu-Ile, Phe-Leu-Glu-Glu-Val, and Phe-Ala-Glu-Glu-Leu all have significant activity whereas Phe-Gly-Glu-Glu-Leu is essentially inactive.[3] At the present time no small peptides with inhibitory properties are known.

TABLE II
PURIFICATION OF VITAMIN K-DEPENDENT CARBOXYLASE[a]

| Step | Carboxylase activity (U) | Protein (mg) | Specific activity (U/mg) |
|---|---|---|---|
| Microsomes | 5.4 | 22 | 0.25 |
| Octylglucoside extraction | 14.6 | 14 | 1.04 |
| Sodium perchlorate extraction | 204.0 | 10 | 20.40 |
| Cholate extraction | 204.0 | 8 | 25.50 |
| Ammonium sulfate precipitation | 188.0 | 5 | 37.60 |

[a] A unit of carboxylase activity is arbitrarily defined as the fixation of 1000 dpm of $H^{14}CO_3^-$ during the standard assay (0.5 m$M$ Phe-Leu-Glu-Glu-Leu) in 30 min at 17°. The data are from a single experiment, and the amount of protein and carboxylase activity in the first step is that contained in 1 ml of the crude microsomal suspension.

### Purification of Rat Liver Carboxylase

The vitamin K-dependent carboxylase has not been purified to a high degree of homogeneity. It is, however, possible to obtain[10] a lipoprotein complex which is 100- to 150-fold increased in specific activity from the starting crude microsomal pellet by the sequential removal of other proteins from the microsomes.

*Removal of Peripheral Protein.* Microsomes are prepared from vitamin K-deficient rats, surface-washed twice with 1 ml of buffer A, suspended in 2 ml of buffer C (1 m$M$ dithiothreitol–25 m$M$ imidazole, pH 7.2) per gram of liver containing 0.1% octylglucoside and 1 m$M$ PMSF (added as a 0.2 $M$ solution in absolute ethanol), and homogenized with six strokes of a Kontes glass homogenizer. This suspension is shaken gently for 30 min at 4° prior to centrifugation for 60 min at 100,000 $g$. This step results in the loss of about 30% of the original microsomal protein.

*Treatment with Chaotropic Ions.* The pellet obtained after centrifugation is surface-washed twice with 1 ml of buffer C and resuspended with gentle homogenization in buffer C (2 ml per gram of liver) containing 0.25 $M$ NaClO$_4^-$. This suspension is frozen rapidly in Corex centrifuge tubes in a slurry of Dry Ice–methanol for 10 min and thawed in a 30° water bath for 4 min. This procedure is repeated, and resulting suspensions are pelleted by centrifugation as above. This fractionation reduces the total amount of protein to about half that originally in the microsomal preparation. Neither freezing and thawing, nor fractionation with sodium perchlorate

[10] L. M. Canfield, T. A. Sinsky, and J. W. Suttie, *Fed. Proc., Fed. Am. Soc. Exp. Biol.* **38**, 710 (1979).

alone, is effective in removing protein from the membrane at this stage. For optimal release of protein, both freeze–thaw cycles are necessary.

*Solubilization by Bile Salts.* The pellet obtained after $ClO_4^-$ treatment is surface-washed twice with 1 ml of buffer C and suspended by homogenization in buffer C (2 ml per gram of liver) to which solid NaCl and 10% recrystallized sodium cholate are added to final concentrations of 0.75 $M$ NaCl and 0.3% sodium cholate. This mixture is incubated at 4° with gentle shaking for 30 min and pelleted by centrifugation as above. This pellet contained about 35% of the original microsomal protein. Treatment with NaCl or sodium cholate alone is insufficient to remove protein from the membrane at this stage.

*Ammonium Sulfate Fractionation.* The pellet obtained from the bile salt solubilization is surface-washed twice with 1 ml of buffer C and resuspended in 80 ml of buffer C per gram of liver originally used. Saturated ammonium sulfate is slowly added to this solution to give a final concentration of 33% ammonium sulfate, and mixing is continued for 30 min at 4° before the pellet is collected by centrifugation for 10 min at 10,000 $g$. Large volumes of buffer are used during this step to remove the sodium cholate effectively. This pellet is resuspended (2 ml per gram of liver) in 1% Triton X-100, 10% glycerol, 1 m$M$ dithiothreitol, 0.5 $M$ KCl, 25 m$M$ imidazole at pH 7.2 and dialyzed against 100 volumes of this solution with one change overnight. The resultant membrane fraction can be used immediately or stored at −70° until ready for use.

### Properties of Purified Preparation

The preparation obtained by this procedure retains at least 50% of its activity for 6 weeks when stored at −70°. It has about a 4-fold increase in the amount of reduced vitamin K required for half-maximal activity and an apparent $K_m$ for Phe-Leu-Glu-Glu-Leu about half that of the solubilized crude microsomal preparation. The data in Table II indicate that the total carboxylase activity is greatly increased during the preparation, and this appears to be due largely to the removal of an inhibitor of the carboxylase in the crude microsomal preparation. The purified preparation retains about 50% of the vitamin K epoxidase activity of the crude microsomes and about 25% of the cytochrome P-450 activity. A portion of the microsomal precursor pool of various vitamin K-dependent proteins is still present in the preparation, as it does carry out a vitamin K-dependent incorporation of $CO_2$ into proteins. It has, however, lost vitamin K reductase activity, as it is completely dependent on reduced vitamin K rather than [vitamin K + NADH]. Electron microscopy of the preparation reveals a lack of membranous elements and the appearance of an aggregated lipoprotein complex.

Section V

# Vitamin A Group

## [24] Colorimetric Estimation of Vitamin A with Trichloroacetic Acid

*By* R. F. Bayfield and E. R. Cole

It is now over 50 years since the color reaction of vitamin A with strong acids, notably sulfuric acid, was first observed.[1,2] The importance of this test lies in the fact that it provided the basis for a general view of the reaction of vitamin A with Lewis acids, including the Carr–Price antimony trichloride reaction,[3] which has tended to dominate colorimetric methods of estimation from that time. However, several unsatisfactory features of the Carr–Price method, particularly interference by moisture and instability of the color, have prompted continued investigations for alternative reagents. Of these, glycerol dichlorohydrin,[4] trifluoroacetic acid,[5] and trichloroacetic acid[6] have received most prominence, but development of methods has proceeded in an empirical manner and only comparatively recently have mechanistic aspects of these reactions received consideration. It is in this context that the trichloroacetic acid method is now discussed with particular reference to estimation of vitamin A in serum and liver tissue.

### Experimental

*Reagents.* All reagents should be freshly prepared. Exposure of samples and reagents to light must be avoided at all times.

Trichloroacetic acid, saturated solution in chloroform. Trichloroacetic acid crystals (50 g), freshly washed with solvent, are dissolved in alcohol-free distilled chloroform (25 ml). This gives about 50 ml of reagent which is stable for 5 hr with daylight excluded. Optical density readings tend to decrease with storage for longer periods, up to 24 hr. No color is given by vitamin A with reagent exposed to bright diffuse sunlight for several hours.

Vitamin A (U.S.P. Reference capsule vitamin A acetate), strong solu-

[1] J. C. Drummond and A. F. Watson, *Analyst* **47**, 341 (1922).
[2] O. Rosenheim and J. C. Drummond, *Biochem. J.* **19**, 753 (1925).
[3] F. H. Carr and E. A. Price, *Biochem. J.* **20**, 497 (1926).
[4] A. E. Sobel and H. Werbin, *J. Biol. Chem.* **159**, 681 (1945); *Ind. Eng. Chem. Anal. Ed.* **18**, 570 (1946).
[5] R. E. Dugan, N. A. Frigerio, and J. M. Siebert, *Anal. Chem.* **36**, 114 (1964).
[6] R. F. Bayfield, *Anal. Biochem.* **39**, 282 (1971).

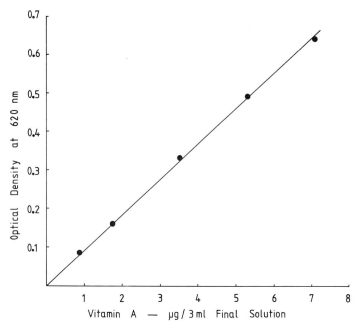

FIG. 1. Relationship between absorbance and vitamin A concentration. From R. F. Bayfield, *Anal. Biochem.* **39**, 282 (1971).

tion.[7] Reference oil (100 mg) is dissolved in chloroform (10 ml). This solution loses about 5% strength after storage for 1 week at $-20°$.

Vitamin A, weak solution. Dilute the strong solution (1 ml) with chloroform (to 50 ml). This solution contains about 7 $\mu$g/ml.

Ethanol and light petroleum (b.p. 60/80°), spectroscopic grade.

*Procedure*

PREPARATION OF STANDARD CURVE. To tubes (1 cm in diameter) add aliquots of the weak solution of vitamin A (0.15–1.0 ml) to obtain a 1–7 $\mu$g range of vitamin. To each tube add chloroform as required to bring the volume to 1 ml. From a fast-delivery pipette, add trichloroacetic acid reagent (2 ml) rapidly, mixing with the vitamin solution. Read the optical density of the solution at 620 nm in a spectrophotometer or colorimeter at *maximum deflection* of the galvanometer spot.

Plot a graph of optical density at maximum deflection vs vitamin A as $\mu$g/3 ml final solution, as illustrated in Fig. 1.

APPLICATION TO SERUM. The serum (15 ml) in a 100-ml stoppered measuring cylinder is shaken with ethanol (15 ml) then with light pe-

---

[7] Alternatively, standard preparations of vitamin A alcohol may be used at equivalent concentrations.

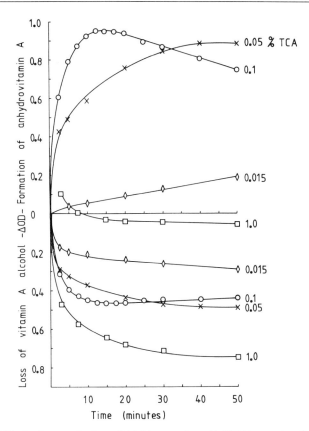

FIG. 2. Effect of concentration of trichloroacetic acid (TCA) on rate of formation of anhydrovitamin A (measured at 377 nm) and loss of vitamin A (measured at 332 nm). Solvent, chloroform.

troleum (30 ml) for 1.5–2 min on a mechanical shaker and allowed to stand to obtain a clear supernatant solution. An aliquot (25 ml) of the light petroleum layer is drawn off, the solvent is removed under vacuum, and the residue is redissolved in chloroform (1.25 ml). This solution (1 ml) is used for color development with trichloroacetic acid reagent as previously described.

APPLICATION TO LIVER TISSUE. The liver (1 g) in a 50-ml beaker with a stout glass rod is finely ground with chromatographic grade silica gel (3–4 g) gradually added, and grinding between additions, until a free-flowing powder is obtained. The mixture is transferred to a 100-ml stoppered measuring cylinder, shaken mechanically for 30 min with acetone/light petroleum, 1 : 1 (50 ml), then allowed to stand 1–1.5 hr to obtain a clear supernatant solution. An aliquot (40 ml) of the solution is drawn off, the solvent is removed under vacuum, and the residue is redissolved in

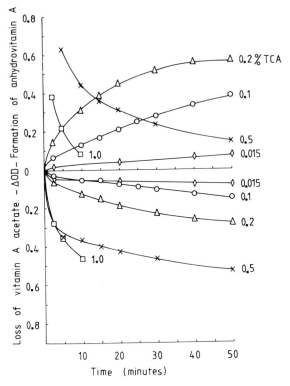

FIG. 3. Effect of concentration of trichloroacetic acid (TCA) on rate of formation of anhydrovitamin A (measured at 377 nm) and loss of vitamin A acetate (measured at 332 nm). Solvent, chloroform.

chloroform (2 ml). The estimation is then completed as described for serum, adjusting the aliquot of chloroform solution used for color development as necessary. In practice it may be an advisable precaution to carry out duplicate extractions of liver tissue if low vitamin A content is suspected.

### Discussion

The colors produced from vitamin A, its acetate and palmitate with the Carr–Price antimony trichloride reagent and trichloroacetic acid reach the same intensity for any given amount of vitamin. However, Fig. 1 shows that color formation is best limited to that from 7 $\mu$g/3 ml final solution for accurate spectrophotometric measurement. The necessity to read at the point of maximum deflection must be emphasized. A fast-measurement

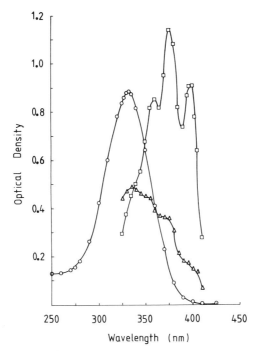

FIG. 4. Effect of concentration of trichloroacetic acid (TCA) on the formation and stability of anhydrovitamin A. O———O, Vitamin A; □———□, anhydrovitamin from 0.1% TCA; △———△, anhydrovitamin from 1% TCA.

accessory and a slave recorder have been described for measurement of the maximum color intensity using the trichloroacetic acid reagent.[8]

Tests with dilute trichloroacetic acid reagent allow dissection of the course of reaction and not only clearly establish that the color must be derived from anhydrovitamin A, but also confirm the sensitivity of vitamin A to dehydration even with acid strengths well below that used in the estimation.

The concentration of trichloroacetic acid controls the rate of formation and stability of anhydrovitamin, which may be correlated with loss of vitamin A or its acetate (Figs. 2 and 3). It is important to observe that change occurs without appearance of blue color in the weaker acid medium. Formation of anhydrovitamin with fine-structure absorption (Fig. 4) follows protonation of vitamin A, and the slower formation from vitamin A acetate reflects differences in the rates of dehydration and deacetoxylation. The catalyzed dehydration of vitamin A has been noted pre-

[8] S. Grys, *Analyst* **100**, 637 (1975).

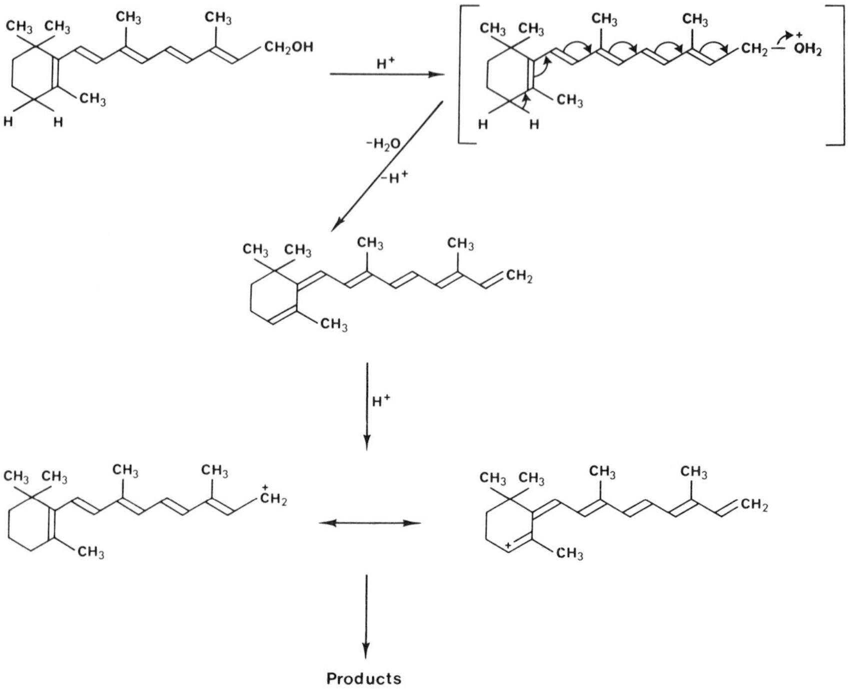

FIG. 5. Sequence of reactions of vitamin A with acids.

viously[9,10] and a method of estimation based on this reaction has been described.[11]

However, it will also be seen (Fig. 4) that there is an area of fine balance between the concentration of trichloroacetic acid catalyzing formation of anhydrovitamin with reasonable stability and that promoting loss in further reactions without appearance of a blue color. The absorbances of anhydrovitamin are derived from the same amount of vitamin A alcohol. Thus, whereas with 0.1% trichloroacetic acid the absorbance of anhydrovitamin at 377 nm is markedly greater than that of vitamin A alcohol at 332 nm, with 1% trichloroacetic acid promotion of further reaction is indicated by the weaker absorbance of the anhydrovitamin. Similar changes occur with vitamin A acetate. Stronger solutions of trichloroacetic acid proceed immediately to protonation of anhydrovitamin, giving the transient blue color of the carbonium ion. Mechanistic aspects of the

[9] J. R. Edisbury, A. E. Gillam, I. M. Heilbron, and R. A. Morton, *Biochem. J.* **26,** 1164 (1932).
[10] E. M. Shantz, G. O. Crawley, and W. D. Embree, *J. Am. Chem. Soc.* **65,** 901 (1965).
[11] P. Budowski and A. Bondi, *Analyst* **82,** 571 (1959).

reaction are indicated in Fig. 5. The spectral properties of this cation and later absorptions in a variety of solvents have been examined,[12] and the necessity for high concentrations of antimony trichloride for even brief stabilization of the colored ion in the Carr–Price reaction has been noted.[13]

Reaction at lower temperatures markedly affects color stability. If the temperature of both vitamin A solution and trichloroacetic acid reagent is lowered before mixing, the time for measurement is increased from about 5 sec at 20° to 19–20 sec at 10° and 28–30 sec at 0°. Incidental to the temperature reduction loss of strength of the reagent occurs, evidenced by crystallization of trichloroacetic acid, but the fact that there is no loss of color shows that the strength of the reagent is still sufficient to sustain the changes discussed above.

Methods of estimation with trifluoroacetic acid as a color-developing reagent using either the acid itself[5] or as a 33% solution in chloroform[14] have been described. The present method employs the less exotic trichloroacetic acid without sacrifice of accuracy.

Attention may be drawn to the anomalous properties of glycerol dichlorohydrin as a reagent with alcoholic groups functioning competitively as Lewis bases, thus accelerating decay of color. The requirement for addition of hydrochloric acid to produce "activated" reagent is thus understood.[4]

With the trichloroacetic acid method, vitamin A concentrations in the range of 1–10 $\mu$g/100 ml have been found in sheep serum, restored to 20–30 $\mu$g/100 ml after grazing the sheep for 1 month.[6] Liver concentrations as low as 0.2 $\mu$g/g have been noted in 2-year-old wethers on a vitamin A-deficient diet and a range 137–453 $\mu$g/g was observed for 1-year-old wethers on green pasture.[15]

[12] P. E. Blatz and D. Pippert, *J. Am. Chem. Soc.* **90**, 1290 (1960).
[13] P. E. Blatz and D. Estrada, *Anal. Chem.* **44**, 570 (1972).
[14] J. B. Neeld and W. N. Pearson, *J. Nutr.* **79**, 454 (1963).
[15] R. F. Bayfield, *Anal. Biochem.* **64**, 403 (1975).

# [25] Indirect Spectrophotometry on Vitamin A Products: Peak Signal Readout

*By* S. GRYS

*General Principle*

The most sensitive reagents known today for determining vitamin A are the Lewis acids antimony(III) chloride, trichloroacetic acid, and trifluoroacetic acid. The maximum absorbance of the blue products occurs within 1.6 sec[1] and fades rapidly. The solvents and reagent impurities (water, HCl, oxidizing agents, etc.) affect the rate at which the absorbance decreases but have less effect on the peak absorbance.[1-4] Thus, to obtain accurate results, maximum absorbance should be noted rather than the measurement at a defined time usually 5 sec after the addition of chromagen. This method describes improvements in reagent preparation and equipment that enable the preparation of reagent of consistent composition and the reading of results within 1 sec.

### Equipment

Spectrophotometer equipped with a linear recorder having a response of less than 0.8 sec at full scale deflection. Some system for the rapid addition and mixing of reagent to the test solution must be devised.

### Reagents

Ethyl alcohol, absolute
n-Hexane
Acetone
Chloroform
Ascorbic acid
Trichloroacetic acid
Magnesium chloride
Silica gel for column chromatography
Vitamin A acetate
β-Carotene

### Preparation of Reagents[1]

Anhydrous chloroform, ethyl alcohol, and phosgene-free: Wash the desired volume of chloroform twice with an equal volume of redistilled water. To 100 ml of washed chloroform add about 10 g of MgCl$_2$ dried overnight in a thin layer at 105° and stored in a tightly closed vessel. Leave the mixture for about 3 hr, mix occasionally, and filter the chloroform before use.

Chromagen reagent: In a 250-ml flask fitted with a ground-glass stopper, mix 100 g of trichloroacetic acid with half its volume of washed

[1] S. Grys, Analyst 100, 637 (1975).
[2] R. F. Bayfield, Anal. Biochem. 39, 282 (1971).
[3] M. S. Kimble, J. Lab. Clin. Med. 24, 1055 (1939).
[4] G. B. Subramanyam and D. B. Parrish, J. Assoc. Off. Anal. Chem. 59, 1125 (1976).

chloroform and about 60 g of dried $MgCl_2$. Take out the stopper occasionally and shake the flask. Liberated HCl is removed with a water aspirator.

*Procedure for Serum*

*Extraction.*[5-7] Place 2 ml of serum in a 10-ml glass-stoppered tube and mix with about 50 mg of ascorbic acid and 2 ml of absolute ethyl alcohol added dropwise. Add 5 ml of *n*-hexane, stopper the tube tightly, shake vigorously for 2 min, and centrifuge at 2500 rpm for 3 min. Transfer the tube to a freezer and, when the lower layer has solidified, decant the upper phase and evaporate to dryness under vacuum at 40° preferably in a vacuum oven. If possible, refill the cabinet with nitrogen, take out and cool the vessel, dissolve the residue in 1 ml of purified chloroform, and stopper.

*Spectrophotometry.*[1,2] Pipette 0.9 ml of chloroform extract into a cell that has been preblanked with chloroform at 620 nm. Insert the spectrophotometer cuvette into the sample beam. Select the 0–0.5 range on the recorder. Set the chart speed to 0.1 cm sec$^{-1}$ and begin recording. Inject 1.8 ml of chromagen reagent into the cell and record the absorbance for 20 sec.

*Procedure for Liver*

*Extraction.*[7,8] Place in a mortar 1.0 g of liver, add 100 mg of ascorbic acid, and grind with a pestle, adding progressively smaller amounts of silica gel until a friable powder is obtained. Transfer the dry homogeneous mixture to a glass-stoppered flask. Rinse the mortar with 50 ml of acetone: *n*-hexane (1 : 1) mixture, pour the solvent mixture into the flask, close tightly, and shake vigorously for 30 min. Centrifuge the contents for 10 min at 3000 rpm, decant the liquid layer, evaporate it to dryness, and dissolve the residue in 2 ml of purified chloroform.

*Spectrophotometry.*[1,7,8] Set the meter range for 0–1.0, place 0.2 ml of the extract into the spectrophotometric cell, add 0.7 ml of chloroform and 1.8 ml of chromagen reagent, and record the absorbance. If the optical density is too low, take a 0.9-ml aliquot and repeat the measurement.

*Recovery*

Recovery curves for the vitamin A alcohol and acetate derivatives added to serum are linear in absorbance and obey Beer's law. Mean

[5] S. R. Targan, S. Merrill, and A. D. Schwabe, *Clin. Chem.* **15**, 479 (1969).
[6] D. W. Bradley and C. L. Hornbeck, *Biochem. Med.* **7**, 78 (1973).
[7] S. Grys, unpublished data.
[8] R. F. Bayfield, *Anal. Biochem.* **64**, 403 (1975).

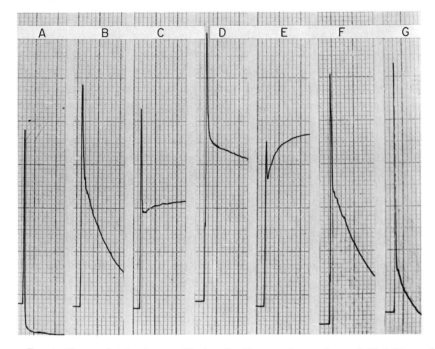

FIG. 1. Changes in absorbance with time for blue reaction products of (B) 2.71 μg of vitamin A ester, (C) 27.72 μg of β-carotene, (F) 0.1 g of chicken liver, and (G) 3.0 ml of swine serum, in 1 ml of chloroform with 2 ml of trichloroacetic acid in chloroform; and of (A) dichloromethane, (D) 2.81 μg of vitamin A ester, and 27.72 μg of β-carotene with trichloroacetic acid in dichloromethane (E).

recoveries for retinol and retinyl acetate without the addition of ascorbic acid as an antioxidant are 92 and 98%, respectively.[6] Recoveries of vitamin A alcohol and acetate added to liver are 87 and 97–101%, respectively. The inclusion of vitamin E before the extraction raises the recovery of retinol to 99%.[8] The substitution of ascorbic acid for α-tocopherol also reduces the destruction of retinol by atmospheric oxygen in the extraction vessel[7] without interfering with color development.

*Characteristic Absorption Spectra and Specificity*

The absorption spectra of 2.71 μg of vitamin A ester (AEC formulation purified by solvent extraction and checked by ultraviolet spectrophotometry), of 27.72 μg of β-carotene (cryst. pure; Koch Light Labs. Ltd), and of chicken liver and swine serum extracts are shown in Fig. 1B, C, F, and G. The changes in absorption with time are quite similar for vitamin A formulations and tissue extracts low in carotene. Thus in most biological samples, except for carotene-rich materials, (e.g., bovine serum), interference is negligible.[1] When the vitamin A concentration is measured in the presence of high levels of carotene, the interfering effect can be eliminated by the measurement of carotene concentration and subtraction of this contribution from the total absorption. Precautions should be taken to ensure carotene purity. Different preparations, even if standardized by spectrophotometry, gave divergent results unless highly purified. The use of the trichloroacetic acid in dichloromethane reagent produces a more stable blue color of slightly greater initial intensity (Fig. 1D). The peak height is, however, less evident, the carotene interference more serious (Fig. 1E), and the blank value (Fig. 1A) 2-fold greater. In this respect the trichloroacetic acid in chloroform reagent is most favorable, although for general use trifluoroacetic acid in dichloromethane is preferred.[4]

## [26] Colorimetric Analysis of Vitamin A and Carotene

### By Lawrence K. Oliver

Circulating levels of vitamin A and carotenoids in serum are decreased significantly in such conditions as chronic nephritis and impairment in fat absorption and in some febrile conditions. In contrast, diabetes mellitus, advanced chronic glomerulonephritis, myxedema, acute hepatic disease, and some dietary conditions may lead to elevated levels of vitamin A and yellow pigmentation of the skin due to accumulation of the carotenoids.

For analysis of vitamin A and carotene in serum, they are first removed from their protein complexes by stripping with ethanolic KOH, then are extracted with petroleum ether and quantitated by measuring the color developed upon treatment with 1,3-dichloro-2-propanol containing acetyl chloride.

### Reagents

Use distilled or deionized water throughout.
Ethanolic KOH: To 1 volume of 10 $N$ KOH, add 9 volumes of denatured ethanol.
Petroleum ether with butylated hydroxytoluene: Dissolve 125 mg of butylated hydroxytoluene in 500 ml of 30–60° petroleum ether.
Chloroform
1,3-dichloro-2-propanol (DCP), 0.32 $M$ acetyl chloride: Use DCP that has been purified by the method below. Store in brown bottles; the activated DCP is stable for at least 2 weeks.

METHODS IN ENZYMOLOGY, VOL. 67

Isooctane

Vitamin A standards, obtained as USP Vitamin A Reference Standard, retinyl (*trans*) acetate in cottonseed oil.

Stock standard: 6 mg/100 ml (20,000 IU/100 ml)[1] in chloroform containing 1% w/v butylated hydroxytoluene

Working standard: 60 $\mu$g/100 ml (200 IU/100 ml); make a 1:100 dilution of the stock with chloroform containing 0.2% w/v of butylated hydroxytoluene. Store at 4° and protect from light.

The 400 and 200 IU/100 ml standards for the assay are prepared by pipetting 2 ml and 1 ml of the working standards into test tubes and evaporating each under nitrogen. The dry tubes are then used in the colorimetry portion of the assay.

Carotene standards

Stock standard: 12 mg/liter in isooctane

Working standards: 1.2 mg/ml and 0.6 mg/ml in isooctane

## Purification of Dichloropropanol

Pour 1 kg of aluminum oxide, basic, activity grade 1, into a clean 5 × 100 cm all-glass chromatography column. Use a glass-wool plug at the bottom of the column and do not use grease on the stopcock. Attach a suction flask with the side arm protected with a drying tube to the bottom of the column by a rubber stopper. Pour 900 ml of dichloropropanol onto the column and allow it to run into the receiver flask. This will take approximately 5 hr and yield 300–400 ml of dichloropropanol, which is slightly turbid owing to suspended fines from the column. Vacuum distill at 10–14 mm Hg (1.33–1.76 $kP_a$) at 70–75°. Do not use grease in the vacuum distillation setup. Discard approximately 10 ml of the forerun (which will be cloudy), then continue to distill until no more distillate is collected. Store the distilled dichloropropanol in brown bottles with plastic-lined caps, where it is stable for at least 4 weeks.

## Assay Procedure

This procedure should be carried out with protection from direct light whenever possible.

Pipette duplicate 1-ml aliquots of each unknown into screw-capped test tubes. Add 1 ml of ethanolic KOH to each tube, cap, mix on a Vortex, then heat in a 60° waterbath for 20 min. Cool to room temperature. To each tube add 4 ml of petroleum ether/butylated hydroxytoluene. Replace

---

[1] One international unit of vitamin A = 0.3 $\mu$g of vitamin A alcohol.

cap, tighten carefully, then mix on a mechanical shaker at high speed for 5 min. Centrifuge at room temperature for 3 min at 700 $g$. Transfer the upper phase without contamination by the aqueous phase to a 15 × 125 mm test tube. Repeat the extraction with a second aliquot of petroleum ether/butylated hydroxytoluene. Evaporate the combined solvents for each sample in a 37° water bath under nitrogen. Rinse down the walls of the tubes with 0.5 ml chloroform and again evaporate to dryness.

## Colorimetry

Zero a spectrophotometer set at 550 nm against water. Prepare a reagent blank of 0.2 ml of chloroform and 0.8 ml of dichloropropanol containing acetylchloride. Read the absorbance between 1.5 and 3 min and record. Treat standards and one of each duplicate sample tubes by adding 0.2 ml of chloroform to dissolve the residue, then adding 0.8 ml of dichloropropanol/acetyl chloride mixture. Read the absorbance against the water blank for 1.5–3 min.

## Carotene Correction

To the second of the duplicate sample tubes, add 1 ml of isooctane to dissolve, mix vigorously, centrifuge at 1200 $g$ for 5 min, then read the absorbance at 450 nm vs isooctane.

## Calculations

IU vitamin A/100 ml = $[(A_x - A_b)/(A_s - A_b)]$ × concentration of standard
$\qquad\qquad$ − (0.4 × carotene concentration in $\mu$g/100 ml)

where $A_x$ = absorbance of the unknown; $A_s$ = absorbance of the standard; $A_b$ = absorbance of the reagent blank.

$$\mu\text{g of carotene/100 ml} = A_x \times 400$$

where $A_x$ = absorbance of the unknown at 450 nm.

## Notes

1. The concentration of the vitamin A standard is verified by measuring the absorbance at the absorbance maximum (approximately 326 nm). The $E_{1\,cm}^{1\%}$ in ethanol is 1548.
2. Carotene has an $E_{1\,cm}^{1\%}$ of 2500 at 450 nm in isooctane.
3. The color developed for vitamin A in dichloropropanol plus acetyl chloride is unchanged ($\Delta A < 0.003$) for 8 min.

4. Carotene levels are usually reported as micrograms per 100 ml. To convert to international units, divide by 0.6.

## Discussion

The Sobel–Snow[2] modification of the Carr–Price[3] procedure for analyzing vitamin A used antimony trichloride as the activating agent for dichloropropanol to generate colored species with the absorbance maximum at 550 nm. Blatz and co-workers[4,5] showed these species to be the retinylic and anhydroretinylic cations, but the mechanism of this reaction was never discovered and remains unknown to this day. Several other activating agents besides antimony trichloride have been found over the years, most of them acting at least as well as antimony trichloride, but sharing the fault of poor stability of the color produced. Naturally this led to poor precision.

We discovered[6] upon investigation of the reaction that the colored species decayed most rapidly when preparations of dichloropropanol were used that contained significant quantities of peroxides. Since vitamin A is highly unsaturated and peroxides are well known to be scavengers of double bonds, we concluded that one of the sources of the instability of the color was the presence of the peroxides in the activating solution. The purification scheme given here removes the peroxides effectively.

Using this purified DCP, we evaluated many different activating reagents. When DCP was activated with antimony trichloride by codistillation as in the original Sobel–Snow work,[2] the reagent was superior (greater color stability) to that obtained from use of unpurified DCP. However, this required a second distillation, since the addition of antimony trichloride to the eluate from the alumina column was inactive upon subsequent distillation. We then chose acetyl chloride for two reasons: (a) relative ease in handling in that it is a liquid that can be measured with a syringe and added to the DCP quite accurately; (b) and more important, the color produced is exceptionally stable. We found that the absorbance changed by less than 0.003 of an absorbance unit in the 8 min after the color was generated.

By scanning the spectrum, we conclude that the colored species are the same as that produced by antimony trichloride. The intensity and position of the maximum absorbance was unchanged when acetyl chloride was used. We were not able to obtain any information that would help in

[2] A. E. Sobel and S. D. Snow, J. Biol. Chem. 171, 617 (1947).
[3] F. H. Carr and E. A. Price, Biochem. J. 20, 497 (1920).
[4] P. E. Blatz and A. Estrada, Anal. Chem. 44, 570 (1972).
[5] P. E. Blatz and D. L. Pippert, J. Am. Chem. Soc. 90, 1296 (1968).
[6] L. K. Oliver and G. Bolz, Clin. Chem. 22, 1541 (1976).

determining what are those colored species. That acetyl chloride is an activating agent leads one to hypothesize that it is acting as an acetylating agent on vitamin A alcohol to generate the acetate, which in turn decomposes to the colored species. That the mechanism is more complex was easily shown by putting vitamin A acetate directly into nonactivated dichloropropanol, a step that produced no color whatsoever.

The precision (1 SD) of the assay using acetyl chloride-activated DCP is approximately 11% at a mean value of 200 IU/100 ml. This precision was established by preparing 121 pairs of specimens, each component of the pair being run on different days, 1 to 3 days apart, no more than two analyses each day.

# [27] Analysis of Geometrically Isomeric Vitamin A Compounds

*By* G. W. T. GROENENDIJK, P. A. A. JANSEN, S. L. BONTING,
and F. J. M. DAEMEN

## I. Introduction

Geometric isomers of vitamin A compounds play an essential role in the visual process. Both the 11-*cis* and all-*trans* configurations of vitamin A aldehyde (retinal, retinaldehyde), vitamin A alcohol (retinol), and vitamin A ester (retinyl ester) have been shown to occur in the visual cycle.[1] In view of the number of substances involved, it is not surprising that important details of this cycle are still unknown or in dispute.[2,3] Since the excellent review of Hubbard, Brown, and Bownds on the methodology of vitamin A compounds in a previous volume of this series,[4] some relatively new analytical methods have been introduced in this field. We shall discuss, in this order: thin-layer chromatography, high pressure liquid chromatography, and the preparation of vitamin A samples for analysis, after their extraction from animal tissues.

[1] G. Wald, *Science* **162,** 230 (1968).
[2] F. Lion, J. P. Rotmans, F. J. M. Daemen, and S. L. Bonting, *Biochim. Biophys. Acta* **384,** 283 (1975).
[3] C. D. B. Bridges, *Exp. Eye Res.* **24,** 283 (1977).
[4] R. Hubbard, P. K. Brown, and D. Bownds, this series, Vol. 18C, p. 615.

METHODS IN ENZYMOLOGY, VOL. 67

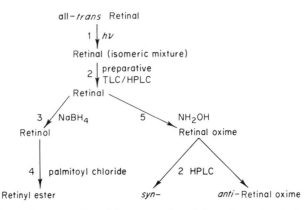

FIG. 1. Schematic presentation of the preparation of the various vitamin A compounds. Numbers refer to description of the various steps in the text.

## II. Reference Compounds

In the chromatographic analysis of vitamin A compounds, the use of authentic reference compounds is almost always necessary. Since only a few of them are commercially available,[5] this section presents methods to obtain pure geometric isomers, as summarized in Fig. 1. All-*trans* retinal is the most convenient starting material.

1. A *mixture of retinal isomers* is prepared by intense illumination of all-*trans* retinal.[4]

2. From this mixture, the pure *retinal* isomers can be isolated, either by preparative TLC (see Section III,D) or, preferably, by HPLC (see Section IV,D). Older methods for the isolation of 11-*cis* and 13-*cis* retinal from an isomeric mixture have used chromatography on $Al_2O_3$ or fractionated crystallization.[4]

3. The reduction of retinal to *retinol* with metal borohydrides has been well described.[4,6] The stereoisomeric configuration is retained during this procedure.

4. *Retinyl esters* (e.g., retinyl palmitate) can be prepared as follows: 5 $\mu$mol of a retinol isomer are solubilized in 500 $\mu$l of toluene and 5 $\mu$l (60 $\mu$mol) of pure, freshly redistilled pyridine are added, followed by 5.5 $\mu$mol of palmitoyl chloride. The mixture is allowed to stand for 15 min at 20°. The reaction must be carried out under anhydrous conditions. Removal of the excess pyridine and palmitoyl chloride is most easily carried out by thick-layer chromatography (PLC-plates, Silica 60, Merck) with

---

[5] All-*trans* and 13-*cis* retinal, and all-*trans* retinol are available from Eastman Organic Chemicals; all-*trans* retinyl palmitate from Sigma Chemical Company.

[6] C. D. B. Bridges, *Exp. Eye Res.* **22**, 435 (1976).

cyclohexane/toluene/ethyl acetate (5/3/2 by volume) as eluent. The retinyl ester is localized by viewing under very weak UV light. The fluorescent spot is scraped off, and the collected adsorbent is extracted with acetone. The yield is nearly quantitative, and the stereoisomeric configuration of the product is usually unchanged. If traces of other isomers are found to be present, purification can be carried out by HPLC, using a LiChrosorb Si 60-5 column (e.g., 250 × 9 mm) with hexane/dioxane (1000 : 1; v/v) as the eluent.

5. *Retinal oxime* is prepared by adding a 50- to 100-fold molar excess of hydroxylamine bicarbonate (pH 6.5) in 70% methanol to the pure retinal isomer. Extraction with dichloromethane yields the oxime in 100% yield. The *syn-* and *anti*-isomers are isolated by preparative HPLC (see Section IV,D).

## III. Thin-Layer Chromatography

### A. General

To obtain reasonably reproducible results, all precautions normally observed in thin-layer chromatography (constant temperature, avoidance of drafts, saturated chambers) are essential.[7] Eluent mixtures should be prepared freshly in view of the different volatilities of the component solvents. To prevent degradation reactions, all manipulations, including chromatography itself, should be carried out in an inert atmosphere (oxygen-free nitrogen or argon). Finally, short wavelength light ($\lambda < 550$ nm) must be avoided in order to prevent photoisomerization reactions. Normal dark-room red light can conveniently be used.

### B. Materials

Silica gel is almost exclusively used for thin-layer chromatography of vitamin A compounds. Its quality and origin do not seem to be very critical. Activation of the plates by heating to 120° or pretreatment with organic solvents rarely results in enhanced resolution. In our experience absolute reproducibility of $R_f$ values is very difficult to obtain with different batches of silica gel. Therefore, the use of authentic reference compounds is always necessary. Recently, so-called HPTLC (high performance thin-layer chromatography) plates have become commercially available, in which the diameter of the silica grains has been reduced to

---

[7] E. Stahl, "Thin-Layer Chromatography." Springer-Verlag, Berlin and New York; and Academic Press, New York, 1969.

TABLE I
DETECTION OF VITAMIN A COMPOUNDS AFTER THIN-LAYER CHROMATOGRAPHY

| Characteristic | Retinol | Retinal | Retinal oxime | Retinyl ester |
|---|---|---|---|---|
| Color | — | Yellow | Pale yellow | — |
| UV fluorescence | Yellow (++) | Absorbs | Orange (+) | Yellow (++) |
| Carr–Price reagent | Blue | Blue-green | Orange-brown | Blue |
| Trifluoroacetic acid | Green-blue | Yellow | Orange | Green-blue |
| Trichloroacetic acid | Blue-green | Brownish | Reddish | Blue-green |

about 5 $\mu$m. In our hands, their main advantage is that the detection limit is reduced by a factor of four, as compared to the conventional plates, to about 0.15 $\mu$g (approximately 0.5 nmol). In addition, their developing time is shortened to only 15 min.

Apolar solvent mixtures should be used for development. A low-boiling alkane [hexane, petroleum ether (b.p. 30–60°), cyclohexane] is invariably the main component, to which variable amounts of slightly more polar solvents like diethyl ether,[8–12] 2-octanone,[9] or ethyl acetate are added. High purity of the solvents is essential where oxidizing contamination is conceivable.

## C. Detection Methods

Standard lipid-spraying reagents,[13] like iodine vapor or charring with strong acids, may be used for the detection of vitamin A compounds following development. More specific detection methods include the very intense colors of retinals and their oximes, the fluorescence under UV light of retinols and retinyl esters and spraying with Carr–Price reagent, a saturated solution of antimony trichloride in chloroform. Recently other Lewis acids, such as trifluoroacetic acid (TFA) and trichloroacetic acid (TCA), have been recommended with dichloromethane as solvent, since they produce more stable colors, are less sensitive to moisture, and are safer.[14] More detailed characteristics of these methods are collected in Table I.

[8] J. P. Rotmans, S. L. Bonting, and F. J. M. Daemen, *Vision Res.* **12**, 337 (1972).

[9] S. Futterman and M. H. Rollins, *J. Biol. Chem.* **248**, 7773 (1973).

[10] A. Futterman and S. Futterman, *Biochim. Biophys. Acta* **337**, 390 (1974).

[11] W. F. Zimmerman, *Vision Res.* **14**, 795 (1974).

[12] D. Oesterhelt, M. Muntzen, and L. Schuhmann, *Eur. J. Biochem.* **40**, 453 (1973).

[13] V. P. Skipski, this series, Vol. 25, p. 396.

[14] G. B. Subramanyan and D. B. Parrish, *J. Assoc. Off. Anal. Chem.* **59**, 1125 (1976).

TABLE II

REPRESENTATIVE $R_f$ VALUES OF GEOMETRIC ISOMERS OF VITAMIN A COMPOUNDS

| | Hexane/ether 92/8 | Hexane/ether 50/50 | Cyclohexane/toluene/ethyl acetate 50/30/20 v/v/v | |
|---|---|---|---|---|
| Retinol | | | | |
| All-*trans* | 0.09 | 0.14 | | 0.21 |
| 9-*cis* | — | 0.17 | | 0.23 |
| 13-*cis* | — | 0.23 | | 0.28 |
| 11-*cis* | 0.12 | 0.28 | | 0.28 |
| Retinal | | | | |
| All-*trans* | 0.27 | 0.47 | | 0.46 |
| 9-*cis* | — | 0.52 | | 0.50 |
| 11-*cis* | 0.47 | 0.58 | | 0.53 |
| 13-*cis* | — | 0.60 | | 0.55 |
| Retinal oxime | | | *anti* | *syn* |
| All-*trans* | — | — | 0.21 | 0.45 |
| 9-*cis* | — | — | 0.23 | 0.40 |
| 11-*cis* | — | — | 0.27 | 0.47 |
| 13-*cis* | — | — | 0.33 | 0.39 |
| Retinyl ester | | | | |
| All-*trans* | 0.88 | >0.90 | | 0.70 |
| 11-*cis* | 0.88 | >0.90 | | 0.70 |

## D. Summary of $R_f$ Values

In all systems used so far, the retinols move slower than the retinals, whereas the retinyl esters move relatively close to the solvent front, even in almost pure hexane.[11] With increasing polarity of the elution solvent (e.g., more diethyl ether), the mobility of all vitamin A compounds increases. Representative $R_f$ values of geometric isomers of vitamin A compounds are collected in Table II.

The data presented in Table II indicate that the common geometric isomers of the vitamin A compounds can in principle be separated by thin-layer chromatography. In general, the method works satisfactorily when a single geometric isomer is to be identified. However, with retinyl esters and with 11-*cis* and 13-*cis* retinal, no reliable separation seems possible. In the latter case, modification with hydroxylamine to retinal oximes appeared to facilitate their identification,[12,15] since each isomer gives two spots, a *syn*- and an *anti*-form, with characteristic $R_f$ values. Meanwhile, we have shown by means of NMR spectroscopy that the slower moving spots represent the *anti*-isomers.[16]

[15] L. Y. Jan, *Vision Res.* **15**, 1081 (1975).
[16] G. W. T. Groenendijk, W. J. de Grip, and F. J. M. Daemen, *Anal. Biochem.*, in press.

TABLE III

ABSORPTION PROPERTIES OF STEREOISOMERS OF VITAMIN A COMPOUNDS IN HEXANE

| Compound | $\lambda_{max}$(nm) | $\epsilon_M(M\ cm^{-1})$ | $\epsilon_{320-350}$ | $\epsilon_{360}$ |
|---|---|---|---|---|
| Retinyl ester[a] | | | | |
| all-*trans* | 326 | — | — | — |
| 11-*cis* | 318 | — | — | — |
| 9-*cis* | 322 | — | — | — |
| 13-*cis* | 328 | — | — | — |
| Retinol[b] | | | $\epsilon_{320}$ | |
| all-*trans* | 325 | 51,800 | 49,500 | — |
| 11-*cis* | 318 | 34,300 | 34,000 | — |
| 9-*cis* | 322 | 42,200 | 41,900 | — |
| 13-*cis* | 328 | 48,500 | 44,700 | — |
| Retinal[b] | | | | |
| all-*trans* | 368 | 48,900 | | 46,700 |
| 11-*cis* | 365 | 26,400 | | 26,000 |
| 9-*cis* | 363 | 39,500 | | 39,000 |
| 13-*cis* | 363 | 38,800 | | 38,200 |
| Retinal oxime | | | $\epsilon_{350}$ | |
| all-*trans* (*syn*) | 357 | 55,500 | 52,500 | 54,900 |
| all-*trans* (*anti*) | 361 | 51,700 | 48,900 | 51,600 |
| 11-*cis* (*syn*)[c] | 347 | 35,900 | 35,700 | 35,000 |
| 11-*cis* (*anti*)[c] | 351 | 30,000 | 30,000 | 29,600 |

[a] Acetate and palmitate, partly from C. D. B. Bridges, *Exp. Eye Res.* **22**, 435 (1976); absorbance data in hexane are not accurately known, but presumably are very similar to those of the corresponding retinols.

[b] Partly (calculated) from R. Hubbard, *J. Am. Chem. Soc.* **78**, 4662 (1956).

[c] Not yet obtained in a crystalline form; therefore, absorption data are not definitive.

Although in a single case even quantitative determination of individual retinals has been achieved after their elution from the plates with pure diethyl ether or ethanol,[9,10] most investigators find thin-layer chromatography to be of limited value with more-complex mixtures. For complete separation and quantitative estimation of individual isomers, high pressure liquid chromatography offers better possibilities, as will be shown in the next section.

## IV. High Pressure Liquid Chromatography

### A. General

In the last few years, high pressure (performance) liquid chromatography (HPLC) has rapidly become a very useful technique in vitamin A analysis. HPLC offers a very sensitive means for fast, nondestructive, and quantitative analysis, which permits separation of all geometric isomers of the various classes of vitamin A compounds mentioned before. The general aspects of HPLC are adequately dealt with in a recent monograph.[17] All precautions mentioned in the preceding section to prevent degradation and isomerization of the vitamin A compounds should always be observed.

Under optimal conditions, quantitative analysis of vitamin A compounds is possible down to the picomole scale, whereas isolation of up to micromoles can be achieved on preparative columns.

## B. Instrumental Aspects

A variety of suitable instruments is available from several manufacturers.[18] We have used, both for analytical and preparative purposes, either a DuPont Model 830 Liquid Chromatograph or an Orlitta Pump (DMP AE 10-4-4). Both were equipped with Valco injection valves (CV 6-UHPA). Alternatively, for analytical purposes only, a Varian 8500 chromatograph with a septumless injector has been used.

Detection is most conveniently accomplished by continuous spectrophotometric analysis of the eluent at the absorbance maxima of the respective compounds (Table III; see also ref. 4). With complex mixtures, the retinol isomers can be determined at 320 nm, the retinals at 360 nm, and retinal oximes either at 350 or 360 nm (Table III). The spectrophotometers (Zeiss PM-2D or Pye Unicam LC3UV) were equipped with a linear potentiometric recorder (Goertz Servogor RE 512). For quantitative determination, all peaks are integrated by means of an electronic integrator (Spectra Physics, Autolab System I).

## C. Adsorbents and Eluents

Various kinds of silica gel have so far turned out to be the most useful type of stationary phase for the analysis of vitamin A compounds. Of all elution solvents tried so far, n-hexane with minor amounts of a slightly less apolar solvent as "modifier" has proved to be most suitable. Table IV presents relevant data on previously published separations of geometric isomers of vitamin A compounds.[6,19-23]

[17] L. R. Snyder and J. J. Kirkland, "Modern Liquid Chromatography." Wiley, New York, 1974.
[18] M. M. McNair, J. Chromatogr. Sci. **14**, 477 (1976).
[19] M. Vecchi, J. Vesely, and G. Oesterhelt, J. Chromatogr. **83**, 337 (1973).
[20] K. Tsukida, A. Kodama, M. Ito, M. Kawamoto, and K. Takahashi, J. Nutr. Sci. Vitaminol. **23**, 263 (1977).
[21] J. P. Rotmans and A. Kropf, Vision Res. **15**, 1301 (1975).
[22] K. Tsukida, A. Kodama, and M. Ito, J. Chromatogr. **134**, 331 (1977).
[23] F. G. Pilkiewitz, M. J. Peltei, A. P. Yudd, and K. Nakanishi, Exp. Eye Res. **24**, 421 (1977).

TABLE IV
HPLC Systems Used in Vitamin A Isomer Analysis

| Vitamin A compounds | Stationary phase | Column Length (cm) | Column I.d. (mm) | Modifier | Analysis time (min)[a] | Reference[b] |
|---|---|---|---|---|---|---|
| Retinyl acetate | | | | | | |
| 13-cis, AT[c] | Corasil II | 300 | ? | 0.1% Dioxane | 15 | 19 |
| 13-cis, AT | Merkosorb Si-60 | 50 | ? | 2% Dioxane | 6 | 19 |
| 9- and 13-cis, AT | $Al_2O_3$ | 50 | ? | 1% Diethyl ether | 7 | 19 |
| Retinyl palmitate | | | | | | |
| 11-cis, AT | μPorasil | 30 | 4 | 2% Ether | 9 | 6 |
| Retinol | | | | | | |
| 11-cis, AT | Corasil II | 200 | 2 | 1% 2-Propanol | 13 | 6 |
| 11-cis, AT | μPorasil | 30 | 4 | 20% Ether | ? | 6 |
| 9,11,13-cis, AT | Zorbax SIL | 25 | 8 | 6% $CH_3COOC_2H_5$, 8% $CH_2Cl_2$ | 21 | 20 |
| Retinal | | | | | | |
| 9,11,13-cis, AT | μPorasil | 30 | 4 | 2% Ether | 19 | 21 |
| 11-cis, AT | μPorasil | 30 | 4 | 12.5% Ether | 7 | 6 |
| 9,11,13-cis, AT | Zorbax SIL | 25 | 8 | 12% Ether | 19 | 22 |
| 9,11,13-cis, AT | μBondapack CN | 30 | 4 | 1% Ether | 15 | 23 |

[a] Retention time of the all-trans isomer.
[b] Numbers refer to text footnotes.
[c] AT = all-trans.

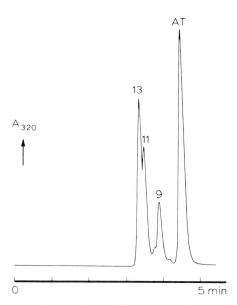

FIG. 2. Chromatogram of a test mixture of four *cis-trans* isomers of retinyl palmitates. (11 = 11-*cis* retinyl palmitate, etc.) Column: packing, Si 60, 5 $\mu$m; length (l), 25 cm; internal diameter (i.d.), 3.0 mm. Eluent: *n*-hexane/dioxane (1000/1, v/v); flow rate (W), 17.5 $\mu$l/sec; pressure (P), 3000 psi. Detection at 320 nm. From J. E. Paanakker and G. W. T. Groenendijk, *J. Chromatogr.* **168,** 125 (1979), with permission.

We have obtained satisfactory results with MicroPak Si-5 (25 × 2 mm), MicroPak Si-10 (25 × 2 mm), Partisil 5 and Partisil 10 (both 25 × 3 mm), LiChrosorb Si 60-5 and silica gel Si-60 (Merck, Darmstadt, GFR). The latter material was sieved to a specific particle size of 5–8 $\mu$m and packed by means of a balanced density procedure, followed by activation.[24] The columns consist of stainless steel, have a length of 25 cm and an inner diameter of either 2–3 mm for analytical purposes or 10 mm for preparative application. For the mobile phase, we prefer *n*-hexane with different concentrations of peroxide-free dioxane as modifier.[25] Dioxane covers a larger polarity range than diethyl ether and has a lower volatility.

Although the reproducibility of elution patterns in terms of the retention times of individual compounds is quite high for a single column and eluent mixture, in normal practice the use of calibration mixtures of authentic compounds is necessary for unambiguous identification and determination.

[24] J. C. Kraak, H. Poppe, and F. Smedes, *J. Chromatogr.* **122,** 147 (1976); J. E. Paanakker, J. C. Kraak, and H. Poppe, *J. Chromatogr.* **149,** 111 (1978).

[25] J. E. Paanakker and G. W. T. Groenendijk, *J. Chromatogr.* **168,** 125 (1979).

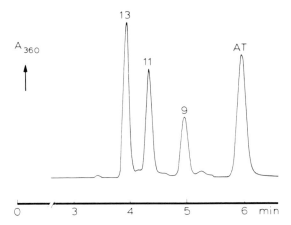

FIG. 3. Chromatogram of a test mixture of four *cis-trans* isomers of retinal. Column; packing: Si 60, 5–8 $\mu$m; 1 = 25 cm; i.d. = 10 mm. Eluent: *n*-hexane/dioxane (975/25); W = 202 $\mu$l/sec; P = 3000 psi. Detection at 360 nm. Amount injected about 0.5 $\mu$g. From J. E. Paanakker and G. W. T. Groenendijk, *J. Chromatogr.* **168**, 125 (1979), with permission.

The small peaks between 11-*cis* and 13-*cis* retinal and between 9-*cis* and all-*trans,* respectively, are presumably 9,13-di-*cis* retinal and 7-*cis* retinal [W. Sperling, P. Carl, C. N. Rafferty, and N. A. Dencher, *Biophys. Struct. Mechanism* **3**, 79 (1977)] and 7-*cis* retinal [M. Denny and R. S. H. Liu, *J. Am. Chem. Soc.* **99**, 4865 (1977)].

The efficiency of the columns decreases after several separation runs under all conditions. Regeneration is usually possible by washing the column consecutively with 50 ml of acetone, 50 ml of diethyl ether, followed by a 1 : 1 mixture of diethyl ether and hexane until a constant base line level at 254 nm is reached. Thereupon, the column is reequilibrated with the appropriate eluent mixture. Ultimately, the columns become silted up and should be repacked with fresh absorbent.

## D. Chromatograms

In this section, we present our results for pure mixtures of vitamin A compounds.[25] With retinyl palmitates the 9-*cis* and all-*trans* esters are completely separated from each other and from the 11-*cis* and 13-*cis* isomers, when 0.1% dioxane in hexane is used as the eluent (Fig. 2). However, the separation of the 11-*cis* and the 13-*cis* ester is not yet ideal. This inconvenience can be overcome by an approach of Bridges,[26] who converts the ester to the corresponding retinol with LiAlH$_4$. Recently the separation of all-*trans* retinyl esters of variable acyl-chain composition has been reported, using reversed-phase HPLC.[27]

[26] C. D. B. Bridges, *Vision Res.* **15**, 1311 (1975).
[27] M. G. M. De Ruyter and A. P. De Leenheer, *Anal. Chem.* **51**, 43 (1979).

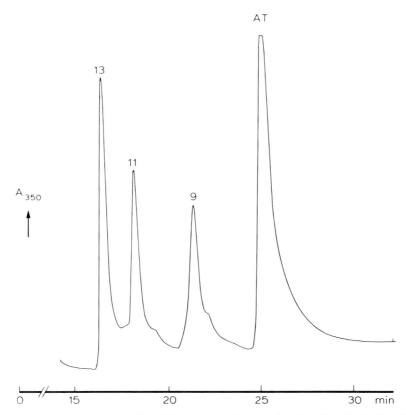

FIG. 4. Preparative separation of four *cis-trans* isomers of retinal. W = 48.4 μl/sec; P = 750 psi. Detection at 350 nm. Amount injected about 0.5 mg. For other conditions, see Fig. 3. From J. E. Paanakker and G. W. T. Groenendijk, *J. Chromatogr.* **168**, 125 (1979), with permission.

With the retinals, a complete base line separation appears possible, using 2.5% dioxane in hexane. Figure 3 demonstrates the analytical separation on a 10-mm (i.d.) column of 0.5 μg of a retinal mixture, whereas Fig. 4 shows the preparative separation of a 1000-fold larger sample (500 μg) on the same column. This separation is made possible by considerably reducing the flow rate.

The elution pattern of the retinol isomers is shown in Fig. 5, using 5% dioxane in hexane as the eluent. Although the separation of the 11-*cis* and 13-*cis* isomers is not optimal, accurate quantitative determination is still possible. It is a remarkable fact that in the retinol series the 11-*cis* isomer is eluted first, whereas in the case of the esters and the aldehydes the 13-*cis* isomers have the shortest retention times. We do not have an explanation for this phenomenon.

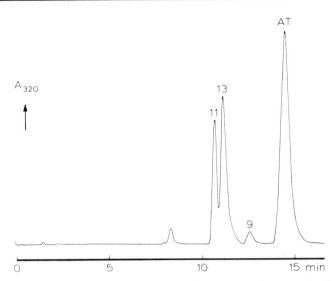

F<sub>IG</sub>. 5. Chromatograms of a test mixture of four *cis-trans* isomers of retinol. Eluent: *n*-hexane/dioxane (95/5); W = 18.1 μl/sec; P = 3500 psi. Detection at 320 nm. For column conditions, see Fig. 2. From J. E. Paanakker and G. W. T. Groenendijk, *J. Chromatogr.* **168**, 125 (1979), with permission.

For reasons to be explained in Section V,C, it may be necessary to convert retinal with hydroxylamine to retinal oxime. In spite of the complication caused by the occurrence of *syn*- and *anti*-isomers, reasonable separation is possible (Fig. 6). Since the ratio of the *syn*- and *anti*-isomers seems to be dependent on the conditions, presently both *syn*- and *anti*-peaks have to be taken into consideration, if quantification is desired. We have been able to isolate pure samples of the *syn*- and *anti*-forms of 11-*cis* and all-*trans* retinal oxime by preparative HPLC. These compounds have been characterized by NMR and mass spectroscopy, and their absorbance maximum and molar absorption in hexane have been determined (Table III). Thus 11-*cis* and all-*trans* retinals, the only isomers of direct importance for the visual process, can be simultaneously assayed as the oximes (see also Section V,C).

The complete analysis in a single run of more complex compounds of vitamin A, comprising esters, aldehydes or oximes, and alcohols, remains difficult. The different classes of vitamin A compounds are well separated, but, depending on the eluent used, the resolution of the isomers is satisfactory only for a single class (Fig. 7). Mixtures containing isomers of retinol, retinal, retinyl ester, retinal oxime and 3-dehydroretinol have been analyzed by programmed high pressure liquid chromatography.[28] The further

[28] C. D. B. Bridges, S. L. Fong, and R. A. Alvarez, *Vision Res.* (1979), in press.

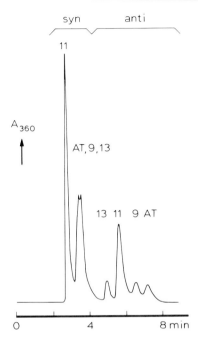

FIG. 6. Chromatogram of a test mixture of four *cis-trans* isomers of retinal oxime; each retinal isomer gives a *syn*- and *anti*-retinal oxime. Column: 1 = 25 cm; i.d. = 2 mm; Packing: MicroPak Si-5. Eluent: *n*-hexane/dioxane (96/4); W = 25.0 μl/sec; P = 2000 psi. Detection at 360 nm. 11(s) = *syn*-11-*cis* retinal oxime, 11(a) = *anti*-11-*cis* retinal oxime, etc.

development of reversed-phase chromatography, introduced in this field,[29] may open up new possibilities as well.

In conclusion, the present state of HPLC allows the separation and quantitative determination of the isomers of a single class of vitamin A compounds. Complete analysis of complex mixtures cannot yet be carried out on a routine basis.

## V. Extraction of Vitamin A Samples for Analysis

### A. General

Vitamin A compounds are apolar lipids and, therefore, lipid extraction is always the first step in their analysis. The more general aspects of lipid extraction procedures are excellently documented in earlier volumes of this series,[30,31] and this presentation assumes that the reader has already

[29] C. A. Frolik, T. E. Tavela, and M. B. Sporn, *J. Lipid Res.* **19**, 32 (1979).
[30] C. Entenman, this series, Vol. 3, p. 299.
[31] N. S. Radin, this series, Vol. 14, p. 245.

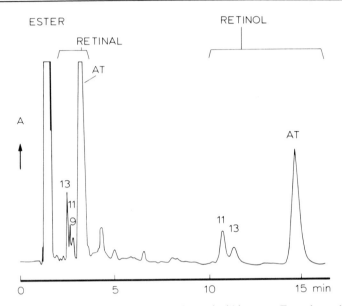

FIG. 7. Separation of a test mixture containing retinoid isomers. For column dimensions and packing, see Fig. 2. Eluent: *n*-hexane/dioxane (95/5); W = 18.1 $\mu$l/sec; P = 3500 psi. Detection at 320 nm (esters and retinols) and 360 nm (retinals).

consulted those chapters. With respect to the extraction of vitamin A compounds, four aspects deserve further consideration. Two of these will be discussed in this section.

It should again be emphasized that the polyunsaturated character of vitamin A compounds makes them very susceptible to a variety of oxidizing conditions.[32] Hence, precautions like working in an inert atmosphere, use of deaerated solvents, and/or addition of antioxidants are essential.[31]

A second aspect concerns the low concentration in which vitamin A compounds occur, even in ocular tissues. When standard lipid extraction methods are used, contamination with a large excess of other lipids, notably phospholipids and cholesterol, is unavoidable. For example, isolated rod outer segment membranes contain about 200 $\mu$g of phospholipids and 12 $\mu$g of cholesterol per microgram of retinal.[33] Although modern absorbents for TLC and HPLC have high intrinsic loading capacities, a prepurification step may be necessary to remove a substantial part of the contaminating lipids. This is conveniently achieved by carrying the extract (containing approximately 500 $\mu$g of lipid) over a small silica column

[32] See, e.g., D. Fischer, F. U. Licht, and J. A. Lucy, *Biochem. J.* **130**, 25g (1972).
[33] W. J. de Grip, F. J. M. Daemen, and S. L. Bonting, this volume [36].

(2 × 0.5 cm) in hexane containing 5% dioxane. Under these conditions, the vitamin A compounds are not absorbed, whereas the phospholipids are completely retained. More recently, we have observed that LiChrosorb Si 60-5 can conveniently be used without prior removal of the phospholipids, since more polar solvents (chloroform, methanol) can be applied to clean the column efficiently afterward.

Two other, more specific aspects of extraction of vitamin A compounds—incomplete extraction of covalently bound retinal and aspecific isomerization of geometric forms—will be discussed in the last section of this chapter.

## B. Extraction of Retinol and Retinyl Esters

Retinol and retinyl esters (i.e., long-chain acyl esters in our case) occur as free compounds and are easily and completely extracted with lipid solvents after homogenization of the tissue. In general, hexane extraction can conveniently be used for the complete extraction of these compounds. Since only part of the membrane phospholipids dissolve readily in this solvent, a favorable fractionation takes place. In addition, hexane does not denature rhodopsin, which obviates the problems with covalently bound retinal in case rod outer segments are to be extracted (see next section).

In our normal procedure tissue fractions are freeze-dried after homogenization and extracted by grinding the material three times for 10 min in a Potter–Elvehjem tube with hexane. The supernatants are collected after low-speed centrifugation, and are concentrated by blowing with a gentle stream of oxygen-free nitrogen. If the freeze-drying step is to be omitted, methanol is added to a homogenized, aqueous suspension to a final concentration of 60% (v/v), and the mixture is extracted with hexane.[34] Alternatively, the tissue can be ground with twice its volume of anhydrous sodium sulfate and is extracted by stirring with two or three portions of hexane.[4]

Organelles rich in vitamin A, like the oil droplets of retinal pigment epithelium cells, can simply be isolated by flotation upon centrifugation of the homogenate.[6] Gentle stirring with an overlying layer of hexane and removal of the organic layer with a pipette or syringe leads to complete extraction of vitamin A compounds, when the procedure is once repeated.[6] The retinal extraction procedure to be described in Section V,C can be conveniently used as well.

[34] D. Bownds, Nature (London) 216, 1178 (1967).

## C. Extraction of Retinal

Retinal is often covalently bound as an aldimine (Schiff base) to primary amino groups in natural products. This applies both to retinal bound as a prosthetic group in rhodopsin and to retinal freed by photolysis or added to tissues that contain hydrophobic domains carrying amino groups, e.g., membranes with phosphatidylethanolamine. This appears to interfere with its complete extraction,[15,25,35,36] although previous lyophilization facilitates the extraction with hexane of retinal not bound in rhodopsin.[4,37]

A second difficulty arises from the fact that the geometric isomers of retinal are, even in darkness, readily susceptible to isomerization, e.g., in the presence of nucleophiles[10] or membrane suspensions[2,38,39] containing phosphatidylethanolamine.[16] Similar phenomena have not been observed with retinol and retinyl esters, unless in trace amounts.

A satisfactory analysis of retinals, therefore, requires complete extraction in the original isomeric form. This can be tested with rhodopsin in rod outer segment membranes, since the amount of retinal present can be independently determined by the assay of rhodopsin.[33] Furthermore, the isomeric state of retinal in rhodopsin is exclusively 11-*cis*, an isomer repeatedly shown to be very susceptible to aspecific isomerization.[2,10,37]

Recently, Pilkiewitz *et al.*[23] have presented a simple nonisomerizing procedure for the identification of protein-linked retinals, using as the extractant dichloromethane with or without the detergent Ammonyx-LO. These authors do not mention their recovery of chromophore. In our hands, this procedure does indeed yield exclusively 11-*cis*-retinal, but the recovery is at best only 30%.

We have combined this method with the modification of the retinal with hydroxylamine.[39] The resulting retinal oximes are extracted with dichloromethane and analyzed by HPLC. In this way it is possible to extract the 11-*cis* chromophore of rhodopsin in the original isomeric state with a recovery of 93% (SD 7%, $n = 9$). Similarly, all-*trans* retinal is quantitatively extracted following illumination of rod outer segments in the presence of hydroxylamine (Fig. 8).

*Procedure.* To a rhodopsin suspension (10–50 $\mu M$) a 1000-fold molar excess of 1 $M$ $NH_2OH$ (pH 6.5) is added, followed by methanol to a final

[35] J. P. Rotmans, G. C. M. v.d. Laar, F. J. M. Daemen, and S. L. Bonting, *Vision Res.* **12,** 1297 (1972).
[36] K. Nakanishi, A. P. Yudd, R. K. Crouch, G. L. Olson, H. C. Cheung, R. Gorindjee, T. G. Ebrey, and D. J. Patel, *J. Am. Chem. Soc.* **98,** 236 (1976).
[37] W. J. de Grip, F. J. M. Daemen, and S. L. Bonting, *Vision Res.* **12,** 1697 (1972).
[38] F. J. M. Daemen, J. P. Rotmans, and S. L. Bonting, *Exp. Eye Res.* **18,** 97 (1974).
[39] G. Wald and P. K. Brown, *J. Gen. Physiol.* **37,** 189 (1953).

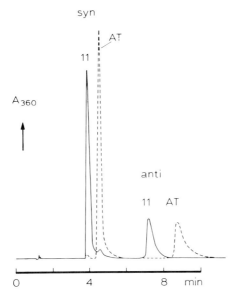

FIG. 8. Chromatograms of extracts of rod outer segment membranes: (a) extracted in the presence of $NH_2OH$ in the dark (——); (b) after photolysis in the presence of $NH_2OH$ (-------). For column dimensions, stationary phase, and mobile phase, see Fig. 6. W = 17 $\mu$l/sec; P = 1750 psi. Detection at 360 nm.

concentration of 70% (v/v). Subsequently water and dichloromethane are added, so that the final ratio water/methanol/dichloromethane becomes 1/1/1 by volume. After thoroughly mixing on a vortex and centrifugation (10,000 g, 1 min) the lower organic layer is collected with a syringe. This procedure is repeated three times. The combined organic layers are evaporated with nitrogen. The residue is dissolved in 2.5% dioxane in hexane (2 $\mu$l/nmol) and analyzed by HPLC.[40]

VI. Conclusion

Although this contribution has concentrated on the analysis of vitamin A compounds in ocular tissues, the techniques can be applied with minor modifications to the analysis of these compounds in liver[41] and plasma.[42] In animal tissues, so far only all-*trans* and 11-*cis* isomers have been found, the 11-*cis* compounds occurring exclusively in the eye. The presence of

[40] G. W. T. Groenendijk, W. J. de Grip, and F. J. M. Daemen, *Biochim. Biophys. Acta,* in press.
[41] R. F. Bayfield, *Anal. Biochem.* **64**, 403 (1975).
[42] J. N. Thompson, P. Erdody, and W. B. Maxwell, *Biochem. Med.* **8**, 403 (1973).

other isomers must be considered to derive from experimental artifacts, in spite of the fact that they are found in various food products.[43] The single case where another isomer, 13-*cis*-retinal, seems to play a physiological role is in bacteriorhodopsin.[44]

[43] D. C. Egberg, J. C. Heroff, and R. H. Potter, *J. Agr. Food Chem.* **25**, 1127 (1977).
[44] D. Oesterhelt, *Angew. Chem. Int.* (*Int. Ed. Engl.*) **15**, 17 (1976); M. J. Pettei, A. P. Yudd, K. Nakanishi, R. Henselman, and W. Stoeckenius, *Biochemistry.* **16**, 1955 (1977).

## [28] High-Pressure Liquid Chromatography of Vitamin A Metabolites and Analogs*

*By* ANNE M. McCORMICK, JOSEPH L. NAPOLI, and HECTOR F. DeLUCA

This chapter describes the high-pressure liquid chromatography methods developed and used in the authors' laboratory for separation of vitamin A metabolites and analogs. Straight-phase and reverse-phase high-pressure liquid chromatography methods are presented, and their application to biochemical problems in the vitamin A field are discussed.

Investigation of vitamin A metabolism has been hampered by the lability of vitamin A compounds (retinoids), which are especially sensitive to oxidation and photoisomerization. In the past, adsorption on silica gel or alumina represented the best methods available for separation of retinoids. However, adsorption chromatography was plagued by poor resolution, incomplete recovery, and artifact production.[1] The application of Sephadex LH-20 chromatography (liquid–gel partition) to the separation of retinoids proved to be a major advance in that quantitative recovery and elimination of artifact formation were achieved.[2] The application of Sephadex LH-20 chromatography to the study of retinol and retinoic acid metabolism revealed several metabolites more polar than retinoic acid.[3] Unfortunately, compounds of similar polarity, including *cis* and *trans* isomers of retinoids, are poorly resolved on Sephadex LH-20 columns. Thus the need for powerful chromatographic methods suited to investigation of vitamin A metabolism remained.

The application of high-pressure liquid chromatography (HPLC) to the

* This work was supported by U.S. Public Health Service Grant AM-14881 and National Institutes of Health Postdoctoral Fellowship DE-07031.
[1] H. F. DeLuca, M. H. Zile, and P. F. Neville, *in* "Lipid Chromatography Analysis" (G. V. Marinetti, ed.), Vol. 2, p. 345. New York, 1969.
[2] Y. L. Ito, M. Zile, H. M. Ahrens, and H. F. DeLuca, *J. Lipid. Res.* **15**, 517 (1974).
[3] Y. L. Ito, M. H. Zile, H. F. DeLuca, and H. M. Ahrens. *Biochim. Biophys. Acta* **369**, 338 (1974).

separation of retinoids has revolutionized vitamin A chromatography and has proved to be an extremely powerful and valuable method for studying vitamin A metabolism. The use of HPLC offers several advantages over conventional methods of vitamin A chromatography, including rapid separation, excellent resolution, quantitative recovery, elimination of artifacts, and the capability of chromatographing a wide range of vitamin A compounds with only minor manipulation of the chromatographic solvent. Reverse-phase HPLC methods have recently been developed[4,5] and used in the isolation and identification of intestinal retinoic acid metabolites,[6-8] determination of tissue levels and tissue distribution of several retinoic acid metabolites (A. McCormick, J. Napoli, and H. DeLuca, unpublished results), determination of 13-cis- and all-trans-retinoic acid in serum,[9] and metabolism of synthetic retinoid analogs.[10] In addition, straight-phase HPLC on microparticulate silica gel columns has been applied to the study of retinol storage and distribution,[11] retinal isomers and analogs,[12,13] determination of serum retinol and retinoic acid,[14,15] and isolation of urinary and fecal metabolites of retinoic acid.[16,17]

### Vitamin A Standard Compounds

Retinol, 13-cis-retinol, retinal, and retinoic acid were obtained from Eastman Kodak Co., Rochester, New York. Retinyl acetate and retinyl palmitate were obtained from Nutritional Biochemicals Corp., Cleveland, Ohio. [11,12-$^3$H]Retinoic acid, 13-cis-retinoic acid, 4-hydroxyretinoic acid, and methyl 4-ketoretinoate were generously supplied by Hoffmann-La Roche Co., Nutley, New Jersey and Basel, Switzerland.

[4] A. M. McCormick, J. L. Napoli, and H. F. DeLuca, Anal. Biochem. 86, 25 (1978).
[5] C. A. Frolik, T. E. Tavela, and M. B. Sporn, J. Lipid Res. 19, 32 (1978).
[6] J. L. Napoli, A. M. McCormick, H. K. Schnoes, and H. F. DeLuca, Proc. Natl. Acad. Sci. U.S.A. 75, 2603 (1978).
[7] A. M. McCormick, J. L. Napoli, H. K. Schnoes, and H. F. DeLuca, Arch. Biochem. Biophys. 192, 577 (1979).
[8] A. M. McCormick, J. L. Napoli, H. K. Schnoes, and H. F. DeLuca, Biochemistry 17, 4085 (1978).
[9] C. A. Frolik, T. E. Tavela, G. L. Peck, and M. B. Sporn, Anal. Biochem. 86, 743 (1978).
[10] A. B. Roberts, M. D. Nichols, C. A. Frolik, D. L. Newton, and M. B. Sporn, Cancer Res. 38, 3327 (1978).
[11] C. D. B. Bridge, Vision Res. 15, 1311 (1975).
[12] T. Ebrey, R. Govindjie, B. Honig, E. Pollack, E. Chan, R. Crouch, A. Yudd, and K. Nakanishi, Biochemistry 14, 3933 (1975).
[13] J. P. Rolmans and A. Kropf, Vision Res. 15, 1301 (1975).
[14] M. G. M. De Ruyter and A. P. De Leenheer, Clin. Chem. 22, 1593 (1976).
[15] C. V. Puglisi and J. A. F. de Silva, J. Chromatogr. 152, 42 (1978).
[16] R. Hänni, F. Bigler, W. Meister, and G. Englert, Helv. Chim. Acta 59, 2221 (1976).
[17] R. Hänni and F. Bigler, Helv. Chim. Acta 60, 881 (1977).

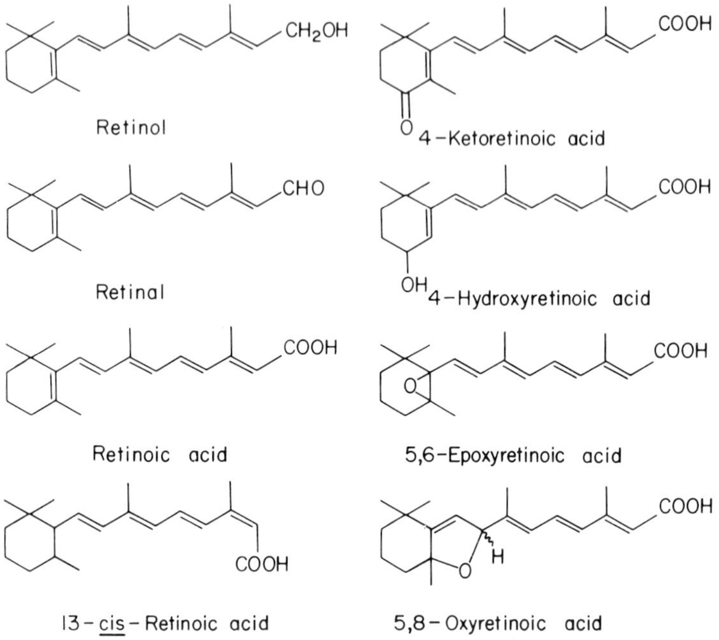

FIG. 1. Structures of vitamin A metabolites and analogs.

Methyl 5,6-epoxyretinoate and methyl 5,8-oxyretinoate were synthesized as previously described.[6,8] 5,6-Epoxyretinoic acid and 5,8-oxyretinoic acid were prepared from the methyl esters by hydrolysis. Retinoid structures are shown in Fig. 1.

### Handling and Storage of Vitamin A Compounds

Vitamin A compounds are extremely sensitive to photoisomerization and oxidation; therefore, certain precautions are necessary during the handling and storage of the compounds. The photoisomerization is alleviated by working under red or yellow lighting. To prevent oxidation, all standard compounds should be stored in degassed solvents. In the authors' laboratory vitamin A compounds are routinely stored under $N_2$ in degassed, distilled methanol at $-70°$. No breakdown of vitamin A compounds is observed if they are handled and stored by the above-described procedures. Radioactive compounds are stored under $N_2$ in toluene at $-70°$.

### High-Pressure Liquid Chromatography Equipment

In the authors' laboratory, a Waters Associates Model ALC/GPC-204 liquid chromatograph was used for HPLC. Detection of compounds was accomplished with a fixed ultraviolet monitor at 313 nm or 340 nm. At 313 nm, 10 ng of retinoic acid are detected routinely; however, the sensitivity is increased to 5 ng at 340 nm. In our laboratory, Waters $\mu$Bondapak $C_{18}$, Whatman ODS-1, and Dupont Zorbax ODS columns are used for reverse-phase HPLC. For straight-phase HPLC, good chromatography of vitamin A compounds is achieved with Waters $\mu$Porasil, Whatman Partisil, and Dupont Zorbax-SIL columns.

### Straight-Phase High-Pressure Liquid Chromatography

The principle of straight-phase HPLC is that of adsorption chromatography. The compounds to be separated are adsorbed to microparticulate silica gel (polar compounds are adsorbed more tightly) and are eluted in the order of least polar to most polar. Good chromatography of neutral and charged retinoids can be achieved with straight-phase HPLC. The advantages of this method include excellent resolution, quantitative recovery, and absence of artifact production. However, straight-phase HPLC is not a satisfactory method for chromatography of acid-sensitive compounds, such as 5,6-epoxyretinoic acid.

### Reverse-Phase High-Pressure Liquid Chromatography

The detailed mechanism of retention in reverse-phase HPLC is not understood. However, the central role of hydrophobic interaction is clear. Solutes are forced out of the mobile phase and are bound with the hydrocarbon ligands of the stationary phase. The driving force for retention of compounds is not the favorable interaction of the compounds with the stationary phase, but the effect of solvent (mobile phase) in forcing the compound to the hydrocarbon stationary phase.[18] Increasing the polarity of the mobile phase increases the retention of substances by the stationary phase. In this method of chromatography compounds are eluted in reverse order compared to straight-phase chromatography (elution order is most polar to least polar). Reverse-phase HPLC columns routinely used for vitamin A chromatography in the authors' laboratory are composed of octadecylsilane molecules (ODS) chemically bonded to microparticulate silica gel. Reverse-phase HPLC has proved to be an extremely powerful tool in vitamin A chromatography. Excellent separation of nonpolar retinoids (retinyl esters, retinol, retinal), acid retinoids (retinoic acid), and esters of acidic polar retinoids has been achieved. In addition to this versatility, reverse-phase HPLC offers quantitative recovery, excellent resolution, elimination of artifact production during chromatography, and good reproducibility.

[18] B. L. Karger and R. W. Giese, *Anal. Chem.* **50**, 1048 (1978).

Recovery of Retinoids from High-Pressure Liquid Chromatography

The recovery of all vitamin A compounds discussed in this chapter was determined by ultraviolet absorption. Both straight-phase and reverse-phase HPLC methods gave recoveries of 95% or greater. In addition, the recovery and stability of retinoic acid was confirmed by chromatography of [11,12-³H]retinoic acid. With reverse-phase and straight-phase chromatography, recoveries were greater than 95% and rechromatography revealed no extraneous peaks, indicating that no isomerization or destruction occurred during HPLC.

Sensitivity

The sensitivity of reverse-phase and straight-phase HPLC methods was determined with all-*trans*-retinoic acid. Monitoring at 340 nm, 5 ng of all-*trans*-retinoic acid are detected routinely.

Separation of Naturally Occurring Vitamin A Compounds (Retinoids)

The separation of several common vitamin A compounds by reverse-phase HPLC is illustrated in Fig. 2. Retinyl palmitate, retinol, retinal, and retinoic acid are well resolved with a solvent system of 10 m*M* ammonium

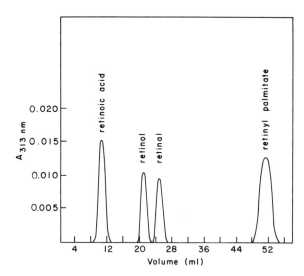

FIG. 2. Reverse-phase HPLC of a mixture of four synthetic vitamin A compounds. Retinoic acid (100 ng), retinol (100 ng), retinal (100 ng), and retinyl palmitate (200 ng) were injected in 10 μl of methanol onto a Waters μC₁₈-Bondapak column (4 mm i.d. × 30 cm). The chromatographic solvent was 10 m*M* ammonium acetate in MeOH : H₂O (80 : 20).

acetate in methanol : water (80 : 20). The addition of salt to the eluting solvent eliminates tailing of the retinoic acid peak encountered with reverse-phase HPLC. Ammonium bicarbonate works equally well in this respect. For chromatography of neutral retinoids, salt is omitted from the chromatographic solvent. All-*trans*- and 13-*cis*-retinol can be separated with methanol : water (80 : 20).[4] Most retinoids are only partially soluble in the methanol : water solvent mixtures used for reverse-phase HPLC. This problem can be circumvented by injecting the compounds to be chromatographed in small volumes of methanol (10–100 $\mu$l).

With minor manipulations of solvent polarity, the chromatography of retinoic acid can be optimized. A 5% increase in the amount of water in the solvent increases the retention of retinoic acid and thereby separates all-*trans*-retinoic acid and its 13-*cis* isomer (Fig. 3).

### Separation of Acidic Polar Retinoids

Two HPLC systems were developed in our laboratory for the chromatography of acidic retinoids, specifically compounds more polar than retinoic acid. The first system is a reversed-phase system using microparticulate ODS columns and methanol : water solvent mixtures. The separation of several biologically significant acidic retinoids on two ODS

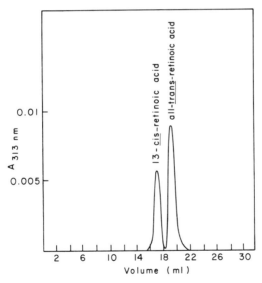

FIG. 3. Separation of all-*trans*-retinoic acid (100 ng) and 13-*cis*-retinoic acid (75 ng) by reverse-phase HPLC. Compounds were injected in 10 $\mu$l of methanol and the chromatographic solvent was 10 m$M$ ammonium acetate in MeOH : H$_2$O (75 : 25).

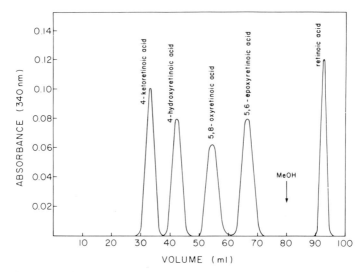

Fig. 4. Reverse-phase HPLC of acidic retinoids. 4-Ketoretinoic acid (1 μg), 4-hydroxy-retinoic acid (1 μg), 5,8-oxyretinoic acid (1 μg), 5,6-epoxyretinoic acid (1.5 μg), and retinoic acid (1 μg) were injected in 50 μl of methanol. Two (0.4 × 30 cm) $\mu C_{18}$-Bondapak columns connected in series were used, and the chromatographic solvent was 10 m$M$ ammonium acetate in MeOH : H$_2$O (60 : 40). Retinoic acid was eluted with methanol.

columns connected in series (total column length 60 cm) is shown in Fig. 4. Note that retinoic acid was eluted by changing the column solvent to methanol since the actual elution volume of retinoic acid with methanol : water (60 : 40) is 255 ml. Similar resolution of these compounds can be achieved with one 0.4 × 30 cm column by increasing the polarity of the solvent, namely to 10 m$M$ ammonium acetate in methanol : water (50 : 50).

Acidic retinoids can also be separated by straight-phase HPLC on a microparticulate silica gel column. The charged retinoids must be eluted with an acidic solvent mixture or the compounds elute in broad, tailing peaks. The separation of retinoic acid, 5,8-oxyretinoic acid, 4-ketoretinoic acid, and 4-hydroxyretinoic acid is depicted in Fig. 5. All the above compounds are stable when chromatographed in this system. However, 5,6-epoxyretinoic acid is converted to 5,8-oxyretinoic acid by the acidic solvent. Thus chromatography of 5,6-epoxyretinoic acid or mixtures of retinoids containing the 5,6-epoxide is best accomplished with the reverse-phase system.

### Separation of Esters of Polar Retinoids

Both the straight-phase and reverse-phase HPLC methods can be adapted for separation of retinoid esters. The methyl esters of several

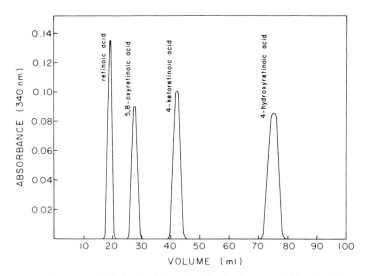

Fig. 5. Straight-phase HPLC of acidic retinoids. Retinoic acid (1 μg), 5,8-oxyretinoic acid (1 μg), 4-ketoretinoic acid (1 μg), and 4-hydroxyretinoic acid (1.5 μg) were separated on a Whatman Partisil 10/50 column with a solvent of hexane : tetrahydrofuran : formic acid (95 : 5 : 0.1). The retinoids were injected in 40 μl of column solvent.

charged retinoids are prepared with diazomethane.[6] The separation of methyl retinoate, methyl 5,8-oxyretinoate, methyl 5,6-epoxyretinoate, and methyl 4-ketoretinoate by reverse-phase HPLC is shown in Fig. 6. The esters are separated on two microparticulate ODS columns connected in series. Note that methyl retinoate is eluted with methanol; the actual elution position in the original column solvent is 210 ml. Similar resolution of the four methyl esters can be obtained with a single column if the polarity of the eluting solvent is increased.

The esters are also well resolved by straight-phase HPLC (Fig. 7). The resolution of methyl 5,6-epoxyretinoate and methyl 5,8-oxyretinoate with the straight-phase system is superior to that obtained by reverse-phase HPLC, although both methods achieve base-line resolution of the compounds.

Biological Applications

*Vitamin A Metabolism.* The study of vitamin A metabolism, in particular, metabolism of retinoic acid in the intestine, has been facilitated by the application of reverse-phase HPLC to the problem. An outline of the method developed in the authors' laboratory for study of retinoic acid metabolism is shown in Fig. 8. Labeled [11,12-³H]retinoic acid of known specific activity is administered intrajugularly in 50–100 μl of distilled

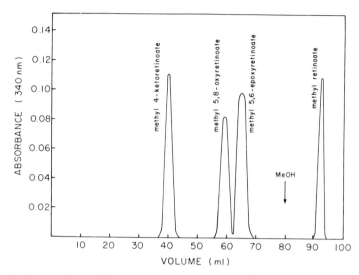

FIG. 6. Reverse-phase HPLC of polar retinoid esters. Methyl 4-ketoretinoate (1 $\mu$g), methyl 5,8-oxyretinoate (1 $\mu$g), methyl 5,6-epoxyretinoate (1 $\mu$g), and methyl retinoate (1 $\mu$g) were separated on two Waters $\mu C_{18}$-Bondapak columns connected in series (total column length equals 60 cm) with MeOH : $H_2O$ (75 : 25). Methyl retinoate was eluted with methanol. The retinoids were injected in 40 $\mu$l of methanol.

ethanol to vitamin A-deficient rats. Three hours after dosing, when radioactivity in the intestinal mucosa is at a maximum, the mucosa is obtained, homogenized, and lyophilized.[7,8] The lyophilized mucosa is extracted twice with 250 ml of distilled methanol (containing 50 $\mu$g of butylated hydroxytoluene per milliliter) under a nitrogen atmosphere. The methanol-soluble metabolites are partitioned between 90% methanol and hexane (250 ml : 250 ml). The methanol phase is evaporated to dryness on a rotary evaporator, and the residue is taken up in 25 ml of acetone : methanol (1 : 1). The soluble metabolites are then chromatographed on a Sephadex LH-20 column (2 × 60 cm) equilibrated with acetone : methanol (1 : 1) to obtain two radioactive fractions designated L1 and L2 [elution volumes are L1 (100 ml) and L2 (220 ml)]. The L1 and L2 fractions were each evaporated to dryness, dissolved in 5 ml of methanol, and applied to separate DEAE-Sephadex (hydroxide form) columns (1 × 25 cm). The DEAE-Sephadex columns are washed with 100 ml of methanol to remove neutral lipids. The charged retinoic acid metabolites are eluted with 0.1 $M$ ammonium acetate in methanol. The L1 fraction yields two radioactive peaks designated D1 and D2 [D1 (40 ml) and D2 (155 ml)] and then L2 fractions have two radioactive components designated D3 and D4 [D3 (50 ml) and D4 (150 ml)]. Reverse-phase HPLC

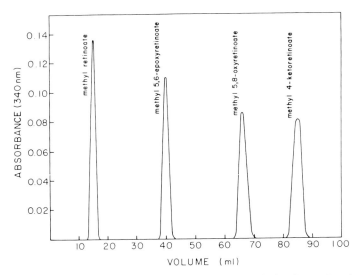

FIG. 7. Straight-phase HPLC of polar retinoid esters. Methyl retinoate (1 μg), methyl 5,6-epoxyretinoate (1 μg), methyl 5,8-oxyretinoate (1 μg), and methyl 4-ketoretinoate (1.5 μg) were separated on a Whatman Partisil 10 : 50 column with hexane : tetrahydrofuran (98 : 2). The retinoids were injected in 40 μl of column solvent.

systems for each metabolite fraction (D1-D4) are shown in Fig. 9. The amounts of each metabolite are calculated from the specific activity of the injected retinoic acid.

## Isolation and Identification of Retinoic Acid Metabolites

*Preparation and Initial Purification of Retinoic Acid Metabolites.* Vitamin A-deficient rats are administered 450 μg each of $[11,12\text{-}^3\text{H}]$retinoic acid (specific activity $8 \times 10^5$ dpm/μg) at 3.5 hr prior to sacrifice. Intestinal mucosa is obtained, homogenized, and lyophilized as described.[7,8] The lyophilized mucosa is extracted with methanol, and the methanol-soluble lipid fraction containing the retinoic acid metabolites is partitioned between methanol : hexane (250 ml : 250 ml). The methanol phase (58% of mucosa radioactivity) contains retinoic acid and polar metabolites. These metabolites are fractionated on a 2 × 55 cm column of Sephadex LH-20 eluted with acetone. A single peak of radioactivity is eluted with acetone (100–150 ml), and two additional radioactive peaks are eluted by changing the solvent to acetone : methanol (1 : 1). The metabolites eluted from the Sephadex LH-20 column with acetone were further purified on a 1 × 20 cm DEAE-Sephadex (hydroxide form) column. Uncharged lipids are eluted with 100 ml of methanol. The charged

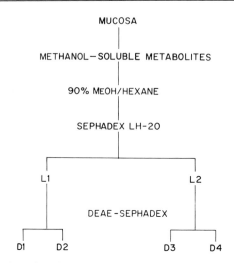

Fig. 8. Fractionation of intestinal metabolites of retinoic acid.

metabolites are eluted with 0.1 $M$ ammonium acetate in methanol as a single peak (D1, elution volume 75 ml).

*HPLC of D1 Metabolites.* The charged D1 metabolites are separated into three components by reverse-phase HPLC on two (0.4 × 30 cm) ODS columns connected in series eluted with 10 m$M$ ammonium acetate in methanol : water (60 : 40) as shown in Fig. 9. The radioactive peak eluted between 68 and 78 ml is 5,6-epoxyretinoic acid, and the peak eluted with methanol is all-*trans*-retinoic acid.[6–8]

*Purification of 5,6-Epoxyretinoic Acid.* 5,6-Epoxyretinoic acid recovered from reverse-phase HPLC is methylated with diazomethane. Methyl-5,6-epoxyretinoate is then purified by reverse-phase HPLC on one microparticulate ODS column with a solvent system of methanol : water (75 : 25). This column separates unreacted 5,6-epoxyretinoic acid (6 ml) from methyl 5,6-epoxyretinoate (54 ml). The final purification of the methylated epoxide is by straight-phase HPLC. On a microparticulate silica gel column (0.4 × 50 cm) eluted with hexane : tetrahydrofuran (99 : 1), methyl 5,6-epoxyretinoate elutes as a single peak of radioactivity and ultraviolet absorbance at 340 nm (25 ml). The ultraviolet spectrum ($\gamma_{max}$ 339 nm with shoulders at 352 and 320 nm) and mass spectrum ($m/e$ 330) demonstrate the compound to be methyl 5,6-epoxyretinoate. The identification of the metabolite as methyl 5,6-epoxyretinoate was confirmed by comigration of synthetic and isolated compounds on reverse-phase and straight-phase HPLC. The reverse-phase system chosen was methanol : water (75 : 25). With two (0.4 × 30 cm) ODS columns connected in series, the compounds coelute at 70 ml. Chromatography on a

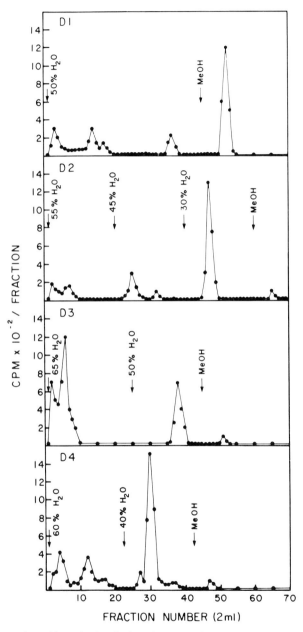

FIG. 9. Separation of intestinal retinoic acid metabolites by reverse-phase HPLC. All metabolite fractions (D1–D4 from Fig. 8) were injected in 100 $\mu$l of methanol. A Waters $\mu$C$_{18}$-Bondapak column was used with the indicated solvents. Metabolite peaks were detected by determining the radioactivity ($^3$H) in the total fraction. Samples were counted in 10 ml of Aquasol (New England Nuclear) with an efficiency of 20–25%.

microparticulate silica gel column (0.4 × 50 cm) with hexane : tetrahydrofuran (99 : 1) elutes the compounds at 32 ml.

## Tissue Distribution of Retinoic Acid and 5,6-Epoxyretinoic Acid

[11,12-³H]Retinoic acid (Hoffman-La Roche Inc., Nutley, New Jersey) is purified on a 1 × 60 cm column of Sephadex LH-20 equilibrated with chloroform : hexane (65 : 35) containing 50 μg of butylated hydroxytoluene (BHT) per milliliter. The retinoic acid peak is adjusted to a specific activity of $4.2 \times 10^7$ dpm/μg. The purity of the retinoic acid is assessed by reverse-phase HPLC on a microparticulate ODS column with a solvent system of 10 m$M$ ammonium bicarbonate in methanol : water (70 : 30) and found to be >99% all-*trans*-retinoic acid. Severely vitamin A-deficient rats (average weight loss 10–20 g/week) are administered 2.2 μg of [11,12-³H]retinoic acid intrajugularly in 50 μl of distilled ethanol. Three hours after dosing, the animals are killed by cardiac puncture; blood and tissues are collected, and 5 tissue specimens are pooled. With a specific activity of $4.2 \times 10^7$ dpm/μg, the detection limits of this experiment are 1 pg/g. Tissue samples are prepared and D1 metabolites are obtained as previously described for intestinal mucosa. The D1 metabolites are fractionated by reverse-phase HPLC on one (0.4 × 30 cm) ODS column eluted with 10 m$M$ ammonium bicarbonate in methanol : water (50 : 50). Unlabeled all-*trans*-retinoic acid (1 μg) and 5,6-epoxyretinoic acid (1 μg) are added to each sample for internal standardization. In this system, 5,6-epoxyretinoic acid elutes at 80–90 ml and retinoic acid elutes at 10–20 ml after the solvent is changed to methanol. Tissue concentrations of retinoic acid and 5,6-epoxide are calculated from the specific activity of the administered [11,12-³H]retinoic acid as follows:

$$\mu g \text{ metabolites} = \frac{\text{total dpm in metabolite peak from HPLC}}{\text{specific activity of } ^3\text{H-retinoic acid (dpm/}\mu g)}$$

The levels of retinoic acid after administration of physiological levels (2 μg) of [³H]retinoic acid to vitamin A-deficient rats are in the range of 1–5 ng per gram of tissue and the levels of 5,6-epoxyretinoic acid are in the range of 50–250 pg per gram of tissue. The sensitivity of this HPLC method is about 1 ng at 340 nm, thus this method could be easily adapted to determine tissue concentrations of unlabeled retinoic acid. However, after the reverse-phase HPLC step, the retinoic peak is radiochemically pure, but not free of all 340 nm absorbing material. Quantitation of retinoic acid by ultraviolet absorbance at 340 nm would require an additional straight-phase HPLC purification step with a solvent of hexane : tetrahydrofuran : formic acid (98 : 2 : 0.1). Quantitation of 5,6-epoxyretinoic acid in this manner is feasible but would require 50–100 g amounts of tissue for reliable results. In addition, purification of 5,6-epoxyretinoic acid by straight-phase HPLC would convert the metabolite to 5,8-oxy-retinoic acid, which has a $\lambda_{max}$ of 298 nm. Thus 5,6-epoxide levels are best quantitated at 313 nm as 5,8-oxyretinoic acid.

### Determination of Radiochemical Purity of Radioactive Vitamin A Compounds

Radioactive vitamin A compounds, especially those of high specific activity (31 Ci/mmol) decompose after prolonged storage. High-pressure liquid chromatography provides a rapid and reliable method for determining the purity of labeled vitamin A compounds. For example, in the authors' laboratory, [$^3$H]- or [$^{14}$C]retinoic acid is routinely checked by reverse-phase HPLC before use. A solvent system of 10 m$M$ ammonium bicarbonate in methanol : water (70 : 30) gives excellent separation of all-*trans*-retinoic acid from any decomposition products (primarily *cis* isomers of retinoic acid). To date, HPLC represents the only method for separation of *cis* and *trans* isomers of retinoic acid. In addition, this solvent system can be used for purification of radioactive retinoic acid samples that show evidence of degradation. In a like manner, unlabeled vitamin A compounds can be assayed for purity and purified by this method.

## [29] Gas Chromatography, Gas Chromatography–Mass Spectrometry, and High-Pressure Liquid Chromatography of Carotenoids and Retinoids

*By* RICHARD F. TAYLOR and MIYOSHI IKAWA

The identification of carotenoids has been of interest to organic chemists and biochemists since the first isolations of carotenes (carotenoids containing only carbon and hydrogen) and xanthophylls (carotenes containing additional atoms of oxygen, nitrogen, etc.) during the early nineteenth century. Such identification has become increasingly more important as the list of naturally occurring carotenoids grows and the functions of carotenoids in living organisms are studied. In addition, the present and future use of carotenoids as acceptable natural food colorings is of great commercial significance.[1] Last, the use of vitamin A de-

---

[1] O. Isler, R. Ruëgg, and U. Schwieter, *Pure Appl. Chem.* **14**, 245 (1967).

METHODS IN ENZYMOLOGY, VOL. 67

## TABLE I
### CAROTENOID TRIVIAL AND SEMISYSTEMATIC NAMES

| Trivial name | Semisystematic name |
| --- | --- |
| $\beta$-Apo-4'-carotenal | 4'-Apo-$\beta$-caroten-4'-al |
| $\beta$-Apo-8'-carotenal | 8'-Apo-$\beta$-caroten-8'-al |
| $\beta$-Apo-10'-carotenal | 10'-Apo-$\beta$-caroten-10'-al |
| $\beta$-Apo-8'-carotenoic acid | 8'-Apo-$\beta$-caroten-8'-oic acid |
| $\beta$-Apo-8'-carotenoic acid ethyl ester | Ethyl 8'-apo-$\beta$-caroten-8'-oate |
| $\beta$-Apo-8'-carotenoic acid methyl ester | Methyl 8'-apo-$\beta$-caroten-8'-oate |
| Astacene | 3,3'-Dihydroxy-2,3,2',3'-tetrahydro-$\beta,\beta$-carotene-4,4'-dione or $\beta,\beta$-carotene-3,4,3',4'-tetrone |
| Azafrin | 5,6-Dihydroxy-5,6-dihydro-10'-apo-$\beta$-caroten-10'-oic acid |
| Bixin | Methyl hydrogen 9'-$cis$-6,6'-diapocarotene-6,6'-dioate |
| Canthaxanthin | $\beta,\beta$-Carotene-4,4'-dione |
| Capsanthin | 3,3'-Dihydroxy-$\beta,\kappa$-caroten-6'-one |
| $\alpha$-Carotene | $\beta,\epsilon$-Carotene |
| $\beta$-Carotene | $\beta,\beta$-Carotene |
| $\gamma$-Carotene | $\beta,\psi$-Carotene |
| $\zeta$-Carotene | 7,8,7',8'-Tetrahydro-$\psi,\psi$-carotene |
| $\beta$-Carotenone | 5,6,5',6'-Diseco-$\beta,\beta$-carotene-5,6,5',6'-tetrone |
| Carotinin | 15,15'-Didehydro-$\beta,\beta$-carotene |
| Crocetin | 8,8'-Diapocarotene-8,8'-dioic acid |
| Cryptoxanthin | $\beta,\beta$-Caroten-3-ol |
| Dehydro-$\beta$-apo-8'-carotenal | 3,4-Didehydro-8'-apo-$\beta$-caroten-8'-al |
| Dehydro-$\beta$-carotene | 3,4-Didehydro-$\beta,\beta$-carotene |
| 4,4'-Diapo-$\zeta$-carotene | 7,8,7',8'-Tetrahydro-4,4'-diapo-$\psi,\psi$-carotene |
| 4,4'-Diapolycopen-4-al | 4,4'-Diapo-$\psi,\psi$-caroten-4-al |
| 4,4'-Diaponeurosporen-4-al | 7',8'-Dihydro-4,4'-diapo-$\psi,\psi$-caroten-4-al |
| 4,4'-Diaponeurosporene | 7,8-Dihydro-4,4'-diapo-$\psi,\psi$-carotene |
| 4,4'-Diaponeurosporen-4-oic acid | 7',8'-Dihydro-4,4'-diapo-$\psi,\psi$-caroten-4-oate |
| 4,4'-Diaponeurosporen-4-oic acid methyl ester | Methyl 7',8'-dihydro-4,4'-diapo-$\psi,\psi$-caroten-4-oate |
| 4,4'-Diapophytoene | 7,8,11,12,7',8',11',12'-Octahydro-4,4'-diapo-$\psi,\psi$-carotene |
| 4,4'-Diapophytofluene | 7,8,11,12,7',8'-Hexahydro-4,4'-diapo-$\psi,\psi$-carotene |
| Diethylcrocetin | Diethyl 8,8'-diapocarotene-8,8'-dioate |
| Dimethylcrocetin | Dimethyl 8,8'-diapocarotene-8,8'-dioate |
| Echinenone | $\beta,\beta$-Caroten-4-one |
| Fucoxanthin | 5,6-Epoxy-3,3',5'-trihydroxy-6',7'-didehydro-5,6,7,8,5',6'-hexahydro-$\beta,\beta$-caroten-8-one 3'-acetate |
| 4-Hydroxy-4,4'-diaponeurosporene | 7',8'-Dihydro-4,4'-diapo-$\psi,\psi$-caroten-4-ol |
| Isocryptoxanthin | $\beta,\beta$-Caroten-4-ol |
| Isozeaxanthin | $\beta,\beta$-Carotene-4,4'-diol |
| Lycopene | $\psi,\psi$-Carotene |

(*continued*)

TABLE I (*continued*)
CAROTENOID TRIVIAL AND SEMISYSTEMATIC NAMES

| Trivial name | Semisystematic name |
| --- | --- |
| Methylazafrin | Methyl 5,6-dihydroxy-5,6-dihydro-10'-apo-$\beta$-caroten-10'-oate |
| Methylbixin | Dimethyl 9'-*cis*-6,6'-diapocarotene-6,6'-dioate |
| Neurosporene | 7,8-Dihydro-$\psi$,$\psi$-carotene |
| Physalien | $\beta$-$\beta$-Carotene-3,3'-diol dipalmitate |
| Phytoene | 7,8,11,12,7',8',11',12'-Octahydro-$\psi$,$\psi$-carotene |
| Phytofluene | 7,8,11,12,7',8'-Hexahydro-$\psi$,$\psi$-carotene |
| Rubixanthin | (3R)-$\beta$,$\psi$-Caroten-3-ol |
| Torularhodin | 3',4'-Didehydro-$\beta$,$\psi$-caroten-16'-oate |
| $\beta$-Zeacarotene | 7',8',Dihydro-$\beta$,$\psi$-carotene |
| Zeaxanthin | $\beta$,$\beta$-Carotene-3,3'-diol |

rivatives, or retinoids, as potential cancer chemotherapeutic agents is currently of great interest,[2] and the analysis of these compounds utilizes methods grounded in and common to carotenoid analysis methods. Thus, the rapid and routine identification of carotenoids and related compounds such as the retinoids is a necessary tool for terpenoid analysis.

Both carotenoids and retinoids are members of the family of terpenoid compounds characterized by their polyene nature. Carotenoids may be defined on the basis of their biosynthetic route, which can involve the condensation of two molecules of either a $C_{15}$ or $C_{20}$ terpenyl pyrophosphate precursor to result in either triterpenoid ($C_{30}$) or tetraterpenoid ($C_{40}$) carotenoids.[3] Enzymic and oxidative modifications of these compounds result in the multitude of naturally occurring carotenoids containing from 20 to 50 carbon atoms. Naturally occurring retinoids arise after oxidative modification of $C_{40}$ carotenoids, usually $\beta$-carotene.

Three methods for carotenoid and retinoid analysis will be discussed here: GC, GC-MS, and HPLC.[3a] GC of carotenoids and retinoids is limited owing to the inherent instability of polyene chains at high temperatures, but the method has been used to determine the size of carotenoid carbon skeletons and the nature of functional groups present in perhydrocarotenoids. GC-MS is an extension of methods developed for the GC analysis of carotenoids in order to further identify the parent carotenoid. HPLC is a rapidly emerging analytical method for both carotenoids and retinoids owing to its nondestructive operating conditions.

[2] M. B. Sporn, N. M. Dunlop, D. L. Newton, and J. M. Smith, *Fed. Proc., Fed. Am. Soc. Exp. Biol.* **35**, 1332 (1976).

[3] B. H. Davies and R. F. Taylor, *Pure Appl. Chem.* **47**, 211 (1976).

[3a] Abbreviations used: GC, gas chromatography; GC-MS, gas chromatography–mass spectrometry; HPLC, high-pressure liquid chromatography.

Standard and Unknown Carotenoids

By recommendation of the Commission on Biochemical Nomenclature,[4] trivial names for carotenoids should always be accompanied by their corresponding semisystematic names. For this reason, all carotenoids hereafter referred to by their trivial names are listed with both their trivial and semi-systematic names in Table I. The rationale for naming $C_{30}$ carotenoids according to IUPAC recommendations has been detailed elsewhere.[5] A summary of the carotenoid parent structures on which the semisystematic nomenclature is based is presented in Fig. 1, as is an example of a $C_{30}$ carotenoid. Retinoids are exempt from this nomenclature system.[4]

Squalene, phytol, cholesterol, cholestane, lanosterol, stigmasterol, lycopene, $\beta$-carotene, retinol, retinyl acetate, retinyl palmitate, retinal, and retinoic acid were all purchased from Sigma Chemical Co., St. Louis, Missouri. 4,4′-Diapocarotenes and diapoxanthophylls were isolated from *Streptococcus faecium* UNH 564P (obtained from Dr. W. R. Chesbro, Department of Microbiology, University of New Hampshire, Durham, N.H.) and *Staphylococcus aureus* 209P (ATCC 6538P) according to methods described elsewhere.[5,6] Phytoene, phytofluene, $\zeta$-carotene, $\beta$-zeaxanthin, and neurosporene were isolated from mutants of *Phycomyces blaksleeanus* by standard methods.[7] Rubixanthin was isolated from rose hips (*Rosa rubrifolia*).[8] All the remaining carotenoids as well as geranylgeraniol, geranyllinalool, and lycopersene were generous gifts from Hoffmann-La Roche through Dr. W. E. Scott (Nutley, New Jersey) and Dr. F. Leuenberger (Basel, Switzerland).

The preparation of unknown carotenoids for analysis should follow standard isolation procedures and partial purification using appropriate column chromatographic techniques, such as those utilizing neutral alumina of varying activity grades (for carotenes and some xanthophylls) and cellulose powders (for highly polar xanthophylls).[5,6] Once carotenoids are partially purified by such methods, they may be analyzed by GC and GC-MS.

All solvents used should be analytical reagent grade or better.[5] Diethyl ether is sodium-dried and then glass-redistilled from reduced iron powder immediately before use to assure it is peroxide-free. Chloroform is glass-redistilled twice, stored in the dark, and used within 1 week to assure that

[4] IUPAC Commission on the Nomenclature of Organic Chemistry and IUPAC–IUB Commission on Biochemical Nomenclature, *Biochemistry* 10, 4827 (1971).
[5] R. F. Taylor and B. H. Davies, *Biochem. J.* 139, 751 (1974).
[6] R. F. Taylor and B. H. Davies, *Biochem. J.* 139, 761 (1974).
[7] B. H. Davies, *Pure Appl. Chem.* 35, 1 (1973).
[8] R. F. Taylor and B. H. Davies, unpublished results.

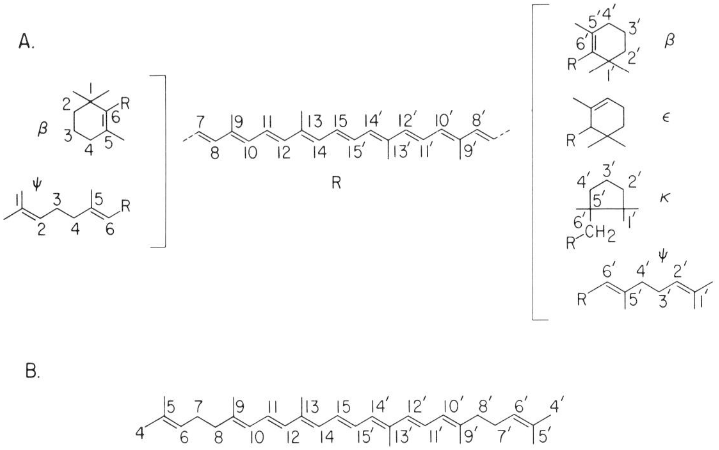

A.

B.

FIG. 1. (A) Structures of the parent carotenes from which all the carotenoids listed in Table I may be derived. The appropriate combinations of each group with the polyene chain (R) illustrate the structures of $\beta,\beta$-carotene ($\beta$-carotene), $\beta,\epsilon$-carotene ($\alpha$-carotene), $\beta,\kappa$-carotene, $\beta,\psi$-carotene ($\gamma$-carotene), and $\psi,\psi$-carotene (lycopene). (B) An example of a symmetrical triterpenoid carotenoid, 7,8,7',8'-tetrahydro-4,4'-diapo-$\psi,\psi$-carotene (4,4'-diapo-$\zeta$-carotene).

no acidic contaminants are present. Pyridine is refluxed over solid KOH for 1 hr and then redistilled through a distilling column fitted with an anhydrous $CaCl_2$ moisture trap. Dry pyridine may be stored over an appropriate molecular sieve, such as Union Carbide 13X. Petroleum ether (b.p. 30–60°) is sodium dried, glass-redistilled, passed through an activated silicic acid column and glass-redistilled once again before use.

All carotenoids and retinoids should be stored under an atmosphere of oxygen-free nitrogen at −40°. The handling of the compounds at room temperature should be rapid and away from strong natural or artificial light. Evaporation *in vacuo* of organic solvents from carotenoid and retinoid solutions should be followed by flushing with oxygen-free nitrogen. Routine ultraviolet-visible absorbance spectra should be determined for all standard carotenoids and retinoids prior to use to assess whether degradation has occurred in storage.

### Preparation of Carotenoid Derivatives

*Hydrogenation.* Routine hydrogenation of carotenoids is carried out using a standard, commercially available microhydrogenation apparatus.[9,10] From 50 to 500 $\mu$g of sample are used, although hydrogenation

9 R. F. Taylor and M. Ikawa, *Anal. Biochem.* **44**, 623 (1971).
10 R. F. Taylor and B. H. Davies, *J. Chromatogr.* **103**, 327 (1975).

of as little as 5 $\mu$g and as much as 20 mg of a compound is possible. The carotenoid is dissolved in 5 ml of chloroform at room temperature, and its absorbance spectrum is determined on a suitable spectrophotometer scanning from 600 to 250 nm. The sample is then placed into a micro-hydrogenation flask and made up to 25 ml in chloroform; 50 mg of platinum dioxide is added to the solution as catalyst. Hydrogenation is carried out for 2–4 hr at room temperature with stirring under a positive hydrogen gas pressure of 100 mm Hg. After this time, the catalyst is removed from the reaction mixture by filtering the solution through a sintered-glass funnel. The catalyst is washed twice with 5-ml volumes of chloroform; the resulting, colorless filtrate is concentrated *in vacuo* to 5 ml, and its absorbance spectrum is determined. A loss of the characteristic absorbance maxima of the polyene prior to hydrogenation indicates the success of hydrogenation. The solution is then reduced in volume under a stream of nitrogen to yield a sample concentration of approximately 10 $\mu$g/ml in chloroform. From 1 to 5 $\mu$l of this solution are subjected to GC analysis.

It has been found that chloroform is the solvent of choice to use for carotenoid hydrogenation. A more polar solvent, however, such as methylene chloride may be required, depending on the solubility characteristics of the polyene. Care should be taken during the removal of reduced platinum dioxide from the reaction solution by filtration. If the (reduced) catalyst is allowed to dry completely on the funnel, spontaneous ignition will result, leading to a potentially hazardous situation. Thus, it is recommended that the catalyst be kept in a small amount of solvent at all times prior to removal from the filter funnel to an appropriate collecting container.

*Acetylation.* Acetylation of xanthophylls,[11] either directly or after hydrogenation, is carried out on 50–100 $\mu$g of compound, although as little as 5 $\mu$g of sample may be used successfully. The compound is dissolved in 1 ml of dry pyridine in a 10-ml, glass-stoppered test tube, and 0.01 ml of acetic anhydride is added. The tube is flushed for 30 sec with nitrogen, then stoppered and stored in the dark at room temperature for 12–24 hr. After this time, 2 ml of distilled water are added to the reaction mixture, and the acetylated product is extracted by adding 3 ml of diethyl ether and vortexting 5 sec. The resulting (upper) ether phase is drawn off, and the (lower) aqueous phase is reextracted twice more using fresh 3-ml volumes of ether. The ether extracts are combined, washed twice with 5-ml volumes of distilled water to remove excess pyridine, and concentrated *in vacuo* to dryness. The resulting residue is dissolved in 0.2–1 ml of chloroform for GC analysis. The completeness of carotenoid acetylation is

[11] A. J. Aasen and S. Liaaen-Jensen, *Acta Chem. Scand.* **20**, 1970 (1966).

followed by thin-layer chromatographic (TLC) analysis of the original carotenoid and the resulting acetylation product on an appropriate system.

*Silylation.* Silylation of hydroxycarotenoids[12] or their perhydro derivatives is routinely carried out on 50–100 μg of compound with a lower limit of 5 μg possible. The compound is dissolved in 1 ml of dry pyridine in a 25- or 50-ml rotary evaporator flask, and 0.5 and 0.25 ml of hexamethyldisilazane and trimethylchlorosilane (both from Pierce, Rockville, Illinois), respectively, are added. The solution is flushed for 30 sec with nitrogen, the flask is stoppered, and the reaction is allowed to take place for 2 hr at room temperature or for 12 hr at 4°. After the reaction is complete, as monitored by TLC before and during the reaction, 10 ml of $CCl_4$ are added and the solution is concentrated *in vacuo* to approximately 1 ml. Additional 10-ml volumes of $CCl_4$ are added, and concentration is repeated until all traces of pyridine are removed (3–5 cycles required). The solution is then concentrated *in vacuo* to dryness, and the residue is extracted with 5-ml volumes of petroleum ether four times. The extracts are combined and reduced in volume to 0.5–1 ml prior to GC analysis. Failure to use purified petroleum ether in this method will result in numerous unidentified elution peaks upon GC analysis.

## GC Apparatus and Columns

While our studies on the GC analysis of carotenoids, retinoids, and related terpenoids have employed either a Barber–Coleman Model 5000[9] or a Pye-Unicam Series 104[10] gas chromatograph, both equipped with dual flame ionization detection systems, the results reported here are limited to those obtained using the latter instrument, since results will vary depending on the size column required by a particular instrument. Either instrument, however, proved satisfactory for terpenoid analysis.

Glass columns, 1.5 m × 4 mm (i.d.) are silylated by washing with 10% (v/v) hexamethyldisilazane in glass-redistilled toluene and then dried at 125° for 8 hr prior to packing. Column packings for the GC analysis of carotenoids and retinoids have routinely used a diatomaceous earth support and a variety of both polar and nonpolar liquid phases (Table II). It is important that the support be acid-washed, rinsed to neutrality, and then treated with a silylating agent such as dimethylchlorosilane to block any polar surface groups capable of interacting with compounds being analyzed and leading to irreversible binding and/or a loss of peak resolution and symmetry. Supports such as Gas Chrom Q (Applied Science Laboratories, Inc., State College, Pennsylvania), Chromosorb W AW-

---

[12] A. M. McCormick and S. Liaaen-Jensen, *Acta Chem. Scand.* **20**, 1989 (1966).

TABLE II

TYPICAL COLUMN PACKINGS USED FOR GC ANALYSIS OF CAROTENOIDS AND RETINOIDS

| Compounds | Packing(s) | Reference |
| --- | --- | --- |
| $C_{40}$ Perhydrocarotenes | 5% Silicone gum rubber on Firebrick | Nicolaides[a] |
| $C_{40}$ Perhydrocarotenes | 5% SE-30 on Chromosorb W | Anderson and Porter[b] |
| Retinoids | 1–3% SE-30 on Gas Chrom P | Dunagin and Olson[c] |
| $C_{40}$ Perhydrocarotenes | 1–3% SE-30 on Gas Chrom Q | Porter[d] |
| $C_{20}$–$C_{40}$ Perhydrocarotenes and xanthophylls | 5% HVG on Chromosorb W | Taylor and Ikawa[e] |
| $C_{30}$–$C_{40}$ Perhydrocarotenes | 5% SE-30 on Chromosorb G | Kushwaha et al.[f] |
| Retinoids | 7% Silicone QF-1 on Chromosorb W-HP | Vecchi et al.[g] |
| $C_{20}$–$C_{50}$ Perhydrocarotenes and xanthophylls | 2% SE-52 on Gas Chrom Q; 2% HVG on Chromosorb W; 3% OV-17 on Universal B | Taylor and Davies[h] |

[a] N. Nicolaides, *J. Chromatogr.* **4**, 496 (1960).

[b] D. G. Anderson and J. W. Porter, *Arch. Biochem. Biophys.* **97**, 509 (1962).

[c] P. E. Dunagin, Jr., and J. A. Olson, this series, Vol. 15, p. 289 (1969).

[d] J. W. Porter, *Pure Appl. Chem.* **20**, 449 (1969).

[e] R. F. Taylor and M. Ikawa, *Anal. Biochem.* **44**, 623 (1971).

[f] S. C. Kushwaha, E. L. Pugh, J. K. G. Kramer, and M. Kates, *Biochim. Biophys. Acta* **260**, 492 (1972).

[g] M. Vecchi, J. Vesely, and G. Oesterhelt, *J. Chromatogr.* **83**, 447 (1973).

[h] R. F. Taylor and B. H. Davies, *J. Chromatogr.* **103**, 327 (1975).

DMCS (Johns-Manville, Denver, Colorado), and Universal B (Phase-Sep Ltd., Flinshire, U.K.) are preferred. Low column loadings of low-polarity, silicone-based liquid phases appear to be the most useful for terpenoid analysis, resulting in rapid elution times, good resolution, and low column bleed in temperature-programmed runs, an important consideration for GC-MS analysis. As an example of the variety of packings that can be used for GC analysis of carotenoids, the results reported below have resulted from the use of four systems:

1. 2% Silicone gum rubber SE-52 (SE-52) on Gas Chrom Q (80-100 mesh)
2. 2% Dow-Corning high vacuum grease (HVG) on Chromosorb W AW DMCS (85–100 mesh)
3. 3% OV-17 on Universal B (85–100 mesh)
4. 2% Dexsil 300 GC (Analabs, Inc., North Haven, Connecticut) on Gas Chrom Q (100–120 mesh).

The SE-52 and OV-17 packings were purchased commercially, whereas the HVG and Dexsil packings were prepared as described previously.[9] To prepare 50 g of packing, 1 g of the liquid phase is dissolved in 200 ml of chloroform and added to 49 g of the solid support in a rotary evaporator flask. The solution is allowed to stand at room temperature for 2 hr with occasional swirling to assure saturation of the solid support by the chloroform solution. After this time, the chloroform is taken off *in vacuo*, and the resulting support is dried in an oven at 80° for 8 hr before being packed into columns.

Columns are packed in the usual manner, using a vibrating tool and low nitrogen gas pressure. The upper 4 cm of a column is plugged with glass wool which has been previously treated by washing with a 10% (v/v) solution of dimethyldichlorosilane in glass-redistilled toluene and dried at 125° for 8 hr. Prior to use, columns are conditioned at 325° for at least 72 hr with a nitrogen carrier gas flow rate of 40 ml/min. We have found that oxygen-free nitrogen is a satisfactory carrier gas for GC analysis of terpenoids provided an appropriate moisture trap is inserted on-line between the gas cylinder and column inlet. The routine injection every 12 hr of column use of Silyl-8 (Pierce, Rockford, Illinois) assured that any remaining or newly appearing active sites on the columns were blocked and also improved column performance and life. In our hands, the columns and packings described were stable indefinitely when used at temperatures below 325°, only requiring occasional replacement of the upper 10 cm of the column packing with fresh packing material due to the build up of charred material.

## GC Analysis Conditions

After extensive studies on the use of isothermal conditions for GC analysis of carotenoids,[10] it was concluded that programmed runs are the most useful in separating a variety of carotenoids in the same extraction sample. Thus, we routinely use a program proceeding from 225° to 300° with a 3°/min rate of rise after an initial isothermal period of 3 min with a nitrogen carrier gas flow rate of 60 ml/min. The column injector and detector oven temperatures are set at 325 and 350°, respectively. Utilizing such conditions, a mixture of perhydrocarotenoids and their derivatives ranging in size from 20 to 50 carbon atoms can be separated in approximately 45–60 min depending on the column used.

Carotenoid samples (1–5 $\mu$l of 1–5 mg/ml solutions in chloroform) are injected directly into the silylated glass-wool plugs in the columns. Each sample is taken up in a syringe already containing 1 $\mu$l of a 5 mg/ml solution of standard squalene in chloroform, which acts as the internal standard. Zero time, the point of reference for calculation of retention times, is at injection.

## Results of GC Carotenoid Analysis

Results from the analysis of perhydrocarotenoids and other terpenoids in the systems described above are presented in Table III. The data are reported as the relative retention time ($R_t$) of any compound with respect to squalene, the internal standard, and calculated as follows:

$$R_t = \text{perhydrocarotenoid } (t)/\text{squalene } (t)$$

where $t$ is the elution time measured from the point of injection to the center of the elution peak for any compound.

*Choice of Liquid Phase.* While all four of the liquid phases utilized are methyl silicone derivatives, each differs in respect to its retention and separation selectivity of perhydrocarotenoids. HVG is the least polar of the phases and contains only methyl groups on the silicone polymer chains. SE-52 and OV-17 contain 5 and 50% phenyl group substitution, respectively, on the silicone polymer chains and are thus increasingly more polar than HVG. Dexsil 300 GC contains a $m$-carborane unit on the methylsilicone polymer chains and has a polarity intermediate between SE-52 and OV-17. Of the four phases, SE-52 was the best suited for routine analysis of all compounds; the next most useful, in order, were HVG, OV-17, and Dexsil 300 GC. Se-52 columns also have the lowest bleed rate in programmed runs, whereas the OV-17 and Dexsil 300 GC columns tend to have high bleed rates even after weeks of conditioning and use.

TABLE III
GC ANALYSIS OF CAROTENOIDS AND RELATED TERPENOIDS[a]

| Compound and molecular size | $R_t$ on system[b,c] | | | |
|---|---|---|---|---|
| | 1 | 2 | 3 | 4 |
| Geranyllinalool ($C_{20}$) | 0.15 | 0.16 | 0.13 | 0.09 |
| Phytol ($C_{20}$) | 0.18 | 0.17 | 0.10 | 0.09 |
| Geranylgeraniol ($C_{20}$) | 0.21 | 0.22 | 0.15 | 0.13 |
| Squalene ($C_{30}$) | 1.00 | 1.00 | 1.00 | 1.00 |
| Cholestane ($C_{27}$) | 1.00 | 1.00 | 1.00 | 1.00 |
| Cholesterol ($C_{27}$) | 1.69 | 1.46 | 1.67 | 1.27 |
| Ergosterol ($C_{28}$) | 1.98 | 1.65 | 1.80 | 1.48 |
| Stigmasterol ($C_{29}$) | 2.14 | 1.82 | 1.95 | 1.63 |
| Lanosterol ($C_{30}$) | 2.31 | 1.91 | 2.11 | 1.75 |
| Lycopersene ($C_{40}$) | 3.95 | 3.15 | 2.76 | 2.86 |
| *Hydrogenation products of* | | | | |
| Retinol ($C_{20}$) | 0.11 | 0.13 | 0.07 | 0.09 |
| Retinaldehyde ($C_{20}$) | 0.10 | 0.10 | 0.06 | 0.08 |
| Crocetin ($C_{20}$) | 0.13 | 0.11 | 0.13 | 0.10 |
| Dimethylcrocetin ($C_{22}$) | 0.34 | 0.33 | 0.40 | 0.30 |
| Diethylcrocetin ($C_{24}$) | 0.45 | 0.45 | 0.50 | 0.39 |
| Bixin ($C_{25}$) | 1.14 | 1.09 | 1.20 | 0.95 |
| Methylbixin ($C_{26}$) | 1.08 | 1.08 | 1.18 | 1.05 |
| $\beta$-Apo-10'-carotenal ($C_{27}$) | 0.42 | 0.52 | 0.41 | 0.45 |
| Azafrin ($C_{27}$) | 2.08 | 1.75 | 1.80 | 1.55 |
| Methylazafrin ($C_{28}$) | 1.70 | 1.49 | 1.73 | 1.44 |
| Squalene (i.e., squalane, $C_{30}$) | 0.66 | 0.73 | 0.54 | 0.72 |
| 4,4'-Diapophytoene ($C_{30}$) | 0.67 | 0.73 | 0.53 | 0.71 |
| 4,4'-Diapophytofluene ($C_{30}$) | 0.67 | 0.74 | 0.54 | 0.72 |
| 4,4'-Diapo-$\zeta$-carotene ($C_{30}$) | 0.67 | 0.73 | 0.53 | 0.73 |
| 4,4'-Diaponeurosporene ($C_{30}$) | 0.67 | 0.73 | 0.53 | 0.73 |
| 4,4'-Diaponeurosporen-4-oic acid ($C_{30}$) | 2.40 | 2.27 | 2.16 | 1.97 |
| 4,4'-Diaponeurosporen-4-oate methyl ester ($C_{31}$) | 1.41 | 1.33 | 1.27 | 1.32 |
| 4-Hydroxy-4,4'-diaponeurosporene ($C_{30}$) | 0.75 | 0.80 | 0.60 | 0.77 |
| $\beta$-Apo-8'-carotenol ($C_{30}$) | 0.82 | 0.85 | 0.70 | 0.76 |
| 3,4-Dehydro-$\beta$-apo-8'-carotenal ($C_{30}$) | 0.82 | 0.85 | 0.69 | 0.75 |
| $\beta$-Apo-8'-carotenoic acid ($C_{30}$) | 1.57 | 1.48 | 1.40 | 1.53 |
| $\beta$-Apo-8'-carotenoic acid methyl ester ($C_{31}$) | 1.61 | 1.54 | 1.46 | 1.56 |
| $\beta$-Apo-8'-carotenoic acid ethyl ester ($C_{32}$) | 1.77 | 1.59 | 1.50 | 1.60 |
| $\beta$-Apo-4'-carotenal ($C_{35}$) | 2.10 | 1.84 | 1.44 | 1.71 |
| Lycopersene (i.e., lycopersane, $C_{40}$) | 3.15 | 2.67 | 1.76 | 2.40 |
| Phytoene ($C_{40}$) | 3.14 | 2.64 | 1.77 | 2.40 |
| Phytofluene ($C_{40}$) | 3.15 | 2.66 | 1.75 | 2.42 |
| $\zeta$-Carotene ($C_{40}$) | 3.13 | 2.66 | 1.78 | 2.41 |
| Neurosporene ($C_{40}$) | 3.14 | 2.68 | 1.77 | 2.41 |
| Lycopene ($C_{40}$) | 3.15 | 2.66 | 1.75 | 2.40 |
| $\gamma$-Carotene ($C_{40}$) | 3.66 | 2.83 | 2.24 | 2.71 |
| $\beta$-Zeacarotene ($C_{40}$) | 3.66 | 2.83 | 2.24 | 2.71 |

*(continued)*

TABLE III (*continued*)
GC ANALYSIS OF CAROTENOIDS AND RELATED TERPENOIDS[a]

| Compound and molecular size | $R_t$ on system[b,c] | | | |
|---|---|---|---|---|
| | 1 | 2 | 3 | 4 |
| $\beta$-Carotene ($C_{40}$) | 3.81 | 2.88 | 2.43 | 3.03 |
| $\alpha$-Carotene ($C_{40}$) | 3.81 | 2.87 | 2.43 | 3.02 |
| Dehydro-$\beta$-carotene ($C_{40}$) | 3.81 | 2.88 | 2.43 | 3.02 |
| Carotinin ($C_{40}$) | 3.81 | 2.88 | 2.42 | 3.03 |
| Rubixanthin ($C_{40}$) | 4.15 | 2.94 | 2.55 | 2.99 |
| Echinenone ($C_{40}$) | 4.65 | 3.59 | 3.19 | 4.64 |
| Canthaxanthin ($C_{40}$) | 5.90 | 4.20 | 3.66 | 5.43 |
| $\beta$-Carotenone ($C_{40}$) | 5.33 | 4.12 | 3.09 | 5.33 |
| Torularhodin ($C_{40}$) | 4.23 | 2.99 | 2.58 | 4.02 |
| Capsanthin ($C_{40}$) | 5.13 | 3.22 | 2.81 | 4.16 |
| Astacene ($C_{40}$) | 5.79 | 3.58 | 3.00 | 4.66 |
| Physalien ($C_{40} + 2C_{16}$) | | | | |
|   $C_{40}$ fragment | 3.90 | 2.92 | 2.42 | 3.67 |
|   Acyl fragment (?) | 0.17 | 0.14 | 0.13 | 0.11 |
| Cryptoxanthin ($C_{40}$) | 4.14 | 3.33 | 2.55 | 3.40 |
| Cryptoxanthin, Ac[d] | 4.10 | 3.28 | 2.53 | 3.29 |
| Cryptoxanthin, TMS | 4.12 | 3.31 | 2.54 | 3.35 |
| Isocryptoxanthin ($C_{40}$) | 3.92 | 2.91 | 2.53 | 2.99 |
| Isocryptoxanthin, Ac | 3.78 | 2.81 | 2.44 | 2.75 |
| Isocryptoxanthin, TMS | 4.07 | 3.02 | 2.51 | 3.01 |
| Zeaxanthin ($C_{40}$) | 4.68 | 3.45 | 3.30 | 3.41 |
| Zeaxanthin diAc | 5.04 | 3.50 | 2.98 | 3.47 |
| Zeaxanthin, diTMS | 5.07 | 3.55 | 3.05 | 3.58 |
| Isozeaxanthin ($C_{40}$) | 4.16 | 3.00 | 2.59 | 3.15 |
| Isozeaxanthin diAc | 4.31 | 3.14 | 2.47 | 3.22 |
| Isozeaxanthin, diTMS | 4.35 | 3.17 | 2.53 | 3.31 |
| Dimethoxyzeaxanthin ($C_{42}$) | 5.15 | 3.67 | 3.44 | 3.93 |
| Dimethoxisozeaxanthin ($C_{42}$) | 4.02 | 2.88 | 2.37 | 3.01 |
| Fucoxanthin ($C_{42}$) | 6.24 | 3.49 | 3.42 | 4.51 |
| Decapreno-$\beta$-carotene ($C_{50}$) | 7.54 | 5.58 | 5.32 | 5.89 |
| *Retention time* (*min*) | | | | |
| Squalene | 5.50 | 8.65 | 11.15 | 10.69 |

[a] R. F. Taylor and B. H. Davies, *J. Chromatogr.* **103**, 327 (1975); R. F. Taylor, unpublished results.

[b] Column systems: 1, SE-52; 2, HVG; 3, OV-17; 4, Dexsil 300 GC. See text for details.

[c] Retention times relative to squalene in a temperature program proceeding from 225° to 300° with a 3°/min rate of rise after an initial isothermal period of 3 min.

[d] Ac, acetate of perhydromonohydroxycarotenoid; diAc, diacetate of perhydrodihydroxycarotenoid; TMS, TMS ether of perhydromonohydroxycarotenoid; diTMS, TMS ether of perhydrodihydroxycarotenoid.

A detailed comparison of these liquid phases as applied to GC analysis of carotenoids has been presented elsewhere.[10] In summary, the following comments may be made concerning the behavior of perhydrocarotenoids on each phase.

Perhydrocarotenes, -xanthophylls, and -xanthophyll derivatives have the lowest retention times and elute with the sharpest peaks from the SE-52 column. While this makes the SE-52 column useful over the entire carotenoid molecular size range, problems may arise during the separation of $C_{20}$–$C_{35}$ carotenoids owing to rapid elution. Such rapid elution is not a problem with the HVG column, and, in fact, most compounds have nearly twice the retention time on HVG columns in comparison to SE-52 columns. While these increased retention times make the HVG column very useful in the $C_{20}$–$C_{35}$ range, high molecular weight carotenes and especially xanthophylls tend to elute as broad and tailing peaks from such columns. Dexsil 300 GC columns behave in a manner similar to HVG columns, but with the additional disadvantage that most perhydrocarotenoids have their longest retention times on such columns.

When the behavior of carotenoids differing with respect to structural and functional groups, but not necessarily molecular size, is compared (see below) on the four systems, it is found that SE-52, HVG, and Dexsil 300 GC columns tend to separate first by molecular weight and then by molecular shape, symmetry, and nature of functional group, whereas the OV-17 column is more selective in separating perhydrocarotenoids according to factors other than molecular weight. This can be illustrated by comparing the $R_t$ values for a number of compounds that have the same number of carbon atoms but differ with respect to double bonds, functional groups, or molecular symmetry (Table IV). Thus, the OV-17 column best separates compounds containing carbon-carbon unsaturation from their saturated derivative (such as squalene and squalane); acyclic, monocyclic, and bicyclic perhydrocarotenes; and hydroxycarotenoids from their derived acetates and trimethylsilyl (TMS) ethers. While the ratio method used in Table IV for the direct comparison of liquid phases can be useful for the selection of the phase to be used, the method must be used carefully since it assumes that all the phases compared have equal resolving powers. Thus, for example, while the OV-17 column is very useful for separating perhydrocarotenes and derivatives of mono- or bifunctional xanthophylls, it does not have high resolving capacity for mixtures of high molecular weight and multifunctional group perhydrocarotenoids.

*Structure and Retention Time.* The phase comparison ratios in Table IV illustrate the value of perhydrocarotenoid GC analysis for identification of basic structural features in an unknown compound. Perhydrocarotenes

TABLE IV

SEPARATION CHARACTERISTICS OF GC PACKINGS IN RELATION TO CAROTENOID STRUCTURE[a]

| Compounds | $R_t$ Ratio[b] on: | | | |
| --- | --- | --- | --- | --- |
| | SE-52 | HVG | DEXSIL 300 GC | OV-17 |
| Squalene : squalane | 100 : 66 | 100 : 73 | 100 : 72 | 100 : 54 |
| Lycopersene : lycopersane | 100 : 80 | 100 : 85 | 100 : 84 | 100 : 64 |
| *Perhydro derivatives of* | | | | |
| β-Carotene : γ-carotene : lycopene | 100 : 96 : 83 | 100 : 98 : 92 | 100 : 89 : 79 | 100 : 92 : 72 |
| Zeaxanthin : isozeaxanthin | 100 : 89 | 100 : 87 | 100 : 90 | 100 : 78 |
| Dimethoxyzeaxanthin : zeaxanthin disilyl ether : zeaxanthin diacetate : zeaxanthin | 100 : 98 : 98 : 91 | 100 : 97 : 95 : 94 | 100 : 91 : 88 : 89 | 100 : 89 : 87 : 96 |

[a] $R_t$ data taken from Table III.

[b] The $R_t$ ratio was calculated by dividing the $R_t$ of each component in a comparison series by the largest $R_t$ in that series and multiplying by 100.

are found to have retention characteristics such that elution time follows the sequence

$$\text{acyclic} < \text{monocyclic} < \text{bicyclic}$$

indicating that the presence of a cyclic end group increases column retention time. An example of perhydrocarotene GC analysis is illustrated in Fig. 2, which also emphasizes the value of GC analysis for determination of carbon chain length in a perhydrocarotene. Perhydroketocarotenoids, such as perhydroechinenone and perhydrocanthaxanthin, are less volatile than their corresponding perhydrohydroxycarotenoid analogs, i.e., perhydroisocryptoxanthin and perhydroisozeaxanthin, respectively. Since catalytic hydrogenation does not reduce carbonyl double bonds, this may indicate that enolization of a keto group in a perhydroketocarotenoid results, effectively, in an olefinic bond and a resulting decrease in volatility with respect to the corresponding perhydrohydroxycarotenoid.

Positional isomers of perhydrohydroxycarotenoids can also be separated by GC. For example, the perhydro derivatives of the mono- and di-4-hydroxycarotenoids isocryptoxanthin and isozeaxanthin are each more volatile than their corresponding mono- and di-3-hydroxycarotenoid isomers, perhydrocryptoxanthin and perhydrozeaxanthin. This volatility difference may be due to more effective shielding of the oxygen sub-

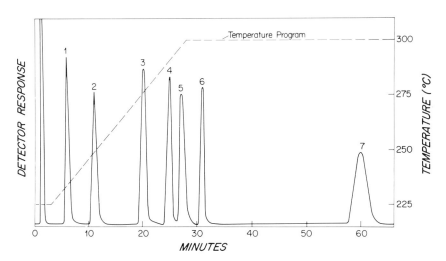

FIG. 2. Gas chromatography (GC) analysis of a mixture of perhydrocarotenes, squalene, and lycopersene on a column of 3% OV-17 on Universal B using a temperature program from 225° to 300°C at 3°/min increase after an initial isothermal period of 3 min. 1, Perhydro-4,4'-diaponeurosporene; 2, squalene; 3, perhydrolycopene; 4, perhydro-γ-carotene; 5, perhydro-β-carotene; 6, lycopersene; 7, perhydrodecapreno-β-carotene.

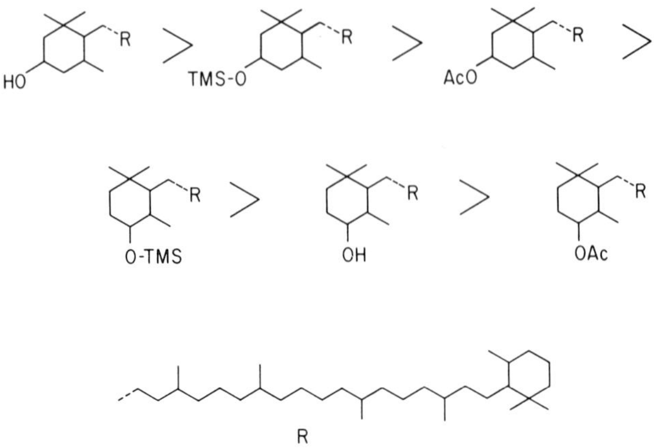

FIG. 3. Structural effects on retention times of perhydrocryptoxanthin, perhydro-isocrypyoxanthin, and their acetylation and silylation derivatives. The compounds are ordered with respect to decreasing retention times (increasing GC volatility).

stituent at $C_4$ of the cyclic end groups by intramolecular folding than if such substitution is at $C_3$. Acetylation and silylation of perhydrohydroxy-carotenoids also results in changes in retention times with respect to the parent compound and can reveal useful information concerning the location of a hydroxyl group. For example, comparison of the retention times for perhydrocryptoxanthin, perhydroisocryptoxanthin, and their acety-lated and silylated derivatives on all four phases, reveals a distinct elution pattern which, again, appears related to the position of the hydroxyl group of the cyclic end group (Fig. 3). More complex patterns are observed when dihydroxy compounds such as perhydrozeaxanthin and perhydro-isozeaxanthin are derivatized but, again, lead to further differentiation of these isomers.

An example of the GC analysis of a variety of perhydroxanthophylls is illustrated in Fig. 4.

*Separation of $C_{30}$ and $C_{40}$ Carotenoids.* The utility of GC analysis to determine major carotenoid structural features is illustrated by the behavior of a number of $C_{30}$ perhydrocarotenoids on the HVG column. The occurrence of $C_{30}$ carotenoids as a new class of carotenoids was established from our initial[13] and later[3,5,6] studies on *Streptococcus faecium* and *Staphylococcus aureus*. In summary, all the carotenes and xanthophylls found in these two bacteria appear to be biosynthesized via a pathway similar to squalene, resulting in pigments with a symmetrical triterpenoid carbon skeleton. The first indications of this unique $C_{30}$ structure resulted from GC analysis of the bacterial carotenes and the rather startling observation that the perhydrocarotenes coeluted not with perhydrolycopene, as

[13] R. F. Taylor, M. Ikawa, and W. Chesbro, *J. Bacteriol.* **105**, 676 (1971).

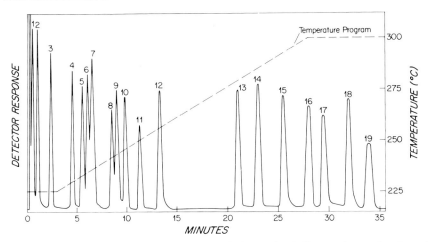

FIG. 4. GC analysis of a mixture of perhydroxanthophylls, squalene, phytol, and perhydroretinol using a column of 2% SE-52 on Gas Chrom Q and the same temperature program described in Fig. 2. 1, Perhydroretinol; 2, phytol; 3, perhydro-β-apo-10′-carotenal; 4, perhydro-β-apo-8′-carotenal; 5, squalene; 6, perhydromethylbixin; 7, perhydrobixin; 8, perhydro-β-apo-8′-carotenoid acid; 9, perhydro-β-apo-8′-carotenoate methyl ester; 10, perhydro-β-apo-8′-carotenoate ethyl ester; 11, perhydroazafrin; 12, perhydro-4,4′-diaponeurosporen-4-oic acid; 13, perhydroisocryptoxanthin; 14, perhydrocryptoxanthin; 15, perhydroechinenone; 16, perhydrocapsanthin; 17, perhydro-β-carotenone; 18, perhydrocanthaxanthin; 19, perhydrofucoxanthin.

expected, but with perhydrosqualene. Subsequent examinations revealed that not only the carotenes were triterpenoids, but the xanthophylls as well, the xanthophylls being derived from the major end product of carotene biosynthesis in the bacteria, 4,4′-diaponeurosporene. Thus, in our studies on the various xanthophylls in both bacteria, a number of carotenoids, including those listed in Table V, were isolated and characterized by GC. Examination of Table V illustrates that GC can be a valuable aid in gaining initial and rapid information concerning the molecular size of a carotenoid as well as the nature of any functional groups present. For example, the perhydro derivative of the carotenoid aldehydes in *Streptococcus faecium,* 4,4′-diapolycopen-4-al or 4,4′-diaponeurosporen-4-al are the most volatile of the xanthophylls on a HVG column, whereas the perhydro derivative of the major xanthophyll in *Staphylococcus aureus,* 4,4′-diaponeurosporen-4-oate, is the least volatile. Chemical modification of the latter acid to its methyl ester results in a marked increase in volatility, as expected. Silylation and acetylation of perhydro-4-hydroxy-4,4′-diaponeurosporene results in decreased volatility in comparison to the free hydroxy parent compound. While not listed in Table V, the perhydro derivative of the major xanthophyll in *Strep-*

TABLE V

GC ANALYSIS OF THE PERHYDRO DERIVATIVES OF 4,4'-DIAPONEUROSPORENE, DERIVED
XANTHOPHYLLS, AND XANTHOPHYLL DERIVATIVES

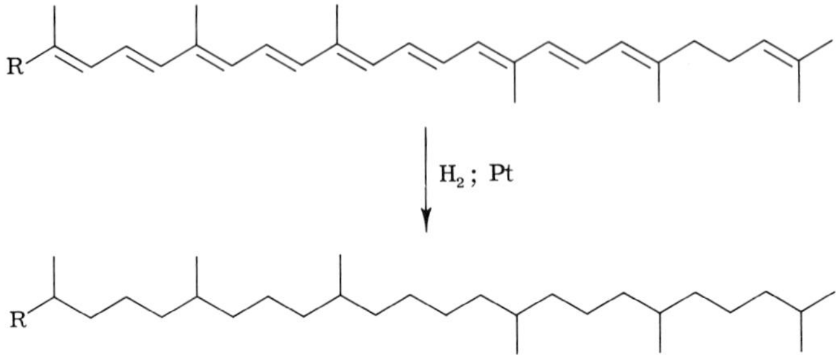

| R | Perhydroderivative of[a] | $R_t$[b] |
|---|---|---|
| H$_3$C- | 4,4'-Diaponeurosporene | 0.73 |
| OHC- | 4,4'-Diaponeurosporen-4-al | 0.76 |
| HOH$_2$C- | 4-Hydroxy-4,4'-diaponeurosporene | 0.80 |
| (H$_3$C)$_3$Si-OH$_2$C- | 4-Hydroxy-4,4'-diaponeurosporene silyl ether | 0.82 |
| H$_3$CCO-OH$_2$C- | Acetylated 4-hydroxy-4,4'-diaponeurosporene | 0.91 |
| H$_3$COOC- | 4,4'-Diaponeurosporen-4-oic acid methyl ester | 1.33 |
| HOOC- | 4,4'-Diaponeurosporen-4-oic acid | 2.27 |

[a] For semisystematic names, see Table I.
[b] Conditions: standard programmed run on 2% HVG on Chromosorb W. All times relative to squalene.

*tococcus faecium,* 4-D-glucopyranosyloxy-4,4'-diaponeurosporene, has also been analyzed on the HVG column. At least four peaks resulted upon each analysis, possibly indicating degradation/rearrangement of the glycosidic perhydrocarotenoid on the GC column. Figure 5 illustrates the GC separation of a number of perhydroapocarotenoids, including the bacterial triterpenoid carotenoids.

GC-MS Analysis of Carotenoids

The successful GC analysis of perhydrocarotenoids has also led to our development and use of combined GC-MS analysis of these compounds. Such GC-MS analysis of perhydrocarotenoids (and their derivatives) together with MS analysis of the original carotenoid prior to hydrogenation provides confirming proof of structural identity.

GC-MS of perhydrocarotenoids may be carried out on any commercially available instrument. Our studies were carried out on an AEI-MS30/DE-50 computerized mass spectrometer linked by a silicone rubber

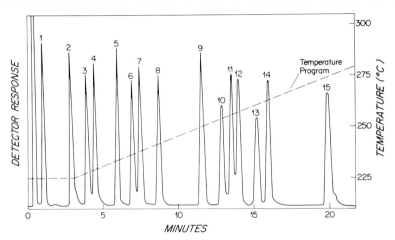

Fɪɢ. 5. GC analysis of a mixture of perhydroapocarotenoids and squalene using a column of 2% HVG on Chromosorb W and the same temperature program described in Fig. 2. 1, Perhydrocrocetin; 2, perhydrodimethylcrocetin; 3, perhydrodiethylcrocetin; 4, perhydro-β-apo-10'-carotenal; 5, perhydro-4,4'-diaponeurosporene; 6, perhydro-4-hydroxy-4,4'-diaponeurosporene; 7, perhydro-β-apo-8'-carotenal; 8, squalene; 9, perhydro-4,4'-diaponeurosporen-4-oate methyl ester; 10, perhydro-β-apo-8'-carotenoic acid; 11, perhydro-β-apo-8'-carotenoate methyl ester; 12, perhydro-β-apo-8'-carotenoate ethyl ester; 13, perhydroazafrin; 14, perhydro-β-apo-4'-carotenal; 15, perhydro-4,4'-diaponeurosporen-4-oic acid.

separator to a Pye-Unicam Series 104 gas chromatograph. The conditions we found to be effective for analysis in this particular system should be used only as a guide in defining the conditions necessary for analysis in other systems.

Prior to GC-MS, studies must be carried out to choose a column and conditions that achieve good separation of the perhydrocarotenoids with a low column bleed rate. While the use of a linked computer can compensate for nominal liquid phase bleed during a GC-MS run, extensive bleed will detract from analysis and especially the analysis of minor components in a mixture. We routinely use the SE-52 column which, as mentioned above, has a very low bleed during programmed runs. In our system, the standard temperature program for 225° to 300° is used. Injection port and sample lines are maintained at 325° and the silicone separator at 230°. Helium is used as the carrier gas in GC-MS, and flow rate is maintained at 40 ml/min. The electron current of the mass spectrometer is maintained at 24 eV with a source temperature of 225°.

The first successful use of GC-MS for carotenoid identification was applied to the triterpenoid xanthophylls from *Staphylococcus aureus* 209P[14] and confirmed the identity of 4,4'-diaponeurosporen-4-oate and a

[14] R. F. Taylor and B. H. Davies, *Int. Symp. Carotenoids, 4th, Abstr.* Berne, p. 66 (1975).

number of its derivatives. Additional studies have also been performed on all the $C_{30}$ carotenes from *Streptococcus faecium* and *Staphylococcus aureus* and illustrate the utility of GC-MS analysis in determining carotenoid structure. For example, when any of the perhydrocarotenes from the bacteria are analyzed by GC-MS, a fragmentation pattern results identical to that of perhydrosqualene (squalane). This pattern is similar to those found upon MS analysis of branched-chain, saturated hydrocarbons[15] and is characterized by clusters of peaks with the corresponding peaks of each cluster 14 mass units apart. The most intense peak of each cluster is defined as a $C_nH_{2n+1}$ fragment, and these clusters begin in the M-47 region and continue for the full spectral span. The $C_{30}$ perhydrocarotenes also contain M-15 and M-27 peaks indicating fission at their terminal methyl branches. As expected with branched-chain compounds, chain fission is favored at branch points, i.e., those bonds adjacent to carbons 5, 9, 13, 5', 9', and 13' (see Fig. 1) carrying branch methyl groups. This characteristic fission pattern at branch points can, in fact, be used to determine the symmetry of a $C_{30}$ carotenoid, a question that cannot be resolved by MS analysis alone, and which requires amounts of carotenoids usually in excess of that obtainable from natural sources for determination by proton magnetic resonance spectroscopy. As an example of such a GC-MS symmetry determination, consider results obtained after analysis of perhydro-4,4'-diapo-$\zeta$-carotene, the hydrogenated derivative of the naturally occurring $C_{30}$ heptaene in *Streptococcus faecium* and *Staphylococcus aureus*. Two structures are possible for this compound, differing only in the arrangement of the four internal methyl branch groups (Fig. 6). If the compound is symmetrical, i.e., arises biosynthetically from the condensation of two $C_{15}$ units,[3] then fragmentation from each end of the molecule would favor ion losses of 85, 113, 155, and 183 mass units about carbon atoms 9, 13, 9', and 13'. As a result, the ion intensities of spectral ions at $m/e$ 337 (M-85), 309 (M-113), 267 (M-155), and 239 (M-183) would be similar. On the other hand, if the molecule is unsymmetrical, i.e., formed (presumably) from the degradative loss of a $C_{10}$ fragment from one end of a $C_{40}$ carotenoid, then additional ion losses of 169 and 197 mass units are possible and the intensities of spectral ions at $m/e$ 253 (M-169) and 225 (M-197) would be expected to be similar to those at $m/e$ 267 (M-155) and 239 (M-183). As shown in Table VI, the ion intensities resulting from GC-MS analysis of perhydro-4,4'-diapo-$\zeta$-carotene confirm the symmetrical nature of the parent compound. Similar studies with the other $C_{30}$ carotenes as well as the major $C_{30}$ xanthophylls from *Streptococcus faecium* and *Staphylococcus aureus* confirmed that they, as well, are

---

[15] E. Stenhagen, *in* "Biochemical Applications of Mass Spectrometry" (G. R. Waller, ed.), p. 11. Wiley (Interscience), New York, 1972.

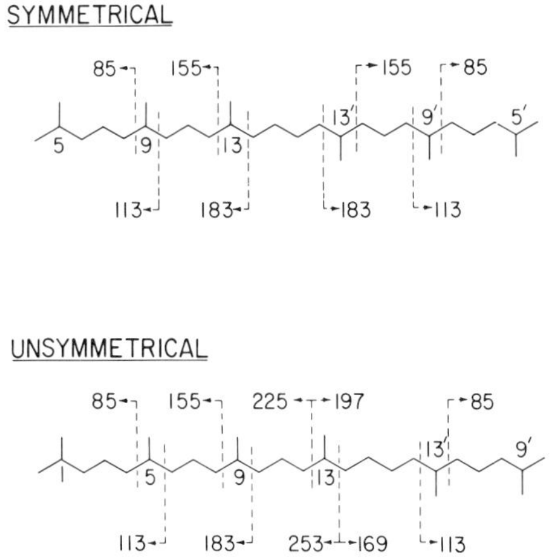

SYMMETRICAL

UNSYMMETRICAL

FIG. 6. Fragmentation patterns expected from fission at methyl branch points during GC-MS analysis of a symmetrical and unsymmetrical perhydrotriterpenoid carotene.

symmetrical, a finding consistent with biosynthetic proof of this symmetry.[3]

## GC of Retinoids

GC analysis of retinol, retinal, and related compounds (retinoids) is limited by the same consideration of heat instability as are the carotenoids. While a standard method has been proposed for the routine GC analysis of retinoids per se,[16] problems with the formation and separation of isomers, on-column degradation, and quantitation remain. The attempted GC analysis of retinoid-TMS ethers has also encountered problems with isomerization and on-column degradation.[17] Our own experience with GC analysis of retinoids per se confirms these problems. While hydrogenation of retinoids prior to GC can be used as described for carotenoids, such hydrogenation limits the usefulness of GC for retinoid identification. For these reasons and also owing to the recent emergence of HPLC for the rapid and nondestructive analysis of retinoids (see below), GC analysis of retinoids appears to be limited as a routine analytical method.

## HPLC of Carotenoids and Retinoids

While GC and GC-MS remain useful tools for the structural characterization of carotenoids and retinoids, the necessity of prior hydrogena-

[16] P. E. Dunagin, Jr., and J. A. Olson, this series, Vol. 15, p. 289.
[17] M. Vecchi, J. Vesely, and G. Oesterhelt, *J. Chromatogr.* **83**, 447 (1973).

TABLE VI

Ion Intensities of Branch Point Fission Fragments Resulting from GC-MS Analysis of Perhydro-4,4′-diapo-ζ-carotene

| $m/e$ | Fragment[a] | Relative intensity[b] |
|---|---|---|
| 337 | M-85 | 28.9 |
| 309 | M-113 | 22.6 |
| 267 | M-155 | 27.6 |
| 239 | M-183 | 21.6 |
| 183 | — | 58.5 |
| 155 | — | 47.2 |
| 113 | — | 65.6 |
| 85 | — | 88.7 |
| 253 | M-169 | 4.2 |
| 225 | M-197 | 5.9 |
| 197 | — | 12.3 |
| 169 | — | 5.6 |

[a] Molecular ion at $m/e$ 422.

[b] Relative intensity calculated with respect to the base peak at $m/e$ 57; Average of three determinations.

tion for such analysis limits these methods. HPLC, however, does not require such prior hydrogenation, and this rapidly emerging method promises to greatly aid the rapid and nondestructive analysis of terpenoid polyene compounds.

The advantages of HPLC for carotenoid and retinoid analysis are obvious: rapid and selective separations resulting in purified compounds which may be collected and further analyzed. This is especially pertinent with respect to the classical methods used to analyze and quantitate retinoids. These methods are based on either colorimetric assay methods,[18] such as the Carr–Price[19] and trichloroacetic acid[20] tests, or on spectrophotometric assay methods of retinoids per se.[21,22] More detailed descriptions of such methods are described in preceding sections of this volume.[23] While such methods will remain of importance for the quantitation of retinol, the increasing study and use of other retinoids require more specific methods for separation and identification of these compounds, and HPLC appears to be suited to this purpose. It is notable, for example,

[18] G. B. Subramanyan and D. B. Parrish, Assoc. Off. Anal. Chem. 59, 1125 (1976).

[19] Official Methods of Analysis, Association of Official Analytical Chemists Washington, D. C., 1975.

[20] R. F. Bayfield, Anal. Biochem. 39, 282 (1971).

[21] J. H. Thompson, P. Erdody, R. Brien, and T. K. Murray, Biochem. Med. 5, 67 (1971).

[22] O. A. Roels and S. Mahadevan, Vitamins 6, 139 (1967).

[23] See this volume [23,24].

that recent studies have confirmed that HPLC analysis matches and sur-
passes classical methods in speed and quantitation for the determination
of retinol in food products.[24,25]

HPLC of carotenoids and retinoids is possible on any of the variety of
commercially available instruments. It is desirable, however, that the
system used be capable of achieving pressures up to at least 3000 psi and
contain a gradient elution capacity. As listed in Table VII, a variety of
column packings and mobile phases have been used for carotenoid and
retinoid HPLC analysis. In general, the most useful packings appear to be
magnesium oxide (carotenes), micro particles (5–10 $\mu$m particle size) of
porous silica such as Zorbax SIL, $\mu$Porasil, Vydac, LiChrosphere, etc.
(retinoids and xanthophylls), and reverse-phase supports such as $\mu$Bon-
dapak $C_{18}$, Vydac ODS, Permaphase-ODS, etc. (most carotenoids and
retinoids). Normal phase solvents routinely contain $n$-hexane (or petroleum ether) with low or increasing concentrations of a polar solvent
whereas reverse-phase systems utilize mixtures of water or buffers in
alcohols or acetonitrile. Detection of most carotenoids is usually at 400–
450 nm, and retinoids may be detected by either UV absorbance between
310 and 350 nm or by exciting between 325 and 365 nm and detecting
fluorescence at 510 nm.[24] HPLC detection limits for both retinoids and
carotenoids are routinely in the low nanogram range.

Due to the possibility that carotenoids and retinoids may exist in samples as isomers or may isomerize during routine handling procedures, it is
advisable to use a nonterpenoid internal standard in each analysis run to
detect such isomerization and also to compensate for variations in column
performance between runs. These standards must, of course, themselves
have absorbance maxima near those of the polyene components being
analyzed. Such standards have included naphthalene, ethyl benzoate,
tetraphenylethylene, and 2,6-di-*tert*-butyl-4-hydroxytoluene.

The preparation of carotenoid and retinoid samples for HPLC analysis
follows routine procedures. Thus, for example, methanol extracts of
serum may be analyzed directly for retinol,[26] whereas food products require prior saponification before analysis of the nonsaponifiable lipid for
retinoids,[24,25] as do citrus fruits prior to carotenoid HPLC analysis.[27] Animal tissues require either saponification,[28] precipitation, or lyophilization[29] prior to extraction of retinoids and carotenoids. In most

[24] D. C. Egberg, J. C. Heroff, and R. H. Potter, *J. Agric. Food Chem.* **25,** 1127 (1977).

[25] D. B. Dennison and J. R. Kirk, *J. Food Sci.* **42,** 1376 (1977).

[26] M. G. M. De Ruyter and A. P. De Leenheer, *Clin. Chem.* **22,** 1593 (1976).

[27] I. Stewart, *J. Agric. Food Chem.* **25,** 1132 (1977).

[28] R. F. Taylor and L. Gaudio, to be published.

[29] A. B. Roberts, M. D. Nichols, C. A. Frolik, D. L. Newton, and M. B. Sporn, *Cancer Res.*
**38,** 3327 (1978).

TABLE VII
TYPICAL HPLC SYSTEMS USED FOR CAROTENOID AND RETINOID ANALYSIS[a]

| Compounds and source | Column packing(s) | Typical mobile phase(s)[a] | Reference |
|---|---|---|---|
| Vegetable carotenes | 1. MgO-Hyflo Super-Cel (1:2)<br>2. Mg(OH)$_2$-Ca(OH)$_2$ (1:6) | 0.5–5% Acetone or $p$-methylanisole in petroleum ether (1 and 2) | Sweeney and Marsh[b] |
| Vegetable/fruit carotenes and xanthophylls | 1. MgO (Sea Sorb 43)<br>2. ZnCO$_3$ (precipitated) | Gradients of $tert$-pentyl alcohol in $n$-hexane containing 1% BHT (1 and 2) | Stewart and Wheaton[c] |
| Fat-soluble vitamins including retinoids | 1. Permaphase ODS (Dupont)<br>2. Zipax HCP (Dupont) | Gradients of 5 to 25% water in methanol or 2-propanol (1 and 2) | Williams et al.[d] |
| Retinoids | 1. Corasil II (Waters)<br>2. Merckosorb SI 60 (E. Merck)<br>3. Al$_2$O$_3$<br>4. Corasil C$_{18}$ (Waters) | 0.1–5% Dioxane in $n$-hexane (1–3); 4% water in methanol (4) | Vecchi et al.[e] |
| Orange juice carotenes and xanthophylls | 1. Al$_2$O$_3$ (Woelm B18)<br>2. Spherisorb (Chromatronix) | Benzene (1) or tetrahydrofuran (2) in $n$-hexane containing 0.01% BHT | Reeder and Park[f] |
| Serum retinol | MicroPak Si-10 (Varian) | Petroleum ether–dichloromethane–2-propanol mixture | De Ruyter and De Leenheer[g] |
| Retinol isomers in foods | 1. Vydac 10 $\mu$m ODS (Separations group)<br>2. Zorbax 5 $\mu$m (Dupont) | Mixtures of acetonitrile in water (1) and hexane–methylene chloride–2-propanol (2) | Egberg et al.[h] |
| Citrus juice carotenes and xanthophylls | 1. MgO<br>2. Pellosil (Reeve Angel) | Gradients of acetone or $tert$-pentyl alcohol in $n$-hexane (1 and 2) | Stewart[i] |
| Retinol in cereal products | $\mu$Porasil (Waters) | 1% Ethanolic chloroform in $n$-hexane | Dennison and Kirk[j] |

| | | | |
|---|---|---|---|
| Retinal isomers | Zorbax SIL (Dupont) | 12% Diethyl ether in n-hexane | Tsukida et al.[k] |
| Plant carotenes and xanthophylls | Bondpak $C_{18}$-Porasil B (Waters) | Gradients of water and diethyl ether in methanol | Eskins et al.[l] |
| Retinol in serum and liver | 1. Permaphase ODS (Dupont) 2. JASCOPAK WC-03 | Mixtures of 2-propanol–ethanol–water (1) and chloroform–isooctane (2) | Abe et al.[m] |
| Retinoids in animal tissues | 5 μm Spherisorb ODS (Spectra Physics) | Acetonitrile in water | Roberts et al.[n] |
| Carotenoids in geological samples | 1. 5 μm Partisil (Whatman) 2. 10 μm Spherisorb (Phase-Sep. Ltd.) | Gradients of acetone in hexane | Hajibrahim et al.[o] |
| Retinoids and retinoid methyl esters | 1. μBondapak $C_{18}$ (Waters) 2. μBondapak Alkyl Phenyl (Waters) 3. μPorasil (Waters) | Gradients of 10 m$M$ sodium acetate in methanol (1 and 2); tetrahydrofuran in hexane (3) | McCormick et al.[p] |

[a] Numbers after a solvent indicate the column with which the solvent was used.
[b] J. P. Sweeney and A. C. Marsh, J. Assoc. Off. Anal. Chem. 53, 937 (1970).
[c] I. Stewart and T. A. Wheaton, J. Chromatogr. 55, 325 (1971).
[d] R. C. Williams, J. A. Schmit, and R. A. Henry, J. Chromatogr. Sci. 10, 494 (1972).
[e] M. Vecchi, J. Vesely, and G. Osterhelt, J. Chromatogr. 83, 447 (1973).
[f] S. K. Reeder and G. L. Park, Assoc. Off. Anal. Chem. 58, 595 (1975).
[g] M. G. M. De Ruyter and A. P. De Leenheer, Clin. Chem. 22, 1593 (1976).
[h] D. C. Egberg, J. C. Heroff, and R. H. Potter, J. Agric. Food Chem. 25, 1127 (1977).
[i] I. Stewart, J. Agric. Food Chem. 25, 1132 (1977).
[j] D. B. Dennison and J. R. Kirk, J. Food Sci. 42, 1376 (1977).
[k] K. Tsukida, A. Kodama, and M. Ito, J. Chromatogr. 134, 331 (1977).
[l] K. Eskins, C. R. Scholfield, and H. J. Dutton, J. Chromatogr. 135, 217 (1977).
[m] K. Abe, K. Ishibashi, M. Ohmae, K. Kawabe, and G. Katsui, Vitamins 51, 275 (1977).
[n] A. B. Roberts, M. D. Nichols, C. A. Frolik, D. L. Newton, and M. B. Sporn, Cancer Res. 38, 3327 (1978).
[o] S. K. Hajibrahim, P. J. C. Tibbetts, C. D. Watts, J. R. Maxwell, G. Eglinton, H. Colin, and G. Guiochon, Anal. Chem. 50, (1978).
[p] A. M. McCormick, J. L. Napoli, and H. F. DeLuca, Anal. Biochem. 86, 25 (1978).

TABLE VIII
HPLC OF VARIOUS RETINOIDS AND CAROTENOIDS[a]

| | Retention time (min) in system[b–d] | | |
|---|---|---|---|
| Compound | 1 | 2 | 3 |
| Retinol | 12.3 | 12.5 | 12.5 |
| Retinal | 11.1 | 13.0 | 14.0 |
| all-*trans*-Retinoic acid | 19.4 | 7.5 | 5.0 |
| 13-*cis*-Retinoic acid | 19.4 | 7.5 | 6.5 |
| Retinyl acetate | sf | 16.5 | 16.0 |
| Retinyl palmitate | sf | 23.0 | 20.0 |
| Methylbixin | r | 5.0 | nd |
| $\beta$-Apo-8'-carotenoic acid ethyl ester | 6.5 | 19.0 | nd |
| $\alpha$-Carotene | 3.5 | 25.0 | nd |
| $\beta$-Carotene | 3.5 | 25.5 | nd |
| Carotinin | 3.5 | 26.0 | nd |
| Echinenone | 9.3 | 22.0 | nd |
| Canthaxanthin | 10.4 | 17.2 | nd |
| $\beta$-Carotenone | 13.0 | 10.5 | nd |
| Zeaxanthin | r | 15.8 | nd |
| Isozeaxanthin | r | 15.8 | nd |

[a] L. Gaudio and R. F. Taylor, unpublished results (1978).
[b] Systems: 1, $\mu$Porasil; 2, $\mu$Bondapak $C_{18}$; 3 $\mu$Bondapak CN. See text for details.
[c] All retention times cited are the means of at least three separate determinations. The standard deviation from the mean for any determination was never more than $\pm 1.8\%$.
[d] r, retained; sf, eluted with the solvent front; nd, not determined.

cases, retinoid and carotenoid analysis on plant and animal tissues is simplified since the extracts, after volume reduction, can be injected directly onto the HPLC column for immediate separation, quantitation, and recovery.

Our experience[28] with HPLC of retinoids and carotenoids is based on studies utilizing two different liquid chromatographic systems with both normal and reverse-phase columns. The instruments used were a Waters Associates Model ALC-202/R-301 W/UV or a Micromeritics Model 701-21204-02 (Micromeritics Corp., Norcross, Georgia), both equipped with full gradient elution and a variable ultraviolet-visible spectrophotometer. In addition, the eluate from the HPLC columns passes through a flow cell in a Varian Model 635 ultraviolet-visible scanning spectrophotometer (Varian, Waltham, Massachusetts), which, by means of a stop-flow device on each chromatograph, allows determination of absorbance spectra on the

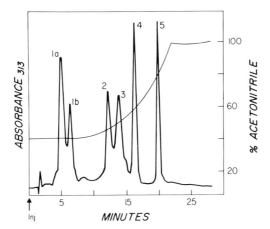

Fig. 7. HPLC of standard retinoids using system 3: μBondapak CN column eluted with a 20-min, concave gradient of 40 to 100% acetonitrile in 50 mM ammonium acetate at 2.0 ml/min. 1a, all-*trans*-Retinoic acid; 1b, 13-*cis*-retinoic acid; 2, retinol; 3, retinal; 4, retinyl acetate; 5, retinyl palmitate.

column eluate at any point during analysis. A fraction collector is used to collect the column eluate for further analysis of the separated carotenoids and retinoids.

The columns are routinely flushed with solvent for 15–30 min prior to use. Carotenoid and retinoid solutions (10–1000 μl samples in hexane, chloroform, etc.) are injected, and the appropriate gradient elution sequence is started. Typically, solvent flow rate is 2 ml/min and detection is at 440 nm (carotenoids) or 313 nm (retinoids).

Table VIII reports preliminary results obtained after analysis of various retinoids and carotenoids in any or all of the three different systems:

1. μPorasil column (3.9 mm i.d. × 30 cm) (Waters Associates, Milford, Massachusetts) using a 35-min, convex mobile phase gradient of 0 to 100% 1% ethanolic chloroform in *n*-hexane
2. μBondapak $C_{18}$ column (3.9 mm i.d. × 30 cm) (Waters Associates) using a 15-min, concave mobile phase gradient of 80 to 100% methanol in water
3. μBondapak CN column (3.9 mm i.d. × 30 cm) (Waters Associates) using a 20-min, concave mobile phase gradient of 40 to 100% acetonitrile in 50 mM ammonium acetate

As shown in Table VIII and in Figs. 7 and 8, these systems provide efficient and reproducible separations of most retinoids and carotenoids within 15–30 min. The normal phase, μPorasil column does not, however,

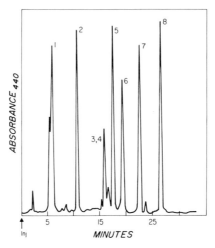

FIG. 8. HPLC of standard carotenoids using system 2: $\mu$Bondapak $C_{18}$ column eluted with a 15-min, concave gradient of 80 to 100% methanol in water at 2.0 ml/min. 1, Methylbixin; 2, $\beta$-carotenone; 3 and 4, zeaxanthin and isozeaxanthin; 5, canthaxanthin; 6, $\beta$-apo-8'-carotenoic acid ethyl ester; 7, echinenone; 8, $\beta$-carotene.

appear to be as useful, since polar compounds such as methylbixin, zeaxanthin, and isozeaxanthin are retained ($>60$ min) on the column while carotenes and retinyl esters elute rapidly with the solvent front. These problems do not occur with reverse-phase columns, and, as expected, the most polar compounds elute first from the reverse-phase columns. Separation of isomers such as all-*trans*- and 13-*cis*-retinoic acid as well as $\alpha$- and $\beta$-carotene is also possible on the reverse-phase columns, but separation of zeaxanthin and isozeaxanthin was not achieved. Separation of the latter compounds may be possible, however, upon modification of the solvent system in system 2 or by utilizing system 3. Thus, while the systems described are useful for separation of the compounds studied, variations on these systems as well as development of new systems should enable more efficient separation of retinoids and carotenoids.

Conclusion

It is concluded that GC and GC-MS are rapid and useful methods for the characterization of carotenoid molecular size. In addition, these methods can distinguish, and aid in the identification of, specific structural features and functional groups present in carotenoids. The requirement, however, that carotenoids be hydrogenated prior to GC analysis limits the use of this method for definitive structural assignment to an unknown polyene. GC analysis of retinoids is also limited due to the thermal instability of these compounds, but both retinoids and carotenoids may be analyzed per se by HPLC. It thus appears that HPLC will become a valuable analysis method for separation, structural characterization, and quantitation of both synthetic and naturally occurring carotenoids and retinoids.

## [30] Separation of Carotenoids on Lipophilic Sephadex

*By* KIYOZO HASEGAWA

For study of the biogenesis of carotenoids using radioisotopes *in vivo,* it is necessary completely to remove steroids from carotenoids. Since both the groups have a common precursor, mevalonic acid, the radioactivity incorporated into steroids may interfere with the precise radioassay for carotenoids. The methods heretofore in use for removing sterols from unsaponifiable material are the digitonide method,[1] the cooling method,[1] and thin-layer chromatography.[2] This chapter deals with the liquid–gel chromatographic separation of carotenes from sterols on Sephadex LH-20.[3]

### Gel Chromatography

Columns of Sephadex LH-20 (Pharmacia Fine Chemicals, Uppsala) are prepared by using the solvent system chloroform–methanol–*n*-hexane[4,5] (65 : 5 : 30) and the column size 1.5 × 140 cm for a long column or 1 × 10 cm for a short column.

Samples are dissolved in a small volume of the same solvent. Elution is carried out at room temperature. Carotenoids are determined spectrophotometrically.[6] Squalene and squalane are analyzed by gas–liquid chromatography.[7] Further, [14]C-labeled squalene also can be used for

[1] E. I. Mercer, B. H. Davies, and T. W. Goodwin, *Biochem. J.* **87**, 317 (1963).
[2] L. K. Lowry and C. O. Chichester, *Phytochemistry* **6**, 367 (1967).
[3] T. Suzuki and K. Hasegawa, *Agric. Biol. Chem.* **38**, 871 (1974).
[4] *n*-Hexane and petroleum ether used were treated with silica gel to remove aromatic hydrocarbons.[5]
[5] M. M. Graff, R. T. O'Connor, and E. L. Skau, *Ind. Eng. Chem. Anal. Ed.* **16**, 556 (1944).
[6] B. H. Davies, *in* "Chemistry and Biochemistry of Plant Pigments" (T. W. Goodwin, ed.), p. 489. Academic Press, New York, 1965.
[7] D. G. Anderson, and J. W. Porter, *Arch. Biochem. Biophys.* **97**, 509 (1962).

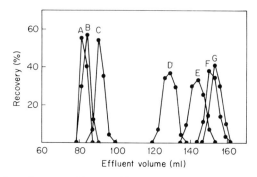

FIG. 1. Separation of carotenoids, steroids, and the related substances on a Sephadex LH-20 column. The column used was 1.5 × 140 cm and was eluted with chloroform–methanol–*n*-hexane (65 : 5 : 30) at a flow rate 6.3 ml/hr/cm². Three-milliliter fractions were collected. From T. Suzuki and K. Hasegawa, *Agric. Biol. Chem.* **38**, 871 (1974).

analysis. Steroids are analyzed by gas–liquid chromatography,[8] and by a colorimetric method.[9] The elution position of each material is also checked by thin-layer chromatography, using a silica gel G plate and petroleum ether–benzene (9 : 1) and MgO-Hyflo Super-Cel (1 : 2) and petroleum ether–benzene (95 : 5) for peak A to C, and silica gel G and benzene–ethyl acetate (5 : 1) for peak D to G (Fig. 1). The elution volumes of the compounds are expressed in terms of RRV, the relative retention volume of the substance to β-carotene.

Figure 1 and the table show part of the results obtained from the experiments performed by using the solvent system chloroform–methanol–*n*-hexane (65 : 5 : 30). In this solvent system, carotenes (RRV, 0.96–1.00) were separated completely free from sterols (RRV, 1.54–1.80), and the contamination of squalene (RRV, 1.07) to carotenes was minimum among the solvent systems tested. An increase of column size to 260 cm caused no better separation than was obtained from the 140-cm column.

Complete separation of carotenes and squalene, a precursor of sterols, was not obtained by the methods generally used, including adsorption chromatography. The present method has an advantage that not only sterols but also squalene can be separated from carotenes.

In the cases in which the removal of squalene is not necessary, a short column, 1 × 10 cm, is sufficient to separate carotenes and sterols.

[8] W. J. A. Van den Heuvel, J. Sjövall, and E. C. Horning, *Biochim. Biophys. Acta* **48**, 596 (1961).
[9] N. Papadopoulos, W. Cevallos, and W. C. Hess, *J. Neurochem.* **4**, 223 (1959).

GEL CHROMATOGRAPHY OF CAROTENOIDS, STEROIDS, AND THE RELATED SUBSTANCES[a]

| Substances[b] | Amount separated (μg) | Peak in which substance was recovered | RRV[d] |
|---|---|---|---|
| Phytoene | 50 | A | 0.96 |
| Phytofluene | 50 | A | 0.96 |
| ζ-Carotene | 30 | A | 0.96 |
| Lycopene | 50 | A | 0.96 |
| β-Carotene | 50 | B | 1.00 |
| α-Carotene | 50 | B | 1.00 |
| Squalene | 100 | C | 1.07 |
| Lanosterol | 1000 | D | 1.54 |
| β-Sitosterol | 1000 | E | 1.70 |
| Ergosterol | 1000 | F | 1.82 |
| Lutein | 150 | G | 1.86 |

[a] From T. Suzuki and K. Hasegawa, *Agric. Biol. Chem.* **38**, 871 (1974).

[b] Phytoene, phytofluene, ζ-carotene, α-carotene, and lutein were isolated from carrot oil (Nutritional Research Associates, Indiana) by a chromatographic method.[c] Other substances were obtained commercially.

[c] K. L. Simpson, T. O. M. Nakayama, and C. O. Chichester, *J. Bacteriol.* **88**, 1688 (1964).

[d] RRV: the relative retention volume of the substance to β-carotene.

Comments

To investigate the mechanism involved in the separation of the materials, several kinds of compounds, which had different polarities, were also subjected to chromatography. They included lutein diacetate,[10] cholesteryl palmitate, squalane, cholestane, vitamin $K_1$, vitamin A acetate, vitamin $D_2$, and cholesterol, RRV values of which were found to be 0.94, 0.94, 1.06, 1.06, 1.11, 1.25, 1.60, and 1.69, respectively. Since nonpolar compounds are eluted before the analogous polar compound, e.g., cholestane (RRV, 1.06) to cholesterol (RRV, 1.69), and carotenes (RRV, 1.00) to lutein (RRV, 1.86), it is suggested that the separation of the compounds on the column is governed mainly by the mechanism of a kind of partition chromatography. Contribution of gel filtration mechanism is not so large as seen in comparison of the elution order of cholesteryl palmitate (RRV, 0.94) and cholestane (RRV, 1.06) or lutein diacetate (RRV, 0.96) and β-carotene (RRV, 1.00). The difference in the elution volumes of vitamin $D_2$ (RRV, 1.60) and ergosterol (RRV, 1.82) suggests that the difference of the structure in part of molecule also participates in the separation. The results described above are in accordance with

[10] P. Karrer and S. Ishikawa, *Helv. Chim. Acta* **13**, 709 (1930).

those concerning the separation on lipophilic Sephadex by Eneroth and Nyström.[11]

In the cases in which methanol is used as a main solvent, e.g., methanol–petroleum ether (7:3) and methanol–acetone–petroleum ether (7:1:2), the complete separation of carotenes from sterols is not obtained. Furthermore, these solvent systems are unsuitable for the purpose of the experiment because of the poor solubility of carotenes in the solvents.

In conclusion, it was found that Sephadex LH-20 chromatography with nonpolar solvent systems, such as chloroform-methanol-$n$-hexane, is one of the convenient methods to separate sterols from carotenes. The method will greatly aid in obtaining a sterol-free carotene preparation from various biological materials.

[11] P. Eneroth and E. Nyström, *Biochim. Biophys. Acta* **144**, 149 (1967).

# [31] Carotenoid Biosynthesis by Cultures and Cell-Free Preparations of *Flavobacterium* R1560

By GEORGE BRITTON, TREVOR W. GOODWIN, DAVID J. BROWN, and NARENDRA J. PATEL

*Flavobacterium* strain R1560 is an orange-yellow, Gram-negative bacterium that produces a high yield of the carotenoid zeaxanthin ($\beta,\beta$-carotene-3,3'-diol). It is also one of the few bacteria to incorporate mevalonic acid efficiently into carotenoids and other isoprenoids, such as ubiquinone. These features make R1560 very effective for the preparation of radioactive zeaxanthin for use as substrate in feeding experiments with marine animals in studies of astaxanthin (3,3'-dihydroxy-$\beta,\beta$-carotene-4,4'-dione) formation. *Flavobacterium* R1560 is also an ideal source of cell-free enzyme systems for studies of carotenoid biosynthesis.

## Culture Conditions

### Medium and Maintenance of Stock Cultures

*Flavobacterium* R1560 is cultured in a medium containing, in grams per liter, glucose (5), yeast extract (10), Tryptone (10), NaCl (30), and $MgSO_4 \cdot 7H_2O$ (5). The pH is adjusted to 7.2 with NaOH, and the medium is transferred to the culture vessels before autoclaving at 15 psi for 15 min. Stock cultures are maintained either in liquid medium (1 ml) in ampoules, frozen and stored in liquid nitrogen, or on slopes prepared from the liquid medium supplemented with agar (3%, w/v). When stocks are maintained

METHODS IN ENZYMOLOGY, VOL. 67

on agar slopes at 22–26° subculturing is necessary every 10 days, but the slopes may be held at 5° for 21 days. Cultures can be stored in liquid $N_2$ for at least 2 years without apparent loss of viability.

## Liquid Cultures

Two methods may be used for culturing *Flavobacterium* R1560 for the preparation of cell-free systems.

*Method 1.* The bacteria are grown in 250-ml conical flasks modified with three or four indentations to act as baffles and thus increase the aeration of the cultures. A liquid inoculum is prepared by inoculating one such flask, containing 50 ml of liquid medium, from a mature agar slope, and incubating in an illuminated (approximately 2000 lux) orbital incubator at 25–27° and 160 rpm for 24 hr. One milliliter of this culture is then used to inoculate each of 20–30 flasks each containing 50 ml of medium. These cultures are incubated as just described for 20–24 hr. The cells are harvested by centrifugation at 10,000 $g$ for 15 min at 0–4°. Lower enzyme activities are obtained if cells cultured for longer than 24 hr are used.

*Method 2.* Larger amounts of bacteria are grown in 10-liter flasks containing 5 liters of medium. An inoculum of 200 ml of 24-hr liquid culture (4 flasks as above) is used. The cultures are incubated at 25° for 24 hr and aerated with a sterile stream of air (flow rate approximately 1 liter/min). The culture is further agitated by magnetic stirring. Excessive frothing is prevented by the inclusion of an antifoaming agent. The cells are harvested either by centrifugation as above, or with an Alpha-Laval Centrifugal Separator operating at a flow rate of approximately 30 liters/hr.

## General Methods of Carotenoid Biochemistry

Carotenoids are unstable to light, heat, oxygen, and acidic and in some cases basic, conditions. A survey of precautions that need to be taken to minimize carotenoid degradation and an account of the general techniques for chromatography, spectroscopic assay, and radioassay of carotenoids is available elsewhere in this series.[1]

## Preparation of [$^{14}$C]Zeaxanthin

### Incubation

*Flavobacterium* R1560 (8 flasks, 50 ml per flask) is grown for 22 hr, and the cells are then harvested by centrifugation, washed with 0.1 $M$

[1] G. Britton and T. W. Goodwin, this series, Vol. 18C, p. 654.

Tris · HCl buffer, pH 7.0, and resuspended in 30 ml of the same buffer. The cell suspension is transferred to a 250-ml culture flask, and potassium mevalonate (440 $\mu$g, 2 $\mu$Ci in 2 ml of aqueous solution), egg-white lysozyme (40 mg), and a solution of $MnSO_4 \cdot 4H_2O$ (4.4 mg), $MgSO_4 \cdot 7H_2O$ (25 mg), and ATP (11.2 mg) in 0.1 $M$ Tris · HCl buffer, pH 7.0 (4 ml) are added. The mixture is incubated for 18 hr at 25–27°, with continuous shaking (160 rpm). Throughout the incubation, air sterilized by passage through a membrane filter is gently bubbled through the cell suspension via a syringe needle (12 cm × 0.2 mm).

## Extraction

At the end of the incubation period, the cells are collected by centrifugation and the lipid material is extracted with acetone or warm methanol (3 × 50 ml). Diethyl ether (200 ml) is added to the yellow extract, and water (100 ml) is added until two phases separate. The colorless lower aqueous phase is removed, and the upper ethereal layer is washed a further three times with water (50 ml). The combined aqueous solution is reextracted with a small volume (30 ml) of diethyl ether, and the combined ethereal extracts are evaporated to dryness on a rotary evaporator at below 30°.

## Purification

The lipid material is chromatographed on thin layers of silica gel G (5 plates, 20 × 20 cm, 0.5 mm layer) with diethyl ether as developing solvent. The main orange band, $R_f$ approximately 0.5, contains the zeaxanthin, which is eluted with diethyl ether and ethanol (approximately 4 : 1) The zeaxanthin is purified further by TLC successively on magnesium oxide : kieselguhr G (1 : 1) with acetone–light petroleum, b.p. 40–60° (1 : 3, v/v), and on silica gel G with diethyl ether as developing solvent. The yield of pure zeaxanthin is approximately 3 mg, containing 0.25 $\mu$Ci of [14]C, i.e., an incorporation of 12.5% of the [14C]mevalonate.

## Preparation of a Cell-Free Enzyme System from *Flavobacterium* R1560

Although freshly cultured and harvested cells can be used for cell disruption, it is usually more convenient to use freeze-dried cells. Substantial quantities of freeze-dried cells may be accumulated for large-scale preparations of the crude enzyme system.

*Freeze-Drying*

The freshly harvested cells are washed three times with ice-cold buffer, (either 0.1 *M* Tris · HCl or 0.1 *M* potassium phosphate) at pH 7.0 and finally collected by centrifugation and rapidly frozen by immersion in liquid nitrogen. The frozen cells are then freeze-dried under vacuum at a temperature not exceeding −20°. The lyophilized cells are stored in a vacuum desiccator below −20° until required. Storage for up to 4 weeks under these conditions is not detrimental.

When required, the freeze-dried cells are resuspended in the pH 7.0 buffer containing the detergent Tween 80 (2%, w/v) (10 ml of buffer per gram of dried cells). Thorough suspension is achieved by use of a loose-fitting Potter–Elvehjem homogenizer. The resuspension is carried out at 0–4°. Alternatively, if freshly cultured and harvested cells are used they are washed three times with pH 7.0 buffer and resuspended in the same buffer containing Tween 80 (2%). In this case, the cells from four 50-ml flask cultures are suspended in 20 ml of buffer.

*Cell Disruption*

Two methods of cell disruption may be employed, namely ultrasonic disintegration and passage through the French pressure cell.

*Ultrasonic Disintegration.* The cell suspension (10 ml) in a thick-walled glass tube is cooled in an ice–NaCl bath and subjected to ultrasonic disintegration with a 100-W instrument at the highest power setting and maximum amplitude of 7 μm peak to peak. Ultrasonication for 5 sec is followed by a 15-sec cooling period. The total sonication time use is 120 sec, i.e., 24 such cycles over a total period of 8 min. The preparation is then centrifuged at 3000 *g* for 30 min at 0–4° to remove intact cells and large particles.

*French Press.* The cell suspension is transferred to a precooled French pressure cell, and the bacteria are ruptured by extrusion through the needle valve at a hydraulic pressure of 10,000 psi. Three passes of the suspension through the cell are made. After each treatment, 1 mg of deoxyribonuclease (dissolved in the minimal volume of 0.15 *M* NaCl) is added, and the cell suspension is mixed thoroughly for 1–2 min before being returned to the French pressure cell. The final crude homogenate is centrifuged at 3000 *g* for 30 min at 0–4°.

The crude 3000 *g* supernatant produced by either method may be used directly to incorporate mevalonate very efficiently into phytoene (30% incorporation in 3 hr in a 0.5-ml incubation) and to some extent into other carotenoid hydrocarbons. Alternatively, the supernatant can be used as a crude enzyme source and subjected to further purification procedures.

### Partial Purification of the Phytoene-Synthesizing Enzyme System

All manipulations are carried out at $0-4°$. The crude $3000\,g$ supernatant solution is centrifuged at $105,000\,g$ for 30 min. The pellet is discarded. To the supernatant (50 ml) is added ammonium sulfate (19.4 g) to bring the solution to 35% saturation. Thirty minutes after addition of the salt, the precipitate is removed by centrifugation for 30 min at $10,000\,g$. The supernatant is then brought to 40% saturation by addition of further ammonium sulfate (2.8 g). After 30 min the pellet is collected by centrifugation at $55,000\,g$ for 30 min, resuspended in 20 ml 0.1 $M$ potassium phosphate buffer, pH 7.0, containing Tween 80 (2%) and dithiothreitol (2 m$M$), and dialyzed for 3 hr against the same buffer. The dialyzed enzyme solution is either used immediately or frozen rapidly in liquid nitrogen and stored in liquid nitrogen or freeze-dried. The lyophilized enzyme preparation may be reconstituted with water to give the appropriate protein concentration.

Approximately 4-fold purification is achieved by this procedure. The partly purified preparation is very active in converting mevalonate into phytoene [approximately 40–50% incorporation of the metabolically active ($3R$) isomer in 3 hr], but little or no incorporation into other carotenoids is obtained.

*Incubation Conditions.* Portions of the crude $3000\,g$ supernatant or the partly purified preparation (350 $\mu$l) are incubated in small tubes (10–15 ml capacity) each with potassium [2-$^{14}$C]mevalonate (0.5 $\mu$Ci, approximately 50 $\mu$mol), ATP (2.0 $\mu$mol), MgCl$_2$ (7.5 $\mu$mol) and MnCl$_2$ (1.5 $\mu$mol) in a final incubation volume of 500 $\mu$l, buffered to pH 7.0. If the enzyme system is prepared from cells disrupted by sonication, nucleotide cofactors (NAD$^+$, NADH, NADP$^+$, NADPH, and FAD, 6 $\mu$mol each) must be included for maximum phytoene synthesis. The mixtures are incubated for 3 hr at 26° with shaking (approximately 100 rpm).

*Extraction and Purification of Products.* Incubations are terminated by the addition of 2 ml of aqueous acetone (1:1), and appropriate unlabeled "carrier" carotenoids are added to facilitate purification. A mixture of acetone and diethyl ether (1:2, v/v; 5 ml) is added, and the tube contents are mixed thoroughly with a vortex mixer. Two phases are separated by centrifugation at $3000\,g$ for 2 min, and the upper etheral layer is carefully removed by a Pasteur pipette. The extraction procedure is repeated four times to ensure complete recovery of carotenoid. The combined ethereal extracts are evaporated to dryness under N$_2$.

The total lipid extract is dissolved in petrol and fractionated by chromatography on a column of neutral alumina (Brockmann grade III, 5 g).[1] Carotenoid hydrocarbons are eluted with 1% diethyl ether in light petroleum (b.p. 40–60°), and zeaxanthin, if required, is eluted with 5% ethanol in diethyl ether. The further purification of zeaxanthin is described above.

The hydrocarbon fraction is separated by TLC on silica gel G with light petroleum as developing solvent. Three bands (I–III) ($R_f$ 0.5, 0.4, and 0.3, respectively) are removed and eluted. Band I is then chromatographed on a thin layer of MgO : kieselguhr G (1 : 1) with 2% acetone in light petroleum to give phytoene ($R_f$ 0.9), phytofluene ($R_f$ 0.7), $\beta$-carotene ($R_f$ 0.4), and $\zeta$-carotene ($R_f$ 0.2). Similarly band II, on MgO : kieselguhr G with 15% acetone in light petroleum, gives $\beta$-zeacarotene ($R_f$ 0.7) and $\gamma$-carotene ($R_f$ 0.4), Band III on MgO : kieselguhr G with acetone : benzene : light petroleum (2 : 2 : 1) gives neurosporene ($R_f$ 0.7) and lycopene ($R_f$ 0.2). The neurosporene, lycopene, $\beta$-zeacarotene, $\gamma$-carotene, and $\beta$-carotene obtained in this way are the all-*trans* isomers and are sufficiently pure for radioassay.

The early biosynthetic intermediates, phytoene, phytofluene and $\zeta$-carotene, may be present as mixtures of the all-*trans* and 15-*cis* isomers, which are separated by column chromatography of each compound, on neutral alumina, Brockmann grade I. 15-*cis*-Phytoene is eluted with 2% diethyl ether in light petroleum (*E/P*) and the all-*trans* isomer with 4% *E/P*. 15-*cis*-Phytofluene is eluted with 5% *E/P*, whereas the all-*trans* isomer requires 70% *E/P*; and 15-*cis*- and all-*trans*-$\zeta$-carotenes are eluted with 10% *E/P* and diethyl ether, respectively.

In experiments in which only phytoene is being examined, the phytoene isomers are obtained by applying a light petroleum solution of the total hydrocarbon fraction from the first chromatography column (grade III) to the grade I column, prewashing with 0.5% *E/P*, and eluting the two phytoene isomers with 2% *E/P* and 4% *E/P*, respectively. Almost all the radioactivity is recovered in the 15-*cis*-phytoene, with only small amounts (usually no more than 5% of the total) in all-*trans*-phytoene and other carotenes.

*Identification and Assay.* The carotenoids are identified by their chromatographic properties, light absorption spectra, and, if possible, mass spectra. Suitable mixtures of authentic carotenoids for use as unlabeled carriers may be obtained from tomatoes[1] or from pigment mutants of the green alga *Scenedesmus obliquus*.[2]

The carotenoids are assayed for radioactivity by liquid scintillation counting; either the colored samples are bleached before counting[1] or corrections are made for color quenching.

[2] R. Powls and G. Britton, *Arch. Microbiol.* **113**, 275 (1977).

Acknowledgments

We are extremely grateful to F. Hoffmann–La Roche and Co. Ltd., Basel, Switzerland for cultures of *Flavobacterium* R1560 and for financial support. We also thank the Science Research Council for a research grant.

## [32] Synthesis and Properties of Glycosyl Retinyl Phosphates*

*By* LARS RASK and PER A. PETERSON

The role of polyprenol phosphate-sugar derivatives in the biosynthesis of bacterial cell wall constituents has been documented in great detail.[1] A mammalian counterpart to the bacterial undecaprenol was first discovered by Caccam et al.,[2] who showed that mannose was transferred to a mannose lipid by crude cell membrane fractions obtained from different tissues. Subsequently Behrens and Leloir[3] showed that the lipid moiety of the glycolipid was dolichyl phosphate. Work from the same laboratory demonstrated unequivocally that glucosyl dolichyl phosphate is an intermediate in the transfer of glucose from UDP-glucose to an oligosaccharide containing dolichyl pyrophosphate.[4] Also mannosyl dolichyl phosphate has been shown to be an intermediate in the transfer of mannosyl to glycoproteins.[5,6]

Vitamin A deficiency affects, among other things, mucous secretion and the biosynthesis of glycoproteins.[7] As retinol is a polyisoprenol, investigations have been carried out to examine whether retinol may fulfill a role analogous to that of dolichol. DeLuca et al. were the first to obtain evidence for the existence of mannosyl retinyl phosphate.[8] Similar observations were obtained in other experimental systems by Helting and Peterson[9] and by Rodriguez et al.[10] From these original reports but a few

* This work was supported by a grant from the Swedish Medical Research Council (Project No. B78-13X-03531-07B)

[1] E. C. Heath, *Annu. Rev. Biochem.* **40**, 29 (1971).
[2] J. F. Caccam, J. J. Jackson, and E. H. Eylar, *Biochem. Biophys. Res. Commun.* **35**, 505 (1969).
[3] N. H. Behrens and L. F. Leloir, *Proc. Natl. Acad. Sci. U.S.A.* **66**, 153 (1970).
[4] N. H. Behrens, A. F. Parodi, and L. F. Leloir, *Proc. Natl. Acad. Sci. U.S.A.* **68**, 2857 (1971).
[5] J. B. Richards and F. W. Hemming, *Biochem. J.* **130**, 77 (1972).
[6] N. H. Behrens, H. Carminatti, R. J. Staneloni, L. F. Leloir, and A. I. Cantarella, *Proc. Natl. Acad. Sci. U.S.A.* **70**, 3390 (1973).
[7] L. DeLuca and G. Wolf, *J. Agric. Food Chem.* **20**, 474 (1972).
[8] L. DeLuca, G. Rosso, and G. Wolf, *Biochem. Biophys. Res. Commun.* **41**, 615 (1970).
[9] T. Helting and P. A. Peterson, *Biochem. Biophys. Res. Commun.* **46**, 429 (1972).
[10] P. Rodriguez, O. Bello, and K. Gaede, *FEBS Lett.* **28**, 133 (1972).

years ago, the field has expanded rapidly. This has to a significant extent been enabled by the utilization of techniques already established for the investigation of other polyprenols.

Materials

All-*trans*-Retinol and all-*trans*-retinoic acid of high purity may be purchased from several manufacturers (e.g., Sigma). It is advantageous to buy these substances in glass ampoules each containing only a small quantity. Opened ampoules can, however, be stored desiccated in a freezer in the dark.

The chemical synthesis of retinyl phosphate has been described by DeLuca and collaborators.[11,12] All-*trans*-Retinol (40 mg) was dissolved in 10 ml of acetonitrile. Trichloroacetonitrile (1 ml) and 10 ml of acetonitrile containing 100 mg of bis (triethylamine) phosphate were added, and the mixture was stirred at room temperature for 1 hr in a stoppered flask under $N_2$. The mixture was neutralized under continuous stirring by the addition of small portions of concentrated $NH_3$. The precipitate formed was removed by filtration. The retinyl phosphate obtained in the filtrate was subsequently purified by chromatography on a DEAE-cellulose column (see below). The yield of retinyl phosphate was about 10%. The whole procedure should be carried out in the dark or under dim red light.

Retinyl phosphate has similar absorption and fluorescence characteristics to those of retinol.[13] Retinyl phosphate is quite labile especially under alkaline conditions and is thereby converted into anhydroretinol. Even on chromatography on Sephadex LH-20 or unbuffered silica, retinyl phosphate may become degraded to yield anhydroretinol and retinol. Anhydroretinol has absorption maxima at 350, 370, and 390 nm and can therefore readily be detected.[13]

[³H]Retinol with a specific activity of 1–5 Ci/mmol can be obtained from New England Nuclear. The vitamin is delivered dissolved in ethanol and can be stored refrigerated under $N_2$ in the dark for shorter periods. After prolonged storage, it is advisable to check the radiochemical purity of the product by thin-layer chromatography or high-pressure liquid chromatography. The last method offers a convenient way of analyzing and quantifying different vitamin A derivatives. An alumina column may be used and the chromatogram developed by a gradient between hexane/dichloromethane (1/1) and dichloromethane/methanol (6/4). The effluent is

[11] G. C. Rosso, L. DeLuca, C. D. Warren, and G. Wolf, *J. Lipid Res.* **16,** 235 (1975).
[12] J. P. Frot-Coutaz and L. DeLuca, *Biochem. J.* **159,** 799 (1976).
[13] L. DeLuca, J. P. Frot-Coutaz, C. S. Silverman-Jones, and P. Roller, *J. Biol. Chem.* **252,** 2575 (1977).

monitored at 360 nm and subsequently collected for radioactivity determination. [³H]Retinyl phosphate was synthesized as described above. [³H]Retinol, 250–300 µCi, was thereby added to the unlabeled retinol.

Unlabeled nucleotide sugars may be purchased from several suppliers (e.g., Sigma). Nucleotide sugars ³H- or ¹⁴C-labeled in the sugar moiety are supplied by New England Nuclear or Amersham/Searle. The radioactive nucleotide sugars are expensive and unfortunately quite labile and should therefore not be purchased in large quantities for prolonged use.

DEAE-cellulose from different producers (e.g., Whatman and Eastman Kodak Co.) has been used with similar results. The DEAE-cellulose gel is converted to the acetate form.[14] The cellulose powder is suspended in 1 $M$ NaOH for 12 hr at room temperature. It is then washed with water and dried by successive washings with 95% ethanol and 99% methanol. After further drying in a desiccator, the material is suspended in glacial acetic acid overnight at room temperature. The acetic acid is washed away with 99% methanol in water and the gel is stored in the latter solution.

DEAE-papers are a product of Whatman. The papers are washed with the same solutions as the DEAE-cellulose powder, but are stored dry.

Silica gel (Bio-Sil HA, 200–325 mesh) is obtained from Bio-Rad Laboratories. Prior to use, it is activated at 100° for 2 hr and thereafter stored desiccated at room temperature.

### Enzyme Preparation

Hitherto no serious attempt has been made to isolate the enzymes catalyzing the formation of glycosylated retinyl derivatives. The enzymes are membrane bound and located in the rough and smooth endoplasmic reticulum and most probably also in the Golgi complex.[15] The occurrence of the enzymes in other intracellular membrane structures or in the plasma membrane has not, to our knowledge, been investigated.

The enzymes may be solubilized at least partially with use of nonionic detergents. They retain most if not all of their activities.[16] Most studies, however, have been performed with a crude membrane preparation as the enzyme source. The tissue (liver, kidney, lung, intestinal mucosa, etc.) is immediately removed from the newly killed animal, chilled on ice, minced with scissors, and homogenized with a Potter–Elvehjem type of

[14] M. Dankert, A. Wright, W. S. Kelley, and P. W. Robbins, *Arch. Biochem. Biophys.* **116,** 425 (1966).
[15] A. Bergman, T. Mankowski, T. Chojnacki, L. DeLuca, E. Peterson, and G. Dallner, *Biochem. J.* **172,** 123 (1978).
[16] T. Helting and P. A. Peterson (1973), unpublished observation.

homogenizer in 4 ml of 0.25 $M$ sucrose buffered with 10 m$M$ Tris-acetate, pH 7.0, per gram of tissue. Some tissues, e.g., mastocytoma tissue, require the homogenization to be performed in a Virtis-type homogenizer. The homogenates are centrifuged at 10,000 $g$ for 10 min to remove nuclei, unbroken cells, and debris. The supernatants are further centrifuged at 105,000 $g$ for 60 min. The pellet, which mainly consists of rough and smooth endoplasmic reticulum, Golgi complex, and plasma membrane, is suspended to a final protein concentration of 5–10 mg/ml with Tris-acetate buffer, pH 7.0, containing 0.25 $M$ sucrose. The synthetases forming the glycosylated retinyl derivatives are stable over periods of several months if kept in the frozen state, preferably at $-70°$.

### Incubation Conditions and Reaction Mechanism

The use of crude microsomal membrane preparations rather than purified enzymes for catalyzing the formation of glycosylated retinyl derivatives makes it difficult to establish optimal incubation conditions for the reaction. First, the membrane preparations contain substrates for the reaction, i.e., retinol, retinyl phosphate, and possibly nucleotide sugars.[10,17,18] Moreover, several enzymes able to degrade both the endogenous and the added substrates are present, e.g., phosphatases, phosphodiesterases. Second, the glycosylated retinyl derivatives are most probably substrates for the transfer of sugar moieties to glycoproteins and/or glycolipids.[9,19,20] What is actually measured is therefore the net outcome between the rate of synthesis and the rate of elimination of the substances. Thus it is impossible to establish true pH optima and $K_m$ values for the synthetic reactions of the glycosylated retinyl derivatives. However, conditions that have been used successfully to demonstrate the synthesis of such substances will be described.

The standard incubation mixture most often employed[11,17,18] is as follows: 0.5–2 mg of membrane protein in 100 $\mu$l of 0.25 $M$ sucrose, 20 m$M$ Tris-acetate, pH 7.0, to which is added, in a volume of 100 $\mu$l of the Tris-acetate buffer, pH 7.0: ATP, 0.2 $\mu$mol; MnCl$_2$, 2 $\mu$mol; EDTA, 0.6 $\mu$mol; nucleotide sugars, $10^{-3}$ to $10^{-2}$ $\mu$mol; all-*trans*-retinol, 0.35 $\mu$mol. Retinol, 10 mg, is dissolved in diethyl ether (5 ml) and mixed with an equal volume of the Tris-acetate buffer containing 4% of a nonionic detergent,

[17] L. DeLuca, N. Maestri, G. Rosso, and G. Wolf, *J. Biol. Chem.* **248**, 641 (1973).
[18] P. A. Peterson, L. Rask, T. Helting, L. Östberg, and Y. Fernstedt, *J. Biol. Chem.* **251**, 4986 (1976).
[19] N. Maestri and L. DeLuca, *Biochem. Biophys. Res. Commun.* **53**, 1344 (1973).
[20] P. A. Peterson, S. F. Nilsson, L. Östberg, L. Rask, and A. Vahlquist, *Vitam. Horm.* (*New York*) **32**, 181 (1974).

e.g., Triton X-100. The ether is removed by flushing with $N_2$. Fifty microliters of the retinol solution are mixed with 50 $\mu$l of Tris-acetate buffer containing the other constituents and then with the membrane protein suspension.

[14]C-Labeled nucleotide sugars, 0.2–2 $\mu$Ci, and/or [3H]retinol, 5–20 $\mu$Ci, are added to the incubation mixture. The radioactive retinol is mixed with its unlabeled counterpart and added in the detergent solution. The incubations are maintained at 37° for 20 min and thereafter interrupted by the addition of 3 ml of ice-cold chloroform/methanol 2/1 (v/v). Subsequently, 0.6 ml of 0.9% NaCl in water is added. After each addition, the reaction mixture is vigorously shaken on a Vortex mixer.

The formation of glycosylated retinyl derivatives seems to be described by the reaction scheme[17,18]:

$$\text{Retinyl phosphate} + \text{XDP-sugar} \rightleftarrows \text{XDP} + \text{retinyl phosphate-sugar}$$

According to this scheme, retinol is phosphorylated prior to its reaction with the dinucleotide sugar. The origin of the phosphate is, however, obscure. ATP is a good candidate as a donor of the phosphoryl group. [32P]Phosphate has been claimed to be transferred to retinol from [$\gamma$-32P]ATP with a liver membrane preparation as the catalyst.[17] This finding has hitherto been difficult to confirm in other systems.[18] However, the phosphorylation of retinol has been well established, i.e., by the finding that retinyl phosphate may be isolated in chemical amounts from hamster small intestinal epithelium.[21] It has also been found that synthetic retinyl phosphate may be used as the substrate for the formation of mannosyl retinyl phosphate in incubation experiments analogous to that described above. Moreover, much lower amounts of retinyl phosphate than of retinol are needed to stimulate the formation of monosaccharide-containing retinyl derivatives, thereby further supporting the suggested reaction mechanism. In such incubation experiments, the retinol is replaced by 10–50 nmol of retinyl phosphate. When retinyl phosphate is used, it seems necessary to add to the incubation mixture either $\alpha$-L-lecithin, 400–800 $\mu$g, or a lipid extract from liver microsomal membranes containing approximately 400 $\mu$g of phospholipids.[11] The reason for this is as yet unknown, but the effect may be due to the more polar nature of retinyl phosphate as compared to retinol. A trivial explanation is that the lipid extract in some way may protect the retinyl phosphate from degradation.

The preparation of a lipid extract suitable as an additive when retinyl phosphate is used has been described.[11] Briefly, 1 ml of a suspension of liver membrane proteins (10 mg) is extracted with 20 ml of chloroform/

[21] J. P. Frot-Coutaz, C. S. Silverman-Jones, and L. DeLuca, *J. Lipid Res.* **17**, 220 (1976).

methanol (2/1) and 4 ml of 0.9% NaCl in water. The lower organic phase is then concentrated to 2–3 ml by flushing with $N_2$.

The presence of ATP in the incubation mixtures is beneficial, probably not as a source of energy but as a protection for the substrates (retinyl phosphate and nucleotide sugars) against unwanted activity of phosphatases and phosphodiesterases. The yield of glycosylated retinyl derivatives may also be improved by the addition of NaF (3–5 m$M$) to the reaction mixture.

Properties of the Reaction Catalyzing the Glycosylation of Retinyl Phosphate

As described above, the reaction catalyzing the glycosylation of retinol is reversible. Accordingly, GDP and to a lesser extent GMP inhibit the formation of mannosyl retinyl phosphate. Likewise UDP (and UMP) impedes the synthesis of galactosyl retinyl phosphate.[17,18] The presence of divalent cations in the reaction mixture is mandatory to ensure maximal production of the glycosylated derivatives. At low concentrations, $Mn^{2+}$ is more effective than $Mg^{2+}$, but similar rate constants may be obtained at higher concentrations of the two cations. The absolute requirement for $Mn^{2+}$ or $Mg^{2+}$ is evidenced by the fact that 50 m$M$ EDTA in the reaction mixture completely abolishes the formation of galactosyl retinyl phosphate.[18]

Several types of detergents will stimulate the synthetase reaction. The reason for this is not well examined. Thus, the detergent may increase the solubility of the exogenously added retinol. Moreover, the unspecific deposition of retinol in the hydrocarbon matrix of the membranes may be prevented. The detergents will also increase the permeability of the membranes, which may improve the accessibility of the substrates for the enzymes.

In summary, if [³H]retinol is used as the substrate in the reaction mixtures described above, the formation of glycosylated retinyl compounds will depend on two separate catalyzing systems. First, the retinol moiety is phosphorylated, possibly by a retinol kinase. Second, the retinyl phosphate serves as the acceptor for a monosaccharide donated by a nucleotide sugar. If, [³H]retinyl phosphate is used as the exogenously added substrate, only the second reaction will be examined.

The details of the synthesis of the glycosyl retinyl phosphates is far from understood. To gain further insight into the molecular reaction mechanisms, the participating enzymes have to be solubilized and preferably separated from inhibitors, endogeneous substrates, and contaminating enzyme activities.

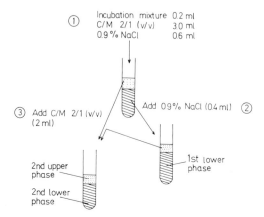

FIG. 1. Separation of glycosyl dolichyl phosphate from glycosyl retinyl phosphate by solvent extraction according to C. S. Silverman-Jones, J. P. Frot-Coutaz, and L. DeLuca [*Anal. Biochem.* **75**, 664 (1976)]. To an incubation mixture of 0.2 ml was added 3 ml of chloroform/methanol (C/M) (2/1, v/v) and 0.6 ml of 0.9% NaCl. The upper and lower phases were separated by low speed centrifugation. The lower phase was washed with 0.4 ml of 0.9% NaCl. The resulting lower organic phase (first lower phase) contains mostly glycosyl dolichyl phosphates. The combined aqueous phases were extracted again with 2 ml of C/M (2/1, v/v). The second upper phase contains approximately 70% of the glycosyl retinyl phosphates in a reasonably pure state.

## Isolation of Glycosyl Retinyl Phosphate

The isolation of glycosyl retinyl phosphate in highly purified form is mostly hampered by the fact that glycosylated dolichyl phosphate derivatives and retinyl phosphate display several characteristics in common with the retinyl sugars. Of the present methods employed no one will achieve the desired separation in a single step. However, by a combination of two methods, the glycosyl retinyl phosphate compounds may be obtained virtually free from contaminants.

A quite efficient separation between glycosyl dolichyl phosphates and glycosyl retinyl phosphates may be accomplished by utilizing the two-phase separation obtained by terminating the incubation reaction with chloroform–methanol and physiological saline (see above). Since dolichyl derivatives generally are less polar than the analogous retinyl compounds, the latter will be enriched in the upper phase. Figure 1 shows a fractionation scheme for the procedure adapted after Silverman-Jones et al.[22]

The phase separation procedure offers a relatively efficient separation of the glycosylated dolichyl and retinyl phosphates. The retinyl compounds are, however, by no means highly purified at this stage and need

[22] C. S. Silverman-Jones, J. P. Frot-Coutaz, and L. DeLuca, *Anal. Biochem.* **75**, 664 (1976).

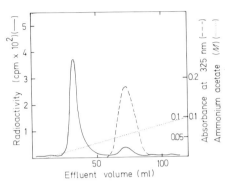

Fig. 2. Purification of glycosyl [³H]retinyl phosphate on a DEAE-cellulose column. The reaction mixture contained 20 μCi of [³H]retinol and 2 mg of crude membrane fraction protein from a mastocytoma [P. A. Peterson, L. Rask, T. Helting, L. Östberg, and Y. Fernstedt, *J. Biol. Chem.* **251**, 4986 (1976)]. After incubation at 37° for 20 min, the incubation was stopped and extracted as described in Fig. 1. The second upper phase was concentrated, mixed with 100 μg of chemically synthesized retinyl phosphate, and applied to a DEAE-cellulose column (10 × 1 cm) equilibrated with 99% methanol. Radioactivity not bound was washed off the column with the equilibrating solvent (not shown in the figure), and the column was eluted with a 100-ml linear ammonium acetate gradient from 0 to 0.1 *M*. Glycosyl [³H]retinyl phosphate was eluted at a salt concentration of about 20 m*M* and retinyl phosphate (absorbance at 325 nm) at a salt concentration of about 60 m*M*.

further purification, preferably by chromatography on DEAE-cellulose columns.[14,17,18] The upper polar phase contains mostly retinyl derivatives, whereas the first lower phase comprises the dolichyl compounds. The second upper phase is dried by flushing with $N_2$. The sample is then dissolved in 2 ml of 99% methanol in $H_2O$ and applied to a column of DEAE-cellulose (1 × 5 cm) in the acetate form (see above). Substances that are not bound are washed off the column with 200–300 ml of 99% methanol in $H_2O$. Polar acidic compounds are then eluted by a linear[17,18] or concave[21] gradient of 0–0.1 *M* ammonium acetate in 99% methanol. The described procedure is applicable also for the isolation of the dolichyl derivatives present in the lower phase. Figure 2 shows a typical separation. The glycosylated dolichyl and retinyl phosphates are eluted at the same salt concentration.[11,17,18,21,23] Dolichyl and retinyl phosphate are eluted somewhat later than the glycosylated compounds, which affords an efficient group separation between the glycosylated and nonglycosylated forms of these substances. Fractions containing the retinyl derivatives are combined and mixed with 2 volumes of chloroform and 0.2 volume of water. The water phase, containing most of the ammonium acetate, is removed and reextracted with chloroform/methanol (2/1). The combined organic phases are dried under $N_2$.[18]

[23] H. G. Martin and K. J. I. Thorne, *Biochem. J.* **138**, 281 (1974).

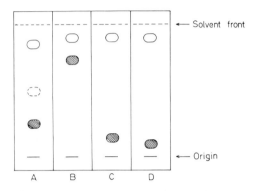

Fig. 3. Schematic picture of chromatography on silica gel thin-layer plates. The following systems were used: (A) chloroform/methanol/$H_2O$ (60/25/4); (B) chloroform/methanol/acetic acid/0.9% NaCl in $H_2O$ (50/25/8/4); (C) benzene/chloroform/methanol (4/1/1); (D) chloroform/methanol/aqueous ammonia (5%) (80/30/4). Symbols: retinol, ○; glycosyl retinyl phosphate, ◍; glycosyl dolichyl phosphate, ◯. The $R_f$ values vary somewhat among different chromatographic runs.

Glycosyl retinyl phosphate may be completely separated from glycosyl dolichyl phosphate by thin-layer chromatography on silica gel plates.[24] The type of resolution obtained is shown in Fig. 3A. The figure demonstrates that retinyl phosphate comigrates with glycosyl retinyl phosphate and that dolichyl phosphate occupies a migration position indistinguishable from that of glycosyl dolichyl phosphate. This fractionation procedure is the simplest and most reliable way to separate the glycosylated retinyl derivatives from their dolichyl counterparts. The identical migration behavior of the glycosylated and nonglycosylated retinyl phosphates in this system does not represent a problem, since the separation between the two types of compounds is achieved by DEAE-cellulose chromatography (see above).

Chromatography on silicic acid columns is an alternative final purification step to obtain the glycosyl retinyl phosphate compounds in highly purified form.[17,18] Silicic acid is activated as described under Materials, suspended in chloroform, and packed into a column (1 × 5 cm). The dried sample is dissolved in chloroform/methanol (8/1) and applied to the column. Elution is initially achieved with chloroform/methanol (8/1) and continued stepwise with chloroform/methanol solutions of increasing polarity. Glycosyl dolichyl phosphate is eluted with chloroform/methanol (2/1) and glycosyl retinyl phosphate with chloroform/methanol (1/1). If the method is used to isolate retinyl phosphate, which has similar chromato-

[24] J. S. Tkacz, A. Herscovics, C. D. Warren, and R. W. Jeanloz, *J. Biol. Chem.* **249,** 6372 (1974).

graphic properties as its glycosylated counterpart, the silica gel has to be titrated to pH 7 with a few drops of $NH_4OH$ in methanol to avoid acid hydrolysis of the retinyl compound. All fractions eluted are also titrated to neutrality.[13] The recovery of the compounds applied to the column is relatively low (10–30%), and the procedure is rather tedious. We therefore prefer, whenever possible, the use of the thin-layer system described above.

When many samples are to be analyzed simultaneously for the quantitation of glycosyl retinyl derivatives, the separation by chromatography on DEAE-cellulose columns is too laborious. However, chromatography on DEAE-papers allows several samples to be handled simultaneously in a short period of time.[18] Glycosyl retinyl phosphate is separated from glycosyl dolichyl phosphate by the two-phase separation described above. The upper phase containing only negligible amounts of glycosyl dolichyl phosphate is dried under $N_2$, dissolved in 99% methanol in $H_2O$, and applied as a narrow zone on a strip of DEAE-paper. The paper is then eluted with 99% methanol in $H_2O$ in a closed, dark jar lined with paper moistened with methanol to prevent evaporation. The methanol front is allowed to reach close to the end of the paper. The paper is dried for a short time in the dark, then chromatography is continued with 50 m$M$ ammonium acetate in 99% methanol. The chromatography is terminated when the front has reached two-thirds of the length of the paper. Figure 4 shows a representative separation obtained on DEAE-paper. By this procedure, it is generally not possible to separate retinyl phosphate from glycosyl retinyl phosphate reproducibly, as it is difficult to control the

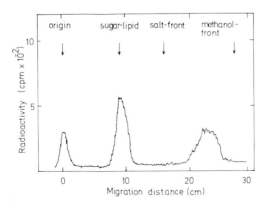

FIG. 4. Separation obtained by the DEAE-paper chromatography method. An incubation of mouse mastocytoma enzyme fraction (1.8 mg of protein) with UDP-[$^{14}$C]galactose (2 $\mu$Ci) was fractionated as outlined in Fig. 1. The second upper phase was applied to the paper, which was eluted as described in the text. After drying, radioactive components were located with a Packard Model 7201 radiochromatogram scanner.

exact salt concentration in the effluent owing to evaporation of the methanol. The method is, however, of use especially when a large number of samples containing glycosyl retinyl phosphate labeled in the sugar moiety are processed.

### Characterization of Glycosyl Retinyl Phosphate

*Thin-Layer Chromatography.* The glycosyl retinyl phosphate derivatives migrated as homogeneous components on thin-layer chromatography in the systems depicted in Fig. 3 (B–D).[17,18] In contrast to the solvent system shown in Fig. 3A, the former systems do not adequately differentiate between glycosyl dolichyl phosphates and glycosyl retinyl phosphates.

*Alkaline Hydrolysis.* Glycosyl retinyl phosphate can be hydrolyzed with alkali under conditions that do not affect glycosyl dolichyl phosphate.[17,18,25] The sample containing the glycosyl retinyl phosphates is dissolved in 300 $\mu$l of chloroform/methanol (1/4) and incubated at 37° for 20 min after the addition of 30 $\mu$l of 1 $M$ NaOH. The sample is then chilled on ice and neutralized with 1 $M$ $CH_3COOH$. This is followed by the sequential addition of 350 $\mu$l of chloroform and 100 $\mu$l of $H_2O$. The lower phase contains retinol and anhydroretinol, and the upper phase contains glycosyl phosphate.

*Acid Hydrolysis.* Owing to the presence of an allylic ester bond in glycosyl retinyl phosphate, this compound is extremely sensitive to acid hydrolysis. At room temperature, 0.1 $M$ HCl achieves almost complete hydrolysis in 10 min.[11,17,18] The products are retinol and dephosphorylated glycose. Glycosyl dolichyl phosphate is hydrolyzed at a significantly lower rate.[11,17,18]

*Catalytic Hydrogenolysis.* Catalytic hydrogenolysis has been used as a means of distinguishing between classes of polyprenyl phosphate sugars that contain saturated $\alpha$-isoprene units from those in which this grouping is allylic. This approach has been used to analyze glycosyl retinyl phosphate.[9,17,18] The sample is dissolved in 95% ethanol to which 2 mg of $PtO_2$ are added. The compound is incubated under 10 psi of $H_2$ at room temperature for 4 hr. Three volumes of chloroform/methanol (2/1) and 1 volume of water are added. Glycosyl phosphate is recovered in the polar phase.[9,18] Results have been presented[26] suggesting that also glycosyl dolichyl phosphate may be cleaved by some type of nonspecific hydrolysis under conditions similar to those described above. Therefore, some cau-

[25] S. Adamo, L. DeLuca, C. S. Silverman-Jones, and S. H. Yuspa, *J. Biol. Chem.* **254**, 3279 (1979).

[26] E. L. Kean, *J. Biol. Chem.* **252**, 5619 (1977).

tion is warranted in the interpretation of results obtained with catalytic hydrogenolysis.

## Biological Significance

Several studies have been devoted to the synthesis of glycosyl retinyl phosphate both *in vivo* and in *in vitro* cultivated cells. Compounds with properties identical to those of glycosyl retinyl phosphates have been isolated from the liver after the prior injection or feeding of the animals with radioactive retinol.[23,27] Mucosal intestinal cells cultivated in the presence of retinol-binding protein containing [³H]retinol[20] or free [¹⁴C]retinol[21] have been found to form polar metabolites of retinol (glycosyl retinyl phosphate and retinyl phosphate). As mentioned, retinyl phosphate was isolated in chemical amounts from hamster intestinal mucosa.[21] Keratinocytes from mouse epidermis cultivated in the presence of [³H]retinol have also been shown to synthesize retinyl phosphate and glycosyl retinyl phosphate.[25] It is, therefore, reasonable to conclude that phosphorylated retinyl derivatives are formed in intact cells both *in vivo* and *in vitro,* not only in cell-free system.

The characterization of the biosynthesis of the retinyl compounds is still at its infancy. The enzymes involved in the various reactions must be solubilized and purified. Competition experiments with use of unrelated isoprenyl lipids, e.g., dolichyl compounds, have to be performed. Such analyses will reveal whether common or separate enzyme systems are responsible for the biosynthesis of the different acidic glycolipids.[18]

The biological role of the glycosyl retinyl phosphate is believed to be the transfer of the sugar moiety to glycoproteins or glycolipids in analogy with other polyprenyl phosphate-sugar derivatives.[9,19,20] However, these reactions are not well documented for the retinyl compounds. In particular, the acceptors of the monosaccharides have to be identified. Moreover, the possibility of a specificity difference between the dolichyl-mediated and the retinyl-mediated monosaccharide transfer reactions has to be elucidated.

---

[27] R. M. Barr and L. DeLuca, *Biochem. Biophys. Res. Commun.* **60,** 355 (1974).

## [33] Fluorescence Assay of Retinol-Binding Holoprotein

*By* J. GLOVER

### Principles

Fluorometric methods of assay are generally more sensitive than colorimetric procedures, and several have been devised for vitamin A. One type depends on extracting the lipids containing the vitamin from plasma or tissues and examining the fluorescence of the extract dissolved in nonpolar solvent.[1] Interference, however, arises from the presence of other fluorescent components in plasma, such as phytoene or tocopherol, or from pigments, such as yellow carotenoids or other lipid fractions, causing quenching. Corrections therefore have to be applied to overcome some of these difficulties.

In assaying retinol in plasma (or serum), it is more useful to determine the physiologically active form of the vitamin bound to retinol-binding protein (holoRBP) separately from any other free or esterified retinol that may be present during absorption from the intestine. A method[2] has been devised that enables this form of vitamin A to be assayed satisfactorily. It takes advantage of three main features of the holoRBP complex. First, retinol attached to its carrier protein fluoresces about 14 times more strongly than free retinol in hydrocarbon solvent. Consequently, direct assay of this holoRBP gives an increased gain in sensitivity. Second, further advantage is obtained by separating the low molecular weight protein complex away from other conjugated plasma proteins using electrophoresis on polyacrylamide gels at low temperatures. This mild procedure removes components whose UV absorption or fluorescence would interfere in the analysis without oxidative or other damage to the holoRBP. Finally, the latter is determined directly in the polyacrylamide gel by fluorescence scanning within a few minutes of the termination of electrophoresis. The process is carried out without detaching retinol from its carrier protein, where it remains stable under the low temperature conditions used. Attempts have also been made to assay retinol-binding holoprotein in serum by examining diluted samples directly in a spectrofluorometer.[3] This does not take into account the variation in fluorescence of pigments bound to albumin and lipoprotein or interference from other conjugated proteins and cannot be regarded as satisfactory.

[1] J. N. Thompson, P. Erdody, R. Brien, and T. K. Murray, *Biochem. Med.* **5**, 67 (1971).
[2] J. Glover, L. Moxley, H. Muhilal, and S. Weston, *Clin. Chim. Acta* **50**, 371 (1974).
[3] S. Futterman, D. Swanson, and R. E. Kalina, *Invest. Ophthalomol.* **14**, 125 (1975).

Reagents, Solutions, and Mixtures

Analytical grade reagents should be used where possible.

Buffer Solutions for Electrophoresis

| Reagent | Upper electrode | Lower electrode | Gel |
|---|---|---|---|
| Tris(hydroxymethyl)-aminomethane | 1 $M$ | 1 $M$ | 1.5 $M$ |
| Glycine | 1 $M$ | — | — |
| Temed ($N,N,N',N'$-tetra-methylethylenediamine) | — | — | 0.02 $M$ (2.3 ml) |
| Hydrochloric acid | — | 0.5 $M$ (45 ml) conc. | 0.24 $M$ (240 ml) NHCl |
| pH adjusted to: | 8.9 | 8.1 | 8.9 |
| Final volume | 1 liter | 1 liter | 1 liter |
| Dilution before use in apparatus | 20 × | 10 × | — |

Buffer Solution for Homogenization

0.25 $M$ sucrose : 25 m$M$ KCl : 5 m$M$ MgCl$_2$ in 10 m$M$ Tris · HCl at pH 7.6
Triton X-100, 10% solution

| Acrylamide Solutions | | Gel Initiator Solutions | Solution A | Solution B |
|---|---|---|---|---|
| Acrylamide (British Drug Houses) | 20 g | Ammonium persulphate | 0.4 g | 0.6 g |
| | | Sucrose | 10 g | 10 g |
| Bis-acrylamide (Serva) | 0.67 g | Volume | 100 ml | 100 ml |
| Final volume | 100 ml | | | |

The above solutions are kept in the refrigerator and in the dark. The persulfate should be used within 7 days.

Polymerization Mixtures (Sufficient for Eight Gels)

| Component | Concentrating gel (large pore) | Resolving gel (small pore) |
|---|---|---|
| Acrylamide | 2 ml | 8 ml |
| Gel buffer | 1 ml | 11 ml |
| Initiator | 3 ml (B) | 11 ml (A) |
| Distilled H$_2$O | 4 ml | — |

Assay Procedure for Homogeneous Solutions of Retinol-Binding
   Holoprotein

*Preparation of Polyacrylamide Gels.* Polacrylamide disc gels are prepared using a procedure similar to that of Davis[4] by filling 6 mm internal
diameter × 7 cm long glass tubes mounted vertically in rubber caps within
2 cm of the top with 5% w/v resolving gel containing 0.75 $M$ Tris · HCl
buffer, pH 8.9, and 3.5% (w/v) sucrose covered with water until the gel
has set. The layer of water is removed. A further 1 cm of 4% (w/v)
concentrating gel containing 0.25 $M$ Tris · HCl at the same pH is placed on
top and covered with water until it sets. This usually takes about 15–20
min; the water is than decanted off, and the gels are transferred to the
standard circular electrophoresis tank (Shandon or similar apparatus), the
lower buffer compartment of which is filled with Tris · HCl at pH 8.1. Care
is taken to ensure that air bubbles are not trapped at the base of the glass
tubes. The upper buffer compartment is then filled with Tris · HCl buffer
pH 8.9 to cover the tops of the gel tubes. The electrophoresis tank is
placed in a cold room or cabinet at +4° and connected to the power unit.

HoloRBP solutions or samples of plasma containing 10% (w/v) sucrose
to make them more dense are applied as a thin layer to the upper buffer–
gel interface using a microsyringe pipette. Sample volumes of 5–20 $\mu$l are
used depending on the expected concentration of holoRBP. It is advisable
also to include a sample of a known standard plasma[5] in the series to serve
as a control check on the performance of the instruments during the assay.
A constant current of 4 mA/tube is applied for 90 min across the tubes, by
which time the holoRBP will have penetrated about 3 cm into the gel.

*Estimation of Fluorescence.* The gels are then removed from the tank
and the glass tubes and placed in turn in a 12 cm long × 0.8 cm interior
width × 1.8 cm high quartz cell of a modified Chromoscan[6] instrument
containing sufficient distilled water to cover the gel to a depth of 3–4 mm.
Gels should be removed from their glass tubes only immediately before
analysis, not be left in water, where losses of RBP by diffusion will occur.
Each gel is placed in the instrument for scanning with a narrow beam of
UV light from the Hg-arc lamp and Woods glass filter (300–400 nm). The
standard 1.0 cm by 1.0 mm (or 0.5 mm) slit as supplied is reduced in height

---

[4] B. J. Davis, *Ann. N. Y. Acad. Sci.* **121**, 404 (1964).
[5] This substandard for checking the instrument can be prepared from a large sample of
   fasting plasma already containing 10% sucrose dispensed in 100-$\mu$l plastic sample tubes
   before snap freezing and keeping at −20°C. A single sample tube can then be removed for
   examination with each series of analyses without disturbing the remaining stock.
[6] Chromoscan instrument obtained from Joyce Loebl & Co. Ltd., Princesway, Team Valley,
   Gateshead, NEll OUJ, U.K. The instrument needs modification as indicated by Glover *et
   al.*[2] for optimal results.

slightly to 7 mm by masking so that the beam just irradiates comfortably the full width of the gel.

The weak background fluorescence of the gel and quartz cell is balanced against the reference beam using either a collimating disk with 2- or 3-mm diameter aperture or an additional neutral gray filter 0.5 D in the slide compartment beneath the phototube and the manually adjustable wedge, so that the recorder trace for a blank gel runs about 5 cm from the right-hand edge of the paper. The most suitable cam for use in the automatic balancing system of the instrument was found to be 5-077D. The optimal setting of the sensitivity potentiometer knob is determined by trial, but usually a value from 3 to 7 suffices and is always fixed at the selected position optimal for maximal signal to background "noise" of the amplifier system for this analysis. The scan gear ratio $1:3$ is used.

After the first scan is completed, the gel is rotated through 180° and scanned a second time so that account may be taken of any uneven application of running of the holoRBP during the electrophoresis.

The peak areas corresponding to the fluorescence of the holoRBP are most easily measured by triangulation. The main area of the peaks from the two scans is linearly related to the concentration of holoRBP in the extract, provided the amount of holoRBP applied to the gel does not exceed 1 $\mu$g (i.e., 20 $\mu$l of a 50 $\mu$g/ml preparation). There is a loss of linearity above this amount due to self-absorption effects. The reproducibility of the assay is $\pm3\%$ SE for replicate analyses done on the same day. Over a long period, variations in the fluorophotometer reference beam energy and background fluorescence of different batches of gels arise so that the overall error rises to $\pm7\%$ SE. It is advisable to switch the instrument on for at least 0.5 hr before use to allow the Hg-arc lamp to reach a steady running temperature. Acrylamide gels as normally prepared contain a low molecular weight bluish UV-fluorescent impurity that migrates in the electrophoresis with the buffer front. This does not interfere, however, with holoRBP. Typical values for some animal plasmas are given in the table.

*Calibration of the Instrument.* A calibration curve for the instrument is most easily prepared by simply plotting double peak areas (i.e., height × base width) against concentration of holoRBP calculated from retinol content of fasting plasma or test fluid. A series of fluorescence results is obtained using 20-$\mu$l portions of large samples (20 ml) of pooled fasting plasma. The retinol in the latter is separated from any residual traces of ester by extracting the total lipids and chromatographing them on a column of weakened alumina.[7] The retinol fraction is then assayed for retinol using the trifluoracetic acid reagent.[8]

[7] H. S. Huang and D. S. Goodman, *J. Biol. Chem.* **240**, 2839 (1965).

[8] J. R. Neeld and W. N. Pearson, *J. Nutr.* **79**, 454 (1963).

NORMAL RANGE OF RETINOL-BINDING HOLOPROTEIN IN PLASMA

| Animal species | HoloRBP ($\mu$g/ml) |
|---|---|
| Human | |
| Child | 20–30 |
| Adult | 30–70 |
| Rat | |
| Weanling | 20–30 |
| Adult | 30–50 |
| Sheep | |
| Anestrus | 35–45 |
| Estrus | 90–110 |
| Chicken | |
| Anestrus | 40–50 |
| Estrus | 60–80 |
| Japanese quail | |
| Anestrus | 100–180 |
| Estrus | 160–260 |

The molecular weight of holoRBP relative to retinol is 77 : 1, hence the concentration of holoRBP can be calculated on the basis that the retinol in fasting plasma is wholly combined with the specific protein in a 1 : 1 molar ratio. Additional checks on the relationship between fluorescence and retinol content may also be carried out using concentrates of holoRBP following chromatography of small samples of plasma on small DEAE–cellulose or arginine–Sepharose columns to separate off any low density lipoproteins that may carry retinol or retinyl esters other than that specifically bound to RBP.

From the calibration curve a factor can be determined to convert fluorescence peak areas directly to holoRBP concentration. This factor will vary from instrument to instrument and should be rechecked following replacement of either the light source or phototubes and after any other readjustment of a component of the optical system. It will also vary with the type of buffer and its concentration in the gel, so once conditions are set they should be maintained. The overall sensitivity of the instrument can, however, be easily checked by using solutions of quinine sulfate in the cell in place of the gel.

First, 10 ml of solution of quinine sulfate (0.1 $\mu$g/ml 0.1 $M$ H$_2$SO$_4$), which provides background fluorescence approximately equivalent to that of the gel, is placed in the cell so that the manual balancing wedge can be adjusted to approximately the same base line as that used in the assay (i.e., with the recording pen at a distance of 5 cm from the right-hand side of the paper). This solution can then be replaced by stronger quinine sulfate solutions up to 1.0 $\mu$g/ml and the response of the recording pen measured for each increase in concentration. The increment in pen deflection per 0.1 $\mu$g increase in quinine sulfate over the linear response range should be constant if the sensitivity of the instrument is unchanged.

Estimation of Retinol-Binding Holoprotein in Animal Tissues

Plasma retinol-binding protein is largely bound to membrane components in the liver and to some extent in the kidney.[9,10] It is possible that in most target tissues the holoprotein will be complexed with plasma membrane receptors. It is necessary therefore to treat tissue homogenates and membrane preparations with a detergent to free the retinol-binding protein, before the RBP can be separated from other proteins by electrophoresis. Triton X-100 (1% solution) was found most suitable, and it does not affect the electrophoretic mobility of the protein in polyacrylamide gel.

*Procedure.* The tissue should be taken from the animal after most of the blood has been removed. It is then washed twice with a small amount of the pH 7.6 Tris · HCl buffer-saline solution used for the homogenization. Five grams of the tissue (blotted to remove surface buffer fluid) are chopped and homogenized in a tissue blender with 10 ml of Tris-buffer saline for 1.5 min at 4°C and filtered through mutton cloth to remove any fibrous tissue. One-tenth volume of 10% Triton X-100 solution (1 ml) is then added to a known volume (10 ml) of the homogenate, and the mixture is allowed to stand for 1 hr at 4°C; it is then spun for 1 hr at 40,000 rpm in a refrigerated centrifuge to remove most particulate material. The supernatant is decanted off, and a suitable aliquot, 20–60 $\mu$l, is applied directly to a polyacrylamide disc gel for electrophoresis and fluorescence analysis as for plasma. The fluorescent zone corresponding to holoRBP has almost the same electrophoretic mobility relative to the buffer front (blue fluorescent zone) as that of holoRBP in plasma (0.57). Other fluorescence peaks nearer the front of the gel should be ignored. In the case of human RBP the same correction factor as for plasma can be used to convert fluorescence areas into holoRBP concentration. In the case of the rat, however, the factor is smaller owing to the lower molecular weight of its RBP. A separate factor should be determined for the RBP of each animal. Factors for the sheep, human, and avian RBPs are very similar.

[9] J. Glover, C. Jay, and G. H. White, *Vitam. Horm.* (*New York*) **32**, 215 (1974).
[10] J. E. Smith, Y. Muto, and D. S. Goodman, *J. Lipid Res.* **16**, 318 (1975).

## [34] Purification of Cellular Retinol and Retinoic Acid-Binding Proteins from Rat Tissue[1]

By DAVID ONG and FRANK CHYTIL

Two intracellular proteins that bind vitamin A-like compounds exist in many animal tissues. The first, called cellular retinol-binding protein (CRBP) binds retinol (vitamin A alcohol) with high affinity and specificity.[2] The second, called cellular retinoic acid-binding protein, or CRABP, binds retinoic acid (vitamin A acid) with high affinity and specificity.[3] CRBP is apparently present in most or all epithelial tissues of the rat, but the amount of binding protein present is small compared to total soluble protein. However, rat liver is a particularly rich source of CRBP. A 10-g liver contains about 150 $\mu$g of CRBP, making liver the tissue of choice for purification of CRBP.

CRABP is present in fewer adult rat tissues than CRBP and is not found in liver. The best source, using criteria of amount of tissue available and level of binding protein, appears to be testis. A 1.5 g testis contains about 4 $\mu$g of CRABP. Testis also contains CRBP (about 5.5 $\mu$g per testis). Since the behavior of CRABP and CRBP in a number of purification techniques is quite similar, the isolation of CRABP requires separation from CRBP. The procedures described here for testis are designed to maximize recovery of CRABP; CRBP can be obtained as a by-product.

### Assay Methods

In both cases detection depends on monitoring the ligands (either retinol or retinoic acid) bound specifically by the proteins. One can utilize radioactive ligands or, if the binding proteins are present in sufficient quantity, monitor the ligands spectrophotometrically.

*Sucrose gradient centrifugation.* This technique is useful for initial extracts when the concentration of the binding proteins is low. Aliquots of extracts are made to 0.3 ml with 50 m$M$ Tris · HCl buffer, pH 7.2. The samples are then incubated with either 100 pmol of all-*trans*-[15-$^3$H]retinol (2.5 Ci/mmol; New England Nuclear) or 20 pmol of all-*trans*-[11, 12-$^3$H]retinoic acid (11.1 Ci/mmol; Hoffmann-La Roche, Nutley, New Jersey) for 4 hr in the dark at 4°. The radioactive ligands are added in 5 $\mu$l of

[1] This work was supported by United States Health Service Grants HD-05384, HD-09195, HL-14214, HL-15341, and CA-20850.
[2] M. M. Bashor, D. O. Toft, and F. Chytil, *Proc. Natl. Acad. Sci. U.S.A.* **70**, 3483 (1973).
[3] D. E. Ong and F. Chytil, *J. Biol. Chem.* **250**, 6113 (1975).

isopropanol containing 1 mg of butylated hydroxytoluene per milliliter. Parallel incubations are carried out with an 100-fold excess of unlabeled ligand, also added in 5 $\mu$l of isopropanol.

Immediately prior to gradient centrifugation the incubations are mixed with 0.2 ml of a charcoal–dextran suspension[4] (for details see this series)[5] to absorb free ligand. After a 5-min incubation the charcoal-coated dextran is removed by centrifugation at 1000 g for 5 min. Aliquots (0.2 ml) of the incubation mixtures are submitted to centrifugation on linear 5 to 20% (w/v) sucrose gradients in 50 m$M$ Tris · HCl, pH 7.2. Details of this method have been described in this series.[6] Centrifugation is for 20 hr at 40,000 rpm in a Beckman SW 50.1 swinging-bucket rotor. The binding proteins are revealed in the radioactive profile of the fractionated gradients as a peak of radioactivity in the 2 S region. The peak is not present in the gradient that contains excess unlabeled ligand. The difference in radioactivity recovered in the 2 S region (specific binding) is used to quantitate the amount of binding protein present, assuming one binding site per molecule of binding protein.

Extracts of liver frequently have sufficient amounts of endogenous free retinol present to interfere with this assay. This problem can be rectified by exposing the extract to a long-wave UV lamp for 10–15 hr (intensity about .05 $\mu$W/cm[2]) This will destroy retinol without detectable damage to the binding protein.

*Spectrophotometry.* After the initial steps in purification, the amount of the binding proteins present can usually be estimated from the absorbance of the bound ligand. Retinol bound to CRBP has a molar extinction coefficient of 50,200 with $\lambda_{max}$ at 350 nm.[7] It should be noted that the spectrum of retinol is considerably different when bound to CRBP, compared to its spectrum in organic solvents; $\lambda_{max}$ shifts from 325–330 nm to 350 nm.[7] When retinoic acid is bound to CRABP, its spectrum does not change significantly and exhibits a molar extinction coefficient of 50,000 with $\lambda_{max}$ at 350 nm.[8]

*Spectrophotofluorometry.* Both retinol and retinoic acid are fluorescent molecules, and their fluorescence is considerably enhanced when they are bound to their respective binding proteins. The fluorescence spectra of the two ligand–protein complexes are quite similar, but the yield of fluorescence for retinol-CRBP (excitation at 350 nm; emission at 485 nm) is about 5 times higher than for retinoic acid-CRABP (excitation at 350

[4] S. G. Korenman, *J. Clin. Endocrinol. Metab.* **28**, 127 (1968).
[5] See this series, Vol. 36 [3].
[6] See this series, Vol. 36 [14].
[7] D. E. Ong and F. Chytil, *J. Biol. Chem.* **253**, 828 (1978).
[8] D. E. Ong and F. Chytil, *J. Biol. Chem.* **253**, 4551 (1978).

nm; emission at 475 nm). Monitoring fluorescence is frequently a convenient technique for determining the elution position of the CRBP–retinol complex from various columns.

## Purification Procedures

The procedures for the purification of CRBP from rat liver and CRABP from rat testes are quite similar in most part (see Tables I and II) and will be described together. All steps are carried out at 4° unless otherwise noted. However, determinations of pH for solutions used for column chromatography are at room temperature. Buffers for ion-exchange chromatography are prepared by dilution from stock buffer solutions.

*Preparation of Buffers.* Measurement of fluorescence during a buffer gradient elution requires low background fluorescence. Consequently, Tris (ultra pure, Schwarz/Mann) is recrystallized from methanol at 20°. Imidazole (grade I, Sigma) is recrystallized twice from benzene, first at 40°, then at 20°. Stock buffer solutions are prepared as follows: Two volumes of 0.5 $M$ Tris are combined with one volume of 0.5 $M$ acetic acid to give a stock 0.33 $M$ Tris-acetate buffer, about pH 8.3. One volume of 1 $M$ imidazole and one volume 1 $M$ acetic acid are combined to give a stock 0.5 $M$ imidazole-acetate buffer, about pH 6.0.

*Step 1. Preparation of Tissue Extracts.* Approximately 1500 g of liver or 4500 g of testes are homogenized with 2 volumes of 10 m$M$ Tris · HCl buffer, pH 7.5. The homogenization is carried out in portions of 200 g for 30 sec in a Waring blender. Debris is removed by centrifugation at 20,000 $g$ for 15 min. The cloudy supernatant liquid is collected and adjusted to pH 5 by dropwise addition of glacial acetic acid with rapid stirring. The resulting mixture is centrifuged at 20,000 $g$ for 15 min to remove precipitated material and provide a clear, cell-free extract.

*Step 2. Batchwise Treatment with CM-Cellulose.* CM-cellulose (CM-52,

TABLE I
PURIFICATION OF CELLULAR RETINOL-BINDING PROTEIN FROM RAT LIVER

| Fraction | Total protein (mg) | Binding protein (mg) | Purification (fold) | Recovery (%) |
|---|---|---|---|---|
| After acid precipitation | 36,000 | 21.3 | — | — |
| CM-cellulose | 20,000 | 19.6 | 1.7 | 92 |
| Sephadex G-75 | 1,800 | 17.8 | 17 | 84 |
| DEAE-cellulose (pH 8.3) I | 25 | 14.0 | 950 | 66 |
| DEAE-cellulose (pH 8.3) II | 15 | 11.6 | 1300 | 54 |
| Sephadex G-50 | 8.9 | 8.6 | 1600 | 40 |

TABLE II
PURIFICATION OF CELLULAR RETINOIC ACID-BINDING PROTEIN FROM RAT TESTES

| Fraction | Total protein (mg) | Binding protein (mg) | Purification (fold) | Recovery (%) |
|---|---|---|---|---|
| After acid precipition | 73,000 | 12.1 | — | — |
| CM-cellulose | 44,000 | 11.0 | 1.5 | 91 |
| Sephadex G-75 | 3,600 | 9.6 | 16 | 79 |
| DEAE-cellulose (pH 8.3) I | 210 | 8.8 | 260 | 71 |
| DEAE-cellulose (pH 8.3) II | 120 | 7.7 | 400 | 64 |
| DEAE-cellulose (pH 6) I | 9.5 | 7.0 | 450 | 58 |
| DEAE-cellulose (pH 6) II | 5.6 | 5.3 | 5800 | 44 |
| Sephadex G-50 | 2.8 | 2.8 | 6000 | 23 |

Whatman) is equilibrated at pH 5.0 (for liver) or pH 5.1 (testes) in 10 m$M$ sodium acetate. The cellulose is then collected by filtration and added as a wet cake to the extract. The ratio of added cellulose to extract should be 1/10 (v/v). The cellulose is suspended by stirring, and the pH is adjusted, if necessary, to pH 5.0 (liver) or pH 5.1 (testes) with 10 $M$ NaOH. The suspension is stirred for 30 min and then filtered to remove the cellulose. The supernatant liquid is titrated to pH 7.2 by dropwise addition of 5 $M$ NaOH. Phenylmethanesulfonyl fluoride in ethanol (2.1 g/100 ml) is added to the extract in the ratio of 1/150 (v/v) to inhibit proteases. The addition of phenylmethanesulfonyl fluoride should be repeated later, after the fractions of interest for each column are combined. The solution is now concentrated 6- to 10-fold by ultrafiltration using an Amicon cell with a PM-10 or UM-10 membrane. The protein concentration after ultrafiltration should be about 40–60 mg/ml.

*Step 3. Saturation of the Binding Proteins with Their Ligand.* To aid in detection and quantitation, the binding proteins are saturated with their respective ligands prior to each chromatographic run. All-*trans*-retinol and/or all-*trans*-retinoic acid (10 m$M$ in isopropanol) are added to the solutions to give a final concentration of 1 $\mu M$. The amount of binding proteins present makes it impractical to saturate them with radioactive ligand. However, they may be labeled sufficiently for detection by adding radioactive ligand (1–2 $\mu$Ci) after dialysis but before application to a column.

*Step 4. Gel Filtration on Sephadex G-75.* The concentrated extract is dialyzed against 0.2 $M$ NaCl in 50 m$M$ Tris · HCl buffer, pH 7.5. Portions (50–100 ml) are then submitted to gel filtration on a column of Sephadex G-75 (5.0 × 75 cm) equilibrated with the same buffer. The flow rate is set at 1 ml/min, and 15-ml fractions are collected. The elution positions of CRBP and CRABP are essentially the same and are determined by monitoring fluorescence of the fractions (excitation at 350 nm; emission at 480 nm). The elution position is centered at approximately fraction 70. Fractions containing the binding protein(s) are combined and concentrated by ultrafiltration as before to a protein concentration of 10–15 mg/ml.

*Step 5. DEAE-Cellulose Chromatography I (pH 8.3).* For preparation of CRBP from liver, the concentrated solutions from the Sepadex G-75 column runs are combined and exhaustively dialyzed against 10 m$M$ Tris-acetate buffer, pH 8.3. This and later buffers are prepared by dilution from the stock buffer solutions. A DEAE-cellulose column (2.6 × 18 cm) is prepared from microgranular cellulose (Whatman, DE-52), equilibrated with the 10 m$M$ Tris-acetate buffer, pH 8.3. The protein solution is applied to the column, and the column is washed first with 50 ml of 10 m$M$ Tris-acetate buffer then two column volumes of 50 m$M$ Tris-acetate. The flow rate is 1 ml/min, and 5-ml fractions are collected. The column is then eluted (see Fig. 1, top) with a linear gradient of the same buffer from 50 m$M$ to 0.33 $M$ (total volume 500 ml). The fractions are monitored for $A_{350}$. The elution position of retinol-CRBP is at about 350 ml of the 500-ml gradient. The appropriate fractions are combined and concentrated by ultrafiltration to a protein concentration of about 3–5 mg/ml.

For preparation of CRABP from testis, the preparation of the DEAE-cellulose column, buffers, and sample are the same as described above for CRBP. After application of the sample and the 50-ml wash with 10 m$M$ Tris-acetate, the column is eluted with 75 ml of 50 m$M$ Tris-acetate, then 125 ml of 100 m$M$ Tris-acetate. The column is then eluted (see Fig. 1, bottom) with a linear gradient of Tris-acetate buffer from 0.1 to 0.3 $M$ (total volume 500 ml). When the fractions are monitored for $A_{350}$, a peak with a prominent shoulder is observed. The shoulder represents retinoic acid-CRABP; the peak, retinol-CRBP. Prior labeling with [³H]retinoic acid helps to resolve the two proteins. The elution position of CRABP is at about 320 ml while CRBP is at 370 ml. The fractions should be combined to maximize recovery of CRABP. This will cause about 30–34% of the CRBP present to be included also.

The fractions are concentrated by ultrafiltration to a protein concentration of about 3–5 mg/ml.

*Step 6. DEAE-Cellulose Chromatography II (pH 8.3).* The next DEAE-cellulose columns (1.6 × 15 cm) are prepared as before. The protein solutions are exhaustively dialyzed against 10 m$M$ Tris-acetate, pH 8.3, then applied to the columns. The columns are first washed with 50 ml of 50 m$M$ Tris-acetate, pH 8.3, then eluted with a linear gradient of NaCl from 0 to 0.1 $M$ in the 50 m$M$ Tris-acetate (total volume, 500 ml). Flow rate is 1 ml/min, and fractions are 5 ml. Fractions are monitored for $A_{350}$ or fluores-

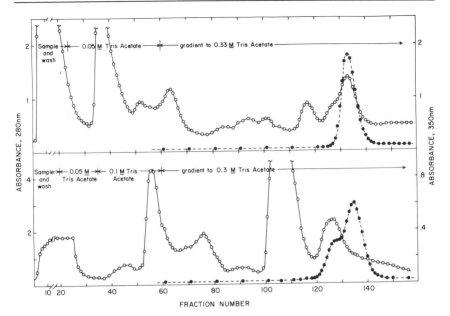

FIG. 1. Chromatography of cellular retinol-binding protein (CRBP) and cellular retinoic acid-binding protein (CRABP) on DEAE-cellulose at pH 8.3. *Top:* chromatography of liver preparation. *Bottom:* Chromatography of testes preparation. O——O, Absorbance at 280 nm; ●——●, absorbance at 350 nm.

cence. CRBP elutes at about 260 ml of the gradient; CRABP at about 220 ml. The appropriate fractions are combined and concentrated to give a protein concentration of about 2 mg/ml.

At this point CRBP from liver should be quite close to homogeneity. This can be monitored by the ratio, $A_{350}/A_{280}$, which is 1.75 for purified CRBP saturated with retinol. Sodium dodecyl sulfate–polyacrylamide electrophoresis is useful to reveal whether a common impurity is present, which has a molecular weight of about 21,000 compared to 14,600 for CRBP. If so, step 9 should be carried out.

*Step 7. DEAE-Cellulose Chromatography I (pH 6.0).* The concentrated solution containing CRABP is adjusted to pH 6.0 with 1 $M$ acetic acid, dialyzed against 10 m$M$ imidazole acetate, pH 6.0, and applied to DEAE-cellulose column (1.6 × 15 cm) prepared and equilibrated in the same buffer. Flow rate is 1 ml/min; fraction size is 5 ml. The column (see Fig 2) is eluted with a linear gradient of imidazole-acetate from 0.01 to 0.10 $M$ (total volume 500 ml). The second protein peak is CRABP, matching the $A_{350}$ peak at about 310 ml. Any CRBP still present elutes at about 270 ml. The fractions containing CRABP are combined, and concentrated

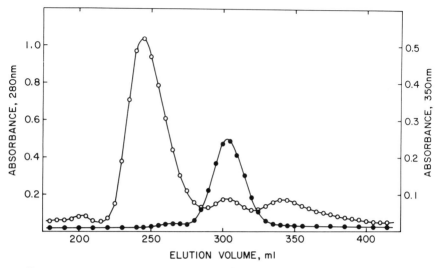

FIG. 2. Chromatography of cellular retinoic acid-binding protein on DEAE-cellulose at pH 6.0. O———O, Absorbance at 280 nm; ●———●, absorbance at 350 nm.

to a protein concentration of about 1 mg/ml. If the solution is to be frozen or stored, the pH should be adjusted to 7.5.

*Step 8. DEAE-Cellulose Chromatography II (pH 6.0).* The chromatography described in step 7 is repeated exactly. The CRABP should now be greater than 95% pure. The remaining impurity is usually a single protein with molecular weight of about 21,000, and can be removed in the next step.

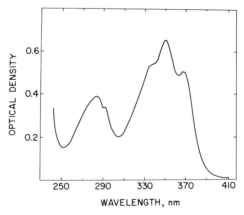

FIG. 3. Absorption spectrum of a homogeneous preparation of cellular retinol-binding protein, 13.5 $\mu M$ in 50 m$M$ Tris · HCl, pH 7.5.

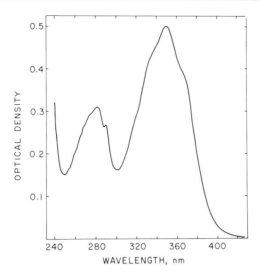

FIG. 4. Absorption spectrum of a homogeneous preparation of cellular retinoic acid-binding protein 10 $\mu M$ in 50 m$M$ Tris · HCl, pH 8.3.

*Step 9. Gel Filtration on Sephadex G-50.* CRBP (liver) or CRABP (testis) in a volume of about 5 ml is dialyzed against 50 m$M$ NaCl in 50 m$M$ Tris-acetate, pH 8.3, then submitted to gel filtration on a column (2.6 × 50 cm) of Sephadex G-50 (fine) prepared and equilibrated in the same buffer. The column is run at a flow rate of 30 ml/hr. $A_{350}$ and $A_{280}$ are determined for each fraction. Fractions with a ratio of $A_{350}$ to $A_{280}$ greater than 1.7 are combined.

Properties

*Storage.* Both binding proteins can be stored at a concentration of 0.5–1.0 mg/ml in 50 m$M$ NaCl and 50 m$M$ Tris · HCl, pH 8.3, at −20° for periods up to a year without loss of binding activity.

*Size.* The molecular weights of the purified proteins have been estimated by SDS-polyacrylamide electrophoresis and gel filtration (CRBP and CRABP)[7,8] and by the sedimentation equilibrium technique (CRBP).[7] Both binding proteins appear to be single polypeptide chains with a molecular weight of about 14,600. The amino acid composition of each has been published.[7,8]

*Ligand Affinity and Specificity.* The dissociation constant (apparent) for retinol and purified CRBP is 16 n$M$.[7] The binding specificity has been investigated using a partially purified preparation of CRBP and various analogs and isomers of retinol.[9] The binding studies showed an absolute dependence for the -OH group, a preference for the all-*trans* double bond configuration, and some freedom allowed in ring structure.

The dissociation constant apparent for retinoic acid and purified CRABP is $4.2 \times 10^{-9} M$.[8] The binding specificity was investigated using various analogs of retinoic acid and a partially purified preparation of CRABP.[10] The binding protein shows an absolute requirement for the carboxyl group. However, a number of modifications of the ring of retinoic acid still permit effective binding.

*Spectral Properties.* The absorbance spectra of retinol-CRBP and retinoic acid-CRABP are shown in Figs. 3 and 4. Both spectra are dominated by a major peak at 350 nm owing to the bound ligand. The molar extinction coefficient for retinol bound to CRBP is 50,200.[7] The ratio of absorbance at 350 to absorbance at 280 for retinol-CRBP is 1.75. The molar extinction coefficient for retinoic acid bound to CRABP is 50,000.[8] The value of $A_{350}/A_{280}$ for retinoic acid-CRABP is 1.8.

[9] D. E. Ong and F. Chytil, *Nature (London)* **255**, 74 (1975).
[10] F. Chytil and D. E. Ong, *Nature (London)* **260**, 5546 (1976).

## [35] Cellular Retinol-, Retinal-, and Retinoic Acid-Binding Proteins from Bovine Retina

*By* JOHN C. SAARI, SIDNEY FUTTERMAN, GENE W. STUBBS, and LUCILLE BREDBERG

An aqueous extract of retina contains three separable retinoid-binding proteins with specificities directed toward retinol,[1,2] retinoic acid,[1,2] and retinal.[3,4] The latter binding protein may be unique to retina,[5] whereas binding proteins specific for retinol and retinoic acid have been reported in many other tissue extracts.[6,7]

### Assay Method

*Principle.* The presence of retinoid-binding proteins in tissue extracts may be detected by incubation with radiolabeled ligand (retinol, retinal, or

[1] B. N. Wiggert and G. J. Chader, *Exp. Eye Res.* **21**, 143 (1975).
[2] J. C. Saari, S. Futterman, and L. Bredberg, *J. Biol. Chem.* **253**, 6432 (1978).
[3] S. Futterman, J. C. Saari, and S. Blair, *J. Biol. Chem.* **252**, 3267 (1977).
[4] G. W. Stubbs, J. C. Saari, and S. Futterman, *J. Biol. Chem.* **254**, 8529 (1979).
[5] S. Futterman and J. C. Saari, *Invest. Ophthalmol. Vis. Sci.* **16**, 768 (1977).
[6] D. E. Ong and F. Chytil, *J. Biol. Chem.* **250**, 6113 (1975).
[7] B. P. Sani and T. H. Corbett, *Cancer Res.* **37**, 209 (1977).

retinoic acid) followed by gel filtration chromatography.[8] Liquid scintillation counting (or, in the case of retinol and bound retinoic acid, monitoring fluorescence) of the column fractions typically reveals peaks at both the void and included volume of the column in addition to the binding protein peak. Ligand appearing at the void volume, since it is not displaced by the addition of excess unlabeled ligand, is thought to be nonspecifically associated with large lipid and/or protein components present in the extract or to be present as microcrystals of undissolved ligand. Hence, it is important to choose a gel in which the binding proteins are clearly separated from both the void and included volume regions. Owing to the nonequilibrium nature of the gel filtration assay, it cannot be considered to yield quantitative binding data, but does provide a semiquantitative measure of binding activity useful in comparative studies.

*Reagents*

All-*trans*-[15-$^3$H]retinol,    2.39    Ci/mmol,    in    ethanol    containing α-tocopherol, 1 mg/ml[9]

All-*trans*-[15-$^3$H]retinal, prepared and photoisomerized[10] as described by Futterman *et al.*[3]

All-*trans*-[11, 12-$^3$H]retinoic acid,[11] 1.45 Ci/mmol, in ethanol containing α-tocopherol, 1 mg/ml

*Procedure.* To 1 ml of extract is added 2 μl of [$^3$H]retinol, photoisomerized [$^3$H]retinal, or [$^3$H]retinoic acid, representing approximately 1 nmol of ligand. The sample is allowed to stand at room temperature in the dark for 30 min, then 100 mg of sucrose are added and the sample is layered on a 1.5 × 140 cm column of Sephadex G-100 equilibrated with 50 mM Tris, pH 7.5, 0.2 M in NaCl. Gel filtration is accomplished at 5° with a flow rate of 18 ml/hr. Fractions of 3 ml are collected and analyzed for radioactivity. The specific binding activity of the preparation is expressed as the picomoles of ligand bound (total cpm bound ÷ specific radioactivity of ligand) per milligram of protein in the original sample. Protein is determined according to Lowry *et al.*[12]

[8] J. C. Saari, A. H. Bunt, S. Futterman, and E. R. Berman, *Invest. Ophthalmol. Vis. Sci.* **16,** 797, (1977).

[9] Available commercially as Vitamin A₁ (all-*trans*), [1-$^3$H(N)]-, from New England Nuclear Corporation.

[10] *Cis*-isomers of [$^3$H]retinal, which preferentially interact with the retinal-binding protein, are produced from all-*trans*-[$^3$H]retinal by photoisomerization.

[11] Obtained as a generous gift from Dr. W. E. Scott, Hoffmann–La Roche.

[12] O. H. Lowry, N. J. Rosebrough, A. L. Farr, and R. J. Randall, *J. Biol. Chem.* **193,** 265 (1951).

Purification Procedure

The complexes formed on incubation of extracts of retina with retinol, retinal, or retinoic acid are sufficiently stable to allow localization of the binding proteins during subsequent purification procedures. Bound retinol or retinoic acid is detected by monitoring its fluorescence at 458 nm (excitation at 335 nm), or by liquid scintillation counting when radioactive ligand is used. Bound retinal is most easily detected by liquid scintillation counting.

All procedures are carried out at 5° in subdued or red illumination unless otherwise noted. The purification procedure has been satisfactorily performed starting with 50–300 cattle retinas. The volumes given below correspond to a preparation starting with 200 retinas.

*Step 1. Extraction.* Two hundred frozen cattle retinas[13] are thawed, homogenized[14] in 50 mM Tris, pH 7.5, 0.2 M in NaCl, and 1 mM in phenylmethane sulfonyl fluoride (32 ml) and centrifuged at 110,000 g for 1 hr in an International B 60 centrifuge, rotor A-211. The supernatant is normally used immediately but may be stored frozen at −20°.

*Step 2. Gel Filtration.* The extract (62 ml) is incubated with 10 μl of photoisomerized [3H]retinal for 30 min at room temperature and applied to a 5 × 100 cm column of Sephadex G-100 previously equilibrated in 50 mM Tris, pH 7.5, 0.2 M in NaCl. Fractions of 20 ml are collected at a flow rate of 60 ml/hr. A radioactive peak representing retinal-binding protein elutes soon after the void volume of the column ($K_d = 0.28$).[15] Retinol- and retinoic acid-binding proteins coelute from this column[2]; hence, it is sufficient to monitor the fluorescence due to endogenous bound retinol and retinoic acid to locate these binding fractions ($K_d = 0.52$).[15]

*Step 3. Adsorption and Elution from DEAE-Cellulose*

RETINOL- AND RETINOIC ACID-BINDING PROTEINS. Fluorescent fractions from step 2 are pooled, dialyzed against 20 mM Tris, pH 7.5, and incubated with 20 μl of [3H]retinoic acid for 30 min at room temperature before application to a 0.9 × 60 cm column of DEAE-cellulose[16] previously equilibrated with dialysis buffer. The relatively large volume of sample (250 ml) is pumped onto the column at a flow rate of 36 ml/hr, followed by 1–2 column volumes of equilibration buffer or a sufficient volume to bring the absorbance at 280 nm to a base line value. Elution of the binding proteins is accomplished by linearly increasing the concentra-

---

[13] Obtained from the George A. Hormel and Co.

[14] Fifty retinas are homogenized at a time in 8 ml of specified buffer using 8–10 passes of a glass-to-glass, 40-ml capacity, Ten Broeck tissue homogenizer.

[15] $K_d = (V_e - V_o)/(V_i - V_o)$

[16] Whatman DE-52 cellulose was used.

tion of NaCl from 0 to 0.13 $M$ in the above buffer (total gradient volume, 500 ml). Fractions of 4 ml are collected during the procedure and analyzed for radioactivity and fluorescence. The peaks of retinoic acid-binding activity (radioactivity and fluorescence) and retinol-binding activity (fluorescence) are eluted at approximately 60 and 75 m$M$ NaCl, respectively.

CELLULAR RETINAL-BINDING PROTEIN. Fractions containing bound [$^3$H]11-*cis*-retinal from step 2 are pooled, dialyzed for 4 hr against 3 liters of 20 m$M$ Tris, pH 7.5, 1 m$M$ in dithiothreitol, and applied to a 0.9 × 60 cm column of DEAE-cellulose[16] previously equilibrated with the same buffer. The cellular retinal-binding protein is eluted with a linear 500 ml gradient from 75 to 250 m$M$ in NaCl containing 20 m$M$ Tris, pH 7.5, and 1 m$M$ dithiothreitol at a flow rate of 1 ml/min. Fractions of 5 ml each are collected and analyzed for radioactivity.

*Step 4. Adsorption and Elution from Calcium Phosphate Gel*

CELLULAR RETINOIC ACID-BINDING PROTEIN. Radioactive fractions from step 3 are pooled (approximately 38 ml) and pumped onto a 0.9 × 60 cm column of calcium phosphate gel[17] previously equilibrated with 20 m$M$ Tris, pH 7.5, 50 m$M$ in NaCl. After washing the column with equilibration buffer to bring the absorbance at 280 nm back to a base line value, the binding protein is eluted with a linear gradient from 0 to 0.1 $M$ sodium phosphate, pH 7.5, in equilibration buffer (total gradient volume, 500 ml). Fractions of 4 ml are collected at a flow rate of 35 ml/hr. The peak of eluted radioactivity occurs 2–3 fractions before the corresponding peak of 280-nm absorbance. Fractions from the tailing side of the absorbance peak contain apo-binding protein whereas fractions taken from the leading side of the peak are a mixture of holo- and aporetinoic acid-binding protein.[2]

CELLULAR RETINOL-BINDING PROTEIN. Fluorescent fractions from step 3 are pooled (approximately 34 ml) and pumped onto a column of calcium phosphate gel.[17] Elution from the column is accomplished as described for the retinoic acid-binding protein except that the linear gradient is from 0 to 50 m$M$ sodium phosphate.

CELLULAR RETINAL-BINDING PROTEIN. Fractions containing radioactivity from step 3 are pooled, concentrated to 10 ml by ultrafiltration (Amicon UM-10), centrifuged at 110,000 $g$ for 20 min, and applied to a 0.9 × 60 cm column of calcium phosphate gel[17] previously equilibrated with 20 m$M$ Tris, pH 7.5, 150 m$M$ NaCl, and 1 m$M$ dithiothreitol. The binding protein is eluted with a 400-ml gradient of 0 to 0.1 $M$ sodium phosphate in the same buffer. Fractions of 3.3 ml each are collected and assayed for radioactivity.

[17] BioGel HTP, available from Bio-Rad Laboratories, was used.

Properties

*Purity.* The retinoic acid-binding activity, purified through step 3, shows a single band on polyacrylamide gel electrophoresis in the presence of sodium dodecyl sulfate (SDS). However, in the absence of SDS, gel electrophoresis reveals the presence of two protein components, only one of which binds retinoic acid.[2] After step 4 the preparation is homogeneous as judged by gel electrophoresis in the presence or the absence of SDS.

The major contaminant of fractions with retinol-binding activity after step 3 is a low molecular weight component (apparent molecular weight, 10,000), which is incompletely resolved from the retinol-binding protein in step 4. Fractions taken from the leading side of the fluorescent peak from step 4 are homogeneous when analyzed by polyacrylamide gel electrophoresis in the presence or the absence of SDS.

The major contaminant associated with the cellular retinal-binding protein, purified through step 3, has an apparent molecular weight of approximately 41,000. This component is largely removed by a single passage through calcium phosphate gel (step 4), but frequently a second passage is required for its complete removal.

*Molecular Properties.* The cellular retinol-, retinoic acid-, and retinal-binding proteins are acidic proteins with apparent molecular weights of 16,600, 16,300, and 33,000, respectively, as determined by polyacrylamide gel electrophoresis in the presence of SDS.[2,4] Each appears to form a one-to-one complex with its respective ligand. Interaction of retinol with the cellular retinol-binding protein results in the appearance of fine structure in the absorption spectrum of the ligand and a red shift in the observed absorption maximum.[2] Interaction of retinoic acid with its binding protein induces fluorescence of the ligand.[2] Illumination of 11-*cis*-retinal, bound to its binding protein, results in isomerization of the chromophore and a shift in the absorbance maximum from 425 nm to 375 nm.[4]

*Specificity.* All-*trans*-[³H]retinoic acid, bound to its binding protein, is displaced by a 100-fold molar excess of unlabeled all-*trans*-retinoic acid, but not by the all-*trans*-isomers of retinol, retinal, retinyl palmitate, retinyl acetate, or retinyl phosphate. All-*trans*-[³H]retinol, bound to its binding protein, is displaced by a 100-fold molar excess of unlabeled all-*trans*-retinol, but not by the all-*trans*-isomers of retinoic acid, retinal, retinyl phosphate, retinyl acetate, or retinyl palmitate. The cellular retinal-binding protein interacts preferentially with 11- and 9-*cis*-retinal and not at all with all-*trans*-retinol and retinoic acid.[3] 11-*cis*-Retinol also binds to the cellular retinal-binding protein, but it is readily displaced by 11-*cis*-retinal.[4]

## [36] Isolation and Purification of Bovine Rhodopsin

*By* W. J. DE GRIP,[1] F. J. M. DAEMEN, and S. L. BONTING

### Introduction

Rhodopsin is the general name for the visual pigments of both the rod cells in the vertebrate retina and the invertebrate photoreceptor cells. The two types of visual pigment differ considerably in properties,[2] and their methodology should be treated separately. The results described in this section are obtained with bovine rhodopsin, but the techniques can be equally well applied to other vertebrate rhodopsins.

Rhodopsin is an intrinsic membrane glycoprotein, confined to the membranous structures (photoreceptor membranes) of a special part of the rod cell, the rod outer segment (ROS),[2a] which is highly specialized for light reception.[3] The early knowledge and methodology of rhodopsin biochemistry, largely developed by Wald[3] and co-workers have been excellently compiled by Hubbard *et al.* in an earlier volume of this series.[4] Futterman[5] subsequently presented a more detailed description of an isolation procedure that was frequently used at that time. However, insights and techniques in membrane biochemistry have developed to such an extent in the last decade that it is important to update these earlier accounts.

[1] Part of the unpublished results were obtained during the first author's stay at the Biological Laboratories, Department of Biology, Harvard University, Cambridge, Massachusetts, supported by a NATO fellowship provided by the Dutch organization for the advancement of basic research (ZWO) and an NIH grant to Professor G. Wald.

[2] S. E. Ostroy, *Biochim. Biophys. Acta* **463**, 91 (1977).

[2a] *Abbreviations:* ROS, rod outer segment(s); SDS, sodium dodecyl sulfate; PAGE, polyacrylamide gel electrophoresis; DTAB, dodecyltrimethylammonium bromide; TTAB, tridecyltrimethylammonium bromide; CTAB, cetyltrimethylammonium bromide; $C_{10}$DAO, decyldimethylaminoxide; $C_{12}$DAO, dodecyldimethylaminoxide; PIPES, piperazine-$N,N'$-bis(2-ethanesulfonic acid); MOPS, morpholino-$N$-propanesulfonic acid; BSA, bovine serum albumin; DTE, dithioerythritol; EDTA, ethylenediaminotetraacetic acid; NADPH, $\beta$-nicotinamide-adenine dinucleotide phosphate, reduced form; $A_{500}$, absorbance at 500 nm; $\epsilon_{500}$, molar absorption coefficient at 500 nm; HPLC, high pressure liquid chromatography; CMC, critical micelle concentration.

[3] G. Wald, *Science* **162**, 230 (1968).

[4] R. Hubbard, P. K. Brown, and D. Bownds, this series, Vol. 18C, p. 615.

[5] S. Futterman, this series, Vol. 32, p. 306.

Properties of Rhodopsin

As an intrinsic membrane protein, rhodopsin is not soluble in aqueous media, and therefore its assay and purification require the use of detergents. The suitability of various types of detergents depends on the purpose for which they are used, and this aspect will be discussed at the appropriate places.

The spectrum of bovine rhodopsin shows five absorption bands. Three derive from the protein backbone: at 191 nm and 227 nm mainly from peptide bonds[6] and at 280 nm mainly from aromatic side chains. The two bands in the visible region at 350 and 500 nm (Fig. 1) derive from the presence of a chromophoric group, 11-cis-retinal (vitamin A aldehyde). It is covalently bound to an ε-amino group of a lysine residue by means of a protonated azomethine group.[7] The ratios $A_{280}/A_{500}$ and $A_{400}/A_{500}$ offer a crude indication of the purity of rhodopsin preparations. Rhodopsin purified by various methods shows $A_{280}/A_{500}$ ratios of 1.65–1.80 and $A_{400}/A_{500}$ ratios of 0.16–0.19.[8–12] These ratios should not be overemphasized, since they are affected to some extent by scattering phenomena in the micellar solution.

Upon absorption of light, 11-cis-retinal is isomerized to all-trans,[3] which is eventually released from the protein. Simultaneously, the two specific visible absorption bands disappear, while new bands appear at 380 nm or 380 and 450 nm, depending on the conditions.[3,13] Incubation of the resulting apoprotein, opsin, with 11-cis-retinal leads under proper conditions to spontaneous re-formation of rhodopsin.

The high light-sensitivity of rhodopsin (quantum yield is about 0.6[13]) requires that isolation and further handling be performed either in darkness, under dim red light ($\lambda > 630$ nm), or under infrared light with the aid of an infrared converter. A simple assay of rhodopsin is based on the decrease of absorbance at 500 nm following illumination.

Estimations of molecular weight of rhodopsin, obtained by SDS-PAGE,[14] amino acid analysis,[7,15] or ultracentrifugation studies,[16] vary be-

[6] C. N. Rafferty, J. Y. Cassim, and D. G. McConnell, Biophys. Struct. Mech. 2, 277 (1977).

[7] F. J. M. Daemen, Biochim. Biophys. Acta 300, 255 (1973).

[8] J. Heller, Biochemistry 7, 2906 (1968).

[9] H. Shichi, M. S. Lewis, F. Irreverre, and A. L. Stone, J. Biol. Chem. 244, 529 (1969).

[10] K. Hong and W. L. Hubbell, Biochemistry 12, 4517 (1973).

[11] P. J. G. M. Van Breugel, F. J. M. Daemen, and S. L. Bonting, Exp. Eye Res. 24, 581 (1977).

[12] W. J. De Grip, unpublished results.

[13] W. W. Abrahamson and R. S. Fager, Curr. Top. Bioenerg. 5, 125 (1973).

[14] R. N. Frank and D. Rodbard, Arch. Biochem. Biophys. 171, 1 (1975).

[15] F. J. M. Daemen, W. J. De Grip, and P. A. A. Jansen, Biochim. Biophys. Acta 271, 419 (1972).

[16] M. S. Lewis, L. C. Krieg, and W. D. Krik, Exp. Eye Res. 18, 29 (1974).

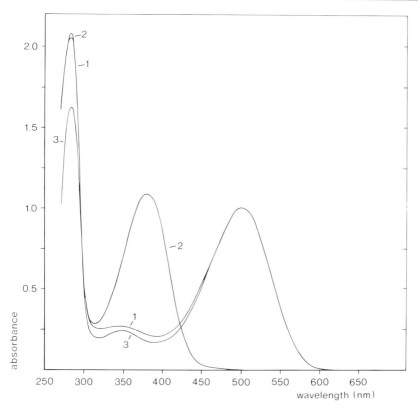

FIG. 1. Absorption spectra in nonylglucose solution ($A_{500}$ set at 1.0). Isolated photoreceptor membranes (1) before illumination and (2) after illumination in the absence of hydroxylamine. 3, Purified rhodospin.

tween 31,000 and 39,000. For most purposes, a value of 36,000 ± 1000 is a satisfactory average.

Assay of Rhodopsin

*Principle*

Rhodopsin is determined by measuring the decrease in absorbance at 500 nm upon illumination, using a molar absorbance of 40,500.[3,15] Under some conditions, the photoproducts also absorb at 500 nm,[4,5] but addition of hydroxylamine will convert all free and bound retinaldehyde into retinal oxime ($\lambda_{max}$ = 365 nm), which has virtually no absorption beyond 450 nm[4]. In detergents like the quarternary ammonium derivatives (DTAB,

TTAB, CTAB) or the amine oxides ($C_{10}DAO$, $C_{12}DAO$), addition of hydroxylamine is not required, since under these conditions retinal is completely released ($\lambda_{max}$ = 370–380 nm) and hardly interferes at 500 nm. Hydroxylamine should be handled with care, since it is highly mutagenic. The assay can be performed on suspensions of photoreceptor membranes, but scattering artifacts may interfere. Most accurate results are obtained after prior solubilization of the membrane in an appropriate detergent, followed by centrifugation or filtration through a membrane filter if some turbidity remains. Most detergents are suitable for this purpose,[4,5] but not SDS, which denatures rhodopsin already at room temperature.[17] We prefer Ammonyx-LO,[18] a commercial mixture of dodecyl- and tetradecyl-dimethylaminoxide, since it has a high solubilizing power and a low absorbance at 280 nm and it obviates the use of hydroxylamine. Normally a 1–2% solution (about 40–80 m$M$) is sufficient to solubilize rhodopsin up to concentrations of 0.2–0.4 m$M$. However, when entire retinas or retinal homogenates are to be extracted, concentrations of 5–7% should be used. In other aspects the method is not different from published procedures.[4,5]

## Reagents

Rhodopsin, either particulate or in detergent-solution
Any buffer at pH 6.0–7.0 (e.g., 0.1 $M$ phosphate or 20 m$M$ PIPES; $Na^+$, $K^+$, $Mg^{2+}$, $Ca^{2+}$ do not interfere)
Ammonyx-LO solution (30% in water)

## Procedure

All procedures involving rhodopsin should be performed in dim red light. A sample of rhodopsin is diluted with buffer and Ammonyx-LO to a final concentration of 2% in Ammonyx-LO and 10–20 $\mu M$ in rhodopsin. This will result in an $A_{500}$ of 0.4–0.8 (1-cm light path). If necessary, centrifugation (10 min; 80,000 $g$, 4°) or filtration through a 0.4-$\mu$m membrane filter will completely clear the solution. The spectrum is recorded between 650 and 250 nm (Fig. 1). The cuvette is illuminated (5 min by a 300-W tungsten-lamp through a heat filter and a filter that cuts off around 500 nm) and rescanned. Alternatively, separate measurements at 650, 500, 490, 480, 400, 380, 350, and 280 nm may be taken. Monitoring at 500 nm only can be hazardous, since artificial photopigments (isorhodopsin, $\lambda_{max}$ = 485 nm) may arise, e.g., during flash-bleaching of rhodopsin.[13] In case the $A_{650}$ increases considerably upon illumination owing to increasing

[17] W. J. De Grip, G. L. M. Van De Laar, F. J. M. Daemen, and S. L. Bonting, *Biochim. Biophys. Acta* **325**, 315 (1973).
[18] T. G. Ebrey, *Vision Res.* **11**, 1007 (1971).

turbidity, either longer centrifugation, a higher detergent concentration, or shorter illumination times should be tried. The rhodopsin concentration can be calculated from the $\Delta A_{500}$ ($A_{500}$ before illumination minus $A_{500}$ after illumination, corrected for differences in $A_{650}$) and the $\epsilon_{500}$.

## Isolation of Rod Outer Segments (ROS)

### Principle

Rhodopsin is the predominant protein of the photoreceptor membrane, accounting for 80–90% of the total membrane protein content.[7,15,19,20] These membranes may therefore be regarded as a fairly pure rhodopsin preparation, the main contaminant being lipid. In view of this and also because solubilization by detergents may introduce unwanted side effects,[6,17] photoreceptor membranes constitute an almost ideal preparation for most types of studies on rhodopsin.

ROS are completely packed with photoreceptor membranes (over 70% of dry weight), and isolation of the latter is therefore most conveniently done by isolating rod outer segments. They can be lysed to remove water-soluble components. Isolation of ROS is facilitated by two facts: first, ROS are connected to the main cell body of the rod cell by a thin connecting cilium only,[21] and thus they break off easily upon mild homogenization of the retina; second, their high membrane content added to the high lipid content of the photoreceptor membrane (see the table) results in a relatively low density (about 1.11–1.12), which permits separation from cells and other cell organelles by means of density centrifugation.

### Available Procedures

In the isolation of ROS from retina, four stages can be distinguished: (1) dissection of retinas from the eye; (2) homogenization of the retinas to break off the ROS; (3) separation of crude ROS fraction from larger fragments; (4) purification of ROS from contaminating cell organelles and membrane material by means of density centrifugation.

For each stage, a variety of techniques has been used. We shall briefly discuss these first and then present a more detailed description of the procedure largely developed and satisfactorily applied in our laboratory.

[19] D. S. Papermaster and W. J. Dreyer, *Biochemistry* **13**, 2438 (1974).
[20] M. Makino, T. Hamanaka, Y. Orii, and Y. Kito, *Biochim. Biophys. Acta* **495**, 299 (1977).
[21] R. W. Young, *Invest. Ophthalmol.* **15**, 700 (1976).

COMPOSITION AND SPECTRAL PROPERTIES OF WATER-WASHED AND LYOPHILIZED, ENRICHED PHOTORECEPTOR MEMBRANES[a]

| Component | Content, on a dry weight base (%) | Specifications | | |
|---|---|---|---|---|
| | | % of protein | $A_{280}/A_{500}$ | $A_{400}/A_{500}$ |
| Total protein[b] | 38 ± 1 | | | |
| Rhodopsin | 8.3 ± 0.2 nmol/mg | 85 ± 3 | 2.0 – 2.2 | 0.20 – 0.25 |
| | | % of lipids | | |
| Total lipid | 51 ± 2 | | | |
| Cholesterol | <3 | <6 | | |
| Retinal | 0.24 ± 0.01 | 0.47 | | |
| Glycolipids | <2 | <4 | | |
| Phospholipids | 41 ± 1 | 81 | % of phospholipids[c] | Mole/mole of rhodopsin |
| Phosphatidylethanolamine (PE) | | | 45 ± 1 | 28 |
| Phosphatidylcholine (PC) | | | 38 ± 2 | 24 |
| Phosphatidylserine (PS) | | | 14 ± 1 | 9 |
| Carbohydrates | 2.2 ± 0.4 | | | |
| Sialic acid | <0.2 | | | |
| $H_2O$[d] | 6 ± 1 | | | |

[a] From W. J. De Grip, Functional groups in rhodopsin and the rod photoreceptor membrane, Thesis, University of Nijmegen, 1974.

[b] Determined by amino acid analysis, since no appropriate protein standard is yet available for colorimetric protein estimations (Lowry, biuret).

[c] These figures are in excellent agreement with data presented by other authors [N. C. Nielsen, A. Fleischer, and D. G. McConnell, *Biochim. Biophys. Acta,* 10 (1970); R. E. Anderson, M. B. Maude, and W. F. Zimmerman, *Vision Res.* 15, 1087 (1975)]. The latter authors provide the most detailed information on fatty acid composition of phospholipids: main classes are in percentage of total: $16:0 = 20.5$, $18:0 = 24.9$, $18:1\omega9 = 4.0$, $20:4\omega6 = 3.0$, $22:5\omega6 = 4.0$, $22:6\omega3 = 34.8$, $24:>4 = 2.3$ (weight base). Our preparations, however, consistently contain a much higher level of unsaturated fatty acyl chains: $16:0 = 15.4$, $18:0 = 19.8$, $18:1 = 2.9$, $20:4 = 4.2$, $22:5 = 4.3$, $22:6 = 48.6$, $24:>4 = 1.8$ ($n = 4$, SD = ±3%).

[d] Can be removed by exposure to $P_2O_5$ *in vacuo*.

In order to obtain a maximal rhodopsin content, the animals should be dark-adapted for at least 3 hr.[4] This is difficult to arrange for cattle, and therefore a method will be given to convert opsin present into rhodopsin by incubation with 11-*cis*-retinal after isolation of the ROS.

Stage 1. Dissection. An incision is made near the edge of the cornea, and cornea and lens are cut away. The vitreous is removed, the eyecup is inverted, the retina is carefully brushed or scraped together and cut at the beginning of the optic nerve.[4,5] Alternatively, when the eyes have not been refrigerated, part of the vitreous can be carefully removed; the connection between retina and pigment epithelium can be loosened at the edges with a spatula, upon which the retina, still sticking to the vitreous, can be carefully guided out and finally cut loose from the optic nerve.[12] The latter procedure generally gives much less contamination from pigment epithelium material. For smaller eyes (frog, rat, rabbit, etc.), the procedure described by Hubbard et al.[1], should be used.

Stage 2. Homogenization. Originally, drastic methods like grinding in a mortar with 40% sucrose solution or homogenization in tightly fitting glass-on-glass Potter–Elvehjem homogenizers have been applied.[4] Later, milder procedures proved to be much more efficient, e.g., homogenization in loose-fitting Teflon on glass Potter–Elvehjem homogenizers, either in isotonic buffer[22] or 40–45% sucrose solution,[4,12] or forceful shaking in sucrose or sucrose–Ficoll solutions,[12,19,23] either by hand or on a Vortex mixer. These methods considerably reduce the amount of contamination, while only moderately lowering the final yield.

Stage 3. Separation from debris. This is most frequently done by one or two centrifugation steps,[4,19] first in a medium containing less than 28% sucrose to remove small and less dense material (ROS float in about 28–30% sucrose), and second in about 40% sucrose to remove larger and more dense material. Sometimes the first centrifugation is omitted. We, however, prefer filtration through a metal or preferentially Teflon screen (70–125 mesh). This is quicker and does not involve packing and resuspending of the crude ROS, which both hampers purification and tends to fragment the ROS.[12,23]

Stage 4. Purification. For final purification of ROS, density centrifugation in various media is employed. The original method, repeated flotation on a sucrose cushion (between 31 and 36%),[4,8] is time-consuming and does not easily yield material of a purity comparable to that obtained by density gradients.[22] It is, therefore, largely replaced by a discontinuous gradient step[19] or continuous gradient centrifugation.[22] In our hands, the continuous gradient technique is faster, more reliable, and more efficient than either the flotation or the discontinuous gradient procedure. Generally, the relatively steep transitions at the boundary of two layers of different density tend to retard material of higher density resulting in loss of ROS and the presence of contamination.

[22] W. J. De Grip, F. J. M. Daemen, and S. L. Bonting, Vision Res. 12, 1697 (1972).
[23] P. P. M. Schnetkamp, A. A. Klompmakers, and F. J. M. Daemen, Biochim. Biophys. Acta 552, 379 (1979).

Most frequently, sucrose is used for preparing density media, since it is cheap and relatively innocuous to biological material. The required solutions, however, are highly hypertonic (0.7 to 1.2 $M$ sucrose) and occasionally other media have been applied. Metrizamide density gradients are approximately isotonic, are less viscous than sucrose, and also appear to be able to separate intact and broken ROS.[24] This has been reported too for BSA density gradients.[25] BSA has little effect on osmolarity but tends to adhere to the membrane.[12] Ficoll-400, a high molecular weight copolymer of sucrose and epichlorohydrin, has been successfully used for the isolation of frog ROS[26,27] We could not, however, obtain highly pure bovine ROS with continuous Ficoll gradients (10 to 25%, w/w).[12] Electron micrographs indicate considerable aggregation of ROS in Ficoll media, which is also observed with erythrocytes,[28] as well as contamination by rod inner segments, at least part of which had not been detached from the outer segments.[12] A density gradient having a constant sucrose concentration with increasing Ficoll concentration, so as to keep the osmolarity constant and as low as possible, gives better results[12,23] and yields ROS that are functionally very well preserved.[23] However, only very fresh eyes yield pure material in this way, and ROS prepared from frozen retinas are highly contaminated.[12]

## Continuous Gradient Procedure

### Reagents

MOPS, 20 m$M$, pH 7.2, containing 2 m$M$ CaCl$_2$ and 3 m$M$ MgCl$_2$ and made about isotonic by addition of sucrose or of 140 m$M$ NaCl or KCl or combinations of both

Sucrose, 23% (w/w), prepared in 1 : 4 diluted MOPS buffer; (0.73 $M$, $d$ = 1.09)

Sucrose, 36% (w/w), prepared likewise; (1.20 $M$, $d$ = 1.15)

Solutions are routinely prepared in 1- to 2-liter volumes, cleared by filtration through 0.2-$\mu$m Nucleopore filters, divided into smaller portions (250 ml), and stored at $-20°$ until use.

---

[24] A. J. Adams, M. Tanaka, and H. Shichi, *Exp. Eye Res.* **27**, 595 (1978).
[25] G. Falk and P. Fatt, *J. Physiol. (London)* **229**, 185 (1973).
[26] R. N. Lolley and H. H. Hess, *J. Cell. Physiol.* **73**, 9 (1969).
[27] D. Bownds, A. Gordon-Walker, A. C. Gaide-Huguenin, and W. Robinson, *J. Gen. Physiol.* **58**, 225 (1971).
[28] T. G. Pretlow, E. E. Weir, and J. G. Zettergren, *Int. Rev. Exp. Pathol.* **14**, 91 (1975).

*Procedure*

All operations should be carried out in dim red light. The following procedure also works well with frozen retinas, which are commercially available.

1. Cattle eyes are collected in a lighttight container, as soon as possible after slaughtering, stored at room temperature, and processed within 2 hr. An incision is made in the sclera, cornea and lens are cut away, and part of the vitreous is carefully removed. At the edges the retina is loosened from the pigment epithelium with a spatula and carefully guided out, while still sticking to the vitreous. When the vitreous has dropped off, the retina hangs down in a little bundle and is cut off at the optic nerve. An adult cattle retina contains 50 to 60 nmol of rhodopsin.

2. Fifty to 60 cattle retinas are collected in 25–30 ml of ice-cold buffer in a Potter–Elvehjem tube. Both MOPS and 23% sucrose can be used for this purpose, since both media give final preparations of comparable yield, purity, and properties. To prevent oxidation of sensitive groups (polyunsaturated lipids; sulfhydryl groups), we routinely add solid DTE to the medium, just before use, to a final concentration of 0.5 to 1 m$M$. Working in a nitrogen or argon atmosphere and addition of $\alpha$-tocopherol (10 nmol per retina) may yield additional protection.[29] The use of EDTA[29] is not required, since DTE also chelates metal ions. In addition, antibiotics or enzyme inhibitors (e.g., benzylsulfonyl fluoride, PMSF, or $\alpha_2$-macroglobulin) may be added when eventually incubation at elevated temperature for longer periods of time is considered, e.g., for enzymic studies. The retinas are then homogenized with 5–10 strokes of a loosely fitting Teflon pestle (clearance 2–3 mm). When frozen retinas are used, they should be completely thawed before homogenization. The ratio $A_{280}/A_{500}$ of the homogenate is 15–25.

3. The homogenate is filtered through a Teflon screen (125 mesh) by gentle swirling or stirring with a plastic spoon. Forty milliliters of filtrate are collected. The isolation can be interrupted at this stage, if so desired, by rapidly freezing the filtrate and storing it at $-70°$. The filtrate contains 60–70% of the rhodopsin originally present. The ratio $A_{280}/A_{500}$ of the filtrate lies between 10 and 15.

4. Linear sucrose gradients are prepared from equal volumes of ice-cold 23% and 36% sucrose in a common, two-leg, plexiglas gradient mixer. Just before mixing, solid DTE is added to both legs (final conc. 0.5–1 m$M$). We routinely fill 3–6 centrifuge tubes simultaneously by means of a peristaltic pump. This takes about 20 min (30 ml per tube). The gradients can be stored overnight at 4° without appreciable loss in separat-

[29] C. C. Farnsworth and E. A. Dratz, *Biochim. Biophys. Acta* **443**, 556 (1976).

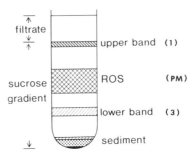

Fig. 2. Band distribution after centrifugation of retinal filtrate on a continuous sucrose (23 to 36% w/w) gradient. Denotations in brackets correspond with gels in Fig. 4b. ROS, rod outer segment(s).

ing efficiency. Subsequently, up to 7 ml of filtrate are layered upon each 30-ml gradient by means of a graduated, polyethylene, 10-ml syringe fitted with teflon tubing (i.d. 3.5 mm). In order not to disturb the gradient, a polypropylene disk (cut out of a bottle) is carefully placed on top of the gradient, and the filtrate is gently ejected on top of the disk, which keeps floating and is easily removed afterward. This guarantees a sharp boundary. Thus, the whole filtrate can be processed within 10 min. Subsequently, the tubes are centrifuged (80,000 $g$, 10°) for at least 1.5 hr. However, centrifugation can also be extended overnight.

In front of a dim red light, the tubes then show the pattern presented in Fig. 2, the bands being detectable owing to their turbidity. A thin band is present at the boundary between upper layer and gradient. It consists mainly of small membrane fragments and vesicles (smooth microsomes), presumably derived from plasma membranes, endoplasmic reticulum (ER), synaptosomes, etc., and it may contain some fragmented ROS and loose discs. The second, heavy band represents almost pure ROS, which are morphologically reasonably intact (Fig. 3). Fragmented ROS in the form of packets of discs band at a slightly lower density above the ROS band.[12,30,31] A small band just below the ROS band contains membrane material of uncertain, possibly Golgi, lysosomal or synaptosomal origin, but very few ROS. This band is considerably more pronounced when retinas are obtained from eyes of young calves and is then often difficult to separate from the ROS band. The sediment consists of three layers. The upper, colorless layer contains ER and lysosomal and mitochondrial material (the presence of divalent cations increases the density of mitochondria, ER, etc.). The middle, brownish one, is very heterogeneous, contain-

[30] W. Krebs and H. Kühn, *Exp. Eye Res.* **25**, 511 (1977).
[31] W. Godchaux III and W. F. Zimmerman, *Exp. Eye Res.* **28**, 483 (1979).

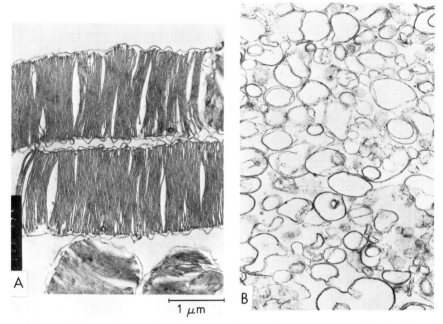

FIG. 3. Electron micrographs of a preparation of rod outer segments (ROS) obtained directly from the sucrose gradient (A) and after subsequent lysis with water (B). Micrographs were obtained in collaboration with Dr. J. Olive and Professor Dr. E. L. Benedetti, Institut de Biologie Moléculaire, Université de Paris VII, France.

ing mitochondria, etc.; the bottom contains melanin granules and red blood cells.

All material above the ROS band is sucked off, and the ROS band is carefully removed with a polyethylene syringe fitted with Teflon tubing. After slow dilution with either 0.5 volume of MOPS or 2 volumes of 23% sucrose, the ROS are sedimented at low speed (3000 $g$ for 10 min at 4°). This removes some contaminating membrane fragments. The ROS can then either be lysed with water or dilute buffer to remove soluble material or can be directly frozen for storage. When stored at $-70°$ under argon or nitrogen, the preparation is stable for at least a year.

The entire isolation can be performed in less than 4 hr. The yield is 40–60%, i.e., 20–30 nmol of rhodopsin per retina, which is quite high as compared to other procedures.[4,5,19,32] After illumination, incubation with a 5-fold excess of 11-*cis*-retinal leads to over 90% regeneration of the original amount of rhodopsin. Both electron micrographs[12] and study of en-

[32] G. J. Sale, P. Towner, and M. Akhtar, *Biochemistry* **16**, 5641 (1977).

zymic activity[12,22,33,34] of the gradient fractions indicate a very high purity of the ROS preparation. In agreement with this, no significant increase in rhodopsin content or changes in enzymic activity are observed upon repeating the gradient centrifugation. Composition, absorption spectrum, and a representative SDS-PAGE pattern of water-washed membranes are presented in the table, Fig. 1, and Fig. 4, respectively.

### Enrichment Procedure

#### Principle

Large amounts of photoreceptor membranes can be conveniently obtained only from cattle retinas. However, since dark adaptation of these animals is not feasible, as much as 50% of the total amount of visual pigment may be in the bleached state. This may seriously interfere with analytical approaches[22] or photolytic studies. We have, therefore, developed a procedure to convert all opsin present into rhodopsin by incubation *in vitro* with an excess of 11-*cis*-retinal.[22] This is performed on the ROS band obtained from the sucrose gradient, since at an earlier stage it would take too much 11-*cis*-retinal because of aspecific isomerization reactions.[35] Excess retinal must be removed afterward. This is best achieved by its reduction to retinol, taking advantage of the presence in the photoreceptor membrane of an intrinsic retinol dehydrogenase.[36] The resulting retinol is removed by a second gradient centrifugation.[22] The enrichment step adds about 3 hr to the isolation procedure.

#### Additional Reagents

A stock solution of 11-*cis*-retinal in hexane is stored at −70° under argon or nitrogen. Just before use, the required amount (1.3–1.5 $\mu$mol) is taken out, blown to dryness with nitrogen, and taken up in 100 $\mu$l of acetone.

#### Procedure

11-*cis*-Retinal can be prepared in 25–40% yield by photoisomerization of all-*trans*-retinal in ethanol or acetonitrile.[35] The resulting mixture of

[33] W. F. Zimmerman, F. J. M. Daemen, and S. L. Bonting, *J. Biol. Chem* 251, 4700 (1976).

[34] A. L. Berman, A. M. Azimova, and F. G. Gribakin, *Vision Res.* 17, 527 (1977).

[35] G. W. T. Groenendijk, P. A. A. Jansen, S. L. Bonting, and F. J. M. Daemen, this volume [27].

[36] S. Futterman, *J. Biol. Chem.* 238, 1145 (1963).

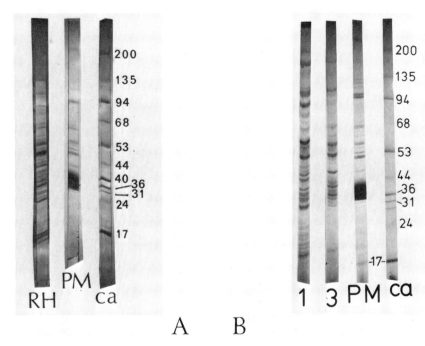

FIG. 4. SDS-PAGE gels of retina homogenate (RH) and isolated photoreceptor membranes (PM) (A) and of the three bands in the sucrose-gradient (B; cf. Fig. 2) both together with a calibration gel (ca) giving molecular weights × $10^{-3}$. The PM gels are overloaded to show the main additional bands (at MW 30,000, 42,000, 48,000, 50,000, 52,000, 53,000, 60,000, 62,000, 94,000, 110,000, 130,000, 190,000, and 230,000). The opsin band around MW 36,000 is often inhomogeneous and apparently split up in two or three "opsins," which behave slightly differently in this system. So far we have not found an explanation for this phenomenon. A gel of purified rhodopsin is not shown, since here overloading results in the appearance of considerable amounts of dimers (MW 65,000) and trimers (MW 95,000). In order to separate a wide range of molecular weights on the gel, a linear polyacrylamide gradient (6 to 20% w/v; 4% cross-linker) is used here. The buffer system was adapted from U. K. Laemlli [*Nature (London)* **227**, 680 (1970)]. Staining was performed with Coomassie Blue R-250 [G. Fairbanks, T. L. Steck, and D. F. M. Wallach, *Biochemistry* **10**, 2606 (1971)].

stereoisomers cannot be used for enrichment owing to the presence of 9-*cis*-retinal, which also combines with opsin, generating isorhodopsin ($\lambda_{max}$ = 487 nm). 11-*cis*-Retinal can be purified on a preparative scale by means of liquid chromatography over $Al_2O_3$[4] or HPLC on silica.[35] DTE accelerates thermal isomerization of 11-*cis*-retinal and should not be added to the sucrose gradients in this case.

The combined ROS bands from the sucrose gradients are added with thorough mixing to a freshly prepared solution of 11-*cis*-retinal (25 nmol per retina) in 100 $\mu$l of acetone. The suspension is incubated for 1 hr under nitrogen at room temperature with occasional shaking. Then, solid NADPH (0.1 mg per retina) and solid DTE (final conc. 1 m$M$) are added, and the incubation is continued for 45 min to reduce excess retinal. Finally, the suspension is diluted with 0.5 volume of ice-cold MOPS and layered upon continuous sucrose gradients, containing 0.5–1 m$M$ DTE. After centrifugation (see preceding section), the gradients show only one band, which contains the ROS. The isolation is then finished as described in the preceding section. Enriched, water-washed and lyophilized photoreceptor membranes have a rhodopsin content of 8.3 ± 0.2 nmol/mg dry weight. The amount of retinol is less than 2% of the molar amount of rhodopsin present. The spectral characteristics are $A_{280}/A_{500} = 2.0–2.2$; $A_{400}/A_{500} = 0.20–0.25$.

## Purification of Rhodopsin: Available Procedures

Purification of rhodopsin requires separation not only from the other membrane proteins, but also from the lipids present. This necessitates the use of detergents so as to fragment the membrane suprastructure into small micellar units and to remove lipids from proteins by exchange with detergent molecules. The suitability of various detergents depends on the purification method applied. In our opinion one should use as mild a detergent as possible in order to minimize changes in protein conformation and properties.[37] Our experience with some common detergents indicates increasing mildness in the order SDS $\ll$ DTAB $\ll$ CTAB $\sim$ $C_{12}$DAO < Emulphogene BC-720 < Triton X-100 < sodium cholate < alkylglucosides < digitonin.[12,17]

Various purification procedures (1–6) have been published.

1. Selective extraction with Tween 80[38] or low SDS-concentration.[39] It is doubtful whether such approaches afford complete purification of rhodopsin.

2. Gel permeation chromatography in CTAB-solution.[8] This technique is also capable of separating rhodopsin from opsin, which obviates the use of enriched ROS. However, separation of rhodopsin from lipids is not easily achieved. Unfortunately, this approach is not very suitable for milder detergents like Triton X-100 or digitonin, presumably owing to the occurrence of too large micelles and too much variation in micellar size.

3. Adsorption chromatography on calcium-phosphate (hydroxyapa-

[37] A. Helenius and D. Simons, *Biochim. Biophys. Acta* **415**, 29 (1975).
[38] M. Zorn and S. Futterman, *Arch. Biochem. Biophys.* **157**, 91 (1973).
[39] N. Virmaux, P. F. Urban, and T. V. Waehneldt, *FEBS Lett.* **12**, 325 (1971).

tite).[9] Nonionic ($C_{12}$DAO, Emulphogene BC-720, digitonin) and cationic (DTAB, TTAB) detergents have been employed with reasonable success.[9,10,18] Rhodopsin and opsin are readily separated. However, complete delipidation is again not easily achieved.

4. Ion-exchange chromatography.[9,38,40] This technique has been applied only with nonionic detergents (Emulphogene,[9] Tween 80,[38] digitonin[40]). Rhodopsin can be separated from opsin, but again this approach does not easily afford complete delipidation. In digitonin two fractions can be obtained, which differ in lipid/rhodopsin ratio.[40]

5. Affinity chromatography on immobilized concanavalin A.[11,14,41,42] Rhodopsin is the only glycoprotein in the photoreceptor membrane[42a], and no glycolipids are present.[12,43] The two saccharide moieties of rhodopsin[44] contain mannose and $N$-acetylglucosamine.[12,45] This allows purification via affinity-binding to lectins like concanavalin A, *Lens culinaris* lectin, or wheat germ agglutinin.[46] Any nondenaturing detergent can be used, but the efficiency of affinity columns is somewhat detergent dependent.[47] Complete delipidation is easily achieved in all detergents, with the exception of digitonin.[12] Opsin appears to bind more strongly to concanavalin A than rhodopsin,[41] and, therefore, this technique also allows separation of rhodopsin from opsin.[12]

6. Ammonium sulfate fractionation in cholate solution.[20] This method separates rhodopsin from opsin, but the purity of the final rhodopsin preparation depends very much on the purity of the starting material, and yields vary between 60 and 90%. The presence of opsin may seriously affect the yield. Complete delipidation is not achieved.

We prefer affinity chromatography for the purification of rhodopsin, since it is versatile and rapid and the purified rhodopsin can be eluted in very high concentration (100–500 $\mu M$). The only disadvantage arises from slow release of concanavalin A from the matrix. Maximal contamination is, however, less than 2% and can easily be removed by passage of the material over a hydroxyapatite or a small Sephadex G-50 column.

[40] H. Shichi, S. Kawamura, C. G. Muellenburg, and T. Yoshizawa, *Biochemistry* **16**, 5376 (1977).

[41] R. Renthal, A. Steinemann, and L. Stryer, *Exp. Eye Res.* **17**, 511 (1973).

[42] J. J. Plantner and E. L. Kean, *J. Biol. Chem.* **251**, 1548 (1976).

[42a] Recently it has been shown that the 230,000 D species is also a glycoprotein [R. S. Molday and L. L. Molday, *J. Biol. Chem.* **254**, 4653 (1979)]. However, in our experience, it does not copurify with rhodopsin.

[43] R. E. Anderson, M. B. Maude, and W. F. Zimmerman, *Vision Res.* **15**, 1087 (1975).

[44] P. A. Hargrave, *Biochim. Biophys. Acta* **492**, 83 (1977).

[45] J. Heller and M. A. Lawrence, *Biochemistry* **9**, 864 (1970).

[46] N. Sharon and H. Lis, *Science* **177**, 949 (1972).

[47] R. Lotan, G. Beattie, W. Hubbell, and G. L. Nicolson, *Biochemistry* **16**, 1787 (1977).

Affinity Chromatography

*Principle*

Concanavalin A (Con A) can be immobilized onto various supports (Sepharose, cellulose, glass) by well established procedures.[48] Con A–Sepharose is also commercially available. For optimal activity, $Ca^{2+}$ or $Zn^{2+}$ and $Mn^{2+}$ or $Ni^{2+}$ are required[46] and must be included in the medium, as well as at least 100 m$M$ NaCl. The pH may vary between 5.5 and 7.5. Although the affinity of Con A is somewhat higher at 25°, the procedure is preferentially carried out at 4° in view of the greater stability of rhodopsin. Most common detergents can be employed, but we prefer to use the $\beta$-alkylglucosides,[49] which are the mildest detergents still permitting complete delipidation.[12] Moreover, they have distinct advantages when reconstitution studies are to follow. Instead of octyl glucose,[49,50] we like to use nonylglucose, which is still somewhat milder and has a lower CMC (7 m$M$ vs 23 m$M$), thus requiring less material.[51] These glucosides have only a low affinity for Con A, as they possess the $\beta$-configuration. The same procedure can, however, be used with other detergents.

*Reagents*

Con A–Sepharose, supplied by Pharmacia as a slurry in buffer solution containing about 10 mg of Con A per milliliter of gel
Photoreceptor membrane preparation
PIPES, 40 m$M$, pH 6.5, containing 2 m$M$ $MgCl_2$, 2 m$M$ $CaCl_2$, 2 m$M$ $MnCl_2$, and 300 m$M$ NaCl
NaCl, 2 $M$
$\beta$-Nonylglucose, 40 m$M$
DTE, 100 m$M$

All solutions are freshly prepared, cleared by filtration through 0.2-$\mu$m Nucleopore filters, and stored in the cold.

*Procedure*

All operations are carried out in dim red light at 4°. The photoreceptor membranes are washed twice with water and once with buffer (PIPES/ $H_2O$/DTE, 50/50/1) and are each time collected by centrifugation (100,000

[48] This series, Vol. 34, entire volume.
[49] C. Baron and T. E. Thompson, *Biochim. Biophys. Acta* **382**, 276 (1975).
[50] G. W. Stubbs, H. G. Smith, Jr., and B. J. Litman, *Biochim. Biophys. Acta* **426**, 46 (1976).
[51] W. J. De Grip and P. H. M. Bovée-Geurts, *Chem. Phys. Lipids* **23**, 321 (1979).

$g$, 30 min, 4°). The required amount of ice-cold nonylglucose is then added under nitrogen (40 m$M$ nonylglucose can dissolve 60–80 nmol of rhodopsin per milliliter). The tube is capped, briefly mixed on a vortex mixer, and incubated for at least 0.5 hr in ice under occasional shaking. Subsequently, an equal volume of PIPES and 1% (v/v) DTE is added, and any remaining turbidity is removed by centrifugation (100,000 $g$, 4°, 15 min). An aliquot of the clear supernatant is diluted in order to measure spectral properties and rhodopsin content, and the remainder is applied to the Con A column. For best results, a small excess of rhodopsin should be used so as to load the column to its maximal capacity. Commercial Con A-Sepharose binds 100–150 nmol of rhodopsin per milliliter of settled gel. Before loading, the column is first washed with 10 volumes of PIPES/ NaCl (1/1) to remove residual carbohydrate and nonimmobilized material, then with 5 volumes of PIPES/H$_2$O (1/1), and finally with 3 volumes of PIPES/β-nonylglucose/DTE (50/50/1). Subsequently, the rhodopsin solution is pumped through. Flow rates of about 2 ml/hr for small columns (1–2 ml volume) up to 10 ml/hr for larger ones may be applied. However, with higher flow rates, part of the rhodopsin passes through without binding and the eluate should be reapplied for maximal loading. Nonbound material is then washed out with PIPES/nonylglucose/DTE (50/50/1). Proteins and excess rhodopsin elute first, immediately followed by lipids (Fig. 5). Lipids can be specifically assayed by phosphate analysis or amino acid analysis (ethanolamine) or, much more rapidly, aspecifically with fluorescamine. When the lipids have been eluted, the bound rhodopsin is eluted with PIPES/nonylglucose/DTE (50/50/1) to which solid α-methylglucose is added to a final concentration of 200 m$M$. Spectra of the eluates are taken, and those containing rhodopsin are combined. Then the column is washed with 3 volumes of PIPES/nonylglucose (1/1) to which solid NaCl is added to a final concentration of 1 $M$, to remove residual aspecifically bound material as well as α-methylglucose, followed by 5 volumes of PIPES/ NaCl (1/1). The column should be stored in the latter medium in the cold and may be used repeatedly until its capacity has decreased too much.

The entire purification takes 5–15 hr depending on column size. The recovery of rhodopsin is 90–100%. Spectra before and after purification are shown in Fig. 1. The purified rhodopsin has an $A_{280}/A_{500}$ ratio of 1.6– 1.8 and an $A_{400}/A_{500}$ ratio of 0.16–0.17, regardless of the properties of the starting material, since opsin is considerably retarded with respect to rhodopsin. If required, salts and α-methylglucoside can be removed by dialysis. Likewise, β-nonylglycose, which is easily dialyzed out,[49,51] can be substituted by digitonin. (Alternatively, nonylglucose may be substituted for any other detergent, when rhodopsin is still bound to Con A-Sepharose). SDS-PAGE shows that Con A, if present, is the only con-

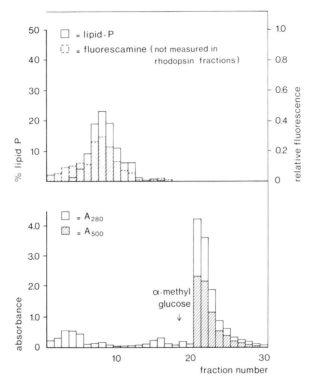

FIG. 5. Purification of rhodopsin by affinity chromatography in 20 m$M$ nonylglucose over Con A-Sepharose. Rhodopsin (600 nmol; 33 nmol/ml) is applied to 8 ml of settled gel. The column is not overloaded to prevent interference of excess rhodopsin with the elution pattern. Flow rate: 6 ml/hr. Fractions of 3.4 ml are collected and checked for $A_{280}$, $A_{500}$, phosphate content, and reaction with fluorescamine.

tamination. Small amounts of Con A can be removed with Sephadex, e.g., by placing the Con A column on top of a small Sephadex G-50 column. Larger amounts are best removed by chromatography over hydroxyapatite. Phosphate analysis indicates that less than 0.5 mol of phosphate is present per mole of rhodopsin, proving virtually complete delipidation. The solution of purified rhodopsin can be stored at $-70°$ for at least half a year.

## Membrane Reconstitution

### Principle

Reconstitution of purified membrane proteins allows their study in their most natural environment, the lipid bilayer, without interference of other proteins. It also allows variation of the lipid constitution almost at will, e.g., for studies of protein–lipid interaction. The first reconstitution studies on rhodopsin[10] elegantly showed that the capacity of rhodopsin to recombine with 11-*cis*-retinal after bleaching, which is lost in most detergents, is recovered upon reconstitution, without significant preference for any class of lipids. Similar results were subsequently reported with respect to thermal stability,[52] and photolytic reaction sequence, with the exception that very saturated lipids tend to slow down the metarhodopsin I–II transition considerably.[52-54] Most reconstitution studies have employed dialysis to remove detergents. This requires detergents with a relatively high CMC ($>1$ m$M$), since otherwise dialysis will take more than a week. This criterion is satisfied by the following detergents, in order of increasing CMC: TTAB $<$ DTAB $\sim$ C$_{12}$DAO $<$ nonylglucose $<$ cholate $<$ C$_{10}$DAO $<$ octylglucose. The rate of dialysis is correlated with the CMC, but the CMC of ionic detergents is strongly affected by the ionic strength of the medium. Other approaches, e.g., gel permeation or density centrifugation, are less suited for preparative purposes, but can be used with more detergents. Reconstitution has also been performed by direct addition of lipids to detergent-solubilized rhodopsin from which bulk detergent is removed by extraction methods.[55,56] Such preparations may, however, contain residual detergent. The techniques mentioned are well outlined in the literature. We will, therefore, only briefly describe the dialysis procedure. This approach is routinely used in our laboratory, as it is easily scaled up and is very suitable using nonylglucose as the detergent. Our attempts to reconstitute rhodopsin while still bound to Con A–Sepharose have not yet been successful.

*Reagents*

Purified rhodopsin in nonylglucose solution
PIPES, 20 m$M$, pH 6.5, containing 1 m$M$ CaCl$_2$, 1 m$M$ MgCl$_2$, and, as required, 150 m$M$ NaCl or other salts
Lipid solution in organic solvent (chloroform/methanol or ethanol)

*Procedure*

For proper reconstitution the lipid/rhodopsin ratio should be at least 50 to 70. The required volume of a concentrated solution of the required

[52] P. J. G. M. Van Breugel, P. H. M. Geurts, F. J. M. Daemen, and S. L. Bonting, *Biochim. Biophys. Acta* **509**, 136 (1978).

[53] M. L. Applebury, D. M. Zuckerman, A. A. Lamola, and T. M. Jovin, *Biochemistry* **13**, 3448 (1974).

[54] D. F. O'Brien, L. F. Costa, and R. A. Ott, *Biochemistry* **16**, 1295 (1977).

[55] A. Darszon, M. Montal, and J. Zarco, *Biochem. Biophys. Res. Commun.* **76**, 820 (1977).

[56] M. Chabre, A. Cavaggioni, H. B. Osborne, and T. Gulik-Krzywicki, *FEBS Lett.* **26**, 197 (1972).

lipid(s) in ethanol ($\sim$100 nmol/$\mu$l) is added under nitrogen to 1 ml of PIPES containing 40 m$M$ nonylglucose and 1 m$M$ DTE. Alternatively, a lipid solution in chloroform/methanol is dried with a stream of nitrogen, and 1 ml of the same buffer is added. After a short incubation under nitrogen the solution should be virtually clear, whereupon it is added to the rhodopsin solution, mixed, incubated first briefly in an ultrasonic bath (10–15 min in ice) and then for another 45–50 min in ice (unsaturated lipids) or as close to the gel–liquid transition temperature as possible (saturated lipids). Subsequently, the solution is placed into dialysis tubing and dialyzed first against 2 volumes of PIPES or any other buffer required (4 hr; one change of buffer) and then against 50–100 volumes of buffer (overnight; one change of buffer). The half-time of dialysis of nonylglucose is about 3 hr. Finally, the recombinants are collected by centrifugation (100,000 $g$ for 30 min at 4°) and stored at $-70°$. They contain less than 1 mol of nonylglucose per mole of rhodopsin and consist of single-walled vesicles together with some liposomal structures. Freeze-fracture electron microscopy, photolytic reaction sequence, and sulfhydryl group reactivity indicate that rhodopsin is indeed incorporated into the lipid bilayer.[12]

For rapid screening, an alternative approach can be applied: dilution of the solution containing rhodopsin and lipids with PIPES to a final nonylglucose concentration of about 3 m$M$ (i.e., clearly below its CMC of 7 m$M$) rapidly leads to formation of reconstituted vesicles which are easily sedimented.[12] The sediment contains 70–80% of the lipid, 90–100% of the rhodopsin, but also variable amounts of detergents.

Dialysis of rhodopsin solutions without previous addition of lipids generates particulate lipid- and detergent-free rhodopsin.[52–54] Mere addition of liposomes or lipid vesicles to this material does not generally lead to incorporation of rhodopsin into the lipid systems.[52,53]

Section VI

# Vitamin D Group

## [37] Colorimetric Determination of Vitamin D$_2$ (Calciferol)

### By SAAD S. M. HASSAN

Few color reactions for vitamin D are known on which the available methods of colorimetric determination of the vitamin are based. The one most widely used and adopted by both British and U.S. pharmacopoeias is that based on reaction with antimony trichloride, in the presence of acetyl chloride.[1] The reaction is relatively sensitive but suffers from interference of vitamin A, which renders it inapplicable to natural vitamin D sources and multivitamin products. This is obviated by a preliminary separation process or treatment with maleic anhydride to form adducts with vitamin A, carotenoids, and polyenes that do not react with antimony trichloride.[2] Furthermore the antimony trichloride reagent is unstable, and the sensitivity of the method decreases with age of the reagent and also with temperature. Small amounts of moisture or alcohol can provoke turbidity, leading to erroneous results. Stannous chloride in acetyl chloride has been also employed, but similar problems are encountered.[3]

Reaction of the vitamin with glycerol dichlorohydrin, in the presence of acetyl chloride ranks second in importance, and gives an immediate yellow color that turns to a stable green.[4] This reaction is affected by several factors and suffers chiefly from relatively low sensitivity. Other chromogenic reactions involving the use of aromatic aldehydes,[5-7] pyrogallol with anhydrous aluminum chloride,[8] and trifluoroacetic acid[9] have also been reported.

An attempt to develop a new chromogen with vitamin D$_2$ was undertaken by the present author.[10] It is desirable to combine simplicity, specificity, particularly with respect to vitamin A, and freedom from fluctuations due to instability of the reagent used.

[1] The United States Pharmacopeia, 18th revision, p. 915. Mack Publ., Easton, Pennsylvania, 1970.
[2] N. Milas, R. Heggie, and J. Raynolds, *Anal. Chem.* **13**, 227 (1941).
[3] H. Tschapke, *Nahrung* **2**, 44 (1958).
[4] J. Campbell, *Anal. Chem.* **20**, 766 (1948).
[5] H. Schaltegger, *Helv. Chim. Acta* **29**, 285 (1946).
[6] J. Buchi and H. Schneider, *Medd. Nor. Farm. Selsk.* **17**, 87 (1955).
[7] D. Laughland and W. Philips, *Anal. Chem.* **28**, 817 (1956).
[8] W. Halden and H. Tzoni, *Nature (London)* **137**, 909 (1936).
[9] S. Gharbo and L. Gosser, *Analyst* **99**, 222 (1974).
[10] S. S. M. Hassan, unpublished work, 1977.

METHODS IN ENZYMOLOGY, VOL. 67

### Reaction with Hydrochloric Acid in Tetrachloroethane

It was found that calciferol reacts with 11 $N$ hydrochloric acid, in the presence of symmetrical tetrachloroethane to develop a greenish-yellow color with maximum absorption at 440–460 nm, with $E_{1cm}^{1\%}$ 55. A broad inflexion at 380 nm with $E_{1cm}^{1\%}$ 120 also appears (Fig. 1). In chlorinated solvents other than symmetrical tetrachloroethane (e.g., chloroform, chlorobenzene, and carbon tetrachloride) the color does not appear. The absorption attained at these wavelengths is rectilinear with the amount of calciferol over the range of 30–800 $\mu$g/ml. The relative standard deviation of the absorption at 450 nm for five identical aliquots containing 50 $\mu$g/ml was 2%. The color is stable for at least 3 hr. This procedure has been satisfactorily applied to the determination of vitamin $D_2$ in mixtures containing vitamin A without any significant interference from the latter vitamin.

### Assay Method

#### *Reagents*

Hydrochloric acid, 11 $N$
Symmetrical tetrachloroethane
Calciferol stock solution, 100 mg in 100 ml of methanol

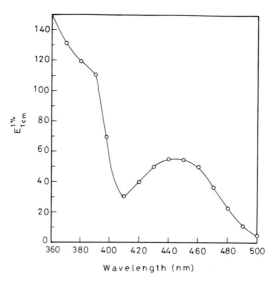

FIG. 1. Absorption spectrum of the reaction product of calciferol with hydrochloric acid and symmetrical tetrachloroethane.

*Procedure.* Introduce 0.5, 1.0, 1.5, 2.0, 2.5, 3.0, 3.5, 4.0, 4.5, and 5.0 ml of the vitamin stock solution into graduated test tubes (20 × 2 cm). Evaporate to dryness in a boiling water bath. Add 1 ml of 11 $N$ hydrochloric acid and 1 ml of symmetrical tetrachloroethane to each tube. Place in a boiling water bath for 10 min with occasional shaking. Cool to room temperature and dilute to 7 ml with acetone. Measure the absorbance directly at 450 nm in a 1-cm cuvette against a blank prepared under similar conditions.

## Reaction with Sulfuric Acid

A recent investigation by Hassan *et al.*[11] revealed that the reaction of steroids with concentrated sulfuric acid can be improved both in sensitivity and stability of the developed chromogen by dilution, which causes significant bathochromic and hyperchromic shifts of the absorption bands. Petersen and Harvey[12] found that the provitamin ergosterol in carbon tetrachloride reacts with sulfuric acid to develop a chromogen with transmittance measurable between 550 and 650 nm. This reaction has been reinvestigated by us, in the light of our findings on the effect of dilution, with vitamin $D_2$. It was found that calciferol reacts with concentrated sulfuric acid to give an absorption spectrum with two maxima at 380 nm ($E_{1cm}^{1\%}$ 290) and 490 nm ($E_{1cm}^{1\%}$ 180). The intensity of the color at both maxima are stable for only 30 min. However, reaction with the acid, followed by graded dilution with ethanol, revealed depression of the band at 380 nm with concomitant amplification of the one at 490 nm with a bathochromic shift to 505 nm, attaining its maximum when the acid concentration is 80% (Fig. 2). Under such conditions, the $E_{1cm}^{1\%}$ equals 205 and 390, respectively. The color is stable for at least 10 hr. This method permits vitamin determination in the range of 5–50 $\mu$g/ml with an accuracy of ±1.5%, but vitamin A interferes.

## Assay Method

*Reagents*

Sulfuric acid, 98%
Ethanol, 96%
Vitamin $D_2$ stock solution, 100 mg in 100 ml of methanol

*Procedure.* Into 20 × 2 cm graduated test tubes, introduce 0.1, 0.3, 0.5, 0.7, 0.9, 1.1, 1.3, and 1.5 ml of the vitamin stock solution. Evaporate

[11] S. S. M. Hassan, M. A. Fattah, and M. T. Zaki, *Z. Anal. Chem.* **281**, 371 (1976).
[12] R. Petersen and E. Harvey, *Anal. Chem.* **16**, 495 (1944).

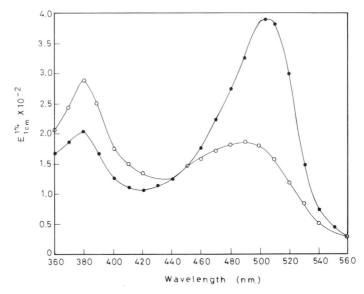

FIG. 2. Absorption spectra of calciferol in (○) 98% sulfuric acid and (●) 80% sulfuric acid.

to dryness in a boiling water bath. Cool to room temperature, add 4 ml of 98% of sulfuric acid, and shake for 2 min. Place the tubes in an ice bath, add 1 ml of 96% ethanol to each, shake, and measure the absorbance at 505 nm in a 1-cm cuvette against a blank prepared similarly.

# [38] Analysis of Vitamin $D_2$ Isomers

*By* KIYOSHI TSUKIDA

## Vitamin $D_2$ Isomers

There exist two important facets of vitamin D analysis. One is an establishment of microanalysis of vitamin D in complex media containing a large excess of other fat-soluble vitamins, such as vitamins A and E.[1-3] The second is the accomplishment of a clear separation as well as a rapid and accurate quantitation of vitamin D and its isomers, (I)–(VIII) (Fig. 1),

[1] K. Tsukida and K. Saiki, *J. Vitaminol.* **16**, 293 (1970).
[2] K. Tsukida and K. Saiki, *J. Vitaminol.* **18**, 165 (1972).
[3] K. Tsukida and K. Saiki, *Int. J. Vitam. Nutr. Res.* **42**, 242 (1972).

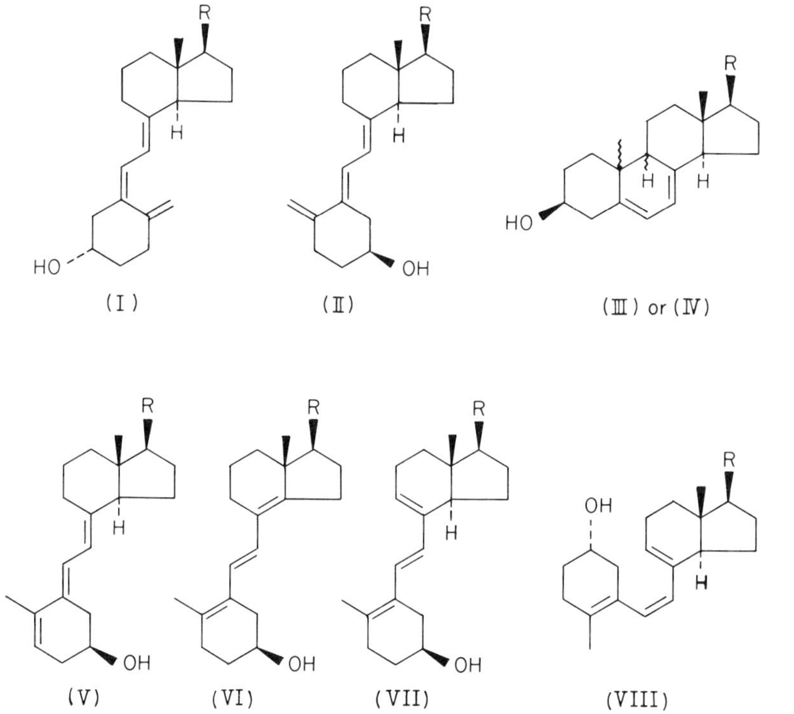

FIG. 1. Structures of vitamin D₂ isomers (see also Table I). (I) Vitamin D₂, (II) 5,6-*trans*-vitamin D₂, (III) ergosterol (9α, 10β), (IV) lumisterol₂ (9β, 10α), (V) isovitamin D₂, (VI) isotachysterol₂, (VII) tachysterol₂, (VIII) previtamin D₂, (Ia–VIIIa) OAc instead of OH; R = C₉H₁₇.

because vitamin D is labile and rapidly isomerizes in a solution upon exposure to light, heat, or chemical reagents. Traditional wet chemical analyses are very tedious and laborious, but no simple method has been developed for many years. Although recent gas–liquid chromatographic technique combined with thin-layer chromatography has obtained some success,[3,4] there is still a need for a faster, simpler, and direct separation of the isomers in a mixture. As novel approaches to this purpose, a relative paramagnetic shift value method employing a nuclear magnetic resonance (NMR) shift reagent[5] and high-performance liquid chromatographic analysis (HPLC)[6] are now presented.

[4] T. Kobayashi and A. Adachi, *J. Nutr. Sci. Vitaminol.* **19**, 303 (1973).
[5] K. Tsukida, K. Akutsu, and M. Ito, *J. Nutr. Sci. Vitaminol.* **22**, 7 (1976).
[6] K. Tsukida, A. Kodama, and K. Saiki, *J. Nutr. Sci. Vitaminol.* **22**, 15 (1976).

*Specimens (Table I)*

Vitamin $D_2$ (I), commercial product

5,6-*trans*-Vitamin $D_2$ (II), prepared from vitamin $D_2$ (0.5 g) by stirring a hexane solution (500 ml) containing iodine (2.5 mg) for 2 hr in a diffused light, followed by chromatographic purification on an alumina column (1.5 × 15 cm) employing ether–hexane (1 : 1) as a developer

Ergosterol (III), commercial product

Lumisterol$_2$ (IV), difficult to obtain the pure specimen in large quan-

TABLE I

Some Characteristics[a] of Vitamin $D_2$ Isomers ($C_{28}H_{44}O$)
and Their Acetates ($C_{30}H_{46}O_2$)[b]

| Compound | Mp (°C) | UV[c] (nm) | RRT[d] |
|---|---|---|---|
| Vitamin $D_2$ (I) | 115–117 | 265 (E 475) | 1.00 + 1.18 |
| Vitamin $D_2$ acetate (Ia) | 88 | 267 | 1.15 + 1.50 |
| 5,6-*trans*-Vitamin $D_2$ (II) | 104–106 | 273 | |
| 5,6-*trans*-Vitamin $D_2$ acetate (IIa) | Oil | 273 | (1.3) |
| Ergosterol (III) | 165 | 294, 281, 270 | 1.32 |
| Ergosteryl acetate (IIIa) | 171–175 | 294, 283, 272 | 1.68 |
| Lumisterol$_2$ (IV) | 120–121 | (281), 272 | 1.03 |
| Lumisteryl$_2$ acetate (IVa) | 100 | (280), 272 | 1.19 |
| Isovitamin $D_2$ (V) | 108–110 | 300, 287, 276 | |
| Isovitamin $D_2$ acetate (Va) | Oil | 300, 287, 276 | (1.7) |
| Isotachysterol$_2$ (VI) | Oil | 302, 290, 280 | |
| Isotachysteryl$_2$ acetate (VIa) | Oil | 301, 289, 279 | 1.89 |
| Tachysterol$_2$ (VII) | Oil | 294, 280, 270 | |
| Tachysteryl$_2$ acetate (VIIa) | Oil | 291, 281, (271) | 1.89 |
| Previtamin $D_2$ (VIII) | Oil | 263 | 1.00 + 1.18 |
| Previtamin $D_2$ acetate (VIIIa) | Oil | 263 | 1.15 + 1.50 |

[a] Authenticity of the specimen can be certified by further spectroscopic information, e.g., by $^1$H-NMR,[e–g] $^{13}$C-NMR,[h] or mass spectroscopic analysis.[g,i]

[b] See also Fig. 1.

[c] (I) and (III)–(VIII), in ethanol; (II) and (Ia)–(VIIIa), in ether.

[d] RRT: relative retention time in gas–liquid chromatography. For analysis of the intact alcohols: 1.5% OV-17, column 240°, $N_2$ 40 ml/min.[e,j] For analysis of the acetates: 1.5% OV-1, column 250°, $N_2$ 55 ml/min.[g]

[e] K. Tsukida and K. Saiki, *J. Vitaminol.* **16**, 293 (1970).

[f] K. Tsukida and K. Saiki, *Int. J. Vitam. Nutr. Res.* **42**, 242 (1972).

[g] K. Tsukida, K. Akutsu, and M. Ito, *J. Nutr. Sci. Vitaminol.* **22**, 7 (1976).

[h] K. Tsukida, K. Akutsu, and K. Saiki, *J. Nutr. Sci. Vitaminol.* **21**, 411 (1975).

[i] K. Tsukida, *Yukagaku* **22**, 575 (1973).

[j] K. Tsukida and K. Saiki, *J. Vitaminol.* **18**, 165 (1972).

tity; may be isolated from ergosterol by irradiating an ether solution with a high-pressure mercury lamp

Isovitamin $D_2$ (V), prepared from vitamin $D_2$ and $BF_3$-etherate in benzene according to the method of Inhoffen *et al.*[7]

Isotachysterol$_2$ (VI), produced from vitamin $D_2$ (0.5 g) by vigorously shaking a hexane solution (100 ml) with 60% $H_2SO_4$ (20 ml) for 10 min, followed by chromatographic purification on an alumina column (1.5 × 15 cm) employing acetone–benzene (1 : 95) as a developer

Tachysterol$_2$ (VII), isolated from ergosterol irradiated with a low-pressure mercury lamp in ether solution.

Previtamin $D_2$ (VIII), produced from vitamin $D_2$ (0.4 g) by refluxing a toluene solution (2 ml) for 2 hr, followed by chromatographic purification on a silica-gel column (1.5 × 15 cm) employing benzene as a developer.

Isomeric acetates (Ia–VIIIa), usually prepared from the parent alcohols in pyridine by adding acetic anhydride and allowing to stand overnight at room temperature in darkness. For preparing (IIa), heating on a boiling water bath for several hours is recommended. The acetate (VIIa) is conveniently obtained from ergosteryl acetate (0.9 g) by irradiating an ether solution with a low-pressure mercury lamp (2 hr, 15–19°, $N_2$). These acetates are purified either by recrystallization or by chromatography on an alumina column.

### Relative Paramagnetic Shift Value ($\nu_{rel}$) Method[5]

*Principle.* The method depends upon the fact that the acetoxyl NMR signals of acetylated vitamin $D_2$ isomers induce paramagnetic shifts with their own $\nu_{rel}$ values when a shift reagent is sequentially added into the sample solution. A relative paramagnetic shift value refers to a relative shift value of an acetoxyl signal of an isomer under consideration against that of vitamin $D_2$ acetate ($\nu_{rel} = 100$), when tris(dipivaloylmethanato)europium(III) [Eu(dpm)$_3$] is used as a shift reagent.

$$\nu_{rel} = \frac{\text{a paramagnetic shift of OAc (sample acetate)}}{\text{a paramagnetic shift of OAc (vitamin } D_2 \text{ acetate)}}$$

Appropriate linearities are confirmed between paramagnetic shift values of acetoxyl protons of the isomer and amounts of the shift reagent. The $\nu_{rel}$ values of the acetoxyl protons of each isomer are simply calculated from NMR charts determined with or without an addition of the reagent (Table II and Fig. 2).

[7] H. H. Inhoffen, G. Quinkert, H.-J. Hess, and H.-M. Erdmann, *Chem. Ber.* **89**, 2273 (1956).

TABLE II
RELATIVE PARAMAGNETIC SHIFT VALUES ($\nu_{rel}$) AND CHEMICAL SHIFTS ($\delta$) OF VITAMIN $D_2$
ISOMERS (60 MHz, $CCl_4$)

| Compound | $\nu_{rel}$ OAc | $\delta$ ppm OAc | 18-Me |
|---|---|---|---|
| Vitamin $D_2$ acetate (Ia) | $100^a$ | 1.96 | 0.55 |
| 5,6-*trans*-Vitamin $D_2$ acetate (IIa) | $71 \pm 1$ | 1.93 | 0.57 |
| Ergosteryl acetate (IIIa) | $86 \pm 1$ | 1.94 | 0.62 |
| Lumisteryl$_2$ acetate (IVa) | $123 \pm 3$ | 1.98 | 0.66 |
| Isovitamin $D_2$ acetate (Va) | 82 | 1.96 | 0.59 |
| Isotachysteryl$_2$ acetate (VIa) | 97 | 1.96 | 0.90 |
| Tachysteryl$_2$ acetate (VIIa) | $100^{a,b}$ | 2.00 | 0.70 |
| Previtamin $D_2$ acetate (VIIIa) | $80 \pm 2$ | 1.95 | 0.73 |

[a] When the shift reagent is added, the 18-Me signal of vitamin $D_2$ acetate induces an anomalous upfield shift and the isolated signal area can be estimated, whereas that of tachysteryl$_2$ acetate moves downfield and is buried in an Me envelope.
[b] $\nu_{rel} = 105 \pm 1$ in $C_6D_6$.

### Reagents

Carbon tetrachloride, guaranteed or spectral grade reagent.

Tetramethylsilane, NMR internal standard ($\delta$ ppm = 0.0)

Eu(dpm)$_3$, *fresh.* This reagent is prepared from 2,2,6,6-tetra-methylheptane-3,5-dione (11 g) in 95% ethanol (30 ml), NaOH (2.4 g) in 50% ethanol (50 ml), and Eu(NO$_3$)$_3$ · 6H$_2$O (8.9 g) in 50% ethanol (50 ml) according primarily to the method of Eisentraut *et al.*[8] Mp. 187–189°.

*Procedure.* A sample containing vitamin $D_2$ isomers is acetylated by a conventional method; its NMR spectra are run in two ways, with or without an addition of the shift reagent. When the reagent is added, downfield-shifted acetoxyl signals should be controlled to be located within a range of $\delta$ 3.3–5.0 ppm, as shown in Fig. 2. When the relative position of each acetoxyl signal is estimated, identification is immediately apparent. If a sample mixture is composed of several isomers and the reference signal is obscure, further addition of some vitamin $D_2$ acetate is recommended. A ratio of the relative contents of the isomers can be determined simultaneously from integrated signal areas of the respective acetoxyl protons. Quantitation of each isomer in a mixture, if necessary, can be performed by employing an appropriate internal standard for de-

[8] K. J. Eisentraut and R. E. Sievers, *J. Am. Chem. Soc.* **87,** 5254 (1965).

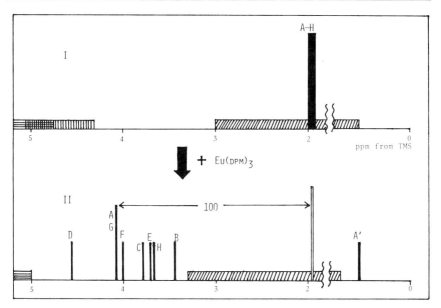

FIG. 2. Schematic diagram of the nuclear magnetic resonance (NMR) separation of vitamin $D_2$ isomers [60 MHz, in $CCl_4$]. I: Normal spectrum of an equimolar mixture of the isomers. II: Spectrum of the same sample solution after addition of $Eu(dpm)_3$. A–H, acetoxyl peaks; A', $C_{18}$-Me signal of vitamin $D_2$ acetate. A, Vitamin $D_2$ acetate (Ia); B, 5,6-*trans*-vitamin $D_2$ acetate (IIa); C, ergosteryl acetate (IIIa); D, lumisteryl₂ acetate (IVa); E, isovitamin $D_2$ acetate (Va); F, isotachysteryl₂ acetate (VIa); G, tachysteryl₂ acetate (VIIa); H, previtamin $D_2$ acetate (VIIIa).

termination or by comparison with an increased signal area resulting from the addition of a definite amount of vitamin $D_2$ acetate.

*Scope and Limitation.* On the basis of a dramatic shift of an acetoxyl signal and a constancy of each $\nu_{rel}$ value, this method permits immediate identification of the isomers and provides valuable information on their stereochemistry. This technique also facilitates simultaneous determination of these isomers in a mixture by estimating each signal area. Applications for vitamin $D_2$ analysis include photochemistry of ergosterol, thermal isomerization of vitamin $D_2$, constitution of vitamin $D_1$, etc.[5,9]

Advantages include the following:

1. The operational procedure is one-stage and simple.
2. An acetoxyl signal appears as a sharp singlet initially in an upper-

[9] The method can also be applied for analysis of isomeric phenols (cresols and tocopherols[10]) and trisubstituted allylic alcohols (vitamin A alcohols, etc.[11]).

[10] K. Tsukida and M. Ito, *Experientia* **27**, 1004 (1971).

[11] K. Tsukida, M. Ito, and F. Ikeda, *J. Vitaminol.* **18**, 24 (1972); *Int. J. Vitam. Nutr. Res.* **42**, 91 (1972); *Experientia* **28**, 721 (1972); *Chem. Pharm. Bull.* **21**, 248 (1973).

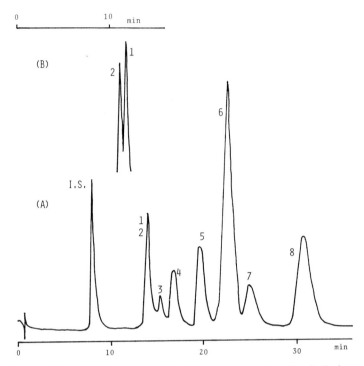

FIG. 3. High-performance liquid chromatographic behavior of vitamin $D_2$ isomers. Experimental conditions: (A) 120 kg/cm², 0.15% methanol + 2% ether in pentane; (B) 80 kg/cm², 55% chloroform (distilled) in pentane. Peak 1, 5,6-*trans*-vitamin $D_2$; peak 2, isovitamin $D_2$; peak 3, previtamin $D_2$; peak 4, lumisterol$_2$; peak 5, isotachysterol$_2$; peak 6, vitamin $D_2$; peak 7, tachysterol$_2$; peak 8, ergosterol; I.S., internal standard ($\alpha$-naphthol).

TABLE III
RELATIVE RETENTION DATA

| Operating condition[a]: | 1 | 2 | 3 |
| | p-Cresol | α-Naphthol | p-Cresol |
| Internal standard: | 1.00 | 1.00 | 1.00 |
|---|---|---|---|
| Vitamin $D_2$ (I) | 2.87 | 2.56 | 1.12 |
| 5,6-*trans*-Vitamin $D_2$ (II) | 1.59 | 1.73 | 0.62 |
| Ergosterol (III) | 3.64 | 3.80 | — |
| Lumisterol$_2$ (IV) | 2.13 | 2.09 | — |
| Isovitamin $D_2$ (V) | 1.62 | 1.73 | 0.58 |
| Isotachysterol$_2$ (VI) | 2.29 | 2.42 | 0.84 |
| Tachysterol$_2$ (VII) | 2.98 | 3.09 | — |
| Previtamin $D_2$ (VIII) | 1.88 | 1.89 | — |

[a] 1: 10% ether in hexane, 120 kg/cm²; 2: 0.15% methanol + 2% ether in pentane, 120 kg/cm²; 3: 55% chloroform (distilled) in pentane, 80 kg/cm².

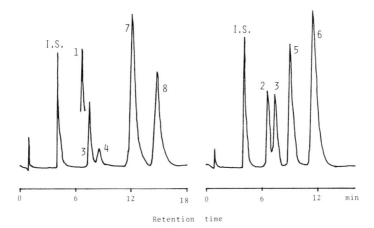

FIG. 4. High-performance liquid chromatographic behavior of vitamin $D_2$ isomers. Experimental conditions: 120 kg/cm², 10% ether in hexane. Peak 1, 5,6-*trans*-vitamin $D_2$; peak 2, isovitamin $D_2$; peak 3, previtamin $D_2$; peak 4, lumisterol₂; peak 5, isotachysterol₂; peak 6, vitamin $D_2$; peak 7, tachysterol₂; peak 8, ergosterol; I.S., internal standard (*p*-cresol).

field region and exhibits the most dramatic shift of all signals. Assignment of each acetoxyl signal is always easy and apparent, even if a sample is contaminated unexpectedly.
3. The total NMR spectrum contains other information valuable for direct identification.
4. High resolution facilitates a simultaneous determination of isomeric members.
5. No analytical obstacles are encountered, except when a hydroxyl- or acetoxyl-bearing compound is included in a sample.

A disadvantage is that this method requires milligram sample size under ordinary NMR running conditions.

High-Performance Liquid Chromatographic Method[6]

*Principle.* Vitamin $D_2$ isomers can be well resolved and determined by respective retention times when a recently developed adsorption mode HPLC is applied.

*Reagents*

Ether and hexane, used as solvents in a mobile phase solution; guaranteed or spectral grade reagent to minimize background absorbance in the UV detector. Usually no special purification is required.

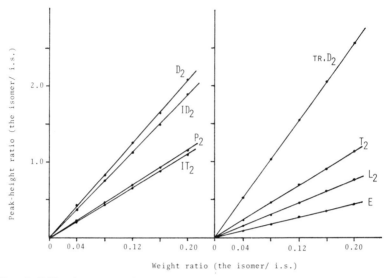

FIG. 5. Calibration curves of vitamin $D_2$ isomers obtained by high-performance liquid chromatography (120 kg/cm², 10% ether in hexane). $D_2$, vitamin $D_2$; TR.$D_2$, 5,6-*trans*-vitamin $D_2$; E, ergosterol; $L_2$, lumisterol₂; $ID_2$, isovitamin $D_2$; $IT_2$, isotachysterol₂; $T_2$, tachysterol₂; $P_2$, previtamin $D_2$; I.S., internal standard (*p*-cresol).

## Apparatus

Shimadzu-DuPont 830 Liquid Chromatograph equipped with a UV-detector (254 nm) system

## Operating Conditions

Column: a stainless steel tube packed with Zorbax SIL, 0.21 × 25 cm
Pressure: 100–120 kg/cm² (flow rate: approximately 0.6 ml/min at 120 kg/cm²).
Temperature: ambient
Mobile phase:
   1. Ether, 10% in hexane for quantitation of ordinary sets of isomers
   2. Methanol, 0.15%, +2% ether in pentane for simultaneous determination
   3. Chloroform (distilled), 55%, in pentane for a separation of isovitamin $D_2$ and 5,6-*trans*-vitamin $D_2$
Sample size: 1 $\mu$l (>1–5 ng), injected with a 10-$\mu$l syringe by stop-flow technique.

*Chromatograms and Internal Standards* (Figs. 3 and 4, Table III). Internal standard should be used to compensate for column characteristics, instrumental variations, and sample introduction technique. For obtaining a similar peak height as that of vitamin $D_2$, 10 times the weight of *p*-cresol or 5 times the weight of $\alpha$-naphthol is required.

*Preparation of Standard Calibrations* (Fig. 5) Standard working solutions are alternately injected into the instrument under the described operational parameters, and the HPLC results are calibrated between the peak-height ratio vs the weight ratio (the isomer/internal standard).

*Advantages.* The proposed method is entirely nondestructive and is sufficient to provide nanogram sensitivities (1–5 ng/$\mu$l); reproducibility of the determination is satisfactory. High speed, base line separation, precise and accurate trace analysis, greater analysis flexibility, no necessity for preparing derivatives, and gentle operating conditions are the main potential advantages. The method is believed to surpass all analytical procedures heretofore proposed.

# [39] Gas–Liquid Chromatography of Vitamin D and Analogs

*By* A. P. De Leenheer and A. A. M. Cruyl

Since the discovery of vitamin D in 1925 by McCollum *et al.*,[1] extensive research has been done to develop appropriate methodology for the analysis of vitamin D, its isomers and metabolites, either in the pure form, in highly concentrated pharmaceutical preparations, or in biological samples. Biological methods are time-consuming and lack specificity. Before the development of high-pressure liquid chromatography (HPLC), gas–liquid chromatography (GLC) proved to be the method of choice compared to other physicochemical techniques, such as ultraviolet (UV) and infrared (IR) spectrophotometry, fluorometry, dienometry, polarography, or the radiochemical methods.

## Problems Related to the GLC of Vitamin D

The secosteroidal structure of vitamin D, incorporating three conjugated double bonds, is fully responsible for the chemical instability. During sample manipulations, thermal and/or photochemical isomerizations are kept under control by standardizing working conditions, i.e., tempera-

[1] E. V. McCollum, N. Simmonds, J. E. Becker, and P. G. Shipley, *J. Biol. Chem.* **65**, 167 (1925).

METHODS IN ENZYMOLOGY, VOL. 67

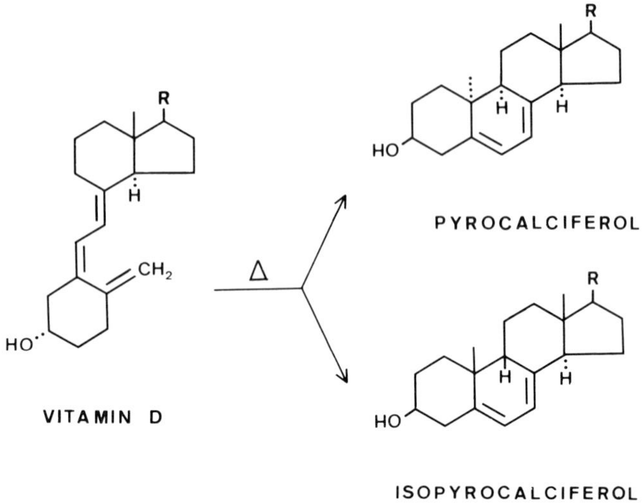

FIG. 1. Thermal isomerization of vitamin D into pyro- and isopyrocalciferol.

ture control and avoidance of acids, daylight, and oxygen. However, cyclization of vitamin D (5,6-cis configuration) into pyro- and isopyrocalciferol (Fig. 1) cannot be avoided at the operating temperature of the gas chromatograph. Although reproducible, this thermal cyclization reaction decreases the sensitivity of the analysis for about a factor of two. On the other hand, the physiological concentrations of vitamin D ranging from 5 to 15 ng per milliliter of plasma require detection at the highest sensitivity attainable. Formation of the trimethylsilyl or methyl ether or of the acetyl, propionyl, trifluoroacetyl, or heptafluorobutyryl ester derivatives does not prevent thermal degradation. Reaction with a potent dienophile such as tetracyanoethylene has no advantage, because two stereoisomeric adducts[2] are formed. The best approach to avoid cyclization is the quantitative conversion of vitamin D into the stable isotachysterol, the all-*trans* isomer of vitamin D. Under the influence of a strong Lewis acid (20% SbCl$_3$ in chloroform)[3], acetyl chloride in ethylene dichloride,[4] or trifluoroacetic acid,[5] vitamin D isomerizes first into isovitamin D, which is then further converted to isotachysterol. The hypothetical mechanism of this reaction has been fully explained by Kobayashi and Adachi.[6] With old batches of heptafluorobutyric anhydride (HFBA), containing traces of the

[2] K. Tsukida, K. Saiki, and M. Ito, *J. Vitaminol.* **18**, 103 (1972).
[3] T. K. Murray, K. C. Day, and E. Kodicek, *Biochem. J.* **98**, 293 (1965).
[4] A. Sheppard, D. E. La Croix, and A. R. Prosser, *J. Assoc. Off. Anal. Chem.* **51**, 834 (1968).
[5] D. Sklan and P. Budowski, *Anal. Biochem.* **52**, 584 (1973).
[6] T. Kobayashi and A. Adachi, *J. Nutr. Sci. Vitaminol.* **19**, 303 (1973).

free acid, vitamin D isomerizes and derivatizes simultaneously into isotachysteryl heptafluorobutyrate.

The alcoholic function in the $3\beta$ position confers to vitamin D a slightly polar character, which hampers the gas chromatographic process. In fact, GLC of underivatized steroids in general is characterized by loss of compound on the column through adsorption and by peak distortion. Adsorption phenomena are minimized by silanization of all glassware, by use of low-loaded columns (stationary phase: 1–2%), and by ether or ester derivatization, which decreases the polarity of the alcohol group(s).

If isomerization of vitamin D is essential for the formation of a compound stable enough and easily amenable for gas chromatographic work, the choice of the derivative should be determined by the separation and the detection limit required for the analysis.

In this chapter two different methods are proposed for the GLC analysis of vitamin D and analogs. The first method involves a two-step conversion: isomerization into isotachysterol by a strong Lewis acid followed by methyl ether derivatization of the alcohol group(s). The derivatives obtained are then submitted to gas chromatography with flame ionization detection. In this way, separation and quantitative analysis of vitamin $D_3$, vitamin $D_2$, and 25-hydroxyvitamin $D_3$ might successfully be applied to commercial pharmaceutical preparations containing high doses of vitamin D and analogs.[7] The second procedure involves the simultaneous isomerization and derivatization into the heptafluorobutyrylisotachysterol structure, which is then submitted to GLC analysis with electron capture or selective ion detection. The latter are suitable for vitamin D analysis in the nanogram range, and the last detection system is especially applicable for determination of vitamin $D_3$ in serum samples.[8]

Derivatization Procedures

*Method A. Preparation of the Methyl Ether Derivatives*

Derivatizations are carried out in conical centrifuge tubes of 10-ml capacity. The two-step derivatization procedure is performed as follows (Fig. 2):

*Isomerization to Isotachysterol.* Adapting the method of Murray *et al.*[3] to a solution containing 50–250 $\mu$g of vitamin D in 0.5 ml of chloroform is added 2 ml of 20% $SbCl_3$ in chloroform; the contents are mixed for 10 sec

[7] A. P. De Leenheer and A. A. M. Cruyl, *J. Chromatogr. Sci.* **14**, 434 (1976).
[8] A. P. De Leenheer and A. A. M. Cruyl, *in* "Quantitative Mass Spectrometry in Life Sciences" (A. P. De Leenheer and R. R. Roncucci, eds.), p. 165. Elsevier, Amsterdam, 1977.

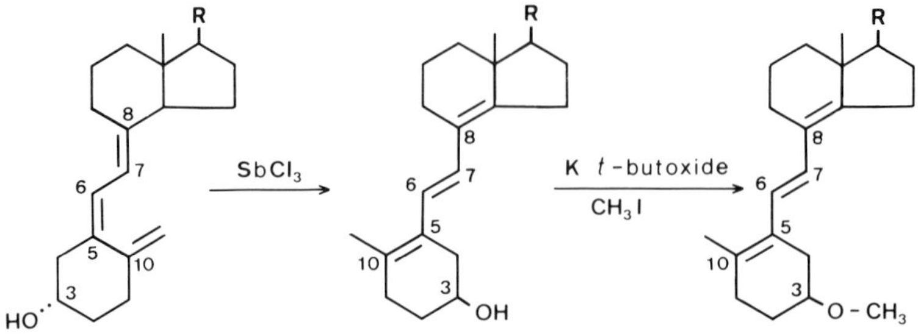

VITAMIN D

$\lambda_{max}^{n-C_6} = 264$ nm

ISOTACHYSTEROL

$\lambda_{max}^{n-C_6} = 279,289,302$ nm

3$\beta$-METHOXY-
ISOTACHYSTEROL

$\lambda_{max}^{n-C_6} = 279,289,302$ nm

FIG. 2. Reaction mechanism of the methyl ether derivatization.

and left for reaction for 1 min at room temperature; the reaction is stopped by addition of 3 ml of 40% tartaric acid in water with continuous mixing for 15 sec; the aqueous phase is removed with the aid of a water jet vacuum pump; the chloroform is filtered over anhydrous $Na_2SO_4$ and dried down with the aid of a stream of purified nitrogen at a temperature below 40°.

*Methyl Ether Formation.* Adapting the method of Clayton,[9] the residue obtained in the isomerization step is dissolved in 1 ml of ether; to the solution 30 mg of potassium *tert*-butoxide is added and immediately suspended by heavily mixing; after a reaction time of 1 min at room temperature, 100 $\mu$l of methyl iodide are added; the reaction is stopped after 5 min by the addition of 4 ml of water; the 3$\beta$-methoxyisotachysteryl derivative is extracted then with 1 × 3 ml and 2 × 2 ml of chloroform; the combined chloroform extracts are filtered over anhydrous $Na_2SO_4$ and evaporated until dryness under a stream of purified nitrogen; the residue is dissolved in an appropriate volume of *n*-hexane for injection into the gas chromatograph.

## Method B. Preparation of the Heptafluorobutyryl Ester Derivatives

Derivatizations (Fig. 3) are done in conical tubes of 3-ml capacity. A few nanograms of vitamin D are dissolved in 200 $\mu$l of benzotrifluoride and derivatized with 5 $\mu$l of heptafluorobutyric anhydride (HFBA) for 30 min at room temperature; the solvent is removed with the aid of a stream of purified nitrogen at a temperature below 40°; the residue is dissolved in an

[9] R. B. Clayton, *Biochemistry* **1**, 357 (1962).

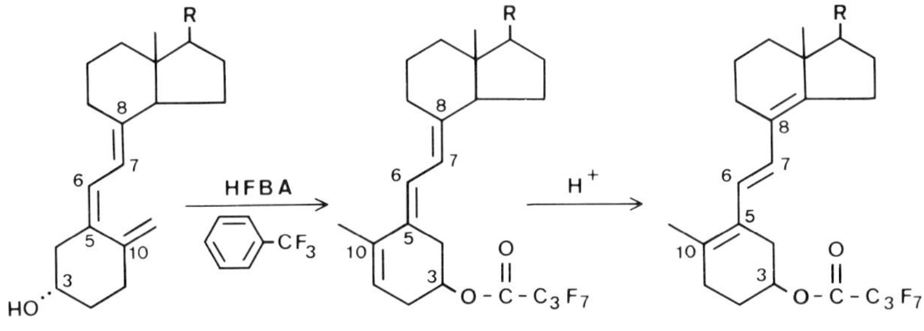

VITAMIN D       ISOVITAMIN D - HFB       ISOTACHYSTEROL-HFB

$\lambda_{max}^{n-C_6} = 264$ nm     $\lambda_{max}^{n-C_6} = 278,298,300$ nm     $\lambda_{max}^{n-C_6} = 280,290,302$ nm

FIG. 3. Reaction mechanism of the heptafluorobutyryl ester derivatization.

appropriate volume of n-hexane (UV spectral grade) for the analysis with electron capture detection or in chloroform (analytical grade) for mass-fragmentographic analysis.

### Gas Chromatographic Conditions

*Method A. Gas–Liquid Chromatography with Flame Ionization Detection*

All separations of the methyl ether derivatives of the isotachysteryl isomers of the vitamin D analogs are carried out on a Hewlett-Packard 5750 G gas chromatograph equipped with a dual flame ionization detector. The analyses are performed on silanized glass columns (1.80 m × 2.00 mm i.d.), coated with different packings of increasing polarity, with high purity nitrogen as carrier gas (linear velocity of 6.0 cm sec$^{-1}$). Taking into consideration the separation, column efficiencies, and peak shapes obtained for the different methoxyisotachysteryl derivatives of vitamin $D_3$, vitamin $D_2$, 25-hydroxyvitamin $D_3$ and for 3$\beta$-methoxydihydro-tachysterol$_2$, the best results are obtained with 3.8% SE-30, 1% OV-1, and 2% QF-1 at an oven temperature of 265, 250, and 225°, respectively. The injector and detector temperatures are set at a few degrees higher than the operating oven temperature.

*Method B.I. Gas–Liquid Chromatography with Electron Capture Detection*

The heptafluorobutyryl derivatives of the vitamin D analogs are chromatographed on a Hewlett-Packard 5750 G gas chromatograph equipped with a pulsed-voltage $^{63}$Ni source. The analyses are carried out on a 1% FFAP Gas Chrom Q (100–120 mesh) silanized glass column (1.80 m × 2.0 mm i.d.) with 5% argon/methane as carrier (31 ± 1 ml min$^{-1}$) and purge (41 ± 1 ml min$^{-1}$) gas. The injector and oven temperatures are kept at 230 and 200°, respectively. The detector is operated in the pulse-sampling mode at 255°, an amplitude of 30 V, a pulse width of 0.75 $\mu$sec, and a pulse interval of 550 $\mu$sec. Contamination by water or oxygen is carefully avoided.

*Method B.II. Gas–Liquid Chromatography with Selective Ion Detection*

All mass-fragmentographic analyses are performed on a LKB 9000S combined gas chromatograph–mass spectrometer (electron-impact mode) equipped with a multiple ion-detection device (MID). The analyses are carried out on a 1% FFAP Gas Chrom Q (100–120 mesh) silanized glass column (2.00 m × 2.00 mm i.d.) with helium as carrier gas (30 ml min$^{-1}$). Temperatures of the injection port and the oven are held at 230–250° and 215°, respectively. The ionization potential is set to 20 eV and the trap current to 60 $\mu$A. The magnet (first channel of the MID unit) is focused on the ion at $m/z$ 580 (M$^+$; isotachysteryl$_3$ heptafluorobutyrate, ISOT$_3$-HFB)

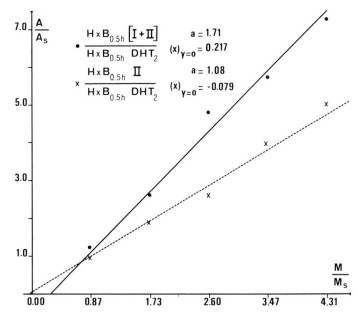

FIG. 4. Calibration graph for the methyl ether derivatives of 25-hydroxyvitamin D$_3$ (0–200 $\mu$g).

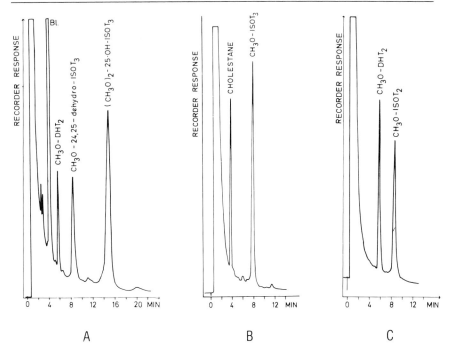

FIG. 5. (A) Gas–liquid chromatography (GLC) (FID) of 3β-methoxydihydrotachysterol₂ (CH₃O-DHT₂: 1 μg) and derivatized 25-hydroxyvitamin D₃ (4 μg), i.e., 3β-methoxy-24,25-dehydroisotachysterol₃ (CH₃O-24,25-dehydro-ISOT₃) and 3β,25-bismethoxyisotachysterol₃ [(CH₃O)₂-25-OH-ISOT₃]. (B) GLC (FID) of cholestane (0.75 μg) and 3β-methoxyisotachysterol₃ (CH₃O-ISOT₃, 3 μg). (C) GLC (FID) of 3β-methoxydihydrotachysterol₂ (CH₃O-DHT₂, 4 μg) and 3β-methoxyisotachysterol₂ (CH₃O-ISOT₂, 4 μg). GLC conditions: A 2% QF-1 column is operated at an oven temperature of 225° and at a linear velocity of nitrogen (carrier gas) of 6 cm sec⁻¹.

whereas for the second ion at $m/z$ 594 (M⁺; dihydrotachysteryl₂ heptafluorobutyrate, DHT₂-HFB) an additional accelerating voltage of 417.25 V is applied. The gain of both channels is adjusted to give a relative intensity of 1:1. The filter setting is 0.6 Hz, and the measuring time is 20 msec.

## Quantitative Measurement

Although subsequent reaction steps are involved in the methyl ether derivatization, the method is easily applicable for quantitative purposes. Though recovery of vitamin D, monitored radioactively, through the complete procedure is not absolute, especially in the nanogram range, it is, however, very reproducible. This is also demonstrated through a linear relationship for vitamin D₃ in the range of 0–250 μg (50 μg interval), first

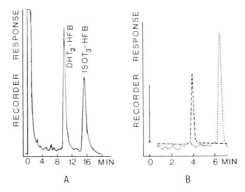

FIG. 6. (A) Gas–liquid chromatography (GLC) (ECD) of 10.8 ng of dihydrotachysteryl₂ heptafluorobutyrate (DHT₂-HFB) and 6.0 ng of isotachysteryl₃ heptafluorobutyrate (ISOT₃-HFB) on a 1% FFAP column (oven temperature, 200°; detector temperature, 255°; pulse interval, 550 μsec). (B) GLC (MID) of 0.8 ng of dihydrotachysteryl₂ heptafluorobuty-rate (---, DHT₂-HFB, M⁺ at $m/z$ 594) and 1.8 ng isotachysteryl₃ heptafluorobutyrate ($\cdots$, ISOT₃-HFB, M⁺ at $m/z$ 580) on a 1% FFAP column (oven temperature, 215°).

for the isomerization reaction alone and then for the entire procedure. Notwithstanding the disadvantage of getting two methylated reaction products for 25-hydroxyvitamin $D_3$, i.e., $3\beta$,25-bismethoxyisotachysterol₃ (II) as the most important, most polar and therefore slowest eluting component on 2% QF-1 and $3\beta$-methoxy-24,25-dehydroisotachysterol₃ (I) as the minor, less polar, and fastest eluting component, the ratio of peak areas for major (II) to minor (I) is constant over the range 0–200 μg of 25-hydroxyvitamin $D_3$. Thus, quantitative measurement is possible either by determining only the major peak (II) area or by summing up the areas of both peaks (I + II) as a measure of the amount of 25-hydroxyvitamin $D_3$ present (Fig. 4). Furthermore, the constancy of the areas' ratio (II/I) is an extra criterion for specificity. Typical examples of gas chromatograms of the methyl ether derivatives are given in Fig. 5.

The methyl ether derivatives can also be applied for quantitative mass-fragmentographic measurements. As internal standard, use can be made of $3\beta$-methoxydihydrotachysterol₂ or of the trideuterated deriva-tives, $3\beta$-methoxy-$d_3$-dihydrotachysterol₂ and $3\beta$-methoxy-$d_3$-isotachy-sterol₃. However, the complete reaction procedure involves many manipulations because for blanks, taken through the entire proce-dure, several peaks are detected at the ions monitored. Therefore the heptafluorobutyric (HFB) derivatives are preferred. Indeed, the higher the mass of the ion monitored; the more selective the analysis becomes. In addition, the HFB derivatives can also be used for electron capture detec-tion. The heptafluorobutyryl derivatization proved to be quantitative for vitamin $D_3$ and $D_2$ in a range of 0–500 ng. Both with $^{63}$Ni electron capture and selective ion detection, a linear response is obtained in the 0–10 ng range by injecting standard solutions on the 1% FFAP column. In both cases dihydrotachysterol$_2$, a structural analog of vitamin $D_3$ is used as a suitable internal standard. Typical examples of gas chromatograms of the HFB derivatives are given in Fig. 6. For electron capture work, quantitation is based on peak area measurement, whereas for mass-fragmentographic analysis, peak height measurement is preferred.

### General Conclusion

Application of GLC for assaying routinely vitamin D in pharmaceutical preparations or in biological samples has become possible. Using different approaches, the major drawback of pyrolysis has been overcome. The irreversible isomerization of vitamin D with a more or less strong acid as catalyst yields the stable all-*trans*-isotachysterol, which possesses good gas chromatographic properties.

The methyl ether derivatives are very useful for the analysis of relatively high doses of vitamin D, but the heptafluorobutyryl derivatives are preferred for the assay of vitamin D in the nanogram range as occurs in plasma samples.

## [40] Gas–Liquid Chromatographic Determination of Vitamin D in Multiple Vitamin Tablets and Their Raw Materials

*By* D. O. Edlund

The method discussed is for the analysis of vitamin $D_2$ or $D_3$ in multiple vitamin tablets or their raw material. Ideally, the form in which the vitamin D is present ($D_2$ or $D_3$) is known and the formulation is available for use in preparing samples. Samples containing known amounts of vitamin D are used to test the procedure for precision and accuracy. If vitamin $D_2$ is present in the sample, vitamin $D_3$ is added (and vice versa) as an internal standard and carried through the purification and extraction procedure. This tends to compensate for losses during the procedure, since vitamins $D_2$ and $D_3$ are very comparable in their chemical and physical properties and chromatograph and extract together. If vitamin E acetate is present in the matrix, an additional cleanup step (chromatography on a phosphate-treated alumina column) is used. In this case also, the vitamin D must be present in an ether insoluble (e.g., gelatin) matrix, since a

METHODS IN ENZYMOLOGY, VOL. 67

preextraction with ether is used to remove most of the vitamin E matrix components prior to solution of the sample.

If this procedure is used, recovery and precision studies should be run using the sample matrix, if possible.

### Preparation of Column for Gas Chromatography

A Pyrex column, 6 ft by $\frac{1}{4}''$ o.d. and 2 mm i.d., packed with 3% DEXSIL 400 on Supelcoport 100/120 is presently used. The packing is solution coated. This can be done by dissolving a known amount (approximately 0.5 g) of liquid phase in a known volume (30 ml) of $CHCl_3$.[1] This solution is then poured through a burette containing the solid support (5 g), allowed to drain, and dried by blowing nitrogen or helium gas through the particles. The column is packed by applying a vacuum to one end (previously closed with glass wool) and pouring packing through a funnel into the other end with occasional tapping on the column. The column is conditioned for at least 2 hr at 300° with a helium gas flow. Coated column packing can be obtained from Supelco, Inc. (Supelco Park, Bellefonte, Pennsylvania).

### GC Parameters

A gas chromatograph with a hydrogen flame detector is used. Column temperature is 265°, injector 280°, detector 300°. The carrier gas is helium at approximately 35 ml/min.

### Reagents and Standards

The silylating agent used is Regisil RC-2 (Regis Chemical Co.; 8210 Austin Avenue, Morton Grove, Illinois.)

Primary and Internal standard solutions are prepared by dissolving USP Reference Standard Vitamin $D_2$ and Vitamin $D_3$ (30–40 $\mu$g/ml of each) in benzene[2] containing 0.01% butylated hydroxytoluene (BHT).

Alcoholic sulfuric acid (2.5 $N$) is prepared by adding slowly, with mixing, 70 ml of concentrated sulfuric acid to approximately 600 ml of 3A alcohol in a 1-liter volumetric flask. The solution is cooled and diluted to volume with 3A alcohol.

Phosphate-treated alumina (PTA) is prepared by placing 250 g of

---

[1] D. O. Edlund and F. A. Filippini, *J. Assoc. Off. Anal. Chem.* **56** No. 6, p. 1374 (1973).

[2] Owing to the hazardous nature of benzene, other solvents may be used, e.g., isooctane or hexane.

Alumina, Basic, Brockman Activity I, 80–200 mesh (Fisher), 20 g of NaH$_2$PO$_4$, and 1600 ml of water in a 3- or 4-liter stainless steel beaker. This is placed on a hot plate and allowed to boil for 15 min. It is then cooled, and the supernatant is decanted. It is washed twice with 1000 ml of water. The water is decanted, and the damp alumina is transferred to a large watch glass or Petri dish, and placed in a 105° oven for at least 4 hr prior to use. Storage of the PTA at 105° prior to use enhances visualization of the band containing vitamin D during the PTA column chromatographic separation.

The PTA column is prepared by placing a glass-wool plug in the bottom of a 7–12-mm i.d. by 300-mm-long column with a Teflon stopper. Petroleum ether (e.g., Skellysolve F) is added to fill the column about halfway to the top. Sea sand is added to give a $\frac{1}{2}''$ layer on the bottom, and the surface is leveled by tapping.

Approximately 3 g of PTA is placed in a 125-ml Erlenmeyer flask. An amount of water corresponding to 4% (v/w) of the PTA is added, and the flask is stoppered tightly with a rubber stopper. The flask is placed on a steam bath for 5 min, shaken to mix, and cooled in cold water (about 5 min). The PTA preparation is then added to the column through a funnel and allowed to settle by gravity. Approximately $\frac{1}{4}$ in. of sea sand is added on top of the PTA mixture. The PTA in the column is kept covered with petroleum ether and is prepared shortly before use.

*Sample Preparation—No Vitamin E*

Samples (approximately 4000 IU of vitamin D) that do not contain vitamin E are ground and transferred to a 500-ml boiling flask. Ten milliliters of a water solution of pyrogallol (5%) and sodium ascorbate (10%) are added. The flask is swirled and allowed to stand for 5–10 min. Fifty milliliters of 3A alcohol and an approximately equivalent amount of internal standard (D$_2$) $\simeq$ (D$_3$) solution are added. Next, 5 ml of 50% KOH are added, an air condensor is attached, and the flasks are placed on a steam bath to reflux for 30 min. Flasks are removed and cooled in cold water. Approximately 100 ml of petroleum ether are added, and flasks are shaken vigorously for approximately 30 sec. Layers are allowed to separate, and the supernatant (ether) layer is decanted to a 250-ml separator. The lower (aqueous) layer is discarded, and ether in the separator is washed with distilled water (50–100 ml) 3 or 4 times until the aqueous phase is no longer basic to phenolphthalein. The aqueous layer is discarded, and the ether is filtered through anhydrous sodium sulfate into a 250-ml, glass-stoppered (24/40) flask. This flask is placed on a rotary vacuum-type evaporator, and the solvent is evaporated to dryness with heat (30–40° water bath) and vacuum. To the residue in the 250-ml flask is added 0.3 ml each of $CHCl_3$ and Regisil-$RC_2$. The flask is stoppered, rotated to dissolve (wet) the sample residue, and allowed to stand 30 min at 40–50°, or overnight at room temperature, before injection (1–5 $\mu$l) to the gas chromatograph.

*Samples Containing Vitamin E Acetate*

These samples are ground, transferred to a boiling flask, and preextracted once with 50 ml of diethyl ether. The ether is allowed to stand on the sample for approximately 2 min with occasional swirling. The ether is then decanted and discarded. Preextraction may not be necessary if the sample contains 10 mg or less of vitamin E acetate. Approximately 20,000 IU of vitamin A acetate is added to the sample to act as a marker on the PTA column. To the sample is added 10 ml of a solution of pyrogallol (5%) plus sodium ascorbate (10%). The sample is swirled and allowed to stand 10 min. Internal standard solution ($D_2$) $\simeq$ ($D_3$) is added and sample is heated gently on a steam bath for 3 min. Pyridine (15 ml) and 2.5 $N$ alcoholic sulfuric acid (40 ml) are added and swirled to mix. An air condensor is attached and the sample is refluxed on a steam bath for 40 min with occasional swirling. The flask is then removed and cooled in cold water. The extraction procedure is similar to that already described. After the sample is extracted and solvent is evaporated, approximately 4 ml of petroleum ether are added to the flask and the flask is swirled to dissolve the residue. This solvent is transferred from the flask through a small funnel to a column of PTA. The flask is rinsed with 5 ml of petroleum ether and added to the column. The column is washed with approximately 20 ml of 8% diethyl ether in petroleum ether to elute the band containing vitamin E. Unnecessary washing at this point may result in loss of vitamin D. The band containing vitamin D is eluted from the column with approximately 50 ml of 20% $CHCl_3$ in petroleum ether. The solvent fraction containing the visible (usually gold-brown) band is collected in a 125- or 250-ml glass-stoppered boiling flask. The solvent is evaporated on a rotary vacuum evaporator. Regisil RC-2 (0.3 ml) and $CHCl_3$ (0.3 ml) are added to the flask as previously described.

Standard solutions are prepared by pipetting equivalent amounts of Primary Standard and Internal Standard solutions to a boiling flask, evaporating the solvent, and adding Regisil and $CHCl_3$ as with the samples.

Response factors are calculated from analysis of the standard solution by measuring the peak height of Pyro $D_3$ and Pyro $D_2$ peaks from a base line drawn tangent to the bottom of the curve. Retention time for the silylated Pyro $D_3$ peak is approximately 12–15 min and the Pyro $D_2$ peak is retained about 2 min longer.

Response factors are used to calculate the sample values. Recoveries of 98–102% with a standard deviation of approximately 2% are usually obtained in statistical studies with this method on our samples.

## [41] Gas–Liquid Chromatographic Determination of Vitamin D

### By TADASHI KOBAYASHI

Gas–liquid chromatographic (GLC) determination of vitamin D ($D_2$ and $D_3$) can be classified according to the following three methods based on the structural forms of vitamin D injected.

1. "Unmodified" vitamin D is injected into a gas chromatograph.[1–5] Since vitamin D is quantitatively isomerized to the thermal cyclized compounds pyro- and isopyrovitamin D at a definite GLC temperature (usually higher than 200°), as shown in Fig. 1, the twin peaks due to the compounds for which the peak area ratio is constant should be observed.

2. A derivative of vitamin D, e.g., trimethylsilyl (TMS) ether or trifluoroacetate, is injected into a gas chromatograph.[6–11] The twin peaks due to derivatives of pyro- and isopyrovitamin D should be also observed in a gas chromatogram, and their peak area ratio is also constant at a definite temperature. An unknown minor peak accompanies the pyro peak, but does not disturb it.

3. Isotachysterol or isovitamin D obtained by isomerization of vitamin D with an acidic reagent, e.g., acetyl chloride or antimony trichloride, is injected into a gas chromatograph.[12–14] Isotachysterol and isovitamin D,

[1] H. Ziffer, W. J. A. Van den Heuvel, E. O. A. Haarhti, and E. C. Horning, *J. Am. Chem. Soc.* **82**, 6411 (1960).

[2] L. V. Avioli and S. W. Lee, *Anal. Biochem.* **16**, 193 (1966).

[3] K. Tsukida and K. Saiki, *J. Vitaminol.* **18**, 165 (1972).

[4] T. Kobayashi and A. Adachi, *J. Nutr. Sci. Vitaminol.* **22**, 41 (1976).

[5] T. Kobayashi and A. Adachi, *J. Nutr. Sci. Vitaminol.* **22**, 209 (1976).

[6] P. P. Nair, C. Bucana, S. DeLeon, and D. A. Turner, *Anal. Chem.* **37**, 631 (1965).

[7] D. O. Edlund and J. R. Anfinsen, *J. Assoc. Off. Anal. Chem.* **53**, 287 (1970).

[8] D. O. Edlund, F. A. Filippini, and J. K. Datson, *J. Assoc. Off. Anal. Chem.* **57**, 1089 (1974).

[9] T. Kobayashi and M. Yasumura, *J. Vitaminol.* **18**, 78 (1972).

[10] T. Kobayashi, A. Adachi, and K. Furuta, *J. Nutr. Sci. Vitaminol.* **22**, 215 (1976).

[11] T. Kobayashi, A. Adachi, and K. Furuta, *Vitamins* **50**, 421 (1976).

[12] T. K. Murray, K. C. Day, and E. Kodicek, *Biochem. J.* **98**, 293 (1966).

[13] T. Panalaks, *Analyst* **95**, 862 (1970).

[14] A. J. Sheppard, D. E. LaCroix, and A. R. Prosser, *J. Assoc. Off. Anal. Chem.* **51**, 834 (1968).

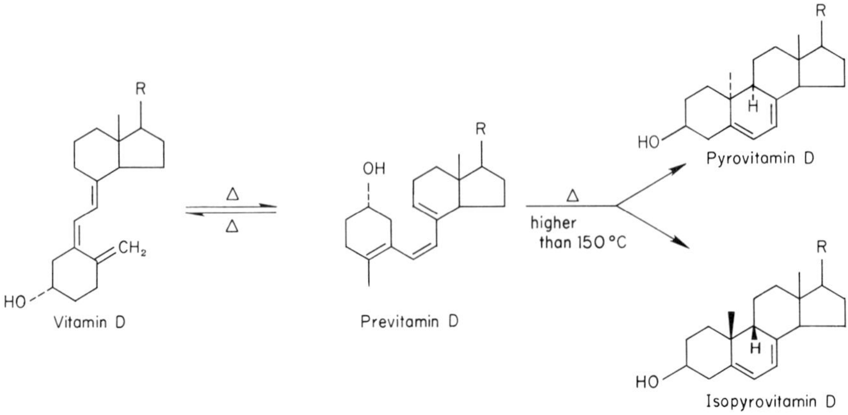

Fig. 1. Thermal isomerization of vitamin D. Vitamin $D_2$ series, R = $C_9H_{17}$; vitamin $D_3$ series, R = $C_8H_{17}$.

which are thermostable at GLC temperatures, give only a single peak in a gas chromatogram.

The third method was previously introduced by Sheppard et al.,[15] so the discussion in this paper is limited to the first and second methods according to the author's experience.[4,5,10,11] As shown in Fig. 1, vitamin D and previtamin D in a solution are thermally interconverted to reach equilibrium at a temperature below 150°, whereas the irreversible thermal cyclization of vitamin D to pyro- and isopyrovitamin D proceeds via previtamin D at a temperature above 150°. A ratio of vitamin D and previtamin D in an equilibrated solution depends on temperature only. A serious problem for the chemical determination of vitamin D is that a ratio of vitamin D and previtamin D originally present in a sample is readily changed during storage or by heating during saponification included as an essential procedure. Keverling Buisman et al.[16] proposed that the sum of vitamin D and previtamin D should be always evaluated as potential vitamin D even though previtamin D itself has little biological potency. Since both vitamin D and previtamin D are thermally cyclized to pyro- and isopyrovitamin D in the same manner, the sum of both compounds can be readily estimated by GLC. Therefore, the term vitamin D means the sum of vitamin D and previtamin D (potential vitamin D) in this chapter.

Since large amounts of vitamin A, E, sterols, and other interfering substances (all of them or parts of them) usually coexist with vitamin D in

[15] A. J. Sheppard and W. D. Hubbard, this series, Vol. 18C, p. 733.
[16] J. A. Keverling Buisman, K. H. Hanewald, F. J. Mulder, J. R. Roborgh, and K. J. Keuning, J. Pharm. Sci. 57, 1326 (1968).

most samples, their removal is necessary before subjecting a sample to GLC. However, complete removal, as in colorimetric methods, is unnecessary because separation of vitamin D from most of interfering substances is possible by GLC if the amounts are not excessive. Clean-up procedures depend on kinds and amounts of interfering substances present in a sample. The procedures that should be applied to the samples investigated by the authors are summarized in the table. Alkali saponification, isolation of unsaponifiable matter, and preparative thin-layer chromatography (TLC) are applied to all the samples. Phosphate-treated alumina column chromatography is useful to eliminate excess amounts of vitamin E, while excess amounts of sterols, e.g., cholesterol and provitamin D (ergosterol and 7-dehydrocholesterol), can be eliminated by application of digitonin–Celite column chromatography. Trimethylsilylation is usually necessary for the samples containing irradiated provitamin D in order to eliminate the influence of lumisterol, and the procedure is essential also for tuna liver oils because some unknown interfering substances cannot be separated from the pyro peak without trimethylsilylation.

## Materials and Reagents

### Multivitamins Preparations I (Usual Preparations)[4]

Vitamin $D_2$ and $D_3$. Commercial grades recrystallized from acetone–water (4 : 1)

ASSAY PROCEDURES FOR THE GAS–LIQUID CHROMATOGRAPHIC DETERMINATION OF
VITAMIN D IN SEVERAL SAMPLES

| Assay procedure | Sample[a] | | | | |
| --- | --- | --- | --- | --- | --- |
| | (1) | (2) | (3) | (4) | (5) |
| Saponification and isolation of unsaponifiable matter | ○ | ○ | ○ | ○ | ○ |
| Phosphate-treated alumina column chromatography | × | ○ | × | × | × |
| Digitonin–Celite column chromatography | × | × | ○ | ○ | ○ |
| Thin-layer chromatography | ○ | ○ | ○ | ○ | ○ |
| Trimethylsilylation | × | × | ○ | ○ | ○ |
| Gas–liquid chromatography | ○ | ○ | ○ | ○ | ○ |

[a] (1) Multivitamins preparations I (usual preparations); (2) multivitamins preparations II (containing excess amounts of vitamin E); (3) tuna liver oils; (4) vitamin $D_3$ resin oils; (5) irradiated "Shiitake" (*Lentinus edodes*). ○, necessary; ×, unnecessary.

Stigmasteryl acetate (SA): commercial grade stigmasterol acetylated according to a conventional method and recrystallized from ethanol

Internal standard solution: stigmasteryl acetate (SA) dissolved in acetone to obtain a concentration of 50 $\mu$g/ml. This is used as the internal standard solution for GLC.

Vitamin D standard solution: Either vitamin $D_2$ or $D_3$ and SA are dissolved in acetone to obtain concentrations of 40 $\mu$g/ml (2000 IU/ml) for vitamin D and 50 $\mu$g/ml for SA, respectively. The vitamin $D_2$ and $D_3$ standard solutions are used for the determination of preparations containing vitamin $D_2$ and $D_3$, respectively.

Reference solution for thin-layer chromatography (TLC): Ten milliliters of either vitamin $D_2$ or $D_3$ solution in ethylene dichloride (500 $\mu$g/ml) are refluxed for 30 min to obtain a mixture of vitamin D and previtamin D. This is used as the reference solution for TLC.

Adsorbent for TLC: commercial grade of Kieselgel $GF_{254}$ from E. Merck Co.

Aldehyde-free ethanol. After adding 5 ml of 50% KOH solution and 5 g of zinc powder to each 1000 ml of ethanol, the preparation is refluxed for 2 hr and then distilled.

$n$-Hexane: guaranteed reagent redistilled

Other guaranteed reagents are used.


*Multivitamins Preparations II (Containing Excess Amounts of Vitamin E)*[5]

Phosphate-treated alumina prepared according to Mulder *et al.*[17] After addition 1600 ml of water and 20 g of disodium hydrogen phosphate ($Na_2HPO_4 \cdot 2H_2O$) to 250 g of alumina (commercial grade of type 1097, E. Merck Co.), the preparation is heated for 30 min on a water bath. The mixture is cooled with gentle swirling, and the upper aqueous layer is decanted. The residue is filtered by suction on a Büchner funnel, then the alumina is activated for 3 hr at 130°. The cooled alumina is stored in a rubber-stoppered container, which is kept in a silica gel desiccator until use.

Petroleum ether reflux over KOH pellets and distilled to collect the fraction between 40 and 60°

Ethyl Ether: Peroxide-free ethyl ether is used.

Other materials and reagents are used as for the multivitamins preparations I.

[17] F. J. Mulder and E. J. De Vries, *J. Assoc. Off. Anal. Chem.* **57**, 1349 (1974).

*Tuna Liver Oils,*[10] *Vitamin D₃ Resin Oils*[10] *and Irradiated "Shiitake"* (*Lentinus edodes*)[11]

Internal standard solution. Stigmasteryl acetate (SA) is dissolved in benzene–pyridine (95 : 5) to obtain a concentration of 100 $\mu$g/ml. This is used as the internal standard solution for GLC. When samples contain vitamin $D_3$ instead of $D_2$, cholesteryl acetate (CA) solution instead of SA solution can be used for this purpose to save GLC time, because the retention time of CA is shorter than that of SA and it is not disturbed by other peaks.

Vitamin D standard solution. Either vitamin $D_2$ or $D_3$ and SA are dissolved in benzene–pyridine (95 : 5) to obtain a concentration of 80 $\mu$g/ml (3200 IU/ml) for vitamin D and 100 $\mu$g/ml for SA, respectively.

Celite: commercial grade of Celite 545 from Johns-Manville

Trimethylsilylation (TMS) reagents. Commercial grades of hexamethyldisilazane and trimethylchlorosilane are used.

Other materials and reagents are used according to the multivitamins preparations I.

Instrumentation

Any gas–liquid chromatograph capable of using a 0.4 cm (i.d.) × 150 cm (length) glass column and equipped with a hydrogen flame ionization detector can be used. Apparatus and analytical conditions used by the author are as follows:

Apparatus: A Shimadzu GC-4BPFE type gas chromatograph equipped with a hydrogen flame ionization detector.

Column: A glass column (0.4 × 150 cm) packed with 1.5% OV-17 on 80–100 mesh Gas Chrom Q (commercial grade of Wako Pure Chem. Ind. Co.). Aging of the column should be performed at 320–330° for more than 24 hours.

Temperature: column, 225°; detector, 300°; injection port, 250°

Flow rate of carrier gas (nitrogen gas): 120 ml/min

Assay Procedures

*Multivitamins Preparations I* (*Usual Preparations*)[4]

This method can be applied to the multivitamins preparations containing vitamin A within 104 of the IU ratio of vitamin A to vitamin D and vitamin E within 2500 of the weight ratio of vitamin E to vitamin D, respectively. Usual commercial preparations are within the limits. Vitamin K does not disturb the assay.

*Saponification and Isolation of Unsaponifiable Matter.* Tablets or capsule samples are pretreated according to a conventional method. A sample containing not less than 2000 IU (50 $\mu$g) of vitamin D is accurately placed in a saponification flask. Twenty milliliters of 5% sodium ascorbate solution is added and swirled gently on a steam bath (80°) for 10 min (this procedure can be omitted for oily and water-dispersed solutions.) After addition of 50 ml of aldehyde-free ethanol, 20 ml of 20% pyrogallol solution in ethanol and 8 ml of 90% (w/v) KOH solution, the mixture is saponified by refluxing on a steam bath for 30 min. It is cooled immediately and exactly 100 ml of benzene are added to isolate the unsaponifiable matter according to Mulder's benzene washing method.[18] It is shaken well and poured into a separatory funnel without rinsing. After adding 40 ml of 6% KOH solution and shaking, the mixture is allowed to stand for the two layers to separate. The lower turbid aqueous layer is discarded, and the upper benzene layer is washed with 40 ml of 3% KOH solution. The aqueous layer is again discarded, and the benzene layer is repeatedly washed with water until the aqueous layer is neutral to phenolphthalein. After standing until both layers are entirely separated, the aqueous layer is discarded and the benzene layer is filtered through a Whatman 1PS filter paper.

*TLC.* Exactly 50 ml of the resulting benzene solution is placed in a round-bottom flask and evaporated under reduced pressure. After dissolving the residue in 1.0 ml of acetone, 0.2 ml of the acetone solution is applied as a zone with a micropipette onto a Kieselgel GF$_{254}$ plate (250 $\mu$m-thick, 20 × 20 cm), which is activated at 105° for 1 hr within 24 hr before use. The reference solution for TLC is spotted on the left side near the zone on the same plate. After developing with a mixture of *n*-hexane–ethyl acetate (4 : 1), the plate is placed under an ultraviolet lamp of 254 nm. Several black-absorbing zones due to compounds separated from the sample and two black-absorbing spots due to vitamin D and previtamin D from the reference solution may be detected against a background of green fluorescene. Zones corresponding to vitamin D and previtamin D assigned from the spots of the reference solution are scraped into a beaker and then combined. After extraction of the adsorbent powder by shaking with 30 ml of acetone and filtering, the remaining powder on the filter paper is repeatedly washed with small quantities of acetone. The washed acetone solutions are combined with the filtrate, and the solvent is evaporated under reduced pressure.

---

[18] F. J. Mulder, E. J. DeVries, and K. J. Keuning, *Pharm. Weekbl.* **100**, 1457 (1965).

*GLC.* After dissolving the resulting residue in 0.5 ml of the internal standard solution, 5 $\mu$l of the solution taken by a microsyringe are applied to the GLC. Five microliters of the vitamin D standard solution are similarly applied to the GLC.

*Calculation.* The peak area ratios of pyrovitamin D to SA are estimated on the gas chromatograms obtained from the sample and standard solution, respectively. The content of vitamin D in a sample is calculated by the following formula:

$$\text{Content of vitamin D (IU/g, tablet or capsule)} = \frac{P_{sa}}{P_{st}} \times \frac{s \times V \times 0.5}{W}$$

$P_{sa}$ = peak area ratio of pyrovitamin D to SA on the gas chromatogram obtained from sample; $P_{st}$ = peak area ratio of pyrovitamin D to SA on the gas chromatogram obtained from vitamin D standard solution; $s$ = concentration of vitamin D in the vitamin D standard solution (IU/ml; 1600 in the above case); $V$ = multiple for dilution ratio (10 in the above case); $W$ = quantity of sample taken for the determination (grams, tablet or capsule).

## *Multivitamins Preparations II (Containing Excess Amounts of Vitamin E)[5]*

This method is used for the multivitamins preparations containing excess amounts of vitamin E (more than 2500 of the weight ratio of vitamin E to vitamin D).

*Deactivation of Phosphate-Treated Alumina and Preparation of Column.* According to Mulder *et al.*,[17] weigh 30 g of phosphate-treated alumina in an Erlenmeyer flask and add 1.5 ml of water, using a pipette. The flask is closed with a rubber stopper, vigorously shaken, and then allowed to stand for 15 min. The deactivated alumina is gradually transferred to a glass tube (2.0 × 30 cm) containing about 50 ml of petroleum ether and allowed to settle. The height of alumina becomes about 12 cm.

*Saponification and Isolation of Unsaponifiable Matter.* A sample containing not less than 2000 IU of vitamin D is treated as for the multivitamins preparations I.

*Phosphate-Treated Alumina Column Chromatography.* Exactly 50 ml of the resulting benzene solution is placed in a round-bottom flask and evaporated under reduced pressure. The residue is dissolved in 5 ml of petroleum ether, then the solution is transferred to the phosphate-treated alumina column with aid of 15 ml of petroleum ether. The column is eluted at a flow rate of 4–5 ml/min with 200 ml of 8% ethyl ether in petroleum ether followed by 100 ml of 30% ethyl ether in petroleum ether. The former eluate is discarded, and the latter eluate is collected in a round-bottom flask. The solvents is evaporated under reduced pressure and the resulting residue is dissolved in 1.0 ml of acetone. This acetone solution is applied to the following TLC.

*TLC, GLC, and Calculation.* The acetone solution is treated and the content of vitamin D in a sample is calculated according to method for the multivitamins preparations I.

*Tuna Liver Oils,*[10] *Vitamin D₃ Resin Oils,*[10] *and Irradiated "Shiitake" (Lentinus edodes)*[11]

These preparations require elimination of sterols and trimethylsilylation.

*Preparation of Digitonin–Celite Column.* The column is prepared according to Sheppard *et al.*[19] Ten grams of Celite 545 dried at 110° for 6 hr are stirred in a flask with 5 ml of aqueous digitonin solution (60 mg/ml). Three grams of the digitonin-treated Celite are transferred to a glass tube (1.0 × 20 cm) with small quantities of *n*-hexane and then allowed to settle while slight pressure is made to the top of tube. The height of the Celite reaches about 12 cm. The stopcock is opened to drain *n*-hexane until the top of the Celite is covered with *n*-hexane to a height of 1 cm.

*Saponification and Isolation of Unsaponifiable Matter:* About 1 g of either tuna liver oil or vitamin D₃ resin oil is weighed accurately in a saponification flask. After adding 50 ml of aldehyde-free ethanol, 20 ml of 20% pyrogallol solution in ethanol, and 8 ml of 90% (w/v) KOH solution are added. On the other hand, when irradiated "Shiitake" is the sample, it is powdered with a mixer and about 10 g of the powder are weighed accurately in a saponification flask. After adding 100 ml of adlehyde-free ethanol, 40 ml of 20% pyrogallol solution in ethanol and 16 ml of 90% (w/v) KOH solution are added. Saponification and isolation of unsaponifiable matter are carried out according to the multivitamins preparations I method.

*Digitonin–Celite Column Chromatography.* An accurate amount of the resulting benzene solution equivalent to about 8000 IU of vitamin D is taken, and the solvent is evaporated under reduced pressure. After dissolving the residue in 3 ml of *n*-hexane, the solution is transferred with 5 ml of *n*-hexane onto the digitonin-Celite column. The elution is started with *n*-hexane at a flow rate of 1–2 ml/min, and the first 30 ml of the eluate are collected in a round-bottom flask. The solvent is evaporated under reduced pressure, and the resulting residue is dissolved in 1.0 ml of acetone. This acetone solution is applied in the following TLC.

[19] A. J. Sheppard, A. R. Prosser, and W. D. Hubbard, this series, Vol. 18C, p. 356.

*TLC.* The acetone solution is applied to the chromatogram according to the multivitamins preparations I method.

*Trimethylsilylation.* The residue obtained by TLC procedure is dissolved in 0.5 ml of the internal standard solution. The resulting solution is trimethylsilylated by mixing with 0.5 ml of hexamethyldisilazane and 0.1 ml of trimethylchlorosilane. The mixed solution is allowed to stand for 10 min at room temperature and then centrifuged for 15 min to eliminate the precipitation formed by adding the TMS reagents. Then 0.5 ml of the vitamin D standard solution is taken and similarly trimethylsilylated.

*GLC.* Five microliters of the resulting supernatant from either sample or vitamin D standard solution is taken with a microsyringe and applied to the chromatogram according to the multivitamins preparations I method.

*Calculation.* The content of vitamin D in a sample is calculated according to the multivitamins preparations I method.

# [42] Thin Layer–Gas Chromatographic Determination of 25-Hydroxycholecalciferol[1]

*By* D. SKLAN

25-Hydroxycholecalciferol (25-HCC) is formed from cholcalciferol in the liver and undergoes further hydroxylation to 1,25-dihydroxy-cholecalciferol, the active metabolite of cholecalciferol, in the kidney.[2] 25-HCC has been separated by small-bore colums of silicic acid[3] by liquid–gel partition chromatography on Sephadex LH-20[4] and by high-pressure liquid chromatography[5] in addition to the thin-layer chromatography (TLC) method described here.

## Principle

In the method developed in our laboratory, determination of 25-HCC is achieved by gas chromatography (GC) after saponification and TLC. Separation from other unsaponifiable materials, such as cholesterol, which are present in amounts manyfold greater then 25-HCC in physiological samples, is critical in obtaining a clean GC determination. Direct GC

[1] The method outlined here has been previously described in *Anal. Biochem.* **56,** 606 (1973).

[2] H. F. DeLuca, *J. Lab. Clin. Med.* **87,** 7 (1976).

[3] J. W. Blunt, H. F. Deluca, and H. K. Schnoes, *Biochemistry* **7,** 3317 (1968).

[4] M. F. Holick and H. F. DeLuca, *J. Lipid Res.* **12,** 460 (1971).

[5] N. Ikekawa and N. Koizumi, *J. Chromatogr.* **119,** 227 (1976).

of 25-HCC results in formation of the iso and pyro ring closure compounds having different retention times. A single peak, however, is obtained after formation of the trifluoroacetate.

## Procedure

### Preparation of the Sample for TLC

Samples should be dispersed, or homogenized, in ethanol containing 50 mg pyrogallol, such that the ratio of solvent to sample is approximately 10 : 1; the sample is brought to boil on a water bath. Aqueous KOH (60% w/v) is added to bring the final KOH concentration to 10% (w/v), and the sample is held under reflux for 30 min. The sample is then cooled rapidly under running water and transferred quantitatively to a separating funnel. Water is added to bring the water : ethanol ratio to about 1 : 1, and the mixture is extracted three times with hexane (0.8 of total volume). The hexane extracts are pooled, washed with water, and dried over anhydrous $Na_2SO_4$. The solvent is then removed under $N_2$ and reduced pressure.

### Thin-Layer Chromatography

Conventional glass plates used in our laboratory are coated with 0.4 mm silica gel G (Merck AG, Darmstadt, containing 15% added $CaSO_4$) and activated at 105° for 60 min after drying at room temperature. Test samples are applied to the plate in chloroform solution, and a 25-HCC standard is applied to the plate, which is developed with chloroform. After drying in a stream of nitrogen, the plate is covered with an additional (clean) glass plate and the area containing the standard is sprayed with a solution of 0.25% rhodamine in ethanol; the standard is located by visualization with shortwave UV light (while the rest of the plate is still covered). The $R_f$ of 25-HCC is 0.18, that cholesterol is 0.35, and that of retinol and of cholecalciferol is 0.45. The corresponding absorbent areas containing the 25-HCC are removed and extracted immediately with methanol. The solution is filtered, the solvent is removed under nitrogen and reduced pressure, and the 25-HCC is taken up in hexane.

### Preparation of the trifluoroacetate

A known amount of 5α-cholestane is added to the hexane containing the 25-HCC to serve as an internal standard, some 10 drops of trifluoroacetic anhydride are added, and the mixture is maintained at 45°

in a heating block for 30 min.[6] The solvent and excess reagent are removed under a stream of nitrogen, and the residue is taken up in hexane. If GC with an electron capture detector is to be carried out, solvent should be added and evaporated several times to remove all traces of halogen-containing reagent.

## Gas Chromatography

The sample is then injected onto a column (2 m long, 3 mm internal diameter) packed with 1% OV-17 (phenyl-methyl silicone, 50% phenyl) maintained at 225° (injection port and detector, 260°) using either a flame ionization or pulsed voltage electron capture detector with carrier flow at 80 ml/min.

The retention time of 25-HCC relative to $5\alpha$-cholestane is 1.31 compared to 1.10 for cholecalciferol trifluoroacetate and 1.94 for cholesterol. The detector response was found to be linear over a range of 1 to 50 ng using the electron capture detector, and 0.5 to 5 $\mu$g with the flame ionization detector. The amount of 25-HCC was determined from the area of the $5\alpha$-cholestane peak, using the flame ionization detector. Recovery as determined by taking samples of plasma with and without added 25-HCC through the whole procedure was 91.5 $\pm$ 2.8 (mean $\pm$ SD from six determinations), and the coefficient of variation obtained by taking the same sample through the procedure was 8.6% (six determinations).

## Comment

As 25-HCC is sensitive to both UV light and to oxygen, the procedure was carried out as much as possible in a darkened room, and exposure of the samples to oxygen, particularly in the dry state, was kept minimal. Alternative developing solvents systems for the TLC plate were examined, and addition of methanol to the developing solvent was found to increase the $R_f$ of all components. Using AgNO$_3$-impregnated silica gel, 25-HCC was also separated well from major unsaponifiable materials other than retinol.

[6] P. P. Nair and S. DeLeon, *Arch. Biochem. Biophys.* **128**, 663 (1968).

---

# [43] High-Performance Liquid Chromatographic Determination of Vitamin D and Metabolites

## By K. T. KOSHY

This chapter is limited to analysis by high-performance liquid chromatography (HPLC) of vitamin D ($D_2$ and $D_3$) and its metabolites in biological samples. Other chapters in this volume cover the determination of vitamins $D_2$ and $D_3$ and their photoisomers by HPLC in nonbiological systems.

During the past decade, there have been dramatic advances in our knowledge of the metabolism and mode of action of cholecalciferol (vitamin $D_3$). The major metabolites have been isolated and identified. It is now well established that vitamin $D_3$ is metabolized in the liver to 25-hydroxyvitamin $D_3$ (25-OH-$D_3$) and to 1,25-dihydroxyvitamin $D_3$ [1,25-$(OH)_2D_3$] in the kidney before it becomes effective in the intestinal absorption of calcium and phosphorus and in the mobilization of these minerals from previously formed bone. The other less active metabolites, whose roles are not so well understood, but have been identified, are 24,25-dihydroxyvitamin $D_3$ [24,25-$(OH)_2D_3$], 25,26-dihydroxyvitamin $D_3$ [25,26-$(OH)_2D_3$] and 1,24,25-trihydroxyvitamin $D_3$ [1,24,25-$(OH)_3D_3$]. The rapid developments in the isolation of these metabolites have been reviewed.[1-5] The current consensus is that vitamin $D_3$ is a prohormone. Figure 1 shows the structures of vitamin $D_3$, the known metabolites of vitamin $D_3$, and the path of its functional metabolism.[4] The metabolism of ergocalciferol (vitamin $D_2$) has also been studied and shown to be identical to that of vitamin $D_3$. The two vitamins differ only in the structure of the side chain.

The initial methods for the determination of vitamin D and its metabolites in human plasma or serum are all based on the principle of competitive protein binding. A number of papers by different authors have appeared in the literature on the use of binding proteins from different sources. These methods present some problems because of lack of precision and lack of specificity due to interference from structurally related compounds, which are characteristic of all *in vitro* binding assay methods.

The rapid improvements in the technology of HPLC have been very beneficial in the development of practical analytical procedures for vitamin D and its metabolites in their bulk products, pharmaceutical dosage forms, biological samples, and animal feeds. Vitamin D and its metabolites have the characteristic *cis*-triene configuration with an extinction coefficient of about 18,000 at 265 nm. They all have considerable absorption at 254 nm, enabling the utilization of the most reliable 254-nm, fixed-wavelength, UV detector for quantitating them by HPLC.

The HPLC methods have many advantages over the potentially useful

[1] H. F. DeLuca, *N. Engl. J. Med.* **289**, 359 (1973).

[2] H. F. DeLuca, *Fed. Proc., Fed. Am. Soc. Exp. Biol.* **33**, 2211 (1974).

[3] D. R. Fraser, *Proc. Nutr. Soc.* **34**, 139 (1975).

[4] H. F. DeLuca, *Clin. Endocrinol.* **5**, Suppl., 97S–108S (1976).

[5] H. F. DeLuca, *Biochem. Pharmacol.* **26**, 563 (1977).

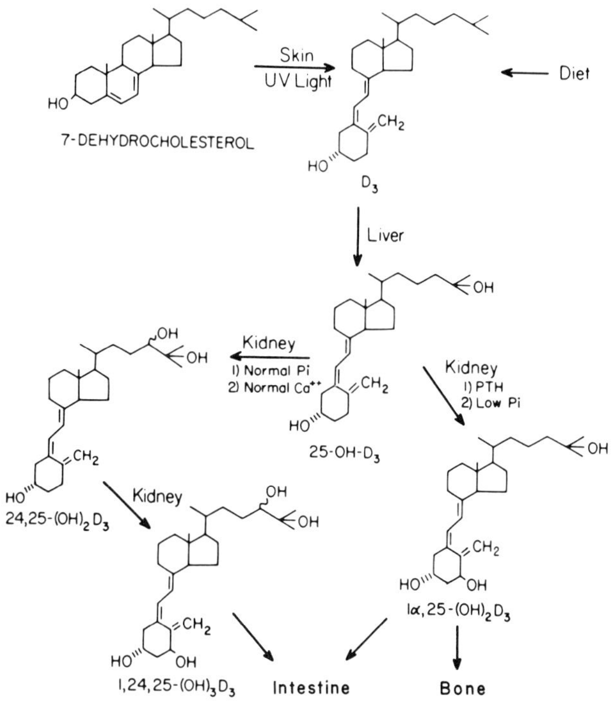

FIG. 1. The functional metabolism of vitamin D. From H. F. DeLuca, *Clin. Endocrinol.* 5, Suppl., 97S–108S (1976).

GLC methods for the analysis of vitamin D and its metabolites: (a) the sample preparation is simpler; (b) the sample does not have to be derivatized or chemically modified; (c) the analysis is performed at ambient temperature and therefore the problems associated with thermal cyclization to the pyro and isopyroforms are avoided; (d) the HPLC system when operated under optimal conditions is more powerful in resolving closely related compounds; (e) the sample fractions can be more easily collected for other purposes, e.g., for determination of radioactivity and for mass spectrometry.

In this discussion it is assumed that the reader is familiar with the basic operational principles of HPLC. One of the very significant advances in HPLC technology in recent years is in the area of column packing material. It may be pointed out that the nature of the column packing materials are such that there is variation from one manufacturer to another, and even batch-to-batch variation with the same manufacturer. Therefore, it may not always be possible to reproduce published methods without making some minor changes in the mobile phase composition or flow rate.

Vitamin D and its metabolites are present in biological samples only in trace amounts. Since these compounds are unstable under adverse conditions, precautionary measures should be taken to avoid unnecessary exposure to heat, light, or harsh chemicals. Glassware cleanliness is also important. Siliconization of all glassware by immersion in a 1% silicone solution (Siliclad, Becton Dickinson and Co., Parsippany, New Jersey) for 5 sec and then heating for 10–15 min is recommended by Lambert et al.[6] Eisman et al.[7] rinsed their glassware with methanol and chloroform prior to use.

## Methods

There are only a few published methods for the HPLC analysis of vitamin D and its metabolites. Mathews et al.[8] showed separation of the major metabolites on a reversed-phase column using a linear solvent gradient from water to methanol, but their system is not suitable for quantitative analysis. Jones and DeLuca[9] and Ikekawa and Koizumi[10] showed isochratic separation of all the known metabolites of vitamin $D_2$ and $D_3$ (see Fig. 1) on a microparticulate silica column. The former group had also shown separation of 25-OH-$D_2$ from 25-OH-$D_3$ under a slightly different HPLC condition. These chromatograms are shown in this volume [44], Figs. 3b and 4. The metabolites of most interest are 25-OH-$D_3$ and 1,25-$(OH)_2D_3$ and to a lesser degree, 24,25-$(OH)_2D_3$. The concentration of 1,25-$(OH)_2D_3$ in biological samples is so low (<1 ng/ml in human blood) that it may not be possible to analyze it by HPLC.

The first HPLC procedure for 25-OH-$D_3$ in a biological sample was on the analysis of cow serum by Koshy and VanDerSlik.[11] The sample volume in this procedure was 25 ml of serum, which is not practical for human serum. Since then, four procedures have appeared in the literature on the analysis of 25-OH-$D_3$ in human serum.[6,7,12,13] There has also been a report on the analysis of 25-OH-$D_3$ in cow liver, kidney, and muscle.[14] The method by Lambert et al.[6] is interesting, as it is a simultaneous

---

[6] P. W. Lambert, B. J. Syverson, C. D. Arnaud, and T. C. Spelsberg, *J. Steroid Biochem.* **8**, 929 (1977).

[7] J. A. Eisman, R. M. Shepard, and H. F. DeLuca, *Anal. Biochem.* **80**, 298 (1977).

[8] E. M. Mathews, P. G. H. Byfield, K. W. Colston, I. M. A. Evans, L. S. Galante, and I. MacIntyre, *FEBS Lett.* **48**, 122 (1974).

[9] G. Jones and H. F. DeLuca, *J. Lipid Res.* **16**, 448 (1975).

[10] N. Ikekawa and N. Koizumi, *J. Chromatogr.* **119**, 227 (1976).

[11] K. T. Koshy and A. L. VanDerSlik, *Anal. Biochem.* **74**, 282 (1976).

[12] K. T. Koshy and A. L. VanDerSlik, *Anal. Lett.* **10**, 523 (1977).

[13] T. J. Gilbertson and R. P. Stryd, *Clin. Chem.* **23**, 1700 (1977).

[14] K. T. Koshy and A. L. VanDerSlik, *J. Agric. Food Chem.* **25**, 1246 (1977).

procedure for the analysis of vitamin D, 25-OH-D, 24,25-(OH)$_2$D and 1$\alpha$,25-(OH)$_2$D from the same sample (the first three were quantitated spectrophotometrically, and the 1$\alpha$,25-(OH)$_2$D was quantitated by a ligand-binding assay).

*Principles of the Methods*

In all the methods for the analysis of 25-OH-D$_3$ in blood, the serum or the plasma is deproteinized and extracted with an organic solvent system. The extract is purified by column chromatography on silicic acid or gel filtration (Sephadex LH-20, Pharmacia Fine Chemicals Co., Inc., Piscataway, New Jersey). It is then analyzed by HPLC on a microparticulate silica column for the separate determination of 25-OH-D$_2$ and 25-OH-D$_3$ or on a C$_{18}$ bonded microparticulate column for determination of total 25-OH-D. A sensitive 254-nm, fixed-wavelength, UV detector is necessary for quantitation owing to the low concentration of 25-OH-D$_3$ in normal human serum, which is in the range of 15–25 ng/ml.

The choice of the solvent system for extraction of the metabolites from the sample seems to be one of individual preference, but will affect the subsequent cleanup step. The solvent systems that have been used are ethanol,[12] methanol–methylene chloride, 2 : 1,[6] ethanol–diethyl ether, 1 : 1.5,[11] methanol–chloroform, 2 : 1,[7] and chloroform–methanol, 2 : 1.[13] Mason and Posen[15] compared the extraction efficiency of ethanol, diethyl ether, and methanol–methylene chloride, 2 : 1, and found ethanol to be the most efficient. Partition chromatography on Sephadex LH-20 column was shown by Holick and DeLuca[16] to be very useful for the separation of metabolites of vitamin D. Adsorption chromatography on silicic acid was used by Lund and DeLuca[17] to separate the metabolites of vitamin D. We have successfully used this adsorbent in purifying human serum,[12] cow serum,[11] and several bovine tissues.[14]

As with the analysis of other trace components in biological samples, it is common practice[6,7,13] to add a known amount of radiolabeled 25-OH-D$_3$ (or any other metabolite of interest) to the sample as an internal standard to monitor the recovery through the entire procedure and then apply a correction factor for loss in the assay. However, we have analyzed a variety of biological samples without the use of labeled tracers and yet achieved good recoveries. But, the use of such an internal standard has definite advantages. It may be cautioned that $^3$H-labeled vitamin D metabolites are less stable than their nonradioactive counterparts and have to

[15] R. S. Mason and S. Posen, *Clin. Chem.* **23**, 806 (1977).
[16] M. F. Holick and H. F. DeLuca, *J. Lipid Res.* **12**, 460 (1971).
[17] J. Lund and H. F. DeLuca, *J. Lipid Res.* **7**, 739 (1966).

be purified at frequent intervals. The use of impure radiotracers will obviously give erroneous results. The following procedure was used by Gilbertson and Stryd[13] for the purification of [³H]25-OH-D₃.

## Purification [³H]25-OH-D₃

Silica gel (230–400 mesh, E. M. Laboratories, Inc., Elmsford, New York) is purified by washing with a generous amount of methylene chloride–ethanol, 92 : 8, followed by hexane. A 0.5 × 2 cm column of washed silica gel is prepared by adding the hexane slurry to a Pasteur disposable pipette containing a small plug of glass wool.

The [³H]25-OH-D₃ is in toluene when received. An aliquot equivalent to about 0.3 μCi is applied to the top of the column and eluted with 2 ml of hexane followed by 7 ml of chloroform–hexane (3 : 2), all of which is discarded. The next 7 ml of chloroform–hexane is collected. This fraction is evaporated under $N_2$ to 0.5 ml, and 4 ml of absolute ethanol is added. It is evaporated to 1.5 ml in a calibrated test tube, and the radioactivity of a 25-μl aliquot is counted to determine recovery. If the recovery is less than 40%, the material is unsuitable for use. If it is suitable, the volume of the solution is adjusted with absolute ethanol so that a 100-μl aliquot contains about 0.01 μCi of the [³H]25-OH-D₃. This solution is stored at 4° and should be used within 4 days.

This procedure is used for purifying both 23,24-[³H]- and 26,27-[³H]25-OH-D₃.

## HPLC Method for 25-OH-D in Blood

Both 25-OH-D₂ and 25-OH-D₃ have the same biological activity. In the clinical application of 25-OH-D₃, it is generally necessary to know only the total 25-OH-D level in the blood, but it is also of interest to have a method specific for 25-OH-D₃ without interference from 25-OH-D₂. The method by Gilbertson and Stryd[13] is specific for 25-OH-D₃. Eisman et al.[7] determined 25-OH-D₂ and 25-OH-D₃ from the same sample, and the sum of the two represented the total 25-OH-D in the blood. Lambert et al[6] determined only total 25-OH-D. The procedure developed in our laboratories[12] is for either total 25-OH-D or specifically for 25-OH-D₃. It is applicable to cow serum and has been adequately tested in our laboratories. The essential details of this procedure are given below.

## Materials

*Glassware.* All glassware is washed with acetone, water, acetone and ethanol prior to use. Generally, pieces are reused without having them washed in the glassware washing area.

*Solvents.* All the solvents, excepting ethanol and diethyl ether, are distilled-in-glass quality (Burdick and Jackson, Inc., Muskegon, Michigan).

### Extraction

A 2.5-ml sample of serum (sample volumes of 1–2.5 ml may be used) is pipetted into a 15-ml screw-cap test tube. (An aliquot of the purified 23,24-[³H]- or 25,26-[³H]25-OH-D₃ equivalent to ≃0.01 μCi is added if it is desirable to use a radiotracer as an internal standard). Five milligrams of NaHCO₃, 5 mg of sodium ascorbate, and 4 ml of 95% ethanol are added, mixed by a Vortex mixer, and centrifuged. The supernatant is transferred to a 20-ml, screw-cap test tube. The residue is reextracted twice more with 2.5 ml each of ethanol, centrifuged, and combined with the first extract. Two milliliters of water and 5 ml of methylene chloride are added to the ethanol extract, and the mixture is shaken for 1 min and centrifuged. The lower phase is removed with a disposable pipette and transferred to a 100-ml, round-bottom flask. The aqueous layer is extracted again with 4 ml of methylene chloride. The combined extract is evaporated to dryness under vacuum in a water bath at 40°. A small quantity of absolute alcohol is added to hasten evaporation of the residual water. The residue is immediately transferred quantitatively into a 2.5-ml tapered test tube using *n*-pentane and concentrated to about 0.5 ml under a stream of N₂.

### Silica Gel Column Chromatography

Two grams of silica gel 60 (230–400 mesh, E. M. Laboratories, Inc., Elmsford, New York), or equivalent, is slurry packed with *n*-hexane–ether (1 : 1) in a 5-ml, disposable, serological pipette (Kimble Products, Toledo, Ohio) with a wad of glass wool at the bottom. The extract from above is transferred to the top of the column using a 9-inch disposable Pasteur pipette (Kimble Products, Toledo, Ohio) and allowed to go down to the top of the packing. The tube is rinsed with small portions of hexane–ether (1 : 1) and transferred to the column, allowing each rinsing to go down to the top of the column before the addition of the next. The column is connected to a glass tube of the same dimension, using a Teflon adaptor to hold about 7–8 ml of the solvent. About 3.5 ml of the solvent is added and eluted under N₂ pressure at a flow rate of about 0.75 ml/min, and the effluent is discarded. The mobile phase is changed to ether–ethyl acetate (9 : 1). The first 4 ml of the effluent are rejected; the next 6 ml are collected in a tapered glass tube, and the solvent is removed under N₂. The residue is reconstituted in ≃100 μl of *n*-pentane.

*Celite Partition Column Chromatography*

In the following steps, Celite 545 (Johns-Manville Product Corp., New York, NY) is the solid support. Methanol–water (80 : 20) containing 0.25% sodium ascorbate is the stationary phase, and *n*-pentane is the mobile phase. The solvents are saturated with each other.

*For Total 25-OH-D Determination.* Two-tenths gram of Celite 545 is mixed with 0.2 ml of methanol–water in a 5-ml test tube. It is packed in a $5\frac{3}{4}$-inch dispo-pipette with a wad of glass wool at the constriction. The extract from the silica column is placed on the column using a 100-$\mu$l syringe (Hamilton Co., Reno, Nevada) followed by 3 × 100 $\mu$l *n*-pentane washings. After the rinsings have gone down the top of the column, the tube is filled with *n*-pentane and the flow is adjusted under very low $N_2$ pressure to a rate of about 0.75 ml/min. The first 0.5 ml of the effluent is rejected, and the next 2.5 ml are collected in a graduated tapered centrifuge tube. The solvent is removed under $N_2$, and the residue is reconstituted in exactly 50 $\mu$l of ethyl acetate or acetonitrile. (If a radiotracer is used as an internal standard, the same quantity of the purified radioactive solution as was added to the sample in the beginning and a 5-$\mu$l aliquot of the final extract are counted in a scintillation spectrophotometer to determine recovery of radioactivity).

*For the Specific 25-OH-D$_3$ Determination.* One gram of Celite is mixed with 0.8 ml of methanol–water in a 15-ml test tube. The column is a 5-ml, disposable, serological pipette with a wad of glass wool at the bottom. The column is packed by collecting small quantities of the wet Celite in the pipette and gently tamping with a glass rod. The extract from the silica column is quantitatively transferred on top of the column, using a 100-$\mu$l syringe, followed by 3 × 100 $\mu$l *n*-pentane rinsings. A glass tube of the same dimension as the pipette is attached with a Teflon adaptor and filled with the *n*-pentane. The column is eluted under low $N_2$ pressure at a flow rate of 0.75 ml/min. The first 3 ml are rejected; the next 6 ml are collected, evaporated under $N_2$, and reconstituted in exactly 50 $\mu$l of ethyl acetate or acetonitrile. (If a radiotracer is used as an internal standard, the same quantity of the purified radioactive solution as was added to the sample in the beginning and a 5-$\mu$l aliquot of the final extract are counted in a scintillation spectrophotometer to determine recovery of radioactivity.)

*Microparticulate Silica Precolumn (Optional)*

This is required only for an occasional sample where the chromatogram after HPLC is not clean enough for quantitation. Unfrozen commercial sera after they have aged fall into this category.

Two hundred and fifty milligrams of microparticulate silica (Partisil 20, Whatman, Inc., Clifton, New Jersey) is slurry-packed with 2.5% isopropanol in hexane in a $5\frac{3}{4}$-inch disposable pipette with a wad of glass wool at the constriction. A pinch of powdered anhydrous $Na_2SO_4$ is added to minimize disturbance of the packing during operation. The solvent for the HPLC step is removed under $N_2$, and the residue is reconstituted in $\simeq 100$ $\mu l$ of 2.5% isopropanol in hexane and applied quantitatively onto the column with small-volume rinsings. The sample application is conveniently performed using a 100-$\mu l$ syringe. Next, a glass tube is attached with a Teflon adaptor to hold 6–7 ml of solvent and filled with 2.5% isopropanol in $n$-hexane. The column is allowed to flow at the rate of about 0.5 ml/min, using very gentle $N_2$ pressure. The first 3 ml are rejected; the next 3 ml are collected, evaporated under $N_2$, and reconstituted in a calculated volume of ethyl acetate for HPLC, taking into consideration the volume of the sample used in the first injection plus the volume of the sample lost in the syringe needle.

## High-Performance Liquid Chromatography

A $C_{18}$ hydrocarbon-bonded microparticulate reversed-phase silica column is used for the determination of the total 25-OH-D, and a nonbonded microparticulate silica column in the adsorption mode is used for the specific determination of 25-OH-$D_3$. The HPLC conditions are shown in Table I.

## Standard Solution

The standard that we used is a crystalline monohydrate (The Upjohn Co., Kalamazoo, Michigan). The first stock solution and dilutions from it

TABLE I
High-Performance Liquid Chromatography Conditions

|  | For total 25-OH-D | For 25-OH-$D_3$ |
| --- | --- | --- |
| Instrument used | Varian 8500[a] | Varian 8500[a] |
| Column | Zorbax ODS[b] | Zorbax Sil[b] |
|  | 2.1 mm × 25 cm | 2.1 mm × 25 cm |
| Mobile phase | $CH_3CN$, $CH_3OH$, $H_2O$, 94 : 3 : 3 | 3% Isopropanol in $n$-Hexane |
| Flow rate | 25 Ml/hr | 45 Ml/hr |
| Pressure | $\simeq 105$ Atm | $\simeq 190$ Atm |
| Detection | UV-254 nm[c] | UV-254 nm[c] |
| Sample size | 5 $\mu l$ | 10 $\mu l$ |
| Sensitivity | 0.005 AUFS | 0.005 AUFS |

[a] Varian Instrument Division, Palo Alto, California.
[b] DuPont DeNemours, Inc., Wilmington, Delaware.
[c] Waters Model 440, Waters Associates, Inc., Framingham, Massachusetts.

are all made in absolute alcohol. All such solutions are prepared in amber volumetric flasks; the head is flushed with $N_2$ and stoppered. These solutions are stable for several weeks in the refrigerator. The working standard is prepared daily from a 10 ng/$\mu$l ethanolic solution of the standard. Usually 25 $\mu$l are transferred to a 2.5-ml tapered test tube, the solvent is removed under $N_2$, and the residue is dissolved in exactly 250 $\mu$l of ethyl acetate or acetnitrile. This 1 ng/$\mu$l standard is adequate for normal samples (10–30 ng/ml).

*Calculations*

The concentrations of 25-OH-D$_3$ and 25-OH-D are calculated by comparison of peak height responses of the standard and the samples. It is possible to calculate the total 25-OH-D concentration from the 25-OH-D$_3$ standard because 25-OH-D$_3$ and 25-OH-D$_2$ have very close molar absorptivity and therefore a comparable response on an ultraviolet detector. On both HPLC systems (Table I) the peak height responses are linear to concentration over a wide range, and therefore it is possible to calculate the concentrations in samples containing varying amounts of 25-OH-D using a single standard. Usually the standard is injected 3 or 4 times to make sure that the HPLC system is working satisfactorily. In our hands, the peak height response of the standard has been very reproducible.

The concentrations in the samples are calculated using the following equation:

$$\text{ng/ml} = \frac{\text{Pk}_{\text{samp.}}}{\text{Pk}_{\text{std.}}} \times C_{\text{std.}} \times V_{\text{std. inj.}} \times \frac{V_{\text{samp.}}}{V_{\text{samp. inj.}}} \times \frac{400.6}{418.6} \times \frac{1}{2.5}$$

If a radiotracer is used as the internal standard

$$\text{ng/ml} = \frac{\text{Pk}_{\text{samp.}}}{\text{Pk}_{\text{std.}}} \times C_{\text{std.}} \times V_{\text{std. inj.}} \times \frac{V_{\text{samp.}}}{V_{\text{samp. inj.}}}$$
$$\times \frac{\text{Added dpm}}{\text{Ext. dpm}} \times \frac{400.6}{418.6} \times \frac{1}{2.5}$$

where $\text{Pk}_{\text{samp.}}$ = peak height of the sample (mm); $\text{Pk}_{\text{std.}}$ = peak height of the standard (mm); $C_{\text{std.}}$ = concentration of the standard (ng/$\mu$l); $V_{\text{std. inj.}}$ = volume of standard injected ($\mu$l); $V_{\text{samp.}}$ = volume of the sample extract ($\mu$l); $V_{\text{samp. inj.}}$ = volume of the sample injected ($\mu$l); Added dpm = radioactivity of tracer added to the sample (dpm); Ext. dpm = total radioactivity of tracer in the sample extract (dpm); 400.6 = molecular weight of 25-OH-D$_3$; 418.6 = molecular weight of 25-OH-D$_3$ · $H_2O$; 2.5 = volume of sample (ml).

*Comments*

Under the HPLC conditions on the reversed-phase system (Zorbax ODS column), 25-OH-D$_3$ and 25-OH-D$_2$ have the same retention time of 6.7 min and are therefore suitable for the determination of the total 25-OH-D in the sample. There is no interference from related compounds. Vitamins D$_3$ and D$_2$ have retention times of $\approx 13$ min. The more polar metabolites, 1,25-(OH)$_2$D$_3$, 24,25-(OH)$_2$D$_3$, 25,26-(OH)$_2$D$_3$, and *trans*-25-OH-D$_3$ all have very short retention times. Comparable metabolites of vitamin D$_2$ behave like the D$_3$ metabolites. Figure 2 shows chromatograms obtained on the Zorbax ODS column.

On the Zorbax SIL column, 25-OH-D$_2$ and 25-OH-D$_3$ are separated, and therefore this column is suitable for the specific determination of 25-OH-D$_3$. The 1.0-g Celite column that is used with this sample removes some of the 25-OH-D$_2$. Figure 3 shows chromatograms of 25-OH-D$_2$ and 25-OH-D$_3$ standards and a chromatogram from a pooled extract.

In general, in our laboratories, the C$_{18}$-bonded, reversed-phase, silica column operated better than the microparticulate colume in the adsorption mode. The former could be operated at a lower flow rate and at a lower pressure. Therefore, the detector noise was lower. We observed about twice the detector response with the reversed-phase column than

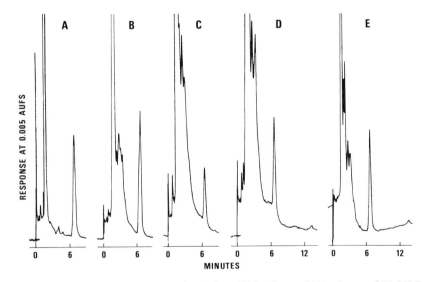

FIG. 2. Liquid chromatograms on the Zorbax ODS column of (A) mixture of 25-OH-D$_2$ and 25-OH-D$_3$ standards, 5 ng each on column; (B, C, D) extracts from a typical fresh serum, and two commercial sera of unknown age, respectively; (E) sample D after additional cleanup on a microparticulate silica precolumn. From K. T. Koshy and A. L. Van Der Slik, *Anal. Lett.* **10**, 523 (1977).

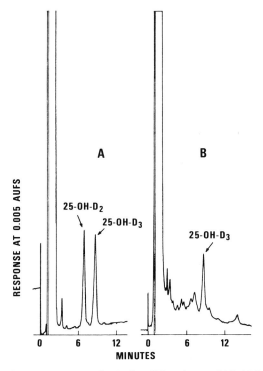

Fig. 3. Liquid chromatograms on the Zorbax SIL column of (A) 25-OH-D₂ and 25-OH-D₃ standards, 10 ng of each on the column; (B), extract from human serum. From K. T. Koshy and A. L. Van Der Slik, *Anal. Lett.* **10**, 523 (1977).

with the adsoprtion column. On both HPLC systems, the peak height responses were linear to concentration and very reproducible for quantitation.

The profiles for the silica gel and Celite partition column chromatographic steps may vary from one laboratory to another and should be established individually. We used a 2-g silica column; it may be possible that a smaller quantity of packing may be sufficient.

*Recovery Data*

Table II shows the results of replicate analysis of pooled serum. Essentially it shows that the endogenous levels of 25-OH-D₃/25-OH-D that we found are in the range of 15–20 ng/ml, which is similar to the range reported using existing methods and with a precision of the assays that is excellent.

Table III shows recovery of 25-OH-D₃ added to the serum. In these

TABLE II
ENDOGENOUS LEVELS (NG/ML) OF 25-OH-D₃ AND TOTAL 25-OH-D IN HUMAN SERUM

| | 25-OH-D$_3$ on Zorbax SIL Column | | | 25-OH-D on Zorbax ODS Column | | |
|---|---|---|---|---|---|---|
| | 1 | 2 | 5 | 3 | 4 | 5 |
| | 15.1 | 21.5 | 14.2 | 17.4 | 12.1 | 16.2 |
| | 18.4 | 21.3 | 14.2 | 17.7 | 12.0 | 18.9 |
| | 15.5 | 20.4 | 15.8 | 18.1 | 12.1 | 18.1 |
| | 16.0 | 21.7 | 15.0 | 22.5 | 13.2 | 19.8 |
| | 14.0 | 21.0 | 14.6 | 17.1 | 12.7 | 20.0 |
| | — | 19.5 | 13.7 | 18.3 | — | 19.5 |
| | — | 19.9 | 14.8 | — | — | — |
| | — | 18.1 | — | — | — | — |
| Mean: | 15.8 | 20.4 | 14.6 | 18.5 | 12.4 | 18.8 |
| SD: | 1.6 | 1.5 | 0.7 | 2.0 | 0.5 | 1.4 |

experiments, known amounts of 25-OH-D$_3$ in ethanol were added to pooled serum prior to extraction and analyzed alongside unfortified blanks. The mean recovery in the six sets of samples ranged from 83 to 90%, which is quite satisfactory. It is possible that the precision and

TABLE III
RECOVERY OF 25-OH-D ADDED TO HUMAN SERUM (NG/ML OR PPB)

| | By the procedure specific for 25-OH-D$_3$[a] | | | | By the procedure for total 25-OH-D[b] |
|---|---|---|---|---|---|
| Added: | 4.8 | 9.5 | 14.3 | 19.0 | 19.0 |
| % Net recovery | 73 | 100 | 106 | 94 | 77 |
| | 100 | 100 | 78 | 86 | 91 |
| | 90 | 110 | 86 | 103 | 92 |
| | 98 | 87 | 78 | 77 | 79 |
| | 81 | 97 | 83 | 72 | 78 |
| | 92 | 76 | 82 | 76 | 80 |
| | — | 78 | 80 | — | 82 |
| | — | 95 | — | — | 87 |
| | — | 67 | — | — | 84 |
| Mean: | 89 | 90 | 84.7 | 84.7 | 83.3 |
| SD: | 10.3 | 14 | 9.8 | 12.0 | 5.6 |

[a] High-performance liquid chromatography (HPLC) on the Zorbax SIL column.
[b] HPLC on the Zorbax ODS column.

accuracy of the assay could be improved if a radiotracer is used as an internal standard.

## Vitamin D and Other More Polar Metabolites in Blood

As mentioned in the introduction, the concentration of $1,25\text{-}(OH)_2D_3$, the most active known metabolite of vitamin D, is so low that it is not possible to determine it by the currently available detectors. However, Lambert *et al.*[6] and Eisman *et al.*[18] have used an HPLC system to obtain a purified fraction of the extract for subsequent analysis of $1,25\text{-}(OH)_2D_3$ by a ligand-binding assay. The former group used an HPLC system that is suitable for the simultaneous determination of vitamin D, 25-OH-D, and $24,25\text{-}(OH)_2D$. They were the first to report HPLC values for vitamin D ($15 \pm 1.8$ ng/ml) and $24,25\text{-}(OH)_2D$ ($2.1 \pm 0.5$ ng/ml). Since no confirmatory tests were performed on the identity of the HPLC peak, these figures have to be verified by other tests.

## 25-OH-D$_3$ in Bovine Tissues

The determination of 25-OH-D$_3$ in bovine liver, muscle, and kidney, was reported from our laboratories.[14] The endogenous levels of 25-OH-D$_3$ in these tissues are much lower (2–10 ng/g) than in the blood. Therefore, a much larger size sample has to be extracted, requiring more extensive cleanup. It is beyond the scope of this article to describe this procedure.

[18] J. A. Eisman, A. J. Hamstra, B. E. Kream, and H. F. DeLuca, *Arch. Biochem. Biophys.* **176**, 235 (1976).

## [44] High-Pressure Liquid Chromatography of Vitamin D Metabolites and Analogs[1]

*By* YOKO TANAKA, HECTOR F. DELUCA, and NOBUO IKEKAWA

It is well established that vitamin D is metabolically converted to many compounds, one of which is accepted as an active and hormonal form of the vitamin, namely 1,25-dihydroxyvitamin D$_3$ [$1,25\text{-}(OH)_2D_3$].[2-4] All

[1] Supported by Grant No. AM-14881 from the National Institutes of Health and U.S.–Japan Cooperative Science Program Grant No. FJ-6037 from the National Science Foundation.
[2] H. F. DeLuca and H. K. Schnoes, *Annu. Rev. Biochem.* **45**, 631 (1976).
[3] H. F. DeLuca, *in* "Hormones and Cell Regulation" (J. Dumont and J. Nunez, eds.), p. 249. Elsevier-North Holland, Amsterdam, 1978.
[4] E. Kodicek, *Lancet* **2**, 325 (1974).

known metabolites of vitamins $D_2$ and $D_3$ and their metabolic pathways are illustrated in Fig. 1. Some of the more important synthetic analogs of the vitamin are illustrated in Fig. 2.

The detection, isolation, and identification of the metabolites of vitamin D have been difficult because of minute amounts found in biological

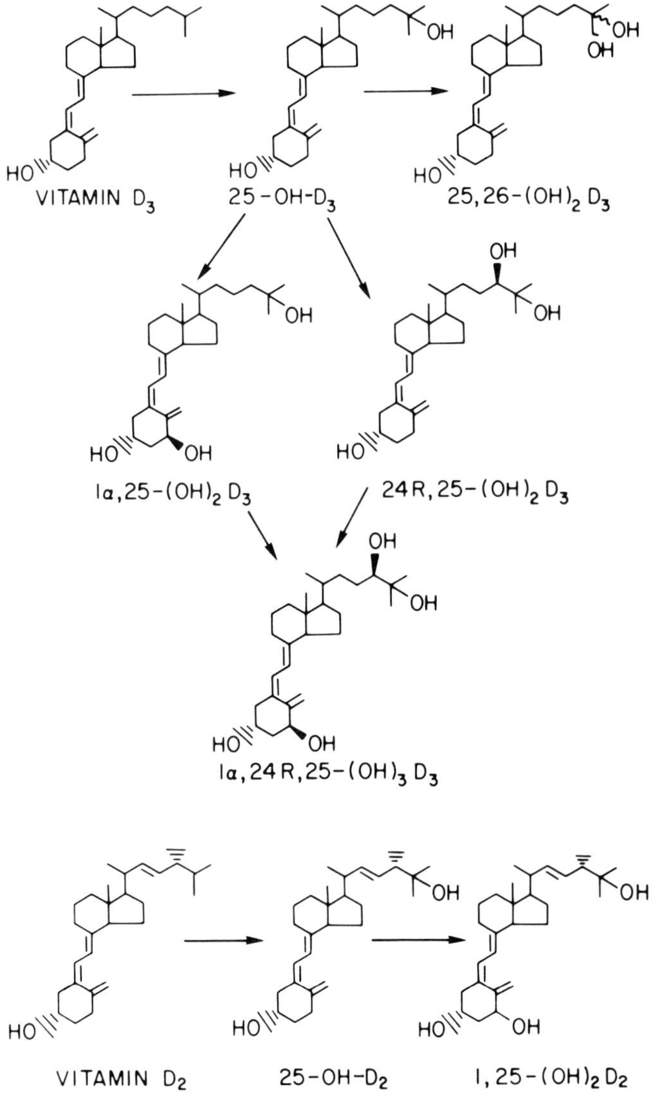

FIG. 1. Metabolic pathway of vitamins $D_2$ and $D_3$. The metabolism of vitamin D is described in this volume [50] by P. S. Yoon and H. F. DeLuca.

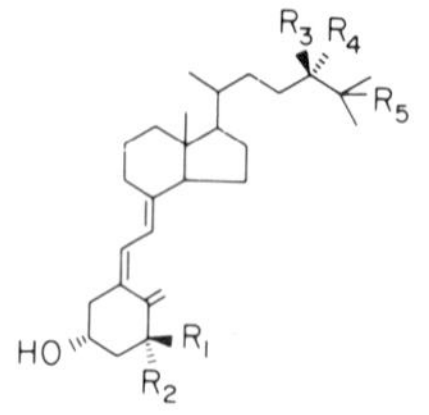

| | | |
|---|---|---|
| 24 R-OH-D$_3$ | R$_3$ = OH, | R$_1$ ,R$_2$ ,R$_4$ ,R$_5$ = H |
| 24 S-OH-D$_3$ | R$_4$ = OH, | R$_1$ ,R$_2$ ,R$_3$ ,R$_5$ = H |
| 24 S, 25-(OH)$_2$D$_3$ | R$_4$ ,R$_5$ = OH, | R$_1$ ,R$_2$ ,R$_3$ |
| 1α, 24 R-(OH)$_2$D$_3$ | R$_1$ ,R$_3$ = OH, | R$_2$ ,R$_4$ ,R$_5$ = H |
| 1α, 24 S-(OH)$_2$D$_3$ | R$_1$ ,R$_4$ = OH, | R$_2$ ,R$_3$ ,R$_5$ = H |
| 1α, 24 S, 25-(OH)$_3$D$_3$ | R$_1$ ,R$_4$ ,R$_5$ = OH, | R$_2$ ,R$_3$ = H |
| 1β, 25-(OH)$_2$D$_3$ | R$_2$   R$_5$ = OH, | R$_1$ ,R$_3$ ,R$_4$ = H |

FIG. 2. Stereochemical isomers of vitamin D$_3$ metabolites and analogs.

materials.[5] Initially the metabolites of vitamin D were separated from vitamin D itself by adsorption chromatography using silicic acid. This method proved to be effective especially for the less polar metabolites, but the columns were long, a large amount of solvent was used, and the method was time consuming. Celite liquid–liquid partition column chromatography has also been employed, but this system is also time consuming and has the added disadvantage of poor reproducibility. Liquid-gel partition chromatography on Sephadex LH-20 was introduced primarily for the separation of polar metabolites, and it rapidly replaced silicic acid adsorption column chromatography in separation of vitamin D metabolites because of markedly reduced time of development and reduced solvent volumes.[6] The Sephadex system is now widely used for the separation and quantitation of vitamin D compounds,[7] the study of regulation of vitamin D metabolism,[8] and the isolation of metabolites for identification.[9] However, there are still difficult-to-resolve metabolites such as 25,26-dihydroxyvitamin D$_3$[25,26-(OH)$_2$D$_3$] and 1,25-(OH)$_2$D$_3$, which can be separated only by Sephadex LH-20 (2 × 40 cm) with a large column eluted with a large amount of solvent.[10] Other separations have also proved to be difficult on Sephadex LH-20. In recent years small-particle

[5] J. Lund and H. F. DeLuca, J. Lipid Res. 1, 739 (1966).
[6] M. F. Holick and H. F. DeLuca, J. Lipid Res. 12, 460 (1971).
[7] Y. Tanaka, H. Frank, and H. F. DeLuca, Science 181, 564 (1973).
[8] M. Garabedian, M. F. Holick, H. F. DeLuca, and I. T. Boyle, Proc. Natl. Acad. Sci. U.S.A. 69, 1673 (1972).
[9] M. F. Holick, H. K. Schnoes, H. F. DeLuca, T. Suda, and R. J. Cousins, Biochemistry 10, 2799 (1971).
[10] M. L. Ribovich and H. F. DeLuca, Arch. Biochem. Biophys. 188, 157 (1978).

columns with a variety of packings have been used for a high degree of resolution. The problem of low solvent flow rate in these columns could be solved with high pressures. This technique, often referred to as "high-performance" or "high-pressure" liquid chromatography has indeed revolutionized the study of vitamin D analysis, metabolism, and identification and is now the most powerful of the methods used in vitamin D chromatography. The ensuing will describe the methods commonly used in the authors' laboratories. To aid in demonstrating the high-pressure liquid chromatographic (HPLC) methods, the following are sources of authentic compounds used in the illustrations: vitamin $D_3$ was purchased from Philips-Duphar, Weesp, The Netherlands; vitamin $D_2$ was purchased from General Biochemicals, Chagrin Falls, Ohio; and 25-hydroxyvitamin $D_3$ (25-OH-$D_3$) was a gift from the Upjohn Company, Kalamazoo, Michigan. 25-OH-$D_2$ and 1,25-dihydroxyvitamin $D_2$ [1,25-(OH)$_2D_2$] were prepared as described by Suda et al.[11] and Jones et al.[12] 1$\alpha$-hydroxyvitamin $D_3$ (1$\alpha$-OH-$D_3$), 24$R$,25-dihydroxyvitamin $D_3$ (24$R$,25-(OH)$_2D_3$], 25,26-(OH)$_2D_3$, 1$\alpha$,25-dihydroxyvitamin $D_3$ [1$\alpha$,25-(OH)$_2D_3$], and 1,24,25-trihydroxyvitamin $D_3$ [1,24,25-(OH)$_3D_3$] were synthesized by Holick et al.,[13] Seki et al.,[14] Lam et al.,[15] Semmler et al.,[16] and Ikekawa et al.[17] respectively. 24$R$,25-(OH)$_2D_3$, 1,25-(OH)$_2D_3$, and 1,24,25-(OH)$_3D_3$ were also a gift from Hoffmann-LaRoche, Nutley, New Jersey. 24-Hydroxyvitamin $D_2$ (24-OH-$D_2$) and 24,25-dihydroxyvitamin $D_2$ [24,25-(OH)$_2D_2$] were isolated from pig plasma (Jones and DeLuca, unpublished data). 1$\alpha$-Hydroxyvitamin $D_2$ (1$\alpha$-OH-$D_2$) was synthesized by Lam et al.,[18] 25-OH-[26,27-$^3$H]$D_3$ was synthesized by the method described by Suda et al.,[19] and 25-OH-[3$\alpha$-$^3$H]$D_3$ was synthesized by Yamada et al.[20] 1,25-(OH)$_2$[26,27-$^3$H]$D_3$ was enzymatically generated from 25-OH-[26,27-$^3$H]$D_3$ using chicken kidney as described by Tanaka et al.[21] 25,26-(OH)$_2$[23,24-$^3$H]$D_3$ was generated from 25-OH-[23,24-$^3$H]$D_3$ enzymatically using kidney from vitamin D-supplemented chickens as described by

[11] T. Suda, H. F. DeLuca, H. K. Schnoes, and J. W. Blunt, *Biochemistry* **8**, 3515 (1969).
[12] G. Jones, H. K. Schnoes, and H. F. DeLuca, *Biochemistry* **14**, 1250 (1975).
[13] M. F. Holick, E. J. Semmler, H. K. Schnoes, and H. F. DeLuca, *Science* **180**, 190 (1973).
[14] M. Seki, N. Koizumi, M. Morisaki, and N. Ikekawa, *Tetrahedron* **1**, 15 (1975).
[15] H.-Y. Lam, H. K. Schnoes, and H. F. DeLuca, *Steroids,* **25**, 247 (1975).
[16] E. J. Semmler, M. F. Holick, H. K. Schnoes, and H. F. DeLuca, *Tetrahedron Lett.* **40**, 4147 (1972).
[17] N. Ikekawa, M. Morisaki, N. Koizumi, Y. Koto, and T. Takeshita, *Chem. Pharm. Bull.* **23**, 295 (1975).
[18] H.-Y. Lam, H. K. Schnoes, and H. F. DeLuca, *Science* **186**, 1038 (1974).
[19] T. Suda, H. F. DeLuca, and R. B. Hallick, *Anal. Biochem.* **43**, 139 (1971).
[20] S. Yamada, H. K. Schnoes, and H. F. DeLuca, *Anal. Biochem.* **85**, 34 (1978).
[21] Y. Tanaka, R. S. Lorenc, and H. F. DeLuca, *Arch. Biochem. Biophys.* **171**, 521 (1975).

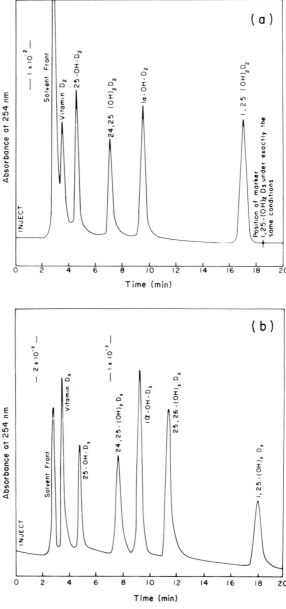

FIG. 3. High-pressure liquid chromatography of vitamins $D_2$ (a) and $D_3$ (b) and their metabolites. HPLC was performed on a DuPont 830LC apparatus fitted with a Waters U6-K injector (Waters Associates, Milford, Massachusetts) with 10% isopropanol in Skellysolve B at 3000 psi pressure and two Zorbax SIL (DuPont) (2.1 mm × 25 cm) columns in series, a flow rate of 0.5 ml/min is achieved. From G. Jones and H. F. DeLuca, *J. Lipid Res.* **16,** 448 (1975).

Tanaka *et al.*[22] Epimers of 24-OH-D$_3$ and 1,24-(OH)$_2$D$_3$ were synthesized by Ikekawa *et al.*[23] and Morisaki *et al.*[24] 24$S$,25-Dihydroxyvitamin D$_3$ [24$S$,25-(OH)$_2$D$_3$] was synthesized by Seki *et al.*,[14] and 1,24$S$,25-(OH)$_3$D$_3$ was synthesized by Ikekawa *et al.*[22] 1$\beta$,25-Dihydroxyvitamin D$_3$ [1$\beta$,25-(OH)$_2$D$_3$] was synthesized by Paaren *et al.*[25]

### Separation of Vitamin D Metabolites and Analogs

Figure 3 illustrates the separation of vitamin D$_3$ and vitamin D$_2$ metabolites on HPLC apparatus equipped with a series of 2 Zorbax SIL (2.1 × 25 cm) DuPont adsorption columns packed with pure, porous silica microparticles (5 $\mu M$) and eluted with a single solvent system of 10% isopropanol in $n$-hexane.[26] The Zorbax SIL has strong adsorptive affinity for the hydroxyl groups of vitamin D compounds, and an increased number of hydroxyl groups results in increased retention of the columns. In addition to the number of hydroxyl groups, 1$\alpha$-hydroxyl seems to interact more strongly with silica than do hydroxyls on the side chain of the molecule. The most distinctive advantage of the HPLC with adsorption columns in the separation of vitamin D$_3$ metabolites is resolution of 25,26-(OH)$_2$D$_3$ and 1,25-(OH)$_2$D$_3$. Previously, the separation had been achieved by only time-consuming Celite liquid–liquid partition column eluted with 500 ml of solvent (CHCl$_3$/Skellysolve B, 20:80)[27] or a lengthy column (1 × 80 cm) of Sephadex LH-20 eluted with 350 ml of hexane/chloroform/methanol (9:1:1).[9] Another advantage is the resolution of vitamin D$_3$ compounds from vitamin D$_2$ metabolites. However, these columns do not resolve vitamin D$_2$ from vitamin D$_3$. With a less polar solvent system often some forms of vitamin D$_3$ can be resolved from vitamin D$_2$ on silica absorbents such as Zorbax SIL. For example, with 2.5% isopropanol in hexane, 24-hydroxyvitamin D$_3$ (24-OH-D$_3$) and 25-OH-D$_3$ can be separated from corresponding D$_2$ compounds even though each has a single hydroxyl group on the side chain (Fig. 4). For resolution of less polar compounds such as vitamin D$_2$ and vitamin D$_3$, a reversed-phase partition column packed with octadecylsilane bonded to the mi-

[22] Y. Tanaka, R. A. Shepard, H. F. DeLuca, and H. K. Schnoes, *Biochem. Biophys. Res. Commun.* **83**, 7 (1978).

[23] N. Ikekawa, M. Morisaki, N. Koizumi, M. Sawamura, Y. Tanaka, and H. F. DeLuca, *Biochem. Biophys. Res. Commun.* **62**, 485 (1975).

[24] M. Morisaki, N. Koizumi, N. Ikekawa, T. Takeshita, and S. Ishimoto, *J. Chem. Soc., Perkin Trans. 1*, 1421 (1975).

[25] H. E. Paaren, H. K. Schnoes, and H. F. DeLuca, *J. Chem. Soc. Chem. Commun.* **8**, 890 (1977).

[26] G. Jones and H. F. DeLuca, *J. Lipid. Res.* **16**, 448 (1975).

[27] T. Suda, H. F. DeLuca, H. K. Schnoes, Y. Tanaka, and M. F. Holick, *Biochemistry* **9**, 4776 (1970).

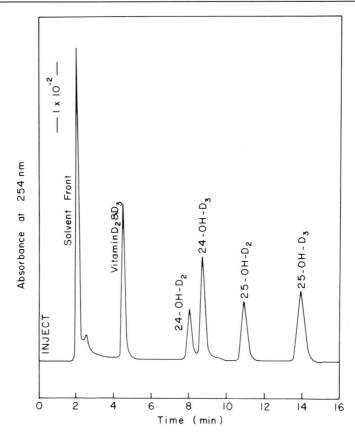

FIG. 4. Separation of 24-OH-D$_2$ and 24-OH-D$_3$, and 25-OH-D$_2$ and 25-OH-D$_3$. HPLC is performed on a DuPont 830LC apparatus with 2.5% isopropanol in Skellysolve B at 4000 psi pressure and two Zorbax SIL (2.1 mm × 25 cm) columns in series and a flow rate of 0.7 ml/min. From G. Jones and H. F. DeLuca, *J. Lipid Res.* **16**, 448 (1975).

croparticulate silica (Zorbax ODS) was used (Fig. 5).[28] The resolution of vitamin D$_3$ metabolites and analogs, including less polar to more polar compounds on a single chromatography system, is achieved by a gradient elution by a solution consisting of 0.02–6% (v/v) methanol–methylene chloride (Fig. 6).[29]

### Separation of Stereoisomers of Vitamin D$_3$ Metabolites and Analogs

Four of the 24-hydroxy-*epi*-vitamin D compounds (Fig. 2) were synthesized by Ikekawa *et al.*,[13,16,22,23] namely, 24-OH-D$_3$, 24,25-(OH)$_2$D$_3$,

[28] R. M. Shepard, R. L. Horst, A. J. Hamstra, and H. F. DeLuca, *Biochem. J.* **182**, 55 (1979).
[29] N. Ikekawa and N. Koizumi, *J. Chromatogr.* **19**, 227 (1976).

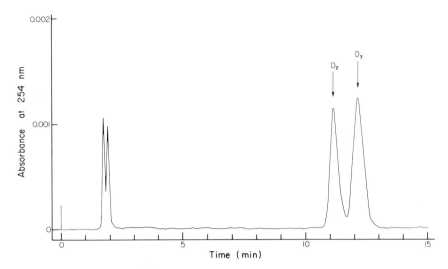

FIG. 5. Separation of vitamins $D_2$ $D_3$ on reversed-phase Zorbax ODS. HPLC is performed on a Waters Model ALC/GPC 204 apparatus with 2% $H_2O$ in methanol at 1200 psi, a Zorbax ODS (4.6 mm × 25 cm) column, and a flow rate of 1.5 ml/min. From R. M. Shepard, R. L. Horst, A. J. Hamstra, and H. F. DeLuca, *Biochem. J.* **182**, 55 (1979).

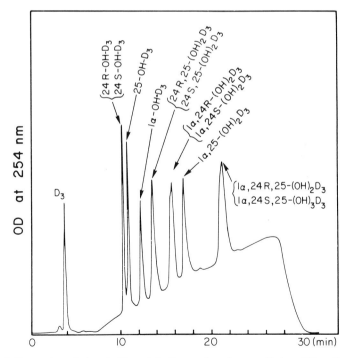

FIG. 6. Separation of vitamin $D_3$ metabolites and analogs on Zorbax SIL with gradient elution from 0.02 to 6% methanol in dichloromethane. HPLC is performed on a Simazu-DuPont 830 with Zorbax SIL (2.1 mm × 25 cm) at 1280 psi pressure and a flow rate of 0.4–0.42 ml/min. From N. Ikekawa and N. Koizumi, *J. Chromatogr.* **19**, 227 (1976).

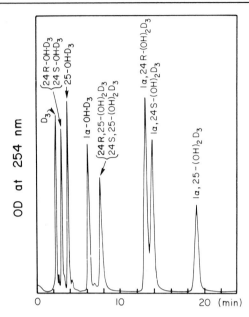

FIG. 7. Separation of vitamin $D_3$ metabolites and analogs on Zorbax SIL with 2% methanol in dichloromethane. HPLC is performed on a Simazu-DuPont 830 with Zorbax SIL (2.1 mm × 25 cm) at 1240 psi pressure and a flow rate of 0.4 ml/min. From N. Ikekawa and N. Koizumi, *J. Chromatogr.* **19**, 227 (1976).

1,24-$(OH)_2D_3$, and 1,24,25-$(OH)_3D_3$. Separation of those stereoisomers is difficult since their polarities are so similar. As shown in Fig. 7, epimers of the compounds that are not 1-hydroxylated are not resolved on Zorbax SIL whereas epimers with $1\alpha$-hydroxyl groups, such as 1,24-$(OH)_2D_3$ and 1,24,25-$(OH)_3D_3$, are partially separated (see next section, Fig. 9, for 1,24,25-$(OH)_3D_3$). In contrast, epimers of the 1-hydroxyl of 1,25-$(OH)_2D_3$ can be separated on a Microparticulate (10 $\mu M$) silica column (0.7 × 30 cm) eluted with 9% isopropanol in hexane at 650 psi pressure with a flow rate of 3 ml/min.[25] In this HPLC system, $1\beta$,25-$(OH)_2D_3$ is eluted at 70 ml and $1\alpha$,25-$(OH)_2D_3$ is eluted at 88 ml. Thus, isomers involving a hydroxyl can be resolved on silica columns, but this depends on how strongly that particular hydroxyl interacts with the absorbent.

### Determination of Stereochemical Configuration of Hydroxyl Groups of Vitamin D Metabolites

Although partial separation of 24-hydroxy epimers which have a $1\alpha$-hydroxyl group can be achieved with Zorbax-SIL as previously de-

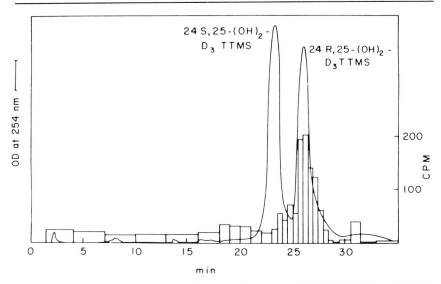

FIG. 8. Separation of tristrimethylsilyl ether derivatives of 24$S$,25-(OH)$_2$D$_3$ and 24$R$,25-(OH)$_2$D$_3$, and cochromatography of the tristrimethylsilyl ether derivative of radioactively labeled, biologically generated 24,25-(OH)$_2$D$_3$ with the 24$R$,25-(OH)$_2$D$_3$ derivative. HPLC is performed on a DuPont 830 equipped with a Zorbax SIL column (2.1 mm × 25 cm), using a pressure of 1140 psi, a solvent system of 2% dichloromethane in hexane, and a flow rate of 0.44 ml/min. The line represents optical density at 254 nm, and the bars represent radioactivity in each 30 sec fraction. From Y. Tanaka, H. F. DeLuca, N. Ikekawa, M. Morisaki, and H. Koizumi, *Arch. Biochem. Biophys.* **170**, 620 (1975).

scribed, the 24 epimers without a 1-hydroxyl group such as 24-OH-D$_3$ and 24,25-(OH)$_2$D$_3$ are not. If the 24-hydroxy epimers are converted to their trimethylsilyl ether (TMS) derivatives, clear resolution is obtained.[30] Separation of the derivative of 24,25-(OH)$_2$D$_3$ epimers on Zorbax SIL is shown in Fig. 8. It is interesting to note that tris-TMS derivative of 24$S$,25-(OH)$_2$D$_3$ is less polar while the free form of 24$S$ isomers of 1,24-(OH)$_2$D$_3$ and 1,24,25-(OH)$_3$D$_3$ are more polar than the corresponding $R$ isomers. When 1-hydroxy compounds of 24-OH-D$_3$ and 24,25-(OH)$_2$D$_3$ are derivatized with TMS in order to get better separation, no resolutions of the 24 epimers are demonstrated, contrary to expectation (N. Ikekawa, unpublished observations). Since the 24 epimers of 24,25-(OH)$_2$D$_3$ have been synthesized and separation of TMS derivatives of the epimers have been successively achieved on HPLC with the Zorbax SIL column, determination of stereochemical configuration of 24-hydroxyl group of naturally produced 24,25-(OH)$_2$D$_3$ was possible.

[30] Y. Tanaka, H. F. DeLuca, N. Ikekawa, M. Morisaki, and N. Koizumi, *Arch. Biochem. Biophys.* **170**, 620 (1975).

Enzymically produced $24,25\text{-}(OH)_2[^3H]D_3$ is mixed with 10 $\mu$g each of the $S$ and $R$ isomers. They are dissolved in 50 $\mu$l of dry hexane to which is added 10 $\mu$l of trimethylsilyl imidazole. The mixture is heated at 70° for 10 min, and 0.3 ml water is then added. This mixture is extracted with $n$-hexane and subjected to HPLC using a Zorbax SIL column (2.1 mm × 25 cm) and a solvent of 2% dichloromethane in hexane. Each 30-sec fraction is collected and counted. As shown in Fig. 8, enzymically generated TMS derivatives of the $24,25\text{-}(OH)_2[^3H]D_3$ comigrate with $24R,25\text{-}(OH)_2D_3$ tris-TMS. The $R$ configuration of biologically produced $24,25\text{-}(OH)_2D_3$ was supported by its biological activity.[30]

The configuration of 24-hydroxyl group of naturally produced $1,24,25\text{-}(OH)_3D_3$ is determined in a similar way except that the compounds are not derivatized. The $1,24,25\text{-}(OH)_3[26,27\text{-}^3H]D_3$ is enzymically generated in the same way as $24,25\text{-}(OH)_2[26,27\text{-}^3H]D_3$ production, except that $1,25\text{-}(OH)_2[26,27\text{-}^3H]D_3$ is used as the substrate.[31] Purified $1,24,25\text{-}(OH)_3[26,27\text{-}^3H]D_3$ is mixed with synthetic epimers of $1,24,25\text{-}(OH)_2D_3$ and subjected to HPLC using a Zorbax SIL column (2.1 mm × 25 cm) eluted with 3.5% methanol in dichloromethane. Although the resolution of the epimers is partial, it is possible to determine the configuration of the 24-hydroxyl group of biologically produced $1,24,25\text{-}(OH)_3D_3$ as $R$ by its clear comigration with synthetic $1,24R,25\text{-}(OH)_3D_3$ (Fig. 9).[32]

### Identification and Quantitation of Known Metabolites of Vitamin D by Cochromatography

One of the applications of HPLC is the identification and quantitation of the metabolites of radioactive vitamin D compounds. Figure 10 illustrates an HPLC profile of a dichloromethane extract of serum obtained from rats fed a rachitogenic diet[33] for 2 weeks and given 325 pmol of $25\text{-}OH[3\alpha\text{-}^3H]D_3$ intravenously 18 hr before sacrifice. Serum is obtained by centrifugation and mixed with 200 ng of each authentic $24R,25\text{-}(OH)_2D_3$, $25,26\text{-}(OH)_2D_3$, and $1,25\text{-}(OH)_2D_3$ dissolved in 50 $\mu$l of 95% ethanol. After 1 hr of incubation at room temperature, the serum (1 ml) is extracted with 10 ml of dichloromethane twice and the combined extracts are dried *in vacuo*. The lipid is dissolved in a solvent system of hexane/chloroform/methanol (9:1:1) and applied to a small batch column of Sephadex LH-20 (7 mm × 14 cm) equilibrated and packed in the same solvent system. The last 18 ml of a total of 27 ml of effluent is dried under

[31] M. F. Holick, A. Kleiner-Bossaller, H. K. Schnoes, P. M. Kasten, I. T. Boyle, and H. F. DeLuca, *J. Biol. Chem.* **248**, 6691 (1973).
[32] Y. Tanaka, L. Castillo, H. F. DeLuca, and N. Ikekawa, *J. Biol. Chem.* **252**, 1421 (1977).
[33] Y. Tanaka and H. F. DeLuca, *Proc. Natl. Acad. Sci. U.S.A.* **71**, 1040 (1974).

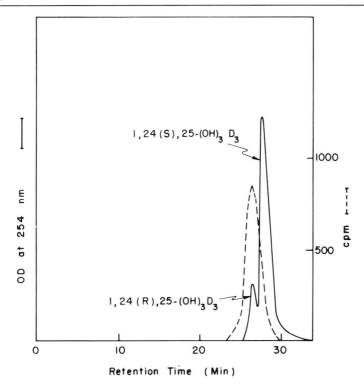

FIG. 9. Cochromatography of synthetically prepared 1,24$R$,25-(OH)$_3$D$_3$ and 1,24$S$,25-(OH)$_3$D$_3$ and *in vitro* generated 1,24,25-(OH)$_3$D$_3$ on Zorbax SIL. Enzymically generated 1,24,25-(OH)$_3$[26,27-$^3$H]D$_3$ was purified with a Sephadex LH-20 column (2 × 18 cm) eluted with a solvent system of chloroform/hexane (75/25). The purified 1,24,25-(OH)$_3$[26,27-$^3$H]D$_3$ is mixed with synthetic isomers of 1,24,25-(OH)$_3$D$_3$ and subjected to HPLC. The cochromatography is performed on Zorbax SIL (2.1 mm × 25 cm) at 1270 psi pressure, with 3.5% methanol in dichloromethane and a flow rate of 0.44 ml/min. ——, Optical density at 254 nm; ---, radioactivity of each 30-sec fraction. From Y. Tanaka, L. Castillo, H. F. DeLuca, and N. Ikekawa, *J. Biol. Chem.* **252**, 1421 (1977).

N$_2$, dissolved in 9% isopropanol in hexane, and subjected to HPLC using a Zorbax SIL column (4.6 mm × 25 cm) developed with 9% isopropanol in hexane at 600 psi pressure and a flow rate of 1.5 ml/min. Each fraction of 0.6 ml is collected and counted. At the same time, the chromatographic profile of the authentic compounds detected by a UV monitor (at 254 nm) is obtained (Fig. 10, solid line). Similar mixtures of authentic 24$R$,25-(OH)$_2$D$_3$, 25,26-(OH)$_2$D$_3$, and 1,25-(OH)$_2$D$_3$ of known concentration are subjected to HPLC using the same system. Standard curves are made by plotting the area under the peak obtained by UV detector vs amount of compound injected. A standard curve of 24$R$,25-(OH)$_2$D$_3$ is shown in Fig. 11. These standard curves are fully reproducible as long as the same

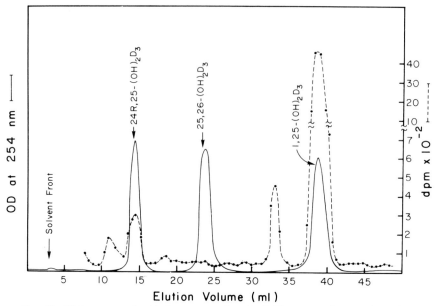

FIG. 10. High-pressure liquid chromatography of a lipid extract of serum from vitamin D-deficient rats given 325 pmol of 25-OH[3α-³H]D₃ 18 hr before sacrifice. Purification of the samples is described in the text. HPLC is performed on a Waters Model ALC/GPC 204 (Waters Associates) equipped with a Zorbax SIL (4.6 mm × 25 cm) column with 9% isopropanol in hexane at 600 psi pressure and a flow rate of 1.5 ml/min ——, Optical density at 254 nm; ---, radioactivity of each 0.6 ml fraction.

chromatographic system is used, although a few points on the standard curve are checked periodically by injecting the standard compounds. Recoveries of the radioactive metabolites in a serum sample are calculated from recoveries of the authentic compounds using the standard curves. Thus, identification of the metabolites by comigration with their authentic compounds is achieved as well as the quantitation of each metabolite. If a conventional column of Sephadex LH-20 with a solvent system of chloroform/hexane (65 : 35) had been used for this sample, it would have resolved the radioactivity into no more than two components as metabolites. Since the amount of authentic compounds added to the sample is large enough to permit a lower sensitivity setting of the UV detector, other UV-absorbing substances extracted from the serum sample do not interfere.

## Use in Measurement of Vitamin D Hydroxylases

Measurement of activities of these enzymes had been achieved by measurement of radioactive metabolites produced from a radioactive sub-

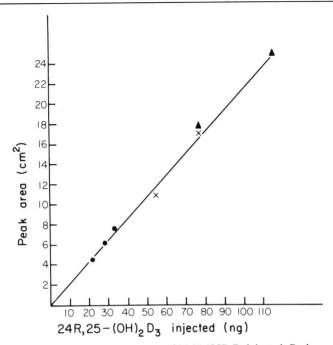

FIG. 11. Relation of peak area to amount of 24,25-(OH)₂D₃ injected. Peak area of 24$R$, 25-(OH)₂D₃ is measured and plotted against amount of 24$R$, 25-(OH)₂D₃ injected. HPLC conditions are as described in Fig. 10. Sensitivity (AFUS): ●, 0.005; ×, 0.01; ▲, 0.02.

strate. The product and substrate are resolved by conventional column chromatography, and the calculation of amount of product is based on the specific activities of the substrate. In addition to superior separation and determination of vitamin D metabolites on HPLC being described here, a sensitive UV detector also provides the possibility of quantitation of non-radioactive metabolites produced *in vitro*. As an example of the assay, Fig. 12 is presented. Kidneys from chickens fed a high-calcium (1.2%), vitamin D-deficient diet[34] for 1 month and given 2.5 $\mu$g of vitamin D per day for the last 4 days are homogenized in ice-cold buffer (pH 7.4) containing 0.19 $M$ sucrose, 15 m$M$ Tris-acetate, and 1.9 m$M$ magnesium acetate. The incubations are carried out in 25-ml Erlenmeyer flasks in duplicate. Each flask contains 1 ml of the 20% (w/v) kidney homogenate (200 mg of tissue), 0.19 $M$ sucrose, 15 m$M$ Tris-acetate, 1.9 m$M$ magnesium acetate, and 25 m$M$ succinate in total volume of 1.5 ml. The incubation medium is flushed with 100% oxygen for 30 sec. One microgram of 25-OH-D₃ dissolved in 25 $\mu$l of 95% ethanol is added. The reaction mixtures are incu-

[34] G. Guroff, H. F. DeLuca, and H. Steenbock, *Am. J. Physiol.* **204**, 833 (1963).

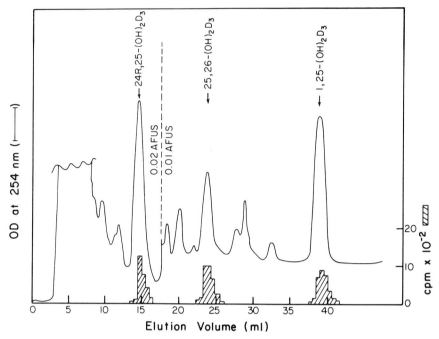

FIG. 12. High-pressure liquid chromatography of a lipid extract of the reaction mixture using kidney homogenate from chickens given vitamin $D_3$. Details of the experiment and purification of the lipid extract are described in the text. HPLC is performed as described in Fig. 10. The solid line represents optical density at 254 nm; the bars represent radioactivity of 0.5-ml fractions, and the arrows show the positions of dihydroxy metabolites on the HPLC system, which previously was calibrated by the authentic compounds.

bated at 37° for 10 min at 120 oscillations per minute. The enzyme activities are determined at initial reaction velocity. The reaction is stopped by addition of 10 ml of dichloromethane, and 3000 cpm of each radioactive 24,25-$(OH)_2D_3$, 25,26-$(OH)_2D_3$, and 1,25-$(OH)_2D_3$ (78 Ci/mmol) (see Fig. 1 legend) is added to the flask. After 1 hr of incubation at room temperature, the dichloromethane phase is separated by Whatman 1ps phase-separating filter paper and dried under nitrogen. The lipid extract is dissolved in chloroform/hexane (65 : 35) and applied to a Sephadex LH-20 column (7 mm × 15 cm) packed and eluted with the same solvent system. The last 20 ml of the total 32-ml effluent is dried in vacuo and redissolved in hexane/chloroform/methanol (9 : 1 : 1). The sample is run through another small-batch column of Sephadex LH-20 (7 mm × 14 cm) packed and eluted with hexane/chloroform/methanol (9 : 1 : 1). The last 18 ml of a total 27-ml effluent is dried, dissolved in 9% isopropanol in hexane, and subjected to HPLC as described in Fig. 10. Fractions of 0.5 ml are collected and counted for the calculation of recovery of the compounds. Peak areas (254 nm) are measured, and total amounts of metabolites produced are calculated by their standard curves (one of them is shown in Fig. 11) and the recovery values. Thus all three 25-OH-D$_3$ hydroxylases (24-, 26-, and 1-hydroxylases) can be determined using one incubation. Since the amount of metabolites produced *in vitro* sometimes requires high sensitivity (0.005 AFUS) of the UV detector, clear-cut base line separation of metabolites from other UV-absorbing substances is necessary. The two-batch Sephadex columns are necessary for a low base line. One investigator claims that the direct use of a preparative microparticulate silica column (2.2 × 6.2 cm) prior to analytical chromatography on HPLC is advantageous.[35] The present authors strongly believe that a batch Sephadex column prepurification of lipid extract is advantageous because the amount of lipid injected into HPLC gradually inactivates the silica particles of column and eventually causes insufficient and irregular resolution of compounds and shortens the life of the expensive columns. Moreover, in order to purify the lipid extracts, all samples must be run through the same preparative HPLC one by one whereas 10–20 of the batch columns can be run simultaneously. Extracts of biological samples purified by the two Sephadex columns are free of lipid contaminants so that samples for the assay can be injected into HPLC equipped with a Zorbax SIL analytical column continuously without washing the column between samples. Thus the use of HPLC for vitamin D hydroxy enzyme assay offers tremendous advantages such as accurate, reproducible separation of metabolites, providing identity of the metabolites, saving solvent, time, radioactive compounds, and counting solutions.

[35] G. Jones, *Clin. Chem.* **24**, 287 (1978).

# [45] Mass Fragmentographic Assay of 25-Hydroxyvitamin D$_3$

*By* INGEMAR BJÖRKHEM and INGER HOLMBERG

The most serious difficulty in the assay of small amounts of 25-hydroxyvitamin D$_3$ in biological fluids is related to the high instability of the compound. Another difficulty in most previous assays of 25-hydroxyvitamin D$_3$ is the lack of specificity. Unless a chromatographic step is included, methods based on competitive protein binding do not discriminate between 25-hydroxyvitamin D$_2$ and 25-hydroxyvitamin D$_3$. Even if a chromatographic step is included in such an assay, it is possible that different endogenous compounds might interfere in the assay. Suc-

METHODS IN ENZYMOLOGY, VOL. 67

cessful recovery experiments with exogenous 25-hydroxy vitamin $D_3$ do not exclude the presence of such interference.

The problems with respect to instability and lack of specificity can be overcome with use of mass fragmentography and deuterium-labeled 25-hydroxyvitamin $D_3$ as internal standard. The internal standard is added to the biological fluid prior to purification and analysis, and it can be assumed that this standard is degraded or decomposed to the same extent as unlabeled 25-hydroxyvitamin $D_3$. In the last step, the amount of unlabeled 25-hydroxyvitamin $D_3$ is determined by combined gas chromatography–mass spectrometry with a multiple ion detector (MID). With the latter detector, a specific ion corresponding to the labeled and unlabeled derivative of 25-hydroxyvitamin $D_3$, respectively, can be followed through a gas chromatogram, and the ratio between the two tracings can be used for calculation of unlabeled 25-hydroxy vitamin $D_3$. This type of assay has been used for determination of 25-hydroxyvitamin $D_3$ in serum[1] and for determination of the rate of 25-hydroxylation of vitamin D in the mitochondrial fraction of rat liver.[2] The type of derivatization used in the latter assay[2] was found to give the most specific assay also in the determination of 25-hydroxyvitamin $D_3$ in serum, and only this latter modification will be reviewed here.

Materials

25-Hydroxy[26-$^2$H$_3$]vitamin $D_3$ can be prepared as summarized in Fig. 1.

3$\beta$-Acetoxy-27-norcholest-5-en-25-one (I) can be synthesized from cholesterol according to Ruzicka and Fischer[3] or obtained commercially from Steraloids (Pawling, New York). 3$\beta$-Acetoxy-27-norcholesta-5,7-dien-25-one (II) is prepared from (I) using the method for introduction of a $\Delta^7$-double bond described by Bernstein et al.[4] A mixture of 1 g and 450 mg of N-bromosuccinimide dissolved in hexane, is refluxed for 4 min under illumination by a photospot lamp (Philips, 17/00, 1000 W). After addition of s-collidine, 0.4 ml, the mixture is rapidly cooled and filtered with suction. The solvent is evaporated from the filtrate in vacuo. The residue is dissolved in a mixture of xylene, 25 ml, and s-collidine, 1 ml, and refluxed in a nitrogen atmosphere for 15 min. The mixture is cooled and filtered, and the solvent is removed from the filtrate in vacuo. The yellow oil ob-

[1] I. Björkhem and I. Holmberg, Clin. Chim. Acta 68, 215 (1976).
[2] I. Björkhem and I. Holmberg, J. Biol. Chem. in press (1978).
[3] L. Ruzicka and W. H. Fischer, Helv. Chim. Acta 20, 129 (1937).
[4] S. Bernstein, L. J. Binovi, L. Dorfman, K. J. Sax, and Y. SubbaRow, J. Org. Chem. 14, 433 (1949).

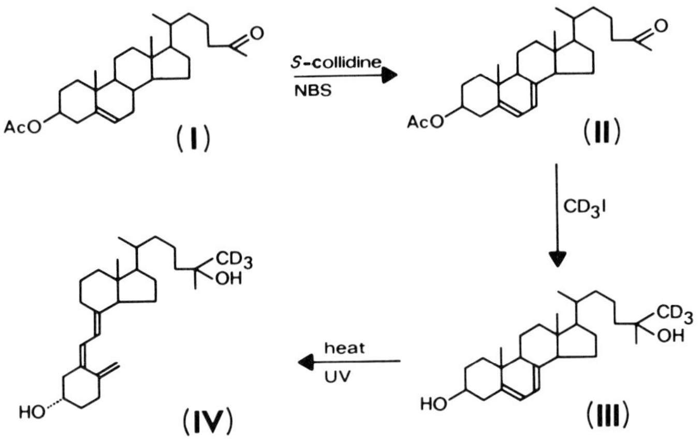

FIG. 1. Synthesis of 25-hydroxy[26-$^2$H$_3$]vitamin D$_3$. The systematic names of the different intermediates in the synthesis are given in the text.

tained is then chromatographed on a column of 50 g of silicic acid containing 0.1 g of silver nitrate per gram of silicic acid. Benzene elutes unchanged (I), and benzene ethyl acetate (9 : 1, v/v) elutes almost pure (II). Crystallization from aqueous acetone yields 100–150 mg of 3$\beta$-acetoxy-27-norcholesta-5,7-dien-25-one with m.p. 132–134°. The identity of the material is further confirmed by finding a $\lambda_{max}$ at 282 nm.

The unchanged (I) recovered in the silicic acid chromatography can be used for preparation of more (II), and thus a total yield of about 20% can easily be obtained in the conversion of (I) to (II).

[26-$^2$H$_3$]Cholesta-5,7-diene-3$\beta$,25-diol (III) is obtained from (II) by Grignard synthesis (cf. Ryer et al.[5]). (II) 200 mg, is refluxed for 3 hr in benzene, 3.5 ml, together with 2.5 ml of a diethyl ether solution containing 103 mg of magnesium and 307 $\mu$l of C$^2$H$_3$I (Radiochemical Centre, Amersham, England). After cooling to 5°, the complex is decomposed by the slow addition of 0.9 ml of ice water and 1.8 ml of 50% aqueous acetic acid solution. After extraction with diethyl ether, the material is purified by preparative thin-layer chromatography using chromatoplates impregnated with silver nitrate (0.04 silver nitrate per gram of silica gel). Benzene/ethyl acetate (3 : 1, v/v) is used as solvent. The yield of (III) is about 30 mg.

25-Hydroxy[26-$^2$H$_3$]vitamin D$_3$ (IV) is obtained from (III) by irradiation with UV light (cf. Ryer et al.[5]). A solution of (III), 90 mg, in diethyl ether, 10 ml, is illuminated at 0° with a UV lamp (Osram, Ultralux, 300 W) under nitrogen for 15 min. The solvent is then removed in vacuo at a temperature below 10°. The residue is immediately dissolved in a mixture

[5] A. I. Ryer, W. H. Gebert, and N. M. Murrill, J. Am. Chem. Soc. 72, 4247 (1950).

of benzene, 10 ml, and methanol, 1 ml, and refluxed for 2 hr. Pure 25-hydroxy[26-$^2$H$_3$]vitamin D$_3$, 3–4 mg, is isolated by thin-layer chromatography using ethyl acetate/chloroform (1 : 3, v/v). Care must be taken that the chromatoplates are developed immediately after the mixture has been applied. The purified 25-hydroxy[26-$^2$H$_3$]vitamin D$_3$, has the same gas chromatographic properties (as trimethylsilyl ether) and the same thin-layer chromatographic properties as authentic unlabeled 25-hydroxy-vitamin D$_3$. In the gas chromatographic analysis, two peaks are obtained in accordance with previous work.[6,7] The mass spectrum of *tert*-butyldimethylsilyl-trimethylsilyl derivative of 25-hydroxy[26-$^2$H$_3$]-vitamin D$_3$ is shown in Fig. 2. The isotopic composition of a typical preparation of 25-hydroxy[26-$^2$H$_3$]vitamin D$_3$ was calculated from the peaks at *m/e* 131–134[8] to be the following: 3% unlabeled molecules, 0% monodeuterated molecules, 5% dideuterated molecules, and 92% trideuterated molecules.[1]

### Reagents and Serum

Ammonium carbonate, chemical powder, (Riedel-De-Haen AG, Seelze-Hannover, West Germany)

*tert*-Butyldimethylchlorosilane imidazole (Applied Science Laboratories Inc., State College, Pennsylvania)

1-Nitroso-2-naphthol and Sudan black B (Merck, Darmstadt)

Sudan black B (Merck, Darmstadt) Solvents, all of analytical grade

Serum collected and frozen immediately. Tubes covered with aluminum foil should be used.

### Analytical Procedure for Determination of 25-Hydroxyvitamin D$_3$ in Serum

In the standard procedure for mass fragmentography, 145 ng of 25-hydroxy[26-$^2$H$_3$]vitamin D$_3$, dissolved in 50 $\mu$l of ethanol, are added to 2 ml of serum. The mixture is allowed to equilibrate for 1 hr at room temperature with continuous shaking in a nitrogen atmosphere. The mixture is protected from light by coating the tubes with aluminum foil. After equilibration, 3 ml of water, 2.5 g of ammonium carbonate, and 2 ml of ethyl acetate are added, and the mixture is shaken thoroughly. Separation of the phases can be achieved by centrifugation for 10–15 min. The organic phase is then collected and the extraction procedure is repeated three times using 2 ml of ethyl acetate in each extraction. Sometimes a

[6] D. Sklan, P. Budowski, and M. Katz, *Anal. Biochem.* **56,** 606 (1973).

[7] J. W. Blunt, H. F. De Luca, and H. K. Schnoes, *Biochemistry* **7,** 3317 (1968).

[8] K. Biemann, "Mass Spectrometry," p. 223. McGraw-Hill, New York, 1962.

foamy layer is formed at the interface; this layer can be removed by gentle stirring with a Pasteur pipette in the upper layer. The combined organic phases are then washed with 1–2 ml of water. After evaporation of the solvent at room temperature under a stream of nitrogen, 50 $\mu$l of the *tert*-butyldimethylchlorosilane reagent is added and the mixture is allowed to stand for 15 min at room temperature. The 3-*tert*-butyldimethylsilyl derivative of 25-hydroxyvitamin $D_3$ is then extracted with hexane. After evaporation of the hexane under nitrogen, the extract is subjected to thin-layer chromatography with toluene/ethyl acetate 3 : 1 (v/v) as solvent. The application of the sample as well as development of the chromato-plates must be performed in a nitrogen atmosphere. The chromatographic zone containing the 3-*tert*-butyldimethylsilyl derivative is detected with the use of 1-nitroso-2-naphthol and Sudan black B as external standards. These compounds have $R_f$ values of 0.68 and 0.40, respectively, under the conditions used. The 3-*tert*-butyldimethylsilyl derivative of 25-hydroxyvitamin $D_3$ has an $R_f$ value of 0.59 and is located in a zone be-tween the 1-nitroso-2-naphthol spot and the Sudan black B spot. This zone should be scraped off immediately after the development of the chromatography and eluted with ethyl acetate. After evaporation of the ethyl acetate under a stream of nitrogen (at room temperature) the 3-*tert*-butyldimethylsilyl derivative of 25-hydroxyvitamin $D_3$ can be con-verted into the 25-trimethylsilyl ether by addition of 400 $\mu$l of a solution containing pyridine/hexamethyldisilazane/trimethylchlorosilane (6 : 4 : 3, v/v/v). The mixture is allowed to stand at room temperature for 1 hr. The solvent and reagents are then removed with a stream of nitrogen at room temperature. Hexane is added and the mixture is treated with sonication for 5 sec. After centrifugation and transfer of the supernatant into a new tube, the solvent is removed under a stream of nitrogen at room tempera-ture, and the residue is dissolved in 20–50 $\mu$l of hexane.

This solution, 4–7 $\mu$l, can then be analyzed by combined gas chromatography–mass spectrometry. In our studies[1,2] we used either an LKB 2091 or an LKB 9000 instrument equipped with an MID unit. A 1.5% SE 30 column (usually 80–100 mesh, 2 mm × 2.5 m) was used. The carrier gas was helium, and the flow rate was 15–30 ml/min. The temperature of the column was 260–275°, and the temperature of the flash heater and ion source was about 40° above this. If the temperature of the column was raised to more than 280°, more than two peaks appeared in the gas chromatogram due to degradation on the column. The electron energy was set to 20 e V and the trap current to 60 $\mu$A. The electron multiplier sensitiv-ity was set to 150 to 200. The first channel of the multiple ion detector was focused on the ion at *m/e* 586 and the second at the ion at *m/e* 589. These ions correspond to the molecular peak in the mass spectrum of the deriva-

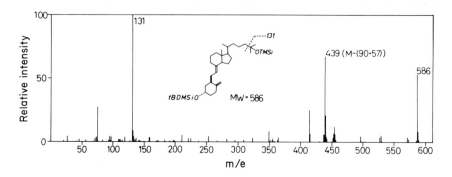

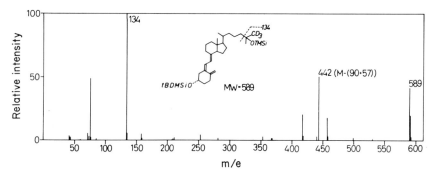

FIG. 2. Mass spectrum of 3-*tert*-butyldimethylsilyl-25-trimethylsilyl derivative of un-labeled (upper spectrum) and 25-hydroxy[26-²H₃]vitamin D₃ (lower spectrum).

tive of unlabeled and $^2H_3$-labeled 25-hydroxyvitamin $D_3$ (Fig. 2). The filter setting was 0.5 Hz for both channels. The amplification used for both channels was 300–900 times. The peak heights of the recordings were measured, as this was found to give more reproducible results than measurements of peak area under the conditions employed. In some experiments, and to check the specificity of the method, the two channels were focused on the ions at *m/e* 131 and at *m/e* 134.

In Fig. 3 a typical MID-recording of trimethylsilyl/*tert*-butyldimethylsilyl derivative of a purified extract of serum is shown. The standard, 25-hydroxy[26-²H₃]vitamin $D_3$, had been added to the serum prior to extraction and purification according to the above procedure. The two peaks obtained in the two tracings correspond to the pyro and isopyro forms of the derivative of 25-hydroxyvitamin $D_3$. Only the first peak, corresponding to the pyro form, should be used in the assay. The amount of unlabeled 25-hydroxyvitamin $D_3$ can be calculated from the ratio between the height of the peak recorded at *m/e* 586 and the height of the peak at *m/e* 589 with use of a standard curve. Figure 4 shows a typical standard

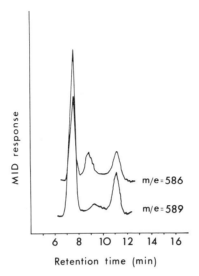

FIG. 3. Multiple ion detector (MID) recordings at *m/e* 586 and *m/e* 589 of *tert*-butyldimethylsilyl/trimethylsilyl derivative of an extract of serum to which 25-hydroxy[26-$^2H_3$]vitamin $D_3$ had been added (145 ng).

curve obtained by analysis of different standard mixtures of unlabeled 25-hydroxyvitamin $D_3$ together with 145 ng of 25-hydroxy[26-$^2H_3$]vitamin $D_3$. The relative standard deviation of the technique was about 3% as calculated from duplicate determinations of serum. The mean values for

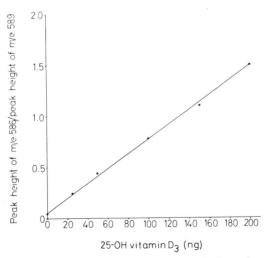

FIG. 4. Standard curve for determination of 25-hydroxyvitamin $D_3$.

concentration of 25-hydroxyvitamin $D_3$ in serum from 12 healthy men and women was 21 ng/ml. The accuracy of the technique was tested by addition of known amounts (25–100 ng) of 25-hydroxyvitamin $D_3$ to a specific serum. The maximum error in these experiments was 4%.

### Analytical Procedure for Determination of Rate of 25-Hydroxylation of Vitamin $D_3$ in Rat Liver Mitochondria

Male rats of the Sprague–Dawley strain weighing about 200 g have been used in our experiments. Liver homogenates, 10% (w/v), are prepared in 0.25 $M$ sucrose. The mitochondrial fraction is isolated by centrifugation of a 800 $g$ supernatant fluid of the liver homogenate at 6500 $g$ for 20 min. The mitochondrial pellet is then washed by resuspension and recentrifugation twice in the same medium. The mitochondrial pellet is finally suspended to the initial volume in phosphate buffer, 0.2 $M$, pH 7.4. The protein content of this mitochondrial preparation should be 1.5–3 mg/ml. In the standard incubation procedure 200 $\mu$g of vitamin $D_3$ dissolved in 50 $\mu$l of 95% ethanol (w/v) are incubated with 1.5 ml of the mitochondrial fraction and 10 $\mu$mol of NADPH in a total volume of 10 ml of 0.2 $M$ potassium phosphate buffer, pH 7.4.

Incubations are performed for 45 min at 37° and are terminated by the addition of 5 ml of toluene together with 145 ng of 25-hydroxy[26-$^2H_3$]vitamin $D_3$. Control incubations should be performed with the same mitochondrial fraction boiled for 10 min. Under the incubation conditions described, the conversion of vitamin $D_3$ into 25-hydroxyvitamin $D_3$ should be linear with time and concentration of mitochondrial protein, and the mitochondrial enzyme should be saturated with respect to substrate and cofactor.[2] The conversion (after correction for blank values) should be of the magnitude 75–300 pmol. In the extraction procedure, the above toluene-containing incubation mixture is shaken thoroughly together with 5 g of ammonium carbonate. The further procedure is exactly the same as that described under the heading Analytical Procedure for Determination of 25-Hydroxyvitamin $D_3$ in Serum.

### Comments

The specificity of the present assay is based on the fact that, to influence the results, a contaminating compound must have the same retention time in gas–liquid chromatography and contain the same ions in its mass spectrum as the derivative of 25-hydroxyvitamin $D_3$. The very specific derivative used in the assay will further increase the specificity. The reagent *tert*-butyldimethylsilylimidazole is sterically hindered for attack in the 25-position, but will attack most other positions in steroids that otherwise eventually may interfere in the assay. In spite of the high specificity in the last step, it was found to be necessary to purify the extract prior to mass fragmentography. Purification by thin-layer chromatography was found to give somewhat better results than purification by Sephadex LH-20 (cf. Björkhem and Holmberg[1]), although the recovery may be better in the latter procedure. The thin-layer chromatography can be performed either with underivatized 25-hydroxyvitamin $D_3$[2] or with the 3-*tert*-butyldimethylsilyl derivative of 25-hydroxyvitamin $D_3$. Preliminary experiments indicate that the latter procedure should be preferred.

The assay described above is designed to determine 25-hydroxyvitamin $D_3$ down to a level of a few nanograms. The sensitivity of the method is limited not by the sensitivity of the instrument, but by the presence of a compound in biological samples with a retention time very similar to that of the derivative of 25-hydroxyvitamin $D_3$ in gas–liquid chromatography (cf. Fig. 3). In a few cases the peak corresponding to this compound was so dominant that, even with the present assay, determination of 25-hydroxyvitamin $D_3$ became difficult or even impossible. If a more purified extract is used, for example after purification by high-performance liquid chromatography, the present assay may easily be modified to determine 25-hydroxyvitamin $D_3$ in a subnanogram range. A smaller amount of 25-hydroxy[26-$^2H_3$]vitamin $D_3$ should be used in such an assay. On the other hand, if a very small amount of 25-hydroxy[26-$^2H_3$]vitamin $D_3$ is used in the assay, the risk for degradation or decomposition during the procedure will increase.

An alternative may be to use an internal standard with a mass more than three units higher than the mass of unlabeled 25-hydroxyvitamin $D_3$. With such internal standard, the ratio between unlabeled 25-hydroxyvitamin $D_3$ and standard can be kept considerably lower than is the case in the present assay. A suitable internal standard with higher mass may be synthesized by substituting $^{13}C^2H_3I$ in the synthesis shown in Fig. 1 for $^{12}C^2H_3I$. The $^{13}C^2H_3I$ is now commercially available.

# [46] Determination of Vitamin D and Its Metabolites in Plasma[1]

*By* RICHARD M. SHEPARD and HECTOR F. DELUCA

In recent years it has been found that vitamin D must be metabolized before its biological activity can be expressed.[2] Vitamin D is hydroxylated

---

[1] Supported by Grant No. AM-14881 from the National Institutes of Health and the Harry Steenbock Research Fund.
[2] H. F. DeLuca and H. K. Schnoes, *Annu. Rev. Biochem.* **45**, 631 (1976).

to 25-hydroxyvitamin D (25-OH-D) in the liver and then further hydroxylated in the kidney to either 1,25-dihydroxyvitamin D [1,25-(OH)$_2$D] or 24$R$, 25-dihydroxyvitamin D (24$R$,25-(OH)$_2$D). 1,25-(OH)$_2$D is now recognized as the active form of vitamin D in bone mineral mobilization and is exclusively responsible for the initiation of active intestinal absorption of calcium and phosphorus.[2] In contrast, a role for 24$R$,25-(OH)$_2$D or 25$R$,26-dihydroxyvitamin D (25$R$,26-(OH)$_2$D, another known metabolite of 25-OH-D) has yet to be established. The 25-hydroxylation reaction of vitamin D is partially feedback regulated, while the metabolism of 25-OH-D to 1,25-(OH)$_2$D or 24,25-(OH)$_2$D is strictly modulated directly or indirectly by serum calcium, serum phosphorus, and parathyroid hormone.[2] Further, the pathogenesis of several metabolic bone diseases is attended by disturbances in the vitamin D metabolic system.[2] In view of the above, a need developed for a multiple assay capable of measuring vitamin D, 25-OH-D, 24,25-(OH)$_2$D, 25,26-(OH)$_2$D and 1,25-(OH)$_2$D in a single 3–5 ml sample of plasma for studies of vitamin D metabolism and its regulation in man, in laboratory research animals, and in the clinical diagnosis of metabolic bone disease states.

The methodology outlined below describes a multiple assay[3] capable of quantitating vitamin D and its metabolites in a single small sample of plasma by means of high-pressure liquid chromatography (HPLC) with an ultraviolet absorbance detector or competitive protein-binding detection. The extraction of vitamin D metabolites from plasma, the purification and separation of the metabolites by conventional column chromatography and HPLC, and their detection by optical absorbance or competitive protein binding are discussed. This method has been put to routine use in our laboratory to measure metabolite levels in plasma samples from humans, rats, chicks, and cows.

### Apparatus and Materials

All HPLC was carried out with a Model LC 204 chromatograph fitted with a Model 6000A pumping system, U6K injection valve, and a Model 440 ultraviolet fixed-wavelength (254 nm) detector (all from Waters Associates, Milford, Massachusetts). A Model 24 spectrophotometer (Beckman Instruments, Inc., Irvine, California) was used to measure concentrations of vitamin D compounds in solution ($\epsilon = 18,200$ at 265 nm). Scintillation counting was performed at room temperature with a Model LS-100C liquid scintillation system (Beckman Instruments, Inc., Fullerton, California) fitted with external standardization.

---

[3] R. M. Shepard, R. L. Horst, A. J. Hamstra, and H. F. DeLuca, *Biochem. J.* **182**, 55 (1979).

*Solvents.* All solvents used for extractions and conventional column chromatography were Fisher Certified ACS grade (Fisher Scientific Co.) and distilled once. All solvents used for HPLC were Fisher HPLC grade.

*Chromatography.* Sephadex LH-20 was purchased from Pharmacia Inc., Piscataway, New Jersey. Lipidex 5000 (similar to hydroxyalkoxypropyl Sephadex, or HAPS) was obtained from Packard Instruments Co., Inc., Downers Grove, Illinois. Stainless steel columns (25 cm × 4.6 mm i.d.), prepacked with microparticulate Zorbax-SIL or Zorbax-ODS, were supplied by DuPont Instruments, Wilmington, Delaware.

*Vitamin D Metabolites.* Crystalline vitamin $D_2$, vitamin $D_3$, and 25-OH-$D_3$ were obtained from Philips-Duphar, Amsterdam, The Netherlands. Crystalline 25-OH-$D_2$ was a gift from the Upjohn Company, Kalamazoo, Michigan. Crystalline 24,25-$(OH)_2D_3$ and 1,25-$(OH)_2D_3$ were gifts from the Hoffmann-La Roche Company, Nutley, New Jersey. 25,26-$(OH)_2D_3$ was synthesized in this laboratory.[4] [$3\alpha$-$^3$H]vitamin $D_2$ (1.9 Ci/mmol) and [$3\alpha$-$^3$H]vitamin $D_3$ (15 Ci/mmol) were synthesized in this laboratory[5] and purified on a 1 × 60 cm hexane : chloroform (90 : 10) Lipidex 5000 column. 25-OH-[$3\alpha$-$^3$H]$D_2$ (1.9 Ci/mmol) was generated biologically from [$3\alpha$-$^3$H]vitamin $D_2$[6] and purified on a 1 × 60 cm chloroform : hexane (50 : 50) Sephadex LH-20 column and a 1 × 60 cm hexane : chloroform (90 : 10) Lipidex 5000 column. 25-OH-[26,27-$^3$H]$D_3$ (80 Ci/mmol) was synthesized in this laboratory[7] and purified on a 1 × 60 cm hexane : chloroform (90 : 10) Lipidex 5000 column. 24,25-$(OH)_2$[26,27-$^3$H]$D_3$ (80 Ci/mmol) was generated biologically from 25-OH-[26,27-$^3$H]$D_3$,[8] and 25,26-$(OH)_2$[23,24-$^3$H]$D_3$ (78 Ci/mmol) was generated biologically from 25-OH-[23,24-$^3$H]$D_3$[9] (78 Ci/mmol; also synthesized in this laboratory[10]). Both of these compounds were purified on a 2 × 40 cm hexane : chloroform : methanol (90 : 10 : 10) Sephadex LH-20 column and then a 1 × 60 cm chloroform : hexane (65 : 35) Sephadex LH-20 column. 1,25-$(OH)_2$[26,27-$^3$H]$D_3$ (80 Ci/mmol) was generated biologically from 25-OH-[26,27-$^3$H]$D_3$[8] and purified on a 1 × 30 cm chloroform : hexane (65 : 35) Sephadex LH-20 column and then a 2 × 40 cm hexane : chloroform : methanol (90 : 10 : 10) Sephadex LH-20 column. Radioactive metabolites that served as internal standards in the assay procedure were further purified on an HPLC Zorbax-SIL column using the

[4] H.-Y. Lam, H. K. Schnoes, and H. F. DeLuca, *Steroids* **25**, 247 (1975).

[5] S. Yamada, H. K. Schnoes, and H. F. DeLuca, unpublished results.

[6] M. H. Bhattacharyya and H. F. DeLuca, *Arch. Biochem. Biophys.* **160**, 58 (1974).

[7] J. L. Napoli, M. A. Fivizzani, and H. F. DeLuca, unpublished results.

[8] Y. Tanaka, R. S. Lorenc, and H. F. DeLuca, *Arch. Biochem. Biophys.* **171**, 521 (1975).

[9] Y. Tanaka, R. M. Shepard, H. F. DeLuca, and H. K. Schnoes, *Biochem. Biophys. Res. Commun.* **83**, 7 (1978).

[10] S. Yamada, H. K. Schnoes, and H. F. DeLuca, *Anal. Biochem.* **85**, 34 (1978).

following solvent systems: [$^3$H]vitamin $D_3$—isopropanol: hexane (1:99); 25-OH-[$^3$H]$D_3$—isopropanol: hexane (4:96); 24,25-(OH)$_2$[$^3$H]$D_3$, 25,26-(OH)$_2$[$^3$H]$D_3$, and 1,25-(OH)$_2$[$^3$H]$D_3$—isopropanol: hexane (10:90).

*Animals.* Male, weanling rats, purchased from the Holtzman Co., Madison, Wisconsin, were fed an adequate calcium, adequate phosphorus, vitamin D-deficient diet for 3 weeks[11] and served as a source of plasma binding protein. One-day-old White Leghorn chickens, obtained from Northern Hatcheries, Beaver Dam, Wisconsin, were fed an adequate calcium, vitamin D-deficient diet for 4 weeks[12] and served as a source of intestinal cytosol binding protein.

Dextran (No. D-4751-clinical grade) and neutralized activated charcoal were purchased from Sigma Chemical Co., St. Louis, Missouri. The charcoal fine particles were removed by suspending the charcoal in 20 volumes of water, stirring well, allowing the charcoal to settle for 10 min, and aspirating off the supernatant. This procedure was repeated four more times, and then the charcoal was filtered on a Büchner funnel and dried in an oven at 100°. $^3$H in fractions from column samples were counted in toluene scintillation solution (2 g of PPO and 100 mg of dimethyl-POPOP per liter of toluene), while that in binding assays was counted in aqueous scintillation solution (5.5 g of PPO and 70 mg of dimethyl-POPOP per liter of 33% Triton X-100 in toluene).

## Procedures

*Extraction of Plasma Samples.* All glassware used at this step and throughout the entire assay procedure must be cleaned by rinsing with methanol and chloroform. To 3–5 ml of plasma are added (as indicated in Fig. 1) the following radioactive internal standards to monitor the analytical recoveries of the assay; 3000 cpm of [$^3$H]vitamin $D_3$, 3000 cpm of 25-OH-[$^3$H]$D_3$, 2000 cpm of 24,25-(OH)$_2$[$^3$H]$D_3$, 2000 cpm of 25,26-(OH)$_2$[$^3$H]$D_3$ and 3000 cpm of 1,25-(OH)$_2$[$^3$H]$D_3$, each in 25 $\mu$l of ethanol. Identical aliquots of each are also placed in counting vials in triplicate to determine recovery totals. The plasma samples are vortexed and incubated for 30 min to ensure thorough equilibration of exogenous and endogenous vitamin D metabolites. Using a modification of the method of Bligh and Dyer,[13] the lipids are extracted by adding 3.75 volumes of methanol: methylene chloride (2:1), shaking vigorously and venting, and

[11] T. Suda, H. F. DeLuca, and Y. Tanaka, *J. Nutr.* **100**, 1049 (1970).
[12] J. A. Eisman, A. J. Hamstra, B. E. Kream, and H. F. DeLuca, *Arch. Biochem. Biophys.* **176**, 235 (1976).
[13] E. G. Bligh and W. J. Dyer, *Can. J. Biochem. Physiol.* **37**, 911 (1959).

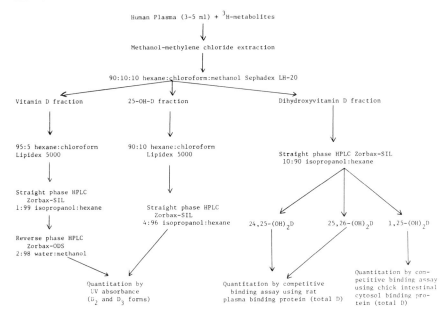

Fig. 1. Outline of the multiple assay procedure for the analysis of vitamins $D_2$ and $D_3$, 25-OH-$D_2$ and 25-OH-$D_3$, 24,25-(OH)$_2$D, 25,26-(OH)$_2$D, and 1,25-(OH)$_2$D in human plasma.

then shaking for 5 min on a horizontal shaker at 3 oscillations per second. The samples are allowed to stand for 15 min, then the phases are separated by adding 1.25 volumes of methylene chloride and shaking them for 1 min, followed by centrifugation at 1500 $g$ for 10 min. The lower methylene chloride layer is collected, and the upper aqueous layer is reextracted with another 1.25 volumes of methylene chloride. The combined methylene chloride layers are evaporated under reduced pressure on a rotary evaporator, adding a small amount of absolute ethanol to clear the solution, and the yellow lipid residue is transferred in several rinses with chloroform to test tubes. The chloroform is evaporated under a stream of nitrogen in a warm water bath, and the residue is solubilized in 0.5 ml of 90 : 10 : 10 hexane : chloroform : methanol. This extraction procedure reliably results in recoveries of 90% or better of ³H-labeled vitamin D or metabolites added to plasma.

*Chromatography of Lipid Extracts.* Before analyses can be done, chromatography of the lipid extracts is necessary to separate the vitamin D metabolites and purify them. Batch elution columns of Sephadex LH-20[14] or Lipidex 5000[15] have been quite useful in providing cleaner samples

[14] M. F. Holick and H. F. DeLuca, *J. Lipid Res.* **12**, 460 (1971).
[15] J. Ellingboe, E. Nyström, and J. Sjövall, *J. Lipid Res.* **11**, 266 (1970).

for HPLC with good recoveries and are quite reproducible. As indicated in Fig. 1, the first step of purification of the lipid extracts is a 90:10:10 hexane:chloroform:methanol Sephadex LH-20 column, which functions to separate the vitamin D metabolites into three fractions of similar polarity ranges—the vitamin D fraction, the 25-OH-D fraction, and the dihydroxyvitamin D fraction—and to remove the bulk of the lipid contaminant from the latter two fractions.

Sephadex LH-20 is thoroughly equilibrated in 90:10:10 hexane:chloroform:methanol (swelled in solvent at least overnight with at least five changes of solvent), and 0.7 × 12 cm columns are poured in the same solvent. The plasma lipid extract is applied in 0.5 ml of solvent, followed by two rinses of 0.5 ml that are used to wash the sample into the column bed. Then, 3.5 ml of solvent is added, and the first 5 ml is collected for the vitamin D fraction. An additional 5.5 ml are added to the column, and the 5.0–10.5 ml region is collected for the 25-OH-D fraction. Finally, 16.5 ml are added, and the 10.5–27.0 ml region is collected for the three dihydroxyvitamin D metabolites. Each of the three column fractions

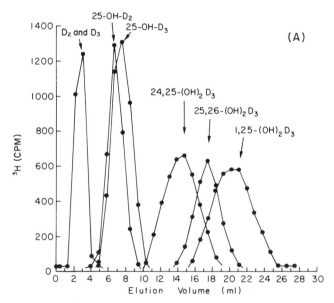

FIG. 2. Purification of vitamin D metabolites by conventional column chromatography. (A) Elution of vitamin D and its major metabolites from a 0.7 × 12 cm Sephadex LH-20 column developed with a solvent system of 90:10:10 hexane:chloroform:methanol. (B) Elution of vitamin $D_2$ and vitamin $D_3$ from a 0.7 × 18 cm Lipidex 5000 column developed with a solvent system of 95:5 hexane:chloroform. (C) Elution of 25-OH-$D_2$ and 25-OH-$D_3$ from a 0.7 × 15 cm Lipidex 5000 column developed with a solvent system of 90:10 hexane:chloroform.

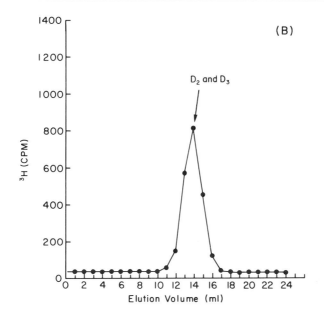

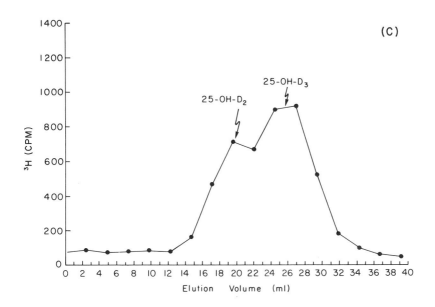

is evaporated under a stream of nitrogen and stored in ethanol under $N_2$ at $-20°$.

A typical chromatogram illustrating the profile of the various vitamin D standards is shown in Fig. 2A. The chromatographic positions of vitamin D and its metabolites are not altered by the presence of plasma lipids, nor is separation of the vitamin $D_2$ and vitamin $D_3$ forms of the various metabolites observed with this column.

*Analysis of Vitamin $D_2$ and Vitamin $D_3$.* Since vitamin $D_2$ and vitamin $D_3$ comigrate with the bulk of the lipid mass on the $90:10:10$ hexane : chloroform : methanol Sephadex LH-20 column, further purification is accomplished by batch elution on a $0.7 \times 18$ cm $95:5$ hexane : chloroform Lipidex 5000 column. The vitamin D fraction is applied to the column in 0.5 ml of $95:5$ hexane : chloroform, followed by two 0.5-ml rinses to wash the sample into the column bed. With the initial 1.5 ml discarded, 9.0 ml of solvent is added and the 1.5–10.5 ml region is also discarded. An additional 7.0 ml is added and the 10.5–17.5-ml region is collected, concentrated under a stream of nitrogen, and transferred to a 3-ml conical tube with two 1-ml rinses of chloroform. This is evaporated under nitrogen, and the residue is redissolved in 50 $\mu$l of $1:99$ isopropanol : hexane. As shown in Fig. 2B, no separation of [³H]vitamin $D_2$ and [³H]vitamin $D_3$ is observed.

Because of the small amounts of vitamin D present in normal plasma samples, further purification is necessary before analysis and is accomplished by HPLC on a Zorbax-SIL column equilibrated in $1:99$ isopropanol : hexane. At a constant flow rate of 2.0 ml/min (900 psi), the sample is injected in 50 $\mu$l of solvent followed by a 50-$\mu$l rinse. The elution region corresponding to vitamin $D_2$ and vitamin $D_3$ (8.3–10.8 min, peak at 9.3 min) is collected and the eluent is concentrated under a stream of nitrogen and transferred to a 3-ml conical tube with chloroform rinsings. This is evaporated under nitrogen and redissolved in 50 $\mu$l of $98:2$ methanol : water. The collection region is previously determined by using a crystalline vitamin $D_3$ standard. As shown in Fig. 3, vitamin $D_2$ and vitamin $D_3$ migrate together in this system.

Final quantitation of plasma vitamin $D_2$ and vitamin $D_3$ is accomplished by reversed-phase HPLC on a Zorbax-ODS column equilibrated in $98:2$ methanol : water at a constant flow rate of 1.5 ml/min (1200 psi). A solution of known concentration (as determined by ultraviolet spectrophotometry)—5 ng vitamin $D_3$ plus 250 cpm [³H]vitamin $D_3$ per 5 $\mu$l—is prepared in $98:2$ methanol : water. Amounts of 5 ng, 10 ng, 20 ng, and 40 ng are injected, and the [³H]vitamin $D_3$ elution region (11.2–13.7 min, peak at 12.2 min) is collected in counting vials, evaporated, and counted in toluene counting solution. Equivalent aliquots of the standard

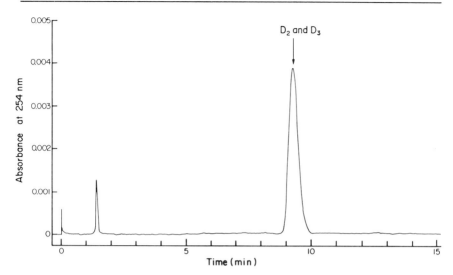

FIG. 3. HPLC with detection by absorbance at 254 nm of vitamins $D_2$ and $D_3$ on a 0.46 × 25 cm Zorbax-SIL silicic acid column developed with a solvent system of 1:99 isopropanol:hexane at a flow rate of 2.0 ml/min—coelution of vitamins $D_2$ and $D_3$ standards.

solution are also counted to determine the percent recovery of each amount. The peak heights at 0.002 AUFS or 0.005 AUFS are measured and divided by the percentage recovery to yield a standard curve relating corrected peak height to nanograms of vitamin $D_3$. A standard of vitamin $D_2$ is injected to determine its elution position (peak at 11.2 min). The plasma samples are injected in 50 $\mu$l of solvent with a 50-$\mu$l rinse; the vitamin $D_3$ region is collected, evaporated, and counted in toluene counting solution along with the initial aliquot of [³H]vitamin $D_3$ to determine the percentage recovery through the overall procedure. The recoveries of vitamins $D_2$ and $D_3$ are assumed to be the same. The peak heights of vitamin $D_2$ and vitamin $D_3$ in the sample are measured at 0.002 AUFS or 0.005 AUFS and divided by the percentage recovery to yield corrected peak heights. The sample peak heights are related to the standard curve to arrive at the total amount of vitamin $D_2$ and vitamin $D_3$ in the original plasma sample. Dividing by the sample volume yields the concentration in nanograms per milliliter.

The 98:2 methanol:water Zorbax-ODS system, as shown in Fig. 4, nearly resolves vitamin $D_2$ and vitamin $D_3$. Peak height in absorbance units of the standards of vitamins $D_2$ and $D_3$, corrected for recovery losses, bear linear, virtually identical, relationships with the amount applied in the range of 0–50 ng (Fig. 5). A typical HPLC profile of a normal human plasma sample (Fig. 6) shows the low levels of vitamin $D_2$ and

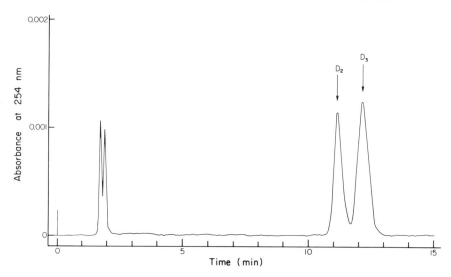

FIG. 4. HPLC with detection by absorbance at 254 nm of vitamins $D_2$ and $D_3$ standards on a 0.46 × 25 cm reversed-phase Zorbax-ODS column developed with a solvent system of 2:98 water:methanol at a flow rate of 1.5 ml/min.

vitamin $D_3$ and the effective removal of all interfering 254 nm absorbing compounds from the vitamins $D_2$ and $D_3$ regions of the Zorbax-ODS chromatogram. The overall recovery of [³H]vitamin $D_3$ added to the plasma sample, after extraction, purification, and analysis on Zorbax-ODS, averages 50.1 ± 7.2% ($n$ = 25). Based on this recovery and assuming a 5-ml plasma sample, the routine limit of detection for vitamins $D_2$ and $D_3$ in plasma is 0.5 ng/ml.

*Analysis of 25-OH-$D_2$ and 25-OH-$D_3$.* Further purification of the 25-OH-D fraction from the initial Sephadex LH-20 column is accomplished on a 0.7 × 15 cm, 90:10 hexane:chloroform Lipidex 5000 column. The 25-OH-D fraction is applied to the column in 0.5 ml of 90:10 hexane:chloroform, followed by two 0.5-ml rinses to wash the sample into the column bed. With the initial 1.5 ml discarded, 12.5 ml of solvent is added and the 1.5–14.0-ml region is also discarded. An additional 20.0 ml is added and the 14.0–34.0-ml region is collected for 25-OH-$D_2$ and 25-OH-$D_3$. The fraction is concentrated under a stream of nitrogen and transferred to a 3-ml conical tube with chloroform rinses, then evaporated under nitrogen; the residue is redissolved in 50 $\mu$l of 4:96 isopropanol:hexane. The chromatogram of 25-OH-[³H]$D_2$ and 25-OH-[³H]$D_3$ in Fig. 2C shows that they are collected together from this column.

Final quantitation of plasma 25-OH-$D_2$ and 25-OH-$D_3$ is accomplished by straight-phase HPLC on a Zorbax-SIL column equilibrated in 4:96

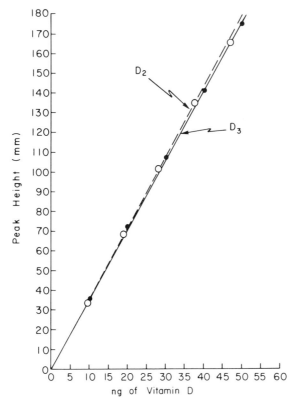

FIG. 5. Standard curves of vitamins $D_2$ and $D_3$, relating corrected peak height to nanograms of vitamin $D_2$ or vitamin $D_3$.

isopropanol : hexane at a constant flow rate of 2.0 ml/min (900 psi). A solution of known concentration (as determined by ultraviolet spectrophotometry)—25 ng 25-OH-$D_3$ plus 250 cpm 25-OH-[$^3$H]$D_3$ per 5 $\mu$l—is prepared in 4 : 96 isopropanol : hexane. Amounts of 25 ng, 50 ng, 100 ng, and 200 ng are injected, and the 25-OH-[$^3$H]$D_3$ elution region (10.4–12.9 min, peak at 11.4 min) is collected in counting vials, evaporated, and counted in toluene counting solution. Equivalent aliquots of the standard solution are also counted to determine the percentage recovery of each amount. The peak heights at 0.005 AUFS or 0.01 AUFS are measured and divided by the percentage recovery to yield a standard curve relating corrected peak height to nanograms of 25-OH-$D_3$. A standard of 25-OH-$D_2$ is injected to determine its elution position (peak at 9.4 min). The plasma samples are injected in 50 $\mu$l of solvent with a 50-$\mu$l rinse; the 25-OH-$D_3$ region is collected, evaporated, and counted in to-

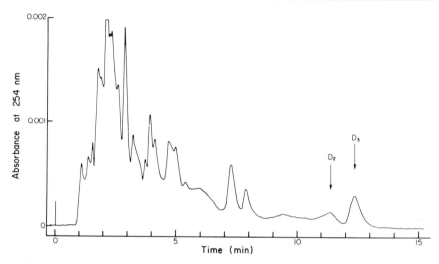

FIG. 6. HPLC with absorbance detection at 254 nm of vitamins $D_2$ and $D_3$ isolated from plasma of normal humans. Five milliliters of plasma were extracted, purified on Sephadex LH-20, Lipidex 5000, and Zorbax-SIL HPLC as described in Procedures, and analyzed using the chromatographic conditions described in Fig. 4. The ultraviolet profiles of normal human plasma extracts showed small absorbance peaks in the vitamin $D_2$ and vitamin $D_3$ regions.

luene counting solution along with the initial aliquot of 25-OH-[$^3$H]$D_3$ to determine the percentage recovery through the overall procedure. The recoveries of 25-OH-$D_2$ and 25-OH-$D_3$ are assumed to be the same. The peak height of 25-OH-$D_2$ and 25-OH-$D_3$ in the sample are measured at 0.005 AUFS or 0.01 AUFS and divided by the percentage recovery to yield corrected peak heights. The sample peak heights are related to the standard curve to arrive at the total amounts of 25-OH-$D_2$ and 25-OH-$D_3$ in the original plasma sample. To correct for different peak shapes, total amounts of 25-OH-$D_2$ determined from a 25-OH-$D_3$ standard curve are multiplied by 0.76. Dividing by the sample volume yields the concentration in nanograms per milliliter.

The 4 : 96 isopropanol : hexane Zorbax-SIL system, as shown in Fig. 7, completely resolves 25-OH-$D_2$ and 25-OH-$D_3$. Peak height in absorbance units of the standards of 25-OH-$D_3$, corrected for recovery losses, bears a linear relationship with the amount applied in the range of 0–200 ng (Fig. 8). Since 25-OH-$D_2$ is a slightly sharper peak than 25-OH-$D_3$ on the Zorbax-SIL column, the amount of 25-OH-$D_2$ represented by a given peak height is only 76% of the amount of 25-OH-$D_3$ represented (Fig. 8). Therefore, 25-OH-$D_2$ determinations are multiplied by 0.76 when 25-OH-$D_3$ is used as a standard. A typical Zorbax-SIL profile of a normal human

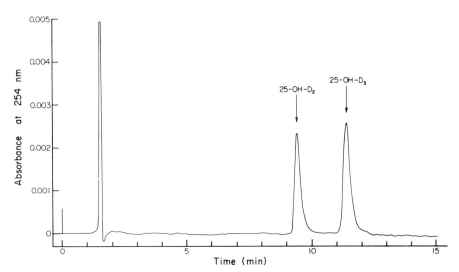

FIG. 7. HPLC with detection by absorbance at 254 nm of 25-OH-D$_2$ and 25-OH-D$_3$ standards on a 0.46 × 25 cm Zorbax-SIL silicic acid column developed with a solvent system of 4:96 isopropanol:hexane at a flow rate of 2.0 ml/min.

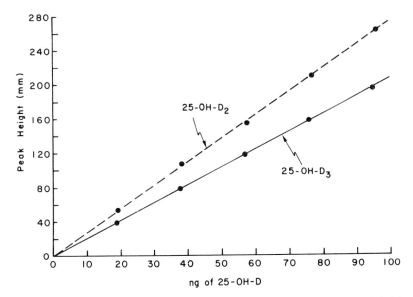

FIG. 8. Standard curves of 25-OH-D$_2$ and 25-OH-D$_3$, relating corrected peak height to nanograms of metabolite. The amount of 25-OH-D$_2$ represented by a given peak height was 76% of 25-OH-D$_3$.

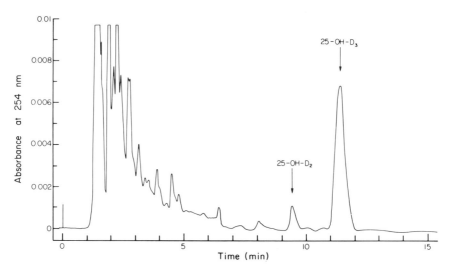

FIG. 9. HPLC with absorbance detection at 254 nm of 25-OH-D$_2$ and 25-OH-D$_3$ isolated from plasma of normal humans. Five milliliters of plasma were extracted, purified on Sephadex LH-20 and Lipidex 5000 as described in Procedures, and analyzed using the chromatographic conditions described in Fig. 7. The ultraviolet profile from normal human plasma showed a larger peak for 25-OH-D$_3$ than for 25-OH-D$_2$.

plasma sample (Fig. 9) shows the presence of 25-OH-D$_2$ and 25-OH-D$_3$ peaks and the effective removal of all interferring 254-nm absorbing compounds from this region. The intra-assay coefficient of variation for the assay of 25-OH-D by this method is 8% ($n = 5$), while the inter-assay coefficient of variation was 10% ($n = 8$). The overall recovery of 25-OH-[$^3$H]D$_3$ added to plasma samples, after extraction, purification, and analysis on the Zorbax-SIL, averaged 74.4 ± 5.0% ($n = 20$). Based on this recovery and assuming a 5-ml plasma sample, the routine limit of detection for 25-OH-D$_2$ and 25-OH-D$_3$ in plasma is 1 ng/ml.

*Separation of Dihydroxyvitamin D Metabolites.* The dihydroxyvitamin D metabolite fraction from the initial Sephadex LH-20 column—containing 24,25-(OH)$_2$D, 25,26-(OH)$_2$D, and 1,25-(OH)$_2$D—is further purified and separated by straight-phase HPLC on a Zorbax-SIL column equilibrated in 10 : 90 isopropanol : hexane at a flow rate of 2.0 ml/min (900 psi). Elution positions and collection regions are determined by injecting 10 μl of a standard ethanol solution containing 5 ng/μl 24,25-(OH)$_2$D$_3$, 5 ng/μl 25,26-(OH$_2$D$_3$, and 7.5 ng/μl 1,25-(OH)$_2$D$_3$. The plasma samples are transferred to 3-ml conical tubes with chloroform rinsings, evaporated under a stream of nitrogen, and redissolved in 50 μl of 10 : 90 isopropanol : hexane. They are injected with a 50-μl rinse; the following regions are collected separately, allowing room for vitamin D$_2$ metabolites:

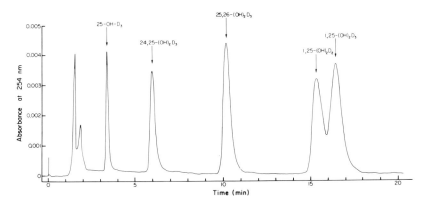

FIG. 10. HPLC with detection by absorbance at 254 nm of the major vitamin D metabolites—standards of 25-OH-$D_3$, 24,25-$(OH)_2D_3$, 25,26-$(OH)_2D_3$, 1,25-$(OH)_2D_2$ and 1,25-$(OH)_2D_3$—on a 0.46 × 25 cm Zorbax-SIL silicic acid column developed with a solvent system of 10 : 90 isopropanol : hexane at a flow rate of 2.0 ml/min.

24,25-$(OH)_2D_3$—5.0–7.5 min (peak at 6.0 min); 25,26-$(OH)_2D_3$—9.2–11.7 min (peak at 10.2 min); and 1,25-$(OH)_2D_3$—14.0–18.7 min (peak at 16.4 min) (Fig. 10). The Zorbax-SIL column is cleaned with methanol before use and after every 10–20 samples to eliminate strongly adsorbed contaminants that might interfere in the binding assays.

While the dihydroxyvitamin D metabolite fraction from the initial Sephadex LH-20 column is separated from most of the lipid impurities, there are still interfering substances present that display binding activity in both the rat plasma protein and chick cytosol receptor competitive binding assays discussed below and must be eliminated by HPLC. As it is known that 24,25-$(OH)_2D_2$ and 1,25-$(OH)_2D_2$ elute slightly ahead of their vitamin $D_3$ analogs in this system,[16] collection of these and 25,26-$(OH)_2D$ is timed to allow for isolation of both analogs from plasma samples. HPLC is not sensitive enough to routinely measure 24,25-$(OH)_2D$, 25,26-$(OH)_2D$, or 1,25-$(OH)_2D$ by ultraviolet absorption, so competitive binding assays are used—Zorbax-SIL being used as a final step of purification and separation.

*Analysis of 24,25-$(OH)_2D$ and 25,26-$(OH)_2D$.* After the separation and purification of the three dihydroxyvitamin D metabolites by HPLC, 24,25-$(OH)_2D$ and 25,26-$(OH)_2D$ are quantitated by a rat plasma protein competitive binding assay modified from the method of Haddad *et al.*[17] In preparation for the assay of plasma samples, several dilutions—1 : 3000,

[16] G. Jones and H. F. DeLuca, *J. Lipid Res.* **16**, 448 (1975).
[17] J. G. Haddad, Jr., C. Min, M. Mendelsohn, E. Slatopolsky, and T. J. Hahn, *Arch. Biochem. Biophys.* **182**, 390 (1977).

1 : 4000, 1 : 5000, and 1 : 6000—of rat plasma in 50 m$M$ sodium phosphate, pH 7.4, are tested using the following amounts of 25-OH-D$_3$ per assay tube: 0.0, 0.1, 0.2, 0.4, 0.8, 1.6, 3.2, and 6.4 ng. The optimal binding curve (that with sensitivity down to 0.1 ng/tube) is selected, and that dilution is used for all subsequent assays. The rat plasma is stored frozen, and diluted rat plasma (usually 1 : 5000) is prepared fresh for each daily assay.

The eluent in the 24,25-(OH)$_2$D and 25,26-(OH)$_2$D fractions collected from the HPLC Zorbax-SIL column are evaporated under a stream of nitrogen and the samples are redissolved in 140 $\mu$l of absolute ethanol. Duplicate 25-$\mu$l aliquots of each are placed in counting vials, evaporated and counted along with the initial aliquots of 24,25-(OH)$_2$[$^3$H]D$_3$ and 25,26-(OH)$_2$[$^3$H]D$_3$ in toluene counting solution to assess percent recovery. Triplicate 25-$\mu$l aliquots of the 24,25-(OH)$_2$D fraction and triplicate 25-$\mu$l aliquots of the 25,26-(OH)$_2$D fraction are pipetted into separate 12 × 75 mm glass tubes. A standard curve is prepared by adding each of the following amounts of 25-OH-D$_3$ in triplicate in 25 $\mu$l ethanol to other tubes—0.0, 0.1, 0.2, 0.4, 0.8, 1.6, 3.2, and 6.4 ng plus 1 $\mu$g (100-fold excess to determine nonspecific binding). 25-OH-[$^3$H]D$_3$ (6000 cpm) is added in 20 $\mu$l of ethanol to all standard and sample tubes on ice; quadruplicate 20-$\mu$l aliquots of 25-OH-[$^3$H]D$_3$ are also placed into counting vials for total radioactivity determinations. To all tubes are added 0.5 ml of diluted rat plasma on ice, and the contents are mixed on a Vortex apparatus. The tubes are covered and allowed to incubate at least 1 hr in an ice-water bath or overnight in the cold room. In a 5-min period, 0.2 ml of dextran-coated charcoal suspension (5% charcoal, 0.5% dextran suspended in the same buffer as above with a magnetic stirrer) is added to all tubes on ice and their contents are vortexed. The tubes are allowed to stand for 30 min on ice, then centrifuged at 4500 rpm for 20 min at 4°. Aliquots (0.5 ml) of the supernatant of all the tubes are mixed with 3.5 ml of aqueous counting solution and counted; 0.5-ml aliquots of buffer are added to the vials relegated to the measurement of total radioactivity, and they are counted in aqueous counting solution.

Recoveries are calculated by using Eq. (1).

$$\% \text{ recovery} = \frac{(\text{cpm-background})100}{(25 \ \mu l/140 \ \mu l)(\text{initial cpm added to sample-background})} \quad (1)$$

The standard curve is prepared by plotting the bound cpm versus the amount of standard on semilog paper. The amounts of 24,25-(OH)$_2$D or 25,26-(OH)$_2$D in each sample tube are determined by relating the bound cpm to the standard curve. The plasma concentration of 24,25-(OH)$_2$D or 25,26-(OH)$_2$D is then calculated using Eq. (2).

$$\text{ng/ml} = \frac{\text{ng in sample tube}}{(25 \ \mu l/140 \ \mu l)(\text{ml of plasma sample})(\% \text{ recovery})} \quad (2)$$

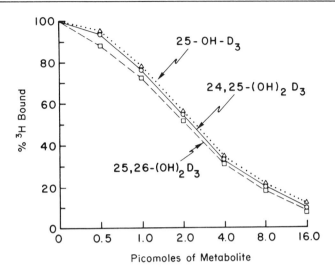

FIG. 11. Competitive displacement of 25-OH-[$^3$H]D$_3$ from rat plasma binding protein by 25-OH-D$_3$ ($\triangle \cdots \triangle$), 24,25-(OH)$_2$D$_3$ (○——○) and 25,26-(OH)$_2$D$_3$ (□---□). Each point represents the mean of three replicates.

A typical binding curve (Fig. 11) shows the displacement of 25-OH-[$^3$H]D$_3$ from rat plasma binding protein by unlabeled 25-OH-D$_3$, 24,25-(OH)$_2$D$_3$, and 25,26-(OH)$_2$D$_3$. Since the three compounds are equipotent in their displacement of 25-OH-[$^3$H]D$_3$, 25-OH-[$^3$H]D$_3$, and unlabeled 25-OH-D$_3$ can be used in the establishment of a standard curve for the assay of 24,25-(OH)$_2$D and 25,26-(OH)$_2$D. As 25-OH-D$_2$ and 25-OH-D$_3$ are equally recognized by the rat binding protein,[18,19] it seems likely that the same is true for 24,25-(OH)$_2$D$_2$ and 25,26-(OH)$_2$D$_2$, should these be present in plasma samples. The intra-assay coefficent of variation for this method is 12% ($n = 6$) and 9% ($n = 6$) for 24,25-(OH)$_2$D and 25,26-(OH)$_2$D, respectively. The inter-assay coefficient of variation is 13% ($n = 3$) and 19% ($n = 3$), respectively. The overall recoveries of 24,25-(OH)$_2$-[$^3$H]D$_3$ and 25,26-(OH)$_2$-[$^3$H]D$_3$ to the point of assay average 71.9 ± 6.2% ($n = 25$) and 75.2 ± 5.6% ($n = 25$), respectively. Since routine sensitivity is 0.1 ng per tube, average recovery is 70% and 18% of the final sample is assayed per tube, the limit of detection of 24,25-(OH)$_2$D and 25,26-(OH)$_2$D for this method is 0.2 ng/ml.

The HPLC Zorbax-SIL system is effective in eliminating binding sub-

[18] M. A. Preece, J. L. H. O'Riordan, D. E. M. Larson, and E. Kodicek, *Clin. Chim. Acta* **54,** 235 (1974).
[19] J. G. Haddad, Jr., L. Hillman, and S. Rojanasathit, *J. Clin. Endocrinol. Metab.* **43,** 712 (1976).

stances interfering in the binding assay of 24,25-$(OH)_2D$ and 25,26-$(OH)_2D$ in normal human and bovine plasma but not in normal rat and chick plasma. In the latter two is present an unknown vitamin D metabolite with binding activity that interferes with the assay of 24,25-$(OH)_2D$. This can be separated from 24,25-$(OH)_2D_3$ by reversed-phase HPLC on a Zorbax-ODS column equilibrated at a flow rate of 2.0 ml/min (3000 psi) with 25:75 water:methanol. With this system, a standard of 24,25-$(OH)_2D_3$ elutes at 17.6 min (collection region 16.5–19.5 min). The sample 24,25-$(OH)_2D$ fraction collected from the Zorbax-SIL column is injected in 50 $\mu$l of methanol with a 50-$\mu$l rinse and the 24,25-$(OH)_2D$ region is collected. This is then subjected to binding analysis as described above.

*Analysis of 1,25-$(OH)_2D$.* After the separation and purification of the three dihydroxyvitamin D metabolites by HPLC Zorbax-SIL, 1,25-$(OH)_2D$ is quantitated by a chick intestinal cytosol protein competitive binding assay modified from the method of Eisman *et al.*[12] In preparation for the assay of plasma samples, chick intestinal cytosol binding protein is prepared from vitamin D-deficient chickens sacrificed by decapitation. The duodenal loops are removed, emptied, and rinsed with 5 ml of ice-cold buffer (25 m$M$ potassium phosphate, 0.1 $M$ potassium chloride, and 1 m$M$ dithiothreitol, pH 7.4); this buffer is used throughout preparation and assay procedures. All subsequent steps are performed either on ice or at 4°. The mucosa is scraped from the serosa and washed twice with 5 volumes (ml/g) cold buffer—suspending in buffer, centrifuging at 4000 rpm (454 g) for 10 min, and discarding the supernatant. The final mucosal pellet is resuspended and homogenized in 2 volumes of the buffer with two passes of a Potter–Elvehjem Teflon–glass homogenizer maintained at 0° with an ice bath. The homogenate is centrifuged at 20,000 rpm (11,352 g) for 75 min in a number of SS-34 Sorvall head of a DuPont Sorvall RC-5 centrifuge. The fine lipid layer is removed with a Pasteur pipette, and the supernatant is collected. This cytosol is frozen in 4- to 6-ml portions on the sides of test tubes in an isopropanol–Dry Ice bath, lyophilized for 24 hr, and stored under nitrogen at −20° until use. The protein concentration of the cytosol is determined by a biuret method with a bovine serum albumin standard, with a yield of binding protein of 20 mg/g wet weight of duodenal mucosa.

The lyophilized cytosol is reconstituted with distilled water, and several diluted protein concentrations—0.6, 0.8, and 1.0 mg protein per ml—are tested using the following amounts of 1,25-$(OH)_2D_3$ per tube: 0.0, 1.5, 3.0, 6.0, 12.0, 24.0, 48.0, 96.0, and 192.0 pg. The optimal binding curve is selected and that dilution is used for all subsequent assays. Diluted chick cytosol (usually 0.8 mg of protein per milliliter) is prepared fresh for each daily assay.

The 1,25-(OH)₂D fraction from the HPLC Zorbax-SIL column is evaporated under nitrogen and redissolved in 210 μl of absolute ethanol. Duplicate 25-μl aliquots are placed in counting vials, evaporated, and counted along with the initial aliquot of 1,25-(OH)₂[³H]D₃ in toluene counting solution to assess percent recovery. Triplicate 50-μl aliquots are pipetted into separate 12 × 75 mm glass tubes. A standard curve is prepared by adding each of the following amounts of 1,25-(OH)₂D₃ in quadruplicate in 50 μl of ethanol to other tubes: 0.0, 1.5, 3.0, 6.0, 12.0, 24.0, 48.0, 96.0, and 192.0 pg plus 15.4 ng (80-fold excess to determine nonspecific binding). 1,25-(OH)₂[³H]D₃ (approximately 400 cpm) is added to each standard tube and three counting vials to compensate for the recovered radioactivity in the sample aliquots. Then, 1,25-(OH)₂[³H]D₃ (3000 cpm) is added in 20 μl of ethanol to all standard and sample tubes on ice and to quadruplicate counting vials for total radioactivity. To all tubes are added 0.5 ml (0.4 mg of protein per tube) of diluted cytosol on ice, and the contents are vortexed. The tubes are covered and incubated in an ice-water bath for 10 min, then incubated for 1 hr at 25° in a water bath at 120 oscillations per minute. The tubes are returned to the ice and 0.2 ml of dextran-coated charcoal suspension (0.5% charcoal, 0.05% dextran suspended in the same buffer as above with a magnetic stirrer) is added in a 5-min period to all tubes and their contents are vortexed. After standing for 10 min on ice, the tubes are centrifuged at 4500 rpm for 20 min at 4°. Aliquots (0.5 ml) of the supernatant of all the tubes are mixed with 3.5 ml of aqueous counting solution and counted; 0.5-ml aliquots of buffer are added to the vials for determination of total radioactivity, and these are similarly counted.

Recoveries are calculated using Eq. (3).

$$\% \text{ recovery} = \frac{(\text{cpm-background})100}{(25\ \mu l/210\ \mu l)(\text{initial cpm added to sample-background})} \quad (3)$$

The standard curve is prepared by plotting the bound cpm versus the amount of standard on semilog paper. The amount of 1,25-(OH)₂D in each sample tube is determined by relating the bound cpm to the standard curve. The plasma concentration of 1,25-(OH)₂D is then calculated using Eq. (4).

$$\text{pg/ml} = \frac{\text{pg in sample tube}}{(50\ \mu l/210\ \mu l)(\text{ml of plasma sample})(\% \text{ recovery})} \quad (4)$$

As with 24,25-(OH)₂D and 25,26-(OH)₂D, the HPLC Zorbax-SIL system is effective and necessary in purifying 1,25-(OH)₂D for detection by binding assay and in removing interfering binding substances. Shown in Fig. 12 is a typical standard binding curve of the displacement of 1,25-

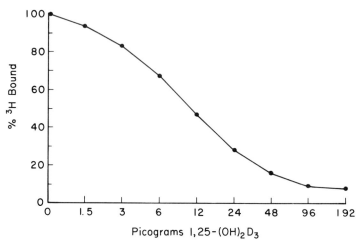

FIG. 12. Competitive displacement of 1,25-(OH)$_2$[$^3$H]D$_3$ from chick intestinal cytosol binding protein by 1,25-(OH)$_2$D$_3$. Each point represents the mean of four replicates.

(OH)$_2$[$^3$H]D$_3$ by unlabeled 1,25-(OH)$_2$D$_3$. Both 1,25-(OH)$_2$D$_2$ and 1,25-(OH)$_2$D$_3$ are equally recognized by the chick intestinal cytosol receptor protein[12] and can be assayed together in this procedure. The intra-assay coefficient of variation is 17% ($n$ = 7), and the inter-assay coefficient of variation is 26% ($n$ = 9). The overall recovery of 1,25-(OH)$_2$[$^3$H]D$_3$ to the point of assay averages 58.4 ± 14.8% ($n$ = 25). Since routine sensitivity is 5 pg per tube, average recovery is 58%; and since 24% of the final sample is asssyed per tube, the limit of detection of 1,25-(OH)$_2$D is 7 pg/ml.

### Concentrations of Vitamin D and Metabolites in Normal Human Plasma (Table)

Total vitamin D determined by the HPLC procedure in plasma samples taken from a group of normal laboratory workers in February was 3.5 ± 2.5 ng/ml ($n$ = 8) with a range of 0.9–7.2 ng/ml. Vitamin D$_2$ in these samples was 1.2 ± 1.4 ng/ml ($n$ = 8) with a range of 0.5–4.6 ng/ml. Vitamin D$_3$ was 2.3 ± 1.6 ng/ml ($n$ = 8) with a range of 0.7–5.7 ng/ml. Total 25-OH-D in normal samples averaged 31.6 ± 9.3 ng/ml ($n$ = 19) with a range of 20.6–45.7 ng/ml. 25-OH-D$_2$ averaged 3.9 ± 3.1 ng/ml ($n$ = 19) and 25-OH-D$_3$ averaged 27.6 ± 9.2 ng/ml ($n$ = 19), with respective ranges of 1.0–15.9 ng/ml and 19.6–41.9 ng/ml. Total 24,25-(OH)$_2$D in normal samples was 3.5 ± 1.4 ng/ml ($n$ = 12) with a range of 1.6–5.8 ng/ml; while total 25,26-(OH)$_2$D was 0.7 ± 0.5 ng/ml ($n$ = 12) with a range of 0.3–1.6 ng/ml. Total 1,25-(OH)$_2$D was 31 ± 9 pg/ml ($n$ = 20) with a range of 20–39 pg/ml.

VITAMIN D METABOLITE LEVELS IN NORMAL MAN

| Metabolite | Plasma concentration[a] |
|---|---|
| Vitamin $D_2$ | $1.2 \pm 1.4$ ng/ml |
| Vitamin $D_3$ | $2.3 \pm 1.6$ ng/ml |
| Total vitamin D | $3.5 \pm 2.5$ ng/ml |
| 25-OH-$D_2$ | $3.9 \pm 3.1$ ng/ml |
| 25-OH-$D_3$ | $27.6 \pm 9.2$ ng/ml |
| Total 25-OH-D | $31.6 \pm 9.3$ ng/ml |
| 24,25-$(OH)_2$D | $3.5 \pm 1.4$ ng/ml |
| 25,26-$(OH)_2$D | $0.7 \pm 0.5$ ng/ml |
| 1,25-$(OH)_2$D | $31 \pm 9$ pg/ml |

[a] Mean $\pm$ SD.

Conclusion

The method outlined above represents a multiple assay procedure for the quantitation of vitamin D and its metabolites that is reasonably fast, accurate, and reproducible. From a single 3–5-ml sample of human plasma, it is possible to measure vitamin $D_2$ and vitamin $D_3$, 25-OH-$D_2$ and 25-OH-$D_3$, total 24,25-$(OH)_2$D, total 25,26-$(OH)_2$D, and total 1,25-$(OH)_2$D. It is possible to streamline the procedure and measure metabolites of particular interest, i.e., 25-OH-D, 24,25-$(OH)_2$D, and 1,25-$(OH)_2$D. The older techniques of organic solvent extraction of plasma lipids and chromatography with Sephadex LH-20 and Lipidex 5000 have been combined with the more recent developments of HPLC[16,20–22] and competitive protein binding assays.[12,17,23] In the development of this multiple assay, it was realized that chromatographic purification of plasma lipid extracts with separation of vitamin D metabolites and removal of interfering contaminants is extremely important in obtaining accurate measurements of plasma concentrations of vitamin D and its metabolites—whether detected by HPLC or competitive protein binding assays. Thus, earlier literature values of metabolites obtained by methods with inadequate chromatography are suspect. The multiple assay procedure outlined here should prove to be a useful tool in the study of clinical disease states related to vitamin D in humans and in studies on vitamin D metabolism in laboratory research animals.

[20] E. W. Matthews, P. G. H. Byfield, K. W. Kolston, I. M. A. Evans, L. S. Galante, and I. MacIntyre, *FEBS Lett.* **48**, 122 (1974).
[21] N. Ikekawa and N. Koizumi, *J. Chromatogr.* **19**, 227 (1976).
[22] Y. Tanaka, and H. F. DeLuca, and N. Ikekawa, this volume [44].
[23] This series, Vol. 36, "Steroid Hormones."

# [47] Assay of 24R,25-Dihydroxycholecalciferol in Human Serum[1]

*By* CAROL M. TAYLOR

It is now well established that under conditions of normo- or hypercalcemia, the hepatic metabolite of vitamin $D_3$ ($D_3$), 25-hydroxycholecalciferol (25-OH-$D_3$), undergoes a second hydroxylation to form 24R,25-dihydroxycholecalciferol [24,25-$(OH)_2D_3$]. The major site of synthesis of this metabolite in both experimental animals[2,3] and man[4] is the kidney, although other sites of synthesis have been suggested, such as cartilage[5,6] and gut.[7]

To complete a picture of vitamin D metabolism in a variety of clinical situations, it was necessary to measure 24,25-$(OH)_2D_3$ concentrations in serum along with the other metabolites of vitamin $D_3$, 25-OH-$D_3$,[8] 1,25-dihydroxycholecalciferol [1,25-$(OH)_2D_3$].[9,10] and 25,26-dihydroxycholecalciferol [25,26-$(OH)_2D_3$].[11] A competitive protein binding assay for this metabolite has therefore been developed.[4,12] As serum 25-OH-$D_3$ concentration is an indicator of the vitamin D status of an animal, this metabolite is assayed along with 24,25-$(OH)_2D_3$, using the methods described below.

[1] This work was supported by a Programme Grant from the Medical Research Council to Professor S. W. Stanbury.

[2] E. B. Mawer, *in* "Clinical and Nutritional Aspects of Vitamin D" (A. W. Norman, ed.), in press. Dekker, New York, 1978.

[3] H. F. DeLuca, *Biochem. Sci. Spec. Publ.* **3**, 5 (1974).

[4] C. M. Taylor, E. B. Mawer, J. E. Wallace, J. St. John, M. Cochran, R. G. G. Russell, and J. A. Kanis. *Clin. Sci. Mol. Med.* **55**, 541 (1978).

[5] M. Garabedian, E. Pezant, M. T. Corval, M. Bailly du Bois, and S. Balsan, *in* Abstracts of the Second International Workshop on Calcified Tissues, Kiriat Anavim, Israel, March, 1976.

[6] M. Garabedian, M. T. Corval, M. Bailly du Bois, M. Liberherr, and S. Balsan, *in* "Endocrinology of Calcium Metabolism" (D. H. Copp and R. V. Talmage, eds.), p. 372, Excerpta Med. Found., Amsterdam, 1978.

[7] H. F. DeLuca and H. K. Schnoes, *in* "Endocrinology of Calcium Metabolism" (D. H. Copp and R. V. Talmage, eds.), p. 178. Excerpta Med. Found., Amsterdam, 1978.

[8] J. G. Haddad and K. J. Chyu, *J. Clin. Endocrinol. Metab.* **33**, 992 (1971).

[9] P. F. Brumbaugh, D. H. Haussler, K. M. Bursac, and M. R. Haussler, *Biochemistry* **13**, 4091 (1974).

[10] J. A. Eisman, A. J. Hamstra, B. E. Kream, and H. F. DeLuca, *Arch. Biochem. Biophys.* **176**, 235 (1976).

[11] T. Suda, H. F. DeLuca, H. K. Schnoes, Y. Tanaka, and M. F. Holick, *Biochemistry* **9**, 4776 (1970).

[12] C. M. Taylor, S. E. Hughes, and P. de Silva, *Biochem. Biophys. Res. Commun.* **70**, 1243 (1976).

Methods

*Preparation of 24R,25-[26,27-$^3$H]Dihydroxycholecalciferol*

Tritium labeled $24R,25\text{-}(OH)_2D_3$ is prepared from 25-[26,27-$^3$H]OH-D$_3$ in chick kidney homogenates from birds treated with ethane-1-hydroxy-1,1-diphosphonate (EHDP).[13] Chicks are maintained on a rachitogenic diet containing 0.6% calcium and 0.5% phosphorus[14] for 3 weeks. An aqueous solution of EHDP (73 m*M*, pH 7.4) is given by subcutaneous injection in a dose of 182 $\mu$mol kg$^{-1}$ day$^{-1}$ from 10 days of age. At 12 days of age, each bird is given 1.6 nmol of cholecalciferol orally in peanut oil. The chicks are killed at 14 days of age, and their kidneys are used in the incubation system described below.

Fifteen grams of kidney are homogenized in buffer (50 m*M*–Tris, 320 m*M* sucrose, pH 7.4) containing Mg$^{2+}$ (2 m*M*), nicotinamide (2 m*M*), and L-malate (1 m*M*) to give a homogenate volume of 40 ml. This is divided among twenty 25-ml conical flasks, each containing 1.2 $\mu$mol of NADP, 12.0 $\mu$mol of glucose 6-phosphate, and 1.2 units of glucose-6-phosphate dehydrogenase in 1.0 ml of buffer. 25-[26,27-$^3$H]OH-D$_3$ (10 $\mu$Ci; 12.2 Ci/mmol; The Radiochemical Centre, Amersham, Bucks, U.K.) is dissolved in 1.0 ml of ethanol and divided among the twenty flasks. Each flask is gassed with oxygen, stoppered, and incubated with continuous gentle agitation at 37° for 2 hr.

A lipid extract of the incubated material is made by homogenizing it with methanol : chloroform : water (2 : 2 : 1 by volume). After centrifugation at 2000 *g* for 20 min, the aqueous layer is washed three times with chloroform; all organic layers are pooled and evaporated to dryness under nitrogen.

The extract is chromatographed on a column of Sephadex LH-20 (20 g, 1.4 × 55 cm) eluted with chloroform : hexane (65 : 35). The 110–190-ml fraction, which contained $^3$H-labeled 24,25-(OH)$_2$D$_3$, is rechromatographed on a second Sephadex LH-20 column (10 g, 0.9 × 40 cm) eluted with methanol. The 25–37 ml fraction, containing the $^3$H-labeled 24,25-(OH)$_2$D$_3$, is evaporated to dryness under nitrogen; the residue is taken up in ethanol and stored at −20° until required.

*Sample Preparation—Extraction and Chromatography*

To monitor recovery through the sample preparation procedure, internal standards of 700 dpm 25-[26,27$^3$H]OH-D$_3$ and 700 dpm 24,25-[26,27$^3$H](OH)$_2$D$_3$ are added to serum samples prior to extraction.

[13] C. M. Taylor, E. B. Mawer, and A. Reeve, *Clin. Sci. Mol. Med.* **49**, 391 (1975).
[14] D. E. M. Lawson, P. W. Wilson, D. C. Barker, and E. Kodicek, *Biochem. J.* **115**, 263 (1969).

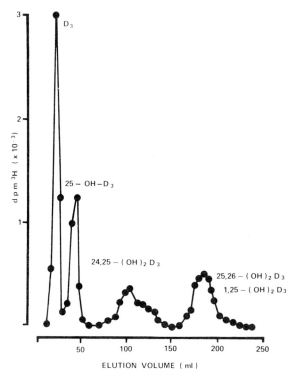

FIG. 1. The elution position of vitamin $D_3$ and its metabolites 25-OH-$D_3$, 24,25-(OH)$_2$D$_3$, 1,25-(OH)$_2$D$_3$, and 25,26-(OH)$_2$D$_3$ from a Sephadex LH-20 column (10 g, 0.9 × 30 cm) eluted with chloroform : hexane (60 : 40).

Vitamin $D_3$ and its metabolites are first extracted from 5.0 ml of serum, using methanol : chloroform : water (2 : 2 : 1) as described above.

The lipid extract is applied to a column of Sephadex LH-20 (10 g, 0.9 × 30 cm), which is eluted with chloroform : hexane (60 : 40) to separate vitamin $D_3$ and its metabolites. Figure 1 shows the elution pattern from such a column. The first fraction from this column (0–70 ml) is rechromatographed on a second column of Sephadex LH-20 (3.5 g, 0.5 × 50 cm) eluted with chloroform : hexane (50 : 50) to separate vitamin $D_3$ and 25-OH-$D_3$. The first 11 ml of eluate are discarded and the next 10 ml, which contains 25-OH-$D_3$, is evaporated to dryness under a stream of nitrogen. The resisue is taken up in 2.50 ml of ethanol; 1.25 ml were taken out and counted for $^3$H activity to estimate the recovery of 25-OH-$D_3$ and the remainder is used in 50-$\mu$l aliquots in the assay of 25-OH-$D_3$.

The next 80-ml fraction, which contains the 24,25-(OH)$_2$D$_3$, is rechromatographed on another Sephadex LH20 column eluted with

methanol as described above. The fraction containing 24,25-(OH)$_2$D$_3$
(25–37 ml) is evaporated to dryness under nitrogen and taken up in 600 $\mu$l
of ethanol. One hundred microliters are counted for $^3$H activity to estimate recovery of 24,25-(OH)$_2$D$_3$, and the remainder is used in 50-$\mu$l
aliquots in the 24,25-(OH)$_2$D$_3$ assay. Recovery through these procedures
is 70 ± 2% for 25-OH-D$_3$ and 78 ± 6% (mean ± SD) for 24,25-(OH)$_2$D$_3$.

### Competitive Protein Binding Assays

*Assay a. 24,25-(OH)$_2$D$_3$.* 24,25-(OH)$_2$D$_3$ is assayed by a competitive
protein binding assay using rachitic rat serum, diluted 1/4000 with
barbital/acetate buffer, pH 8.6 as the binding protein.

Ethanol, 50 $\mu$l containing standard sterol or serum extract is pipetted
into a plastic tube, followed by 25 $\mu$l of ethanol containing approximately
4000 dpm [$^3$H]25-OH-D$_3$. One milliliter of binding protein solution is
added, and the contents of the tubes are mixed on a vortex mixer. After
incubation at 4° for 2 hr, 0.2 ml of dextran-coated charcoal solution is
added (0.0625 g of Dextran 40, 0.625 g of Norit GSX, 100 ml of barbital/
acetate buffer, pH 8.6). The tubes are mixed and allowed to stand for 10
min at 4°, then centrifuged at 800 $g$ for 20 min. Aliquots (0.6 ml) are
transferred to counting vials containing 10 ml of scintillator mixture (toluene : Triton X-100 2 : 1, 0.4% 2,5-diphenyloxazole) for counting on an
LKB Ultrobeta scintillation counter. Each serum sample is assayed in

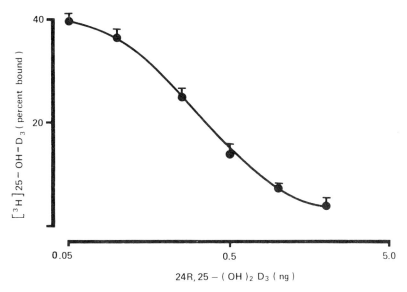

FIG. 2. A typical displacement curve for the 24,25-(OH)$_2$D$_3$ assay.

triplicate, and duplicate blanks (serum extract, [³H]25-OH-D₃, and 1.0 ml of buffer) are included for each sample. Standard curves are constructed over the range 0.05 to 2.00 ng of 24,25-(OH)₂D₃ (Fig. 2). The limit of detection of the assay is 0.1 ng per milliliter of serum.

*Assay b. 25-OH-D₃.* 25-OH-D₃ is assayed by a method exactly analogous to the one described above, rachitic rat kidney cytosol being used as the binding protein.[8] The assay is incubated at 26° for 1 hr, and standard curves are constructed over the range 0.25–4.00 ng of 25-OH-D₃. The limit of detection of the assay is 0.5 ng per milliliter of serum.

### Specificity—Competition in the Assays of Other Vitamin D Sterols

The displacement of tritiated 25-OH-D₃ from both the serum and kidney binding proteins was studied with the following compounds: D₃, 25-OH-D₃, 24R,25-(OH)₂D₃, 24S,25-(OH)₂D₃, 25,26-(OH)₂D₃, and 1,25-(OH)₂D₃. The sterols were dissolved in ethanol to give concentrations between 0.05 and 4.00 ng per 50 μl and used in competition experiments using the methods described above. Figures 3 and 4 show the results of such procedures. In both systems the metabolites 25-OH-D₃, 24R,25-(OH)₂D₃, 24S,25-(OH)₂D₃, and 25,26-(OH)₂D₃ were equipotent in displacing the tritiated 25-OH-D₃, emphasizing the need for an efficient chromatography system separating all these metabolites prior to assay. It is particularly important to separate all 25-OH-D₃ from the 24,25-(OH)₂D₃

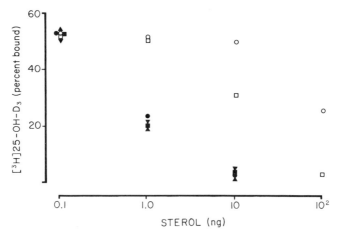

FIG. 3. Competitive displacement of [³H]25-OH-D₃ from rachitic rat serum, diluted 1/4000 with barbital/acetate buffer, pH 8.6, by vitamin D₃ and its metabolites. ○ = D₃, ▲ = 25-OH-D₃, ● = 24R,25-(OH)₂D₃, ■ = 24S,25-(OH)₂D₃, □ = 1,25-(OH)₂D₃, ▼ = 25,26-(OH)₂D₃.

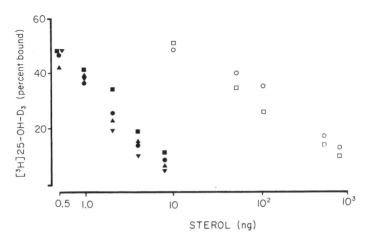

FIG. 4. Competitive displacement of [³H]25-OH-D₃ from the binding protein prepared from rachitic rat kidneys by vitamin D₃ and its metabolites. ○ = D₃, ▲ = 25-OH-D₃, ● = 24R,25-(OH)₂D₃, ■ = 24S,25-(OH)₂D₃, □ = 1,25-(OH)₂D₃, ▼ = 25,26-(OH)₂D₃.

before assaying the dihydroxymetabolite, as 25-OH-D₃ is present in serum at a concentration about ten times that of 24,25-(OH)₂D₃, and their similar behavior in the assay systems could lead to erroneous results.

Applications of the 24,25-(OH)₂D₃ Assay

The assay has been used to determine the serum concentration of 24,25-(OH)₂D₃ in a variety of clinical situations.

It was first established that 24,25-(OH)₂D₃ is present in human serum at a concentration of 6–10% that of the prevailing serum 25-OH-D₃ concentration in subjects with normal renal function. There was a strong positive correlation between the levels of the two metabolites ($r = 0.87, n = 19, y = 0.06x + 0.47$) as shown in Fig. 5.[15] Thus, when measuring a serum 24,25-(OH)₂D₃ concentration, it is important to know the serum 25-OH-D₃ concentration of the subject before any conclusion may be drawn as to whether the concentration measured is normal or not.

In contrast, anephric patients produce no, or very little, 24,25-(OH)₂D₃ even when their serum 25-OH-D₃ levels have been increased by oral administration of 25-OH-D₃.[4] Figure 6 shows the levels of 24,25-(OH)₂D₃ measured in the sera of a group of anephric patients with serum 25-OH-D₃ concentrations ranging from 5 to 80 ng/ml.

Patients with chronic renal failure, but not nephrectomized, do seem

[15] C. M. Taylor, P. de Silva, and S. E. Hughes, *Calcif. Tissue Res.* **22**, Suppl., 40 (1977).

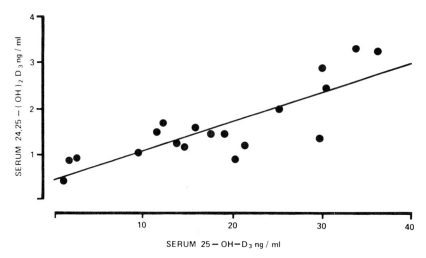

FIG. 5. The relationship between serum concentrations of 25-OH-D$_3$ and 24,25-(OH)$_2$D$_3$ in subjects with normal renal function ($y = 0.06x + 0.47$, $r = 0.87$, $n = 19$).

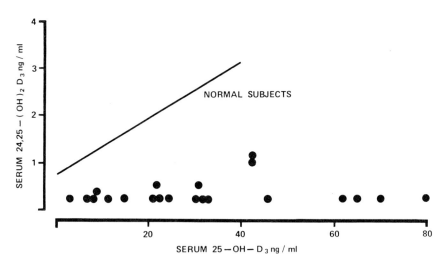

FIG. 6. Serum 25-OH-D$_3$ and 24,25-(OH)$_2$D$_3$ concentrations in anephric patients. The correlation between the two metabolites in subjects with normal renal function is shown as the line $y = 0.06x + 0.47$.

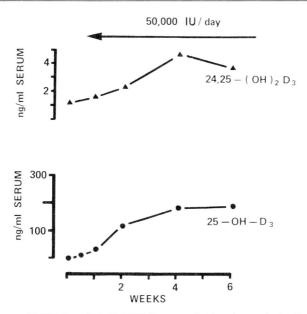

FIG. 7. Serum 25-OH-D$_3$ and 24,25-(OH)$_2$D$_3$ concentrations in a patient with chronic renal failure before and during treatment with 50,000 IU (1.25 mg) of vitamin D per day.

able to produce 24,25-(OH)$_2$D$_3$ from 25-OH-D$_3$, especially if they are being dosed with vitamin D (Fig. 7).[16] Whether or not this formation of 24,25-(OH)$_2$D$_3$ is reduced compared with formation in subjects with normal renal function of comparable vitamin D status has yet to be established.

Parathyroid disorders are also conditions in which abnormal metabolism of vitamin D has been suggested.[17] This has been confirmed in patients with primary hyperparathyroidism who have raised levels of serum 24,25-(OH)$_2$D$_3$ (Fig. 8).[15,16] After parathyroidectomy serum 24,25-(OH)$_2$D$_3$ levels fall rapidly (Fig. 9), accompanied by a transient fall in serum 25-OH-D$_3$.

Conversely, patients with hypoparathyroidism undergoing no therapy with vitamin D or its metabolites have normal serum levels of 24,25-(OH)$_2$D$_3$. Treatment of such patients with 1,25-(OH)$_2$D$_3$, however, leads to an increase in serum 24,25-(OH)$_2$D$_3$ concentration (Fig. 10).

[16] C. M. Taylor, *in* "Vitamin D: Biochemical, Chemical and Clinical Aspects Related to Calcium Metabolism. (A. W. Norman *et al.*, eds.), p. 541. de Gruyter, Berlin, 1977.
[17] E. B. Mawer, J. Backhouse, L. F. Hill, G. A. Lumb, P. de Silva, C. M. Taylor, and S. W. Stanbury, *Clin. Sci. Mol. Med.* **48**, 349 (1975).

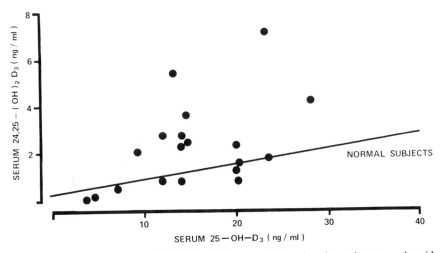

FIG. 8. Serum 25-OH-D₃ and 24,25-(OH)₂D₃ concentrations in primary hyperparathyroid patients. The correlation between the two metabolites in subjects with normal renal function is shown as the line $y = 0.06x + 0.47$.

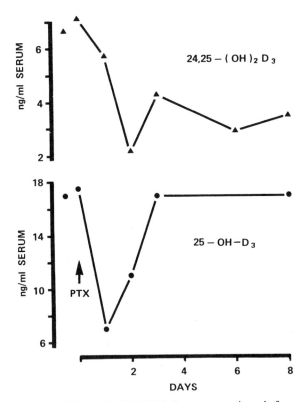

FIG. 9. Serum 25-OH-D₃ and 24,25-(OH)₂D₃ concentrations before and following parathyroidectomy (PTX) in a patient with primary hyperparathyroidism.

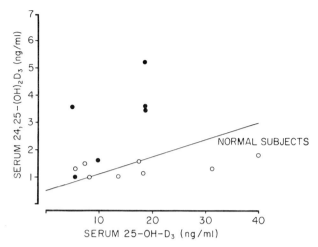

FIG. 10. Serum 25-OH-D$_3$ and 24,25-(OH)$_2$D$_3$ concentrations in hypoparathyroid patients. ○, Untreated patients; ●, patients treated with 1,25-(OH)$_2$D$_3$.

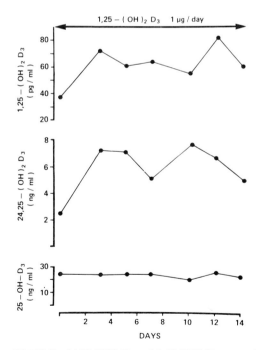

FIG. 11. Serum 25-OH-D$_3$, 24,25-(OH)$_2$D$_3$ and 1,25-(OH)$_2$D$_3$ concentrations in a normal subject dosed with 1,25-(OH)$_2$D$_3$, 1 $\mu$g/day for 14 days.

The observation that 1,25-$(OH)_2D_3$ stimulates the 24-hydroxylation of 25-$(OH)D_3$ has previously been made in the rat[18] and the chick,[19] but not in man. The concomitant measurements of serum 25-OH-$D_3$, 24,25-$(OH)_2D_3$, and 1,25-$(OH)_2D_3$[20] in a normal subject treated with 1,25-$(OH)_2D_3$, 1 $\mu g$/day for 14 days, show that increase in serum 1,25-$(OH)_2D_3$ levels as a result of oral administration of this metabolite does indeed lead to raised blood levels of 24,25-$(OH)_2D_3$[21] (Fig. 11). Thus, the raised serum 24,25-$(OH)_2D_3$ seen in primary hyperparathyroidism may be a direct consequence of the raised levels of 1,25-$(OH)_2D_3$ seen in this condition.[20,22]

Finally, it has recently been discovered that there are metabolites of vitamin $D_2$, ergocalciferol, that can interfere in the assay for 24,25-$(OH)_2D_3$ described here and hence lead to erroneous results.[21] Thus a careful assessment of the diet and the vitamin D therapy received by a subject under investigation should be made before interpreting results obtained using this assay method.

[18] Y. Tanaka and H. F. DeLuca, *Science* **183**, 1198 (1974).
[19] Y. Tanaka, R. S. Lorenc, and H. F. DeLuca, *Arch. Biochem. Biophys.* **171**, 521 (1975).
[20] C. M. Taylor, J. Hann, J. St. John, J. E. Wallace, and E. B. Mawer, *Clin. Chim. Acta* **96**, 1 (1979).
[21] C. M. Taylor, *in* "Vitamin D; Basic Research and Its Clinical Application" (A. W. Norman *et al.*, eds.), p. 197. de Gruyter, Berlin, 1979.
[22] M. R. Haussler, M. R. Hughes, J. W. Pike, and T. A. McCain, *in* "Vitamin D: Biochemical, Chemical and Clinical Aspects Related to Calcium Metabolism (A. W. Norman *et al.*, eds.), p. 473. de Gruyter, Berlin, 1977.

# [48] Enzymic Preparation of [$^3$H]1,25-Dihydroxyvitamin $D_3$

*By* ANTHONY W. NORMAN and JUNE E. BISHOP

Since radioactive 1,25-dihydroxyvitamin $D_3$ [1, 25-$(OH)_2D_3$] is not readily available commercially, the method given below has been developed for the enzymic preparation of the steroid in our laboratory. 1,25-Dihydroxyvitamin $D_3$ *in vivo* is produced in the kidney by hydroxylation of 25-hydroxyvitamin $D_3$.[1,2] This method uses the rachitic chick kidney 25-hydroxyvitamin $D_3$ 1-hydroxylase system to hydroxylate radioactive 25-hydroxyvitamin $D_3$ *in vitro*.[3]

[1] D. R. Fraser and E. Kodicek, *Nature (London)* **288**, 764 (1970).
[2] A. W. Norman, R. J. Midgett, J. F. Myrtle, and H. G. Nowicki, *Biochem. Biophys. Res. Commun.* **42**, 1082 (1971).
[3] This is a slight modification of the incubation system used to measure initial rates of 1-hydroxylation described by H. L. Henry and A. W. Norman, *J. Biol. Chem.* **249**, 7529 (1974).

METHODS IN ENZYMOLOGY, VOL. 67

Methods

*Equipment*

Dissecting scissors, forceps
Analytical balance
Potter–Elvehjem homogenizer
Refrigerated centrifuge, centrifuge tubes
Glass chromatography column, 1 × 80 cm
Volumetric fraction collector
Liquid scintillation counter, counting vials
Pipettes
Beakers
Erlenmeyer flask
Separatory funnel
Test tubes

*Reagents.* All reagents used are analytical grade; water and solvents are glass-distilled.

NaCl, 0.15 $M$
Sucrose, 0.25 $M$
Tris · HCl, 50 m$M$, pH 7.4
MgCl₂, 50 m$M$ in 50 m$M$ Tris · HCl, pH 7.4
Malate, 0.3 $M$ in 50 m$M$ Tris · HCl, pH 7.4
[³H]25-Hydroxyvitamin D₃ (available commercially from Amersham-Searle and New England Nuclear)
Sephadex LH-20 (Pharmacia)
Chloroform
Methyl alcohol
Petroleum ether
Ethanol
Butyl-phenyl-biphenyl-oxadiazole
Toluene

*Animals.* Male, White Leghorn chicks are placed on a rachitogenic diet[4] on the day of hatch. By their fourth week the animals are depleted of vitamin D₃; their serum Ca should be 6–7 mg/100 ml. At this point, they have become rachitic and their kidney 1-hydroxylase system is maximally active.[5]

*Tissue Preparation.* The animals are sacrificed by decapitation, and the kidneys are immediately removed to ice-cold saline. The remaining steps of tissue preparation are carried out at 4°. The kidneys are rinsed with

[4] A. W. Norman and R. G. Wong, *J. Nutr.* **102**, 1709 (1972).
[5] H. L. Henry, R. J. Midgett, and A. W. Norman, *J. Biol. Chem.* **249**, 7584 (1974).

saline to remove the remaining blood, and then a 10% homogenate (w/v) is made in 0.25 $M$ sucrose using a Potter–Elvehjem glass–Teflon homogenizer. The homogenate is centrifuged at 600 $g$ for 10 min. The supernatant fraction is centrifuged at 8500 $g$ for 20 min. The resulting mitochondrial pellet is resuspended in the original volume of 0.25 $M$ sucrose.

*Incubation.* The following reaction mixture is allowed to incubate in 30-ml beakers for 1 hr at 37° in a gently shaking, constant-temperature water bath: 4.0 ml of 10% mitochondrial suspension in 0.25 $M$ sucrose; 5.2 ml of 50 m$M$ Tris · HCl, pH 7.4; 0.4 ml of 50 m$M$ MgCl$_2$ in 50 m$M$ Tris · HCl, pH 7.4; 0.33 ml of 0.3 $M$ malate in 50 m$M$ Tris · HCl, pH 7.4; 2.6 nmol of [$^3$H]25-hydroxyvitamin D$_3$ in 0.1 ml of ethanol.

The reaction is stopped by adding the reaction mixture to three volumes of 2:1, methanol:chloroform in a separatory funnel. With the addition of one volume each of water and chloroform, the organic and aqueous phases separate. After removal of the organic (lower, chloroform) layer, the aqueous (upper, methanol–water) layer is washed twice with two volumes of chloroform. The organic layers are pooled in an Erlenmeyer flask and dried under nitrogen.

*Chromatography.* Sephadex LH-20, (25–30 g) is swelled for at least 3 hr in the chromatography solvent, 65:35, chloroform:petroleum ether. A 1 × 80 cm column is poured, and the reaction mixture is applied in 0.5 ml of chromatography solvent. The sample is washed onto the column with two 0.5-ml rinses of the sample flask. A total of 100 3-ml fractions are collected, and 50-$\mu$l aliquots are removed. The aliquots are air-dried and after addition of 7 ml of butyl-PBD-toluene the radioactivity in each fraction is determined by liquid scintillation counting. The 1,25-(OH)$_2$D$_3$ region (usually fractions 70–90) is pooled, dried under nitrogen, and resuspended in ethanol. After an aliquot is taken to determine final yield, the steroid is stored under nitrogen or argon in the dark at −25°.

## [49] Vitamin D Metabolites: Extraction from Tissue and Partial Purification prior to Chromatography

By Frederick P. Ross and Anthony W. Norman

Numerous studies have shown the role of vitamin D metabolites in the maintenance of calcium and phosphorus homeostasis in higher animals.[1–3]

[1] A. W. Norman and H. Henry, *Recent Prog. Horm. Res.* **30**, 431 (1974).

[2] A. W. Norman, "Vitamin D: The Calcium Homeostatic Steroid Hormone," pp. 1–490. Academic Press, New York, 1979.

[3] H. F. DeLuca, *Fed. Proc. Fed. Am. Soc. Exp. Biol.* **33**, 2211 (1974).

METHODS IN ENZYMOLOGY, VOL. 67

It is known now that vitamin D is metabolized in successive hydroxylations, occurring first in the liver and then in the kidney.[1-3] These processes result in the production of 25-hydroxyvitamin D (25-OH-D) and then 1,25- and/or 24,25-dihydroxyvitamin D [1,25-$(OH)_2D_3$ or 24,25-$(OH)_2D_3$]. Further metabolism of the latter two compounds can occur, at least *in vitro*[4] and possibly *in vivo*,[5] to 1,24,25-trihydroxyvitamin D [1,24,25-$(OH)_3D_3$]. In addition 1,25-$(OH)_2D_3$ undergoes catabolism with at least partial oxidation of the side chain. This reaction is thought to occur in the intestine.[6] Furthermore 25-OH-$D_3$ can be converted to 25,26-dihydroxyvitamin D (25,26-$(OH)_2D_3$),[7] possibly by the liver. There is now good evidence that 1,25-$(OH)_2D_3$ functions as a classical steroid hormone.[2] The physiological importance per se of the other metabolites is more equivocal.

Many of the studies that have led to our understanding of the metabolism and mode of action of these compounds have had as their biochemical basis either the measurement of the concentration of the individual steroids or a study of the tissue or subcellular localization of radiolabeled substrates or metabolites. The recent availability of 25-OH-D of high specific activity (60–100 Ci/mmol labeled with $^3$H), from which it is possible to biosynthesize [$^3$H]1,25-$(OH)_2D_3$[8] or [$^3$H]24,25-$(OH)_2D_3$[9] of similar specific activity has greatly facilitated all studies in this field. Nevertheless, since the concentration of all metabolites is low in those tissues that have been studied, the problem remains of establishing simple and reliable methods for the quantitation of these compounds.

The compounds of interest are secosteroids containing one to four hydroxyl groups, one or more of which may be esterified with long-chain fatty acids. Such structures are best extracted from tissue along with other tissue lipids. The most common and efficient method of extraction is that of Bligh and Dyer.[10] Problems can arise, particularly when a study involves the measurement of radioactivity or metabolite levels in extracts containing large amounts of lipid.

Radioactivity can be measured by conventional means provided that (a) the compound is radiochemically pure and (b) contamination with

[4] E. J. Friedlander and A. W. Norman, *Arch. Biochem. Biophys.* **170**, 731 (1975).

[5] M. F. Holick, A. Kleiner-Bossaller, H. K. Schnoes, P. M. Kasten, and H. F. DeLuca, *J. Biol. Chem.* **248**, 6691 (1973).

[6] P. Kumar and H. F. DeLuca, *Biochem. Biophys. Res. Commun.* **69**, 197 (1976).

[7] T. Suda, H. F. DeLuca, H. K. Schnoes, Y. Tanaka, and M. F. Holick, *Biochemistry* **9**, 4776 (1970).

[8] A. W. Norman and R. G. Wong, *J. Nutr.* **102**, 1709 (1972).

[9] C. M. Taylor, S. E. Hughes, and P. DeSilva, *Biochem. Biophys. Res. Commun.* **70**, 1243 (1976).

[10] E. G. Bligh and W. J. Dyer, *Can. J. Biochem. Physiol.* **37**, 911 (1959).

other compounds that interfere with the photochemical event basic to liquid scintillation counting has been at least reduced to acceptable levels. In short, the presence of unlabeled, colorless, compounds is of no issue when measuring radioactivity in partially purified tissue extracts.

The measurement of metabolite levels by either competitive protein binding assay or by high-pressure liquid chromatography is more exacting. Ideally the compound(s) in question should be both chemically and radiochemically pure. In practice, it is sufficient to obtain no interference with the assay technique.

None of the above goals are easily attained for all the metabolites of interest. Measurement of the levels of the individual compounds has been performed by three basically different methods. These are receptor-bindings assays,[9,11,14,15] bioassays,[16,17] and high-pressure liquid chromatography.[18] All these processes involve prior extraction of the tissue sample and, with the exception of one procedure, the chromatographic purification of the extract either prior to or as part of the procedure. Many of the assay procedures are sensitive to the presence of contaminants of various kinds, and this has resulted in the need for extensive purification procedures including chromatography on silica gel, multibore silicic acid columns, Sephadex LH-20, and high-pressure chromatographic columns. In some instances, more than one chromatographic step is necessary before the assay is performed.

In an attempt to overcome the problems caused by interfering substances and also the difficulty of dealing with large amounts of contaminating lipids, various authors have resorted to extraction procedures using diethyl ether, benzene, or dichloromethane, which selectively remove the compounds of interest while leaving the bulk of the other lipids in the aqueous phase. These procedures are fraught with the problem of variable efficiency of extraction. The majority of authors continue to perform total lipid extraction by the method of Bligh and Dyer.[10] Given the average chemical composition of the most commonly extracted tissues, i.e.,

[11] J. G. Haddad and K. J. Chyu, *J. Clin. Endocrinol. Metab.* **33**, 992 (1971).

[12] R. Belsey, H. F. DeLuca, and J. T. Potts, Jr., *J. Clin. Endocrinol. Metab.* **33**, 554 (1971).

[13] R. Belsey, H. F. DeLuca, and J. T. Potts, Jr., *J. Clin. Endocrinol. Metab.* **38**, 1046 (1974).

[14] P. F. Brumbaugh, M. R. Haussler, and K. M. Bursac, *Biochemistry* **13**, 4091 (1974).

[15] J. A. Eisman, A. J. Hamstra, B. Kream, and H. D. DeLuca, *Arch. Biochem. Biophys.* **176**, 235 (1976).

[16] L. F. Hill, E. B. Mawer, and C. M. Taylor, *in* "Vitamin D and Problems Related to Uremic Bone Disease" (A. W. Norman *et al.,* eds.), p. 755. Berlin, de Gruyter.

[17] P. H. Stern, T. E. Phillips, and S. V. Lucas, *in* "Vitamin D: Biochemical, Chemical and Clinical Aspects Related to Calcium Metabolism" (A. W. Norman *et al.,* eds.), p. 519. de Gruyter, Berlin, 1977.

[18] P. W. Lambert, B. J. Syrenson, C. D. Arnaud, and T. C. Spelsberg, *J. Steroid Biochem.* **8**, 929 (1977).

serum, intestine, liver, and kidney, we have found it possible to apply purification procedures[19] that result in a decrease in the mass of the lipid some 2 to 4-fold, with virtually no loss of the vitamin D metabolites of major interest. When this technique is combined with the effective partition procedure of Mawer and Backhouse,[20] an overall purification of 25 to 40-fold can be obtained. At worst this reduces both the scale and sophistication of the following chromatographic procedures. At best it may allow one to forego preliminary chromatography and make direct application of the sample to the highly efficient process of HPLC.[21]

The method described here has been tested on both a macro (1–3 g of lipid) and a micro (5–10 mg of lipid) scale and is equally effective for both.

## Reagents and Materials

1. Where the samples are to be subjected to HPLC followed by a bioassay procedure, solvents of the highest quality should be used throughout. Generally chromatographic-quality, glass-distilled solvents will suffice.

2. Acetone is purified by refluxing over $KMnO_4$, followed by distillation in glass and storage at $-20°$.

3. Lipids are extracted as described by Bligh and Dyer.[10] Many authors use modifications of this method that result in small amounts of water being left in the organic phase. All such water must be removed by drying over $Na_2SO_4$, since evaporation of such solutions under reduced pressure may cause foaming and loss of sample.

4. The organic phase is evaporated under reduced pressure (for large quantities) or under a stream of nitrogen. Water-bath temperature should not exceed 40°. Small quantities of lipid (up to 100 mg) can be evaporated directly into a clean,[22] 15-ml, conical centrifuge tube. Larger quantities of lipid can be transferred with aliquots of chloroform into a 50-ml polypropylene or glass centrifuge tube equipped with a tight-fitting stopper. With samples larger than 1 g, the material is divided equally among two or more tubes.

5. Chloroform is added to each tube. The amount added is not critical but should not exceed 0.2–0.3 ml in the case of 15-ml tubes or 2–3 ml in the case of the larger tubes. The tube is swirled gently to facilitate solubilization of the sample, and the tube and contents are cooled in ice. To each

[19] M. Kates in "Laboratory Techniques in Biochemistry and Molecular Biology (T. S. Work and E. Work, eds.), Vol. 3, Part II, p. 393. North-Holland Publ., Amsterdam, 1972.

[20] E. B. Mawer and J. Backhouse, Biochem. J. 112, 255 (1969).

[21] G. Jones and H. F. DeLuca, J. Lipid Res. 16, 448 (1975).

[22] All glassware should be cleaned thoroughly with either a strong laboratory detergent or in sulfuric acid–dichromate solution (followed by soaking in dilute bicarbonate solution).

15-ml tube is added 5 ml of acetone cooled to $0°$; to the larger tubes the amount added is 40 ml per tube. A cold solution of $MgCl_2 \cdot 10H_2O$ in methanol (10% w/v) is then added to each tube (100 $\mu$l or 1 ml) and the tube contents are mixed well and left at $0°$ for 1 hr. Centrifugation in a bench-top centrifuge equipped with swing-out buckets yields a white precipitate containing only phospholipids (as shown by TLC on silica gel G plates, 0.25 mm, eluted with hexane : diethyl ether : acetic acid, 70 : 30 : 1), and no vitamin D metabolites (as shown by standard tracer methods using $^3$H-labeled metabolites). The supernatant is removed and retained. The precipitate is washed twice with acetone (1 ml or 4 ml, respectively), and the washings and supernatant are combined. Solvent is removed either under reduced pressure or with a stream of $N_2$. The resulting semisolid mass is transferred with Skellysolve B (three aliquots of 0.5 ml or 5.0 ml, respectively) to a clean tube (15 ml or 50 ml, respectively). More Skellysolve B is added to make the volume 2 ml or 20 ml, respectively, followed by an equal volume of a methanol/water (9 : 1, v/v) mixture. The tubes are shaken vigorously (by capping and shaking in the case of the larger tubes, and by mixing the contents on a vortex mixer for the smaller tubes) and centrifuged in a bench-top centrifuge to separate the two layers. The upper layer is removed and retained; it is replaced with fresh Skellysolve B (1 ml or 5 ml, respectively) and the process is repeated twice. After final separation of the layers, the two phases are stored separately. The upper hydrocarbon layer contains the bulk of the neutral lipid mass, but only vitamin D or vitamin D esters. The methanolic layer contains little mass and the more polar metabolites of vitamin D.

# [50] Purification and Properties of Vitamin D Hydroxylases[1]

By POKSYN S. YOON and HECTOR F. DELUCA

Hydroxylation reactions active in the *in vivo* metabolism of vitamin $D_3$ include at least five known oxidative enzyme systems, three of which have been characterized as mixed-function oxidases that derive reducing equivalents from NADPH for the catalytic incorporation of oxygen derived specifically from molecular oxygen, rather than water.[2-4] The first of these metabolic hydroxylation reactions occurs at the C-25 position of

[1] Supported by National Institutes of Health Program Project Grant AM-14881 and the Harry Steenbock Research Fund.
[2] J. G. Ghazarian, H. K. Schnoes, and H. F. DeLuca, *Biochemistry* 12, 2555 (1973).
[3] T. C. Madhok, H. K. Schnoes, and H. F. DeLuca, *Biochemistry* 16, 2142 (1977).
[4] T. C. Madhok, H. K. Schnoes, and H. F. DeLuca, *Biochem. J.* 175, 479 (1978).

vitamin $D_3$ and is found in the liver microsomes. This vitamin $D_3$ 25-hydroxylase has recently been shown to be sensitive to metyrapone (2-methyl-1,2-di-3-pyridyl-1-propanone) and light-reversible carbon monoxide inhibition, strongly suggesting the participation of a P-450 hemoprotein component. 25-Hydroxyvitamin $D_3$ (25-OH-$D_3$) is further oxidized at the C-26,[5] C-24,[6,7] or C-1 positions when the substrate is incubated *in vitro* using kidney homogenates. Whether oxidation occurs on the A ring or on the side chain appears to be tightly controlled by a number of physiological regulatory factors, such as serum calcium, and phosphorus levels, parathyroid hormone, sex hormones, and 1,25-dihydroxyvitamin $D_3$ [1,25-$(OH)_2D_3$]. Hydroxylated products at either the C-24 or C-26 positions generated *in vitro* by both renal and extrarenal tissue preparations from D-replete animals give rise to metabolites that are less active than 25-OH-$D_3$ in the *in vivo* stimulation of intestinal transport of calcium and phosphorus, as well as the mobilization of these ions from mineralized bone.[5-7] In contrast, C-1 hydroxylation, catalyzed by renal mitochondria from D-deficient animals results in the production of the highly potent hormone 1,25-$(OH)_2D_3$. Hence, recent attention has focused upon the characterization of the components of the liver microsomal 25-hydroxylase and the renal mitochondrial 25-OH-$D_3$ 1$\alpha$-hydroxylase, since both of these reactions are essential for the activation of vitamin $D_3$ *in vivo*. Our knowledge, to date, of the isolated and purified electron-transport components extends only to those enzymes involved in the 1$\alpha$-hydroxylation of 25-OH-$D_3$. This hydroxylase system contains a cytochrome P-450 hemoprotein that is known to catalyze 1$\alpha$-hydroxylation in intact mitochondria, isolated from D-deficient chicks, when mitochondria are supplied with oxidizable Krebs cycle intermediates, magnesium ions, and molecular oxygen, or in calcium-swollen mitochondria in the presence of exogenous NADPH.[8-11] A cytochrome P-450 hemoprotein, detectable by its malate-reduced carbon monoxide minus oxidized carbon monoxide difference spectrum, can be observed in these same mitochondria using special techniques described previously.[9] The detergent-solubilized cytochrome P-450 from vitamin D-deficient chick mitochondria catalyze 1$\alpha$-hydroxylation when incubated with an NADPH-generating system,

[5] Y. Tanaka, R. A. Shepard, H. F. DeLuca, and H. K. Schnoes, *Biochem. Biophys. Res. Commun.* **83**, 7 (1978).

[6] J. C. Knutson and H. F. DeLuca, *Biochemistry* **13**, 1543 (1974).

[7] R. Kumar, H. K. Schnoes, and H. F. DeLuca, *J. Biol. Chem.* **253**, 3804 (1978).

[8] D. R. Fraser and E. Kodicek, *Nature (London)* **228**, 764 (1970).

[9] J. G. Ghazarian, C. R. Jefcoate, J. C. Knutson, W. H. Orme-Johnson, and H. F. DeLuca, *J. Biol. Chem.* **249**, 3026 (1974).

[10] H. L. Henry and A. W. Norman, *J. Biol. Chem.* **249**, 7529 (1974).

[11] J. G. Ghazarian and H. F. DeLuca, *Arch. Biochem. Biophys.* **160**, 63 (1974).

purified bovine adrenocortical ferredoxin reductase (a flavoprotein), bovine adrenal ferredoxin, and molecular oxygen. Similarly, C-1 hydroxylation of 25-OH-$D_3$ is catalyzed by the solubilized cytochrome P-450 in the presence of reduced pyridine nucleotides, an NADPH-diaphorase-containing fraction and an iron-sulfur protein, isolated from chick renal mitochondria. Thus, the renal NADPH–cytochrome P-450 reductase appears to be a two-component system consisting of a flavoprotein and an iron-sulfur-containing ferredoxin.

### Solubilization of Renal Cytochrome P-450

Cytochrome P-450 is isolated from 1.33–1.40 g of mitochondria isolated from 25–30 vitamin D-deficient chicks raised on 1.1% calcium diets for 6 weeks prior to sacrifice. The mitochondria are hypotonically treated by resuspension in 50 m$M$ KCl, 50 m$M$ Tris · HCl (pH 7.4) to a final concentration of 20 mg of protein per milliliter. The suspension is allowed to equilibrate for 30 min at 0° with constant stirring, then divided into aliquots of 120–125 mg of protein. To each aliquot is added 9.0 ml of 0.1 $M$ potassium phosphate buffer (pH 7.4) containing 5% butanol (v/v) and 40 $\mu$l of 10% potassium cholate. After centrifugation at 105,000 $g$ for 60 min, the sedimented pellets are washed twice by resuspension in 10.0 ml of 0.1 $M$ potassium phosphate buffer containing 1 m$M$ dithiothreitol (DTT). Each pellet is then resuspended in 8.0 ml of 0.1 $M$ potassium phosphate buffer (pH 7.4), 25% glycerol (v/v), and 1 m$M$ DTT with the addition of 10% potassium cholate to a final concentration of 0.5 mg cholate per milligram of protein. This ionic detergent treatment solubilizes 70–72% of the total cytochrome P-450[12] from the chick mitochondria when equilibrated at 0° for 20 min. Extraneous proteins are then removed by centrifugation at 105,000 $g$ for 90 min; to the supernatant fraction, solid ammonium sulfate is added to 30% saturation. After centrifugation, additional ammonium sulfate is added to 45% saturation, and the suspension is recentrifuged. The 30–45% fractionated pellets are then resuspended in 1.0–2.0 ml of 0.1 $M$ potassium phosphate buffer (pH 7.4), 25% glycerol (v/v), and 1 m$M$ DTT. Most of the ionic detergent is then removed by chromatography on small Sephadex G-25 columns (0.7 × 25 cm), thereby preventing conversion of the P-450 species to P-420. Further chromatography of the combined peak fractions, concentrated by ammonium sulfate precipitation between 30 and 45% saturation, on a Sephadex G-50 column (2.2 × 60 cm)

---

[12] The concentration of cytochrome P-450 was determined from the carbon monoxide-reduced minus reduced-difference spectrum using a value of 91 cm$^{-1}$ mM$^{-1}$ for the extinction increment between 424 nm and 409 nm according to the method of T. Omura and R. Sato, this series, Vol. 10, p. 556.

equilibrated in the same 25% glycerol, potassium phosphate, DTT buffer system partially separates the two forms of the cytochrome, resulting in higher P-450 to P-420 ratios. This procedure yields preparations with 2.0–2.8-fold purification over crude mitochondria; they are free from detectable amounts of endogenous 25-OH-D$_3$ 1$\alpha$-hydroxylase activity and contain minimal amounts of other carbon monoxide binding chromophores that interfere with spectral quantitative determinations of P-450 concentrations, such as hemoglobin and cytochrome oxidase. The solubilized P-450 hemoprotein is stable over a period of months at temperatures below $-70°$ when stored in the presence of 25% glycerol and potassium phosphate buffer containing DTT.

### Solubilization of Renal Ferredoxin Reductase

The NADPH-reducible component from chick mitochondria, which is believed to be a flavoprotein, is prepared by sonication of osmotically shocked mitochondria and chromatography of the submitochondrial extract on DEAE-cellulose equilibrated in 50 m$M$ Tris · HCl (pH 8.0). The flavoprotein-containing fractions are detected by NADPH-catalyzed reduction of cytochrome $c$, which is dependent upon the addition of purified adrenocortical ferredoxin. This component elutes from the ion-exchange resin at low salt concentrations (0.1–0.3 $M$ KCl) and catalyzes the reduction of DCPIP (2,6-dichlorobenzenoneindophenol) in the presence of NADPH, similar to the NADPH-diaphorase activity observed for the bovine adrenal ferredoxin reductase. Although the chick kidney mitochondrial component has been solubilized, resolved from the other electron-transferring components, and has been shown to substitute for the bovine flavoprotein in the reconstituted 1$\alpha$-hydroxylase system, this component has been only partially purified due to its instability.

### Purification of Renal Mitochondrial Ferredoxin

The following procedure describes the purification of the iron-sulfur component from renal mitochondria of vitamin D-replete chicks, to apparent homogeneity, through solubilization by hypotonic treatment and sonication, chromatography on DEAE-cellulose, and separation by discontinuous preparative and disc gel electrophoresis.

*Assay Method*

The iron-sulfur component from chick kidney mitochondria will transfer electrons from NADPH-reduced bovine adrenocortical ferredoxin reductase to either solubilized chick cytochrome P-450 or to an artificial electron acceptor such as cytochrome *c*. Although the chick kidney flavoprotein will substitute for the bovine adrenocortical flavoprotein in the NADPH-catalyzed reduction of cytochrome *c*, the protein from beef adrenocortical glands is more readily available in pure form and can, therefore, be used routinely in the following standard assay. The bovine adrenal ferredoxin reductase is prepared from the $S_2$ fraction of Omura *et al.*[13] and purified according to the procedure described by Sugiyama and Yamano,[14] in which extracts obtained from the high speed supernatant fractions of sonicated mitochondria are precipitated by ammonium sulfate between 30 and 60% saturation, dialyzed against 70 m$M$ NaCl, 10 m$M$ potassium phosphate (pH 7.4), centrifuged at 47,000 *g* for 30 min, then chromatographed on an adrenal ferredoxin-Sepharose affinity column. The beef adrenocrotical flavoprotein is eluted with 0.4 $M$ NaCl, 10 m$M$ potassium phosphate buffer (pH 7.4) and concentrated by ammonium sulfate precipitation between 30 and 70% saturation. Further chromatography on a Sephadex G-50 column (1.8 × 59 cm) equilibrated with 0.3 $M$ KCl, 50 m$M$ Tris · HCl (pH 7.5) yields flavoprotein preparations with OD 272 nm/OD 450 nm ratios between 6.9 and 7.8. The bovine adrenal flavoprotein is stable at liquid nitrogen temperatures without loss of activity.

Adrenocortical flavoprotein-dependent cytochrome *c* reductase activity is monitored spectrophotometrically in semimicrocuvettes at 25° by measuring the increase in absorbance of the reduced cytochrome *c* at 550 nm under conditions in which the reaction rate is linear with respect to renal ferredoxin concentration. The standard reaction mixture contains 10 $\mu$mol of potassium phosphate buffer (pH 7.4), 30.5 nmol of cytochrome *c* (Sigma, type III or VI), 92.0 pmol of bovine adrenal flavoprotein, and 100 nmol of NADPH freshly prepared in 25 m$M$ sodium borate (pH 8.1), for a total reaction volume of 1.0 ml. The rate of cytochrome *c* reduction is calculated using an extinction coefficient of 21.0 cm$^{-1}$ m$M^{-1}$ at 550 nm.[15]

Crude mitochondrial extracts containing respiratory chain membrane fragments that may contain contaminating NADPH–cytochrome *c* reductase activity are assayed using a dual-beam Cary-14 spectrophotometer. The extract is incubated in the presence of potassium phosphate buffer (pH 7.4), 10 m$M$ rotenone, and 5 m$M$ sodium cyanide in a total volume of 1.9 ml for 5 min at 25°. After the addition of 200 nmol of NADPH, equal amounts of the reaction mixture are divided between the reference and sample compartments until a stable base line is reached. The rate of cytochrome *c* reduction, which is independent upon the presence of ad-

[13] T. Omura, E. Sanders, and R. W. Estabrook, *Arch. Biochem. Biophys.* **117**, 660 (1966).
[14] T. Sugiyama and T. Yamano, *FEBS Lett.* **52**, 145 (1975).
[15] C. H. Williams and H. Kamin, *J. Biol. Chem.* **237**, 587 (1962).

renal ferredoxin reductase, is then determined by the addition of 92.0 pmol of bovine flavoprotein to the sample cuvette with an equivalent volume of buffer added to the reference compartment.

## Preparation of Mitochondria

Male, 6–8-week-old, white Plymouth Rock × Cornish cockerel chicks raised on a vitamin D replete, 1.0% calcium diet are killed by decapitation and exsanguination. Kidney tissue is removed, placed in chilled isolation media containing 0.25 $M$ sucrose, 15 m$M$ Tris-acetate (pH 7.4), 1 m$M$ DTT, and 0.1 m$M$ EDTA, and cleaned of adhering fat and connective tissue. Mitochondria are prepared by differential centrifugation from a 20% (w/v) homogenate, which is subfractionated by sedimentation of cellular debris, nuclei, and erythrocytes at 400 $g$ for 15 min using a swinging bucket rotor. The pellets are resuspended once in isolation media and recentrifuged. Mitochondria are then sedimented from the combined postnuclear supernatants by centrifugation at 7000 $g$ for 15 min. Mitochondrial pellets are washed twice by resuspension and centrifugation to minimize microsomal contamination. After the final wash, the pellets are resuspended in 50 m$M$ KCl, 50 m$M$ Tris · HCl (pH 7.5) to a concentration of approximately 50–60 mg of protein per milliliter prior to freezing in liquid nitrogen and storage at −70°.

## Purification Procedure

The purification procedure outlined in Table I can be utilized for the preparation of renal mitochondrial ferredoxin from vitamin D-replete animals raised on 1.0% dietary calcium. All steps are carried out at 0–5° unless otherwise indicated.

*Step 1. Solubilization by Hypotonic Treatment and Sonication.* Mitochondrial pellets containing 18–20 g of mitochondrial protein are resuspended in 50 m$M$ KCl, 50 m$M$ Tris · HCl (pH 7.5) to a final concentration of 20 mg of protein per milliliter. The suspension is allowed to equilibrate at 0° for 30 min with constant stirring. Aliquots (100 ml) are then sonicated for nine 30-sec intervals with a Branson Sonifier W-182 at 6 A output with intermittent cooling to maintain the temperature below 8°. The sonicated suspension is then centrifuged at 105,000 $g$ for 3 hr to yield a deep yellow supernatant containing the solubilized NADPH–cytochrome P-450 reductase components. The iron-sulfur protein from this extract is stable at temperatures below −70° over a period of weeks.

*Step 2. Chromatography on DEAE-Cellulose Column.* The supernatant fraction from the high speed centrifugation of sonicated mitochondria is

TABLE I

PURIFICATION OF RENAL FERREDOXIN FROM VITAMIN D-REPLETE CHOCK MITOCHONDRIA

| Step | Total protein (mg) | Total activity | Specific activity[a] | Yield (%) |
|---|---|---|---|---|
| Sonicated mitochondria | 18666 | 31680 | 1.69 | 100 |
| DEAE-cellulose pooled fractions | 190 | 12628 | 66.40 | 39.9 |
| Preparative electrophoresis pooled fractions | 16 | 5648 | 355.23 | 17.8 |
| DEAE-cellulose pooled fractions | 0.125 (0.100–0.340)[b] | 715 | 5721.60 (2102.9–7150.0) | 2.2 (2.1–3.3) |
| Disc polyacrylamide electrophoresis | 0.100 (0.085–0.124) | 676 | 6758.00 (5451.61–7952.94) | 1.8 (1.6–2.0) |

[a] Nanomoles of cytochrome $c$ reduced per minute per milligram protein.

[b] The degree of purity obtained from this step is highly dependent upon which peak fractions are combined. Values given in parentheses indicate the range of variation observed during separate purification experiments.

then loaded onto a DEAE-cellulose (Whatman No. 52, microgranular) column (3.5 × 17 cm) equilibrated with 50 m$M$ Tris · HCl (pH 8.0) at a flow rate of 60 ml/hr. The column is then washed with 250 ml of low ionic strength buffer containing 0.18 $M$ KCl, 50 m$M$ Tris · HCl (pH 7.8), which removes most of the extraneous proteins contained in the submitochondrial extract. Continuous gradient elution from 0.18 to 0.45 $M$ KCl in 10 m$M$ Tris · HCl (pH 7.5) using 200 ml of each elution buffer results in the separation of the ferredoxin component, as detected by its adrenal flavoprotein-dependent NADPH–cytochrome $c$ reductase activity, from other mitochondrial or microsomal contaminating activities. Fractions containing the highest amount of ferredoxin-catalyzed cytochrome $c$ reductase activities are pooled and frozen by swirling the combined fractions in an acetone–Dry Ice bath. The frozen sample is then lyophilized under vacuum at −50 to −60°, and the dried powder is dissolved in 28–30 ml of chilled distilled water. The concentrate is then dialyzed against two changes of 1.5 liters of 50 m$M$ Tris · HCl (pH 8.0) to remove salts prior to subsequent electrophoresis.

*Step 3. Discontinuous Preparative Electrophoresis.* A preparative polyacrylamide electrophoresis system consisting of a 3.8% stacking gel, 9.7% separating gel, and 9.7% supporting gel are prepared at room temperature immediately before use. Details of the apparatus used are described elsewhere.[16] The system consists of a 3 × 20 cm cylindrical glass

[16] P. W. Ludden, Doctoral Thesis, University of Wisconsin-Madison (1977).

chamber fitted with two microstopcocks 3 cm from the base, which has a 2-mm lip to prevent slippage of the gels during electrophoresis. The 9.7% support gel extends from the base to the level of the microstopcocks. A space between the separating and supporting gels is created during polymerization by layering a 30% sucrose (w/w) solution above the bottom gel before pouring the upper two gel phases. The sucrose solution is later replaced by elution buffer containing 50 mM Tris · HCl (pH 7.8), which flows continuously during separation through the microstopcocks. Thirty-five milliliters of 9.7% acrylamide gel and 12 ml of the 3.8% gel are then sequentially polymerized in the chamber. All gels are prepared in a final buffer concentration of 0.15 M Tris HCl (pH 8.9) and are prerun, following equilibration at 5°, in the presence of the same buffer for 2.5 hr at 100 V. The upper chamber buffer is replaced by 500 ml of 84 mM Tris-glycine (pH 8.3), and the lower portion of the gel system is immersed in 500 ml of 45 mM Tris-glycine (pH 8.5). Electrophoresis is carried out at 5° to prevent denaturation of the iron-sulfur protein and to minimize heating during electrophoresis. If all glassware is properly acid-washed prior to use, no distortion of the gels upon cooling should be observed. The dialyzed concentrate from the preceding step is made 10% (w/w) with sucrose and carefully layered above the stacking gel. Slow loading of the sample at 50 V over a period of 10–12 hr minimizes precipitation of the sample. After the concentrate enters the gel, voltage is increased to 100 V and protein bands can be seen migrating toward the anode. The microstopcocks are then opened, allowing flow of elution buffer at 120 ml/hr. A high flow rate is necessary for the separation of closely migrating bands; therefore, the renal ferredoxin elutes over a wide range of fractions in a large elution volume. The protein is concentrated on a small DEAE-cellulose column (1.4 × 6.5 cm) equilibrated with 0.2 M KCl, 50 mM Tris · HCl (pH 8.0) at a flow rate of 30 ml/hr and eluted with 0.45 M KCl, 50 mM Tris · HCl (pH 7.8).

*Step 4. Disc Gel Electrophoresis.* Protein aliquots of (9–10 μg) of the concentrated peak fractions are then simultaneously electrophoresed on a series of 8-cm, 14% disc gels, which are prepared and developed in the presence of 0.5 M Tris-glycine (pH 8.3) at 3.0 mA per tube for 2 hr. One gel is stained with 0.04% Coomassie-Blue G-250 in perchloric acid,[17] and all other gels are immediately fractionated into 1-mm slices using a Gilson Automatic Gel Crusher, and proteins are allowed to elute from the gels into 25 mM Tris · HCl (pH 7.5) at 5° for 12 hr. After enzymic and spectral analysis, corresponding fractions are pooled and concentrated by lyophilization. The final electrophoresis step effectively separates a soluble $b_5$-type cytochrome contaminant from the renal ferredoxin.

[17] A. H. Reisner, P. Nemes, and C. Bucholtz, *Anal. Biochem.* **64**, 509 (1975).

This purification procedure yields chick renal ferredoxin preparations from 3600–4000-fold purification, in 1.6–2.0% yield, from crude mitochondrial extracts isolated from vitamin $D_3$-replete animals. The specific content of this iron-sulfur component is estimated to be 24 pmol per milligram of mitochondrial protein based upon comparison of the value of the double integral of its electron paramagnetic resonance spectrum (EPR) in the reduced state with the value of the double integral of a copper-EDTA standard of known concentration.

### Properties

*Purity, Stability, and Molecular Weight.* Polyacrylamide electrophoresis of the purified renal ferredoxin in the presence of 0.1% sodium dodecyl sulfate (SDS) on 14% gels reveals that the iron-sulfur protein, purified to apparent homogeneity, has an estimated molecular weight of 11,900 based upon comigration with marker proteins. The protein is stable under conditions of high ionic strength (0.2–0.4 $M$ KCl) at pH levels above 7.5, and will retain both spectral properties and enzymic activity when stored at temperatures below $-70°$. Lyophilization of dilute ferredoxin-containing solutions at low temperatures is the most effective method for concentrating this component with minimal loss of activity.

*Spectral Properties.* The optical absorption spectrum of the oxidized form exhibit maxima at 412 nm and 454 nm, both of which are diminished upon reduction by sodium dithionite. Low temperature EPR spectra of the reduced renal ferredoxin yield prominent signals with $g$ values at $g_z = 2.02$ and $g_y \cong g_x = 1.94$. Further characterization of the iron-sulfur center contained in the chick ferredoxin has demonstrated that the active site of electron transfer contains $Fe_2S_2$, rather than $Fe_4S_4$, type clusters.[18]

*Reconstitution of 25-Hydroxyvitamin $D_3$ 1$\alpha$-Hydroxylase Activity.* In addition to its ability to transfer electrons to cytochrome $c$ from NADPH-reduced adrenal ferredoxin reductase, as utilized in the standard assay, the chick renal ferredoxin isolated from vitamin D-replete animals will reduce solubilized cytochrome P-450 from vitamin D-deficient chick mitochondria in the reconstitution of 25-OH-$D_3$ 1$\alpha$-hydroxylase activity. Incubations are performed in 25-ml Erlenmeyer flasks containing 3.3 $\mu$mol of glucose 6-phosphate, 100 nmol of NADPH, 1.0 $\mu$mol of magnesium acetate, and 1.5 enzyme units of yeast glucose-6-phosphate dehydrogenase. The concentrations of protein components were 0.25–0.26 nmol of cytochrome P-450, 4.40 nmol of renal ferredoxin, 0.16 nmol of adrenal ferredoxin reductase, and 4.51 nmol of cytochrome $b_5$, as indicated in

[18] P. S. Yoon, J. Rawlings, W. H. Orme-Johnson, and H. F. DeLuca, *Biochemistry* (in press).

TABLE II

RECONSTITUTION OF 25-HYDROXYVITAMIN $D_3$ $1\alpha$-HYDROXYLASE

| Incubation No. | Components in reaction system[a] | Activity (mol 1,25-$(OH)_2D_3$ formed mol P-450$^{-1}$ × $10^3$)[b] |
|---|---|---|
| 1 | Cytochrome P-450 | 0.00 |
| 2 | + Renal ferredoxin | 0.00 |
| 3 | + Adrenocortical ferredoxin reductase | 0.01 |
| 4 | + Renal ferredoxin + adrenocortical ferredoxin reductase | 11.30 |
| 5 | + Cytochrome $b_5$ + adrenocortical ferredoxin reductase | 0.02 |
| 6 | + Cytochrome $b_5$ + adrenocortical ferredoxin reductase + renal ferredoxin | 11.38 |

[a] The amounts of protein components added were 0.25–0.26 nmol of cytochrome P-450, 4.40 nmol of renal ferredoxin, 0.16 nmol of adrenocortical ferredoxin reductase, and 4.51 nmol of cytochrome $b_5$, where indicated. At these concentrations the chick renal ferredoxin component was rate-limiting. All incubations contained buffer, cofactors, and salt solutions as described in the text.

[b] Activity was calculated based upon comigration of the reaction product with authentic vitamin $D_3$ metabolite markers on HPLC.

Table II. All cofactor and salt solutions are prepared immediately before use. 4-Morpholine propane sulfonic acid (Mops), 25 $\mu M$ at pH 7.4, is added to a final concentration of 18 m$M$ for a total reaction volume of 1.1 ml. The mixture is gassed for 1.0 min with 100% oxygen and equilibrated at 37° for 3 min prior to initiation of the reaction with 19.8 pmol of 25-OH-[$3\alpha$-$^3$H]$D_3$ in 3 $\mu$l of 95% ethanol. All solvents are glass-distilled before use. After 60 min of incubation at 37°, with gentle agitation, the reaction is terminated by the addition of 10 ml of 2 : 1 methanol : chloroform mixture and extracted according to the method of Bligh and Dyer.[19] A two-phase system is formed by the addition of 2.5 ml of distilled water and 5 ml of chloroform at 5°. The lower chloroform phase is removed, and the upper aqueous phase is reextracted with 5 ml of chloroform. The organic phases are then combined and evaporated to dryness. The residue is redissolved in 100 $\mu$l of 10% isopropanol in $n$-hexane. Twenty-five microliters of extract are removed for determining extraction and chromatographic recoveries, and an additional 30 $\mu$l are subjected to high-pressure liquid chromatography (HPLC) using a Waters U6K injector fitted with a Zorbax-silica column (0.46 × 25 cm). The sample is co-injected with nonradioactive synthetic vitamin $D_3$ markers and eluted with 10% iso-

[19] E. G. Bligh and W. J. Dyer, *Can. J. Biochem. Physiol.* 37, 911 (1959).

propyl alcohol in $n$-hexane at a pressure of 900 psi of nitrogen gas yielding a flow rate of 2.4 ml/min. Twenty 2.5-ml fractions are collected; the retention times of 25-OH-$D_3$, 24,25-$(OH)_2D_3$, 25,26-$(OH)_2D_3$, and 1,25-$(OH)_2D_3$ are monitored by measuring the amount of absorbance of each internal standard at 254 nm. The solvent from each fraction is evaporated under an air stream, and 5 ml of counting solution[6] are added. Each sample is counted using a Packard Model 3255 liquid scintillation spectrometer, and the amount of 25-OH-$[3\alpha\text{-}^3H]D_3$ converted to 1,25-$(OH)_2D_3$ is determined by comigration with the known standards. Recoveries from the HPLC columns average 80–85% of injected radioactivity. As shown in Table II, hydroxylation at the C-1 position requires the presence of adrenal ferredoxin reductase, renal ferredoxin, and solubilized cytochrome P-450. Requirements for other components in the reaction mixture have been described in preliminary reconstitution experiments.[20] The soluble $b_5$-type cytochrome fails to support 1$\alpha$-hydroxylation when substituted for the renal ferredoxin at equimolar concentrations; nor does the $b_5$ hemoprotein affect the rate of the renal ferredoxin-catalyzed reaction. Under these reconstitution conditions, analysis of the hydroxylation products reveals that only the 1,25-$(OH)_2D_3$ product is formed.[21]

## Comments

While the purification procedure and reconstituted enzymic activity of the chick renal ferredoxin described above demonstrate that vitamin D supplementation does not appear to affect either the enzymic properties or the specific content of this electron transport component when isolated from mitochondria of normal animals, the source of the cytochrome P-450 hemoprotein is known to be dependent upon vitamin D status.[20] Hence, the solubilization procedures described herein for the P-450 component is applicable only for preparations from vitamin D-deficient chick mitochondria.

Significant amounts of up to 90 pmol per milligram of mitochondrial protein of a low molecular weight $b_5$-type cytochrome appeared consistently during purification of the chick renal ferredoxin until the final electrophoresis step. Although this cytochrome does not affect the rate of 1$\alpha$-hydroxylation of 25-OH-$D_3$, determinations of renal ferredoxin concentrations in impure preparations require the use of alternative methods of detection, such as EPR, rather than optical absorption spectroscopy.

[20] J. I. Pedersen, J. G. Ghazarian, N. R. Orme-Johnson, and H. F. DeLuca, *J. Biol. Chem.* **251**, 3933 (1976).
[21] P. S. Yoon and H. F. DeLuca, *J. Biol. Chem.* (in preparation).

# [51] 25-Hydroxylation of Vitamin D₃ in Liver Microsomes and Their Smooth and Rough Subfractions

*By* SOREL SULIMOVICI, MARTIN S. ROGINSKY, and ROBERT F. PFEIFER

In 1971, when "Methods in Enzymology" last dealt with vitamin D, isolation of the plasma metabolites 25-hydroxyergocalciferol (25-OH-$D_2$) and 25-hydroxycholecalciferol (25-OH-$D_3$) was described.[1] Subsequently, the production of 25-OH-$D_3$ from $D_3$ was demonstrated in liver,[2,3] kidney, and small intestines.[4] The enzyme responsible for this reaction is calciferol 25-hydroxylase, which occurs in the microsomal fraction of the endoplasmic reticulum.[5] This step requires NADPH and molecular oxygen and is similar to hydroxylations involved in the biosynthesis of steroid hormones. All these reactions are catalyzed by enzymes designated as mixed-function oxidases that activate molecular oxygen and cause the incorporation of one atom of the oxygen into the substrate while the other oxygen atom is presumably reduced to water in the presence of NADPH.[6]

In 1973, Bhattacharyya and DeLuca[2] described an assay for calciferol 25-hydroxylase activity in liver homogenates. The activity of this enzyme in D-deficient chicken[7] and rat liver homogenates,[2,3] and in rat liver microsomes has been studied.[5] This chapter describes a modification of the above procedure to study the properties of the enzyme system in rat liver microsomes and in smooth- and rough-surfaced subfractions.

## Preparation of Microsomes

Male albino rats (Holzman Co., Madison, Wisconsin) from D-deficient mothers are used for all organelle preparations. It is essential that from birth the animals be shielded from ultraviolet light and fed a low D diet (Nutritional Biochemicals, Cleveland, Ohio) with water supplemented ad libitum. At age 6–8 weeks, the rachitic animals are sacrificed without anesthesia; the livers are immediately removed, cleaned free of connective tissue, and kept on ice. Within 30 min of sacrifice, the minced livers are homogenized in 4 volumes of ice-cold 0.25 $M$ sucrose using a motor-

[1] H. F. DeLuca and J. W. Blunt, this series, Vol. 18C, p. 709.

[2] M. H. Bhattacharyya and H. F. DeLuca, *J. Biol. Chem.* **248**, 2969 (1973).

[3] M. H. Bhattacharyya and H. F. DeLuca, *J. Biol. Chem.* **248**, 2974 (1973).

[4] G. Tucker, III, R. E. Gagnon, and M. R. Hausler, *Arch. Biochem. Biophys.* **155**, 47 (1973).

[5] M. H. Bhattacharyya and H. F. DeLuca, *Arch. Biochem. Biophys.* **160**, 58 (1974).

[6] H. S. Mason, *Adv. Enzymol.* **19**, 79 (1957).

[7] M. H. Bhattacharyya and H. F. DeLuca, *Biochem. Biophys. Res. Commun.* **59**, 734 (1974).

driven homogenizer fitted with a Teflon pestle. Careful homogenization is important, and no more than five or six passes are made. The resulting suspension is centrifuged in the cold at 600 g for 15 min; the precipitate is discarded, and the supernatant fraction is recentrifuged at 9000 g for 20 min. The supernatant is carefully decanted and further centrifuged at 105,000 g for 60 min in a refrigerated ultracentrifuge to sediment the microsomes. The microsomal pellet is washed with 0.25 M sucrose and recentrifuged at the same speed for an additional 60 min. This washed microsomal pellet can either be assayed for 25-hydroxylase activity or further subfractionated into the smooth and rough membranes.

### Isolation of Smooth and Rough Microsomes

The crude microsomal pellet is resuspended in ice-cold 0.25 M sucrose containing 20 mM CsCl (10–12 mg of microsomal protein per milliliter) using hand homogenization. This suspension is then subjected to ul-trasonication at 0° for 30 sec using a sonifier cell disruptor operating at 20 kHz. Sodium deoxycholate (0.01%) is added, and the mixture is incubated at 4° for 30 min. Immediately after this, the suspension (4 ml) is layered on top of 3 ml of 1.4 M sucrose solution containing 15 mM CsCl[8] and centrifuged for 2 hr at 200,000 g in a Beckman Model L ultracentrifuge using a 50 type rotor (20° tube angle). After centrifugation, the clear reddish upper phase is removed with a Pasteur pipette and discarded. The fluffy layer (smooth microsomes) formed at the boundary of the two sucrose phases is aspirated and resuspended in 0.25 M sucrose. The lower 1.4 M sucrose phase is discarded, and the residue, a dark brown pellet (rough microsomes), is also resuspended in 0.25 M sucrose. Both subcellular membranes are centrifuged at 200,000 g for 45 min, and the resulting pellets are washed once with 0.25 M sucrose. These smooth and rough microsomes are resuspended in ice-cold 0.1 M $K_2HPO_4$, pH 7.4, and must be assayed immediately for 25-hydroxylase activity.

### Calciferol 25-Hydroxylase Assay

*Principle*. The conversion of $[^3H]D_3$ to $[^3H]25$-OH-$D_3$ over a standard period of time is the basis of the assay. Quantitation of the $[^3H]25$-OH-$D_3$ produced by the particular microsomal or submicrosomal fraction provides a measure of enzymic activity.

*Reagents and Materials*. The following are used in this assay: sucrose, $MgSO_4$, KCl, nicotinamide, ATP, $NADP^+$, glucose, glucose-6-phosphate dehydrogenase, glucose 6-phosphate, cesium chloride (CsCl), sodium

---

[8] G. Dalner, *Acta Pathol. Microbiol. Scand. Suppl.* **166,** 1 (1963).

deoxycholate, toluene, $K_2HPO_4$, chloroform, methanol, Skellysolve B, Sephadex LH-20, Tween 80, [$^3$H]D$_3$, 2,5-diphenyloxazole (PPO), 1,4-bis-2-(5-phenyloxazolyl)benzene (POPOP).

*Preparation of [$^3$H]D$_3$ Standard.* [$^3$H]D$_3$(2.5 $\mu$Ci(12.3 Ci/mmol), Amersham Searle) is mixed with 100 $\mu$l of Tween 80 and 4.9 ml of freshly prepared 0.1 $M$ $K_2HPO_4$, pH 7.4, to yield a solution of about 100,000 dpm per 100 $\mu$l of buffer. Storage at $-20°$ in small aliquots to avoid frequent freezing and thawing allows the compound to be stable for a month.

*Incubation Procedure.* A portion of the microsomes or of the smooth or rough membrane subfractions equivalent to 5–10 mg of protein is resuspended in 0.1 $M$ $K_2HPO_4$ buffer, pH 7.4. To 1 ml of this microsomal suspension in a 25-ml Erlenmeyer flask is added a mixture of 5.0 m$M$ $MgSO_4$, 0.1 $M$ KCl, 0.16 $M$ nicotinamide, 20 m$M$ ATP, 20 m$M$ glucose, 0.4 m$M$ NADP$^+$, 22.4 m$M$ glucose 6-phosphate, 2.5 IU of glucose-6-phosphate dehydrogenase, and 0.05 $\mu$Ci [$^3$H]D$_3$ (4.1 pmol) in 0.1 $M$ $K_2HPO_4$ buffer, pH 7.4, to a total volume of 3 ml. Incubations are performed in triplicate at 37° for 2 hr in air using a Dubnoff incubator with constant shaking. A blank containing the substrate [$^3$H]D$_3$ in buffer alone is carried through the entire procedure.

*Extraction and Separation of D$_3$ Metabolites.* The addition of 15 ml of methanol : chloroform (2 : 1, v/v) terminates the incubation period, and the combined contents are decanted into a small separatory funnel. After gentle shaking for 1 hr at 4°, 2.5 ml of water and 5 ml of chloroform are added and the two-phase system is allowed to separate. The lower organic phase is removed and the aqueous phase is reextracted with an additional 10 ml of chloroform.[9] Combined chloroform extracts are taken to dryness at 38° under a stream of nitrogen. Separation of [$^3$H]D$_3$ from [$^3$H]25-OH-D$_3$ is carried out on a 1 × 52 cm column of 17 g of Sephadex LH-20 slurried in chloroform : Skellysolve B (60 : 40, v/v).[10] The dry residue is redissolved in 1 ml of the same solvent and applied to the prepared Sephadex LH-20 column. Elution with 75 ml of the column solvent at a flow rate adjusted to deliver 0.5 ml/min is carried out. A total of eighteen 4-ml fractions are collected directly into counting vials. These are taken to dryness, redissolved in 9 ml of scintillation mixture (5 g of PPO and 100 mg of POPOP in 1 liter of dry toluene containing 5% methanol), and counted in a liquid scintillation spectrometer. Total radioactivity of the two separate peaks representing [$^3$H]D$_3$ and [$^3$H]25-OH-D$_3$ is determined for the pooled vials.

*Calculation of the Enzymic Activity.* Quantitation of enzymic activity is derived from the extent of conversion of added [$^3$H]D$_3$ to [$^3$H]25-OH-D$_3$

[9] E. G. Bligh and M. J. Dyer, *Can. J. Biochem. Physiol.* **37**, 911 (1959).
[10] M. F. Holick and H. F. DeLuca, *J. Lipid Res.* **12**, 460 (1971).

per unit weight (milligrams of protein content) of the individual microsomal or submicrosomal fractions. It is assumed that the fractions examined for 25-hydroxylase activity contain negligible endogenous $D_3$ in these rachitic animals. Results are expressed as picomoles of 25-OH-$D_3$ formed per milligram of protein per minute (calculated best from the specific activity of the substrate [$^3$H]$D_3$).

## Comments

The principle of this method utilizing the conversion of a radioactive substrate to an end product is applicable to the study of any enzymic step in the metabolism of D. Although activity can be demonstrated only in preparations obtained from rats maintained on D-deficient diet and allowed to become rachitic, the procedure described provides a consistently reproducible and sensitive means for determining microsomal 25-hydroxylase activity, as well as that of the smooth and rough microsomal subfractions.

The identity of the product of the enzymic reaction [$^3$H]25-OH-$D_3$ is established by additional chromatography on silicic acid,[11] Celite liquid–liquid partition,[12] and high-pressure liquid chromatography.[13] Confirmation of this identity is further established by comparative binding of the isolated metabolite and pure standard [$^3$H]25-OH-$D_3$ to its serum transport protein.[14]

The reaction follows a linear time course for 1 hr and continues to increase for an additional hour. At that time, the conversion of [$^3$H]$D_3$ to [$^3$H]25-OH-$D_3$ is 6–8% per 10 mg of microsomal protein for total microsomes. The relationship between enzyme concentration and the amount of product formed is linear between 3 and 10 mg of microsomal protein.

The most common difficulty encountered in the assay is the instability of the enzyme system. In most cases, preparations of smooth and rough vesicles that are left in ice overnight show no detectable hydroxylase activity when assayed the following day. It is essential that the incubation step be performed immediately following the preparation of the membrane subfractions.

[11] M. A. Preece, J. L. H. O'Riordan, D. E. M. Lawson, and E. Kodicek, *Clin. Chem. Acta* **54**, 235 (1974).
[12] J. W. Blunt, H. F. DeLuca, and H. K. Schnoes, *Biochemistry* **7**, 3317 (1968).
[13] J. A. Eisman, R. M. Shepard, and H. F. DeLuca, *Anal. Biochem.* **80**, 298 (1977).
[14] S. Sulimovici and M. S. Roginsky, *Biochem. Biophys. Res. Commun.* **71**, 1078 (1976).

## [52] Measurement of the Chicken Kidney 25-Hydroxyvitamin D$_3$ 1-Hydroxylase and 25-Hydroxyvitamin D$_3$ 24-Hydroxylase

### By HELEN L. HENRY

In the kidney 25-hydroxyvitamin D$_3$ is hydroxylated at the C-1 position by 25-hydroxyvitamin D$_3$ 1-hydroxylase (1-hydroxylase) or alternatively at the C-24 position by 25-hydroxyvitamin-D$_3$ 24-hydroxylase (24-hydroxylase). The activity of the 1-hydroxylase is highest in the completely vitamin D-deficient state, somewhat lower when vitamin D is present and dietary calcium is low, and very low when both vitamin D and high levels of dietary calcium are present. As shown in Table I, the activity of the 24-hydroxylase is highest under conditions of vitamin D repletion and high dietary calcium.

Properties of 25-Hydroxyvitamin D$_3$ 1-Hydroxylase and
  25-Hydroxyvitamin D$_3$ 24-Hydroxylase

The 25-hydroxyvitamin D$_3$ 1-hydroxylase enzyme complex is located exclusively in the mitochondria. It consists of cytochrome P-450[1,2] and a nonheme ion protein (ferredoxin) similar to the adrenodoxin component of adrenal mitochondrial steroid hydroxylases.[3]

The $K_m$ for the substrate, 25-hydroxyvitamin D$_3$, is 1 to $2 \times 10^{-7} M$.[1] NADPH is the required source of reducing equivalents, and it can be supplied intramitochondrically by oxidation of either malate or succinate. In isolated mitochondria the enzyme is mardekly inhibited by Mn$^{2+}$ ($K_i = 3 \times 10^{-6} M$), calcium ($K_i = 2 \times 10^{-5} M$), and strontium ($K_i = 3 \times 10^{-5} M$). Inorganic phosphate ($K_i = 7 \times 10^{-3} M$) also inhibits the 1-hydroxylase.[1]

The measurable activity of the 1-hydroxylase in isolated mitochondria is greater in a hypotonic than in an isotonic incubation medium,[4] even when malate is the cofactor.

The 25-hydroxyvitamin D$_3$ 24-hydroxylase is also localized in the kidney mitochondria and requires NADPH and molecular oxygen.[5] The $K_m$

[1] H. Henry and A. W. Norman, *J. Biol. Chem.* **249**, 7529 (1974).

[2] J. G. Ghazarian, C. R. Jefroate, J. C. Knutson, W. H. Orme-Johnson, and H. F. DeLuca, *J. Biol. Chem.* **249**, 3026 (1974).

[3] J. I. Pedersen, J. G. Ghazarian, N. R. Orme-Johnson, and H. F. DeLuca, *J. Biol. Chem.* **251**, 3933 (1976).

[4] H. L. Henry and A. W. Norman, *Arch. Biochem. Biophys.* **172**, 582 (1976).

[5] J. C. Knutson and H. F. DeLuca, *Biochemistry* **13**, 1543 (1974).

TABLE I
CONCENTRATIONS OF 25-HYDROXYVITAMIN D$_3$ 1- OR 24-HHDROXYLASES UNDER VARIOUS
CONDITIONS

| Dietary condition | | Picomoles/min/mg kidney homogenat protein | |
|---|---|---|---|
| Vitamin D | Calcium | 1-Hydroxylase | 24-Hydroxylase |
| − | Low or high | 2–3 | 0 |
| + | Low (0.2%) | 0.6–1.5 | 0.02–0.10 |
| + | High (1.2%) | 0.1–0.3 | 0.1–0.5 |

for 25-hydroxyvitamin D$_3$ is similar to that of the 25-hydroxyvitamin D$_3$
1-hydroxylase ($10^{-7}$ $M$).[6]

Assay Method

*Principle.* The conversion of tritium-labeled 25-hydroxyvitamin D$_3$ to
1,25-dihydroxyvitamin D$_3$ or to 24,25-dihydroxyvitamin D$_3$ is measured
by chromatographic separation of the product(s) from the substrate in a
lipid extract of incubation samples. Although there are a number of variations that can be introduced, the following procedure results in reproducible initial rates of enzyme activity. It involves the separate handling of
several timed samples for each assay, but since the length of time over
which the reaction is linear can vary, it is preferable to a single time-point
assay.

*Reagents*

Tris · HCl, 50 m$M$, pH 7.4
Sodium malate, 0.3 $M$ (in Tris · HCl)
MgCl$_2$, 50 m$M$ (in Tris · HCl)
25-Hydroxy[26,27-$^3$H]vitamin D$_3$, approximately 20 Ci/mol, 20 $\mu M$
   (in ethanol)

$^3$H-25-Hydroxyvitamin D$_3$ may be purchased at a specific radioactivity
of 9–12 Ci/mmol and diluted with a known amount (determined by absorbance at 264 nm, $E$ = 16.3 × 10$^3$) of nonradioactive 25-hydroxyvitamin
D$_3$ (presently available only from Dr. John Babcock of the Upjohn Company). The specific activity of the 25-hydroxyvitamin D$_3$ to be used as
substrate must be accurately determined by measuring the absorbance at
264 nm for total concentration of 25-hydroxyvitamin D$_3$ and measuring the
disintegrations per minute in an aliquot containing a known amount of the
steroid.

[6] H. L. Henry, M. V. Icenogle, and A. W. Norman, unpublished data (1975).

*Tissue Preparation.* Homogenates of chick kidney (10%, w/v) are prepared in 0.25 $M$ sucrose, maintaining the temperature at 4° or below. The homogenate may be assayed directly or mitochondria may be prepared by the following procedure. The homogenate is centrifuged at 600 $g$ for 10 min, the supernatant fraction is set aside, and the pellet is resuspended in the original volume of 0.25 $M$ sucrose. After a second 600 $g$ centrifugation, the two supernatant fractions are combined and centrifuged at 8500 $g$ for 20 min. At least 90% of the original 1-hydroxylase activity will be present in the pelleted mitochondria, which are then resuspended in the original volume of 0.25 $M$ sucrose. The 1-hydroxylase is not stable to freezing and should be assayed within 90 min of removal of the kidney.

*Incubation Procedure.* A 10-ml incubation mixture in a 20-ml beaker is prepared by adding 7.27 ml of Tris · HCl, 0.33 ml malate (final concentration, 10 m$M$), 0.4 ml MgCl$_2$ (final concentration, 2 m$M$), and 2.0 ml homogenate or mitochondrial suspension. After warming the mixture to 37°, the reaction is initiated by the addition of 0.05 ml of [$^3$H]25-hydroxyvitamin D$_3$; a 2.0-ml sample is immediately removed for a zero time point. Samples of 2.0 ml are subsequently removed from the same beaker at 4, 8, and 12 min after beginning the assay. As each 2.0-ml sample is removed, it is added to 4 ml of methanol: CHCl$_3$, 2 : 1 (v/v).

*Lipid Extraction.* In a 30-ml separatory funnel, a second 4 ml of methanol: CHCl$_3$, 2 : 1, is added to the sample, followed by the addition of 1.0 ml of H$_2$O and 1.0 ml of CHCL$_3$ and thorough agitation of the mixture. When both the CHCL$_3$ (lower) and methanol–water phases are clear (10–15 min), the CHCl$_3$ layer is removed and saved. The methanol–water layer must be back-extracted two times, using 1 ml of CHCl$_3$ each time and adding the CHCl$_3$ phases to the original one. Without these back extractions, the recovery of [$^3$H]1,25-diydroxyvitamin D$_3$ may be as low as 70%.

*Chromatography.* COLUMN PREPARATION. If only 1,25-dihydroxyvitamin D$_3$ production is of interest, columns of 0.8 × 15 cm may be used. If 24,25-dihydroxyvitamin D$_3$ is also being determined, a column length of 23 cm is required. In either case, Sephadex LH-20 is equilibrated with the elution solvent (hexane: CHCl$_3$, 35:65), glass columns with Teflon stopcocks are partially filled with solvent (glass wool is an inexpensive and adequate alternative to fritted-glass columns), the slurry of Sephadex LH-20 is added to the column, and the stopcock is opened. Continued additions of the slurry are made until the desired column height is achieved. Depending on the amount of lipid in each sample, columns can be reutilized 5–15 times. It is convenient to run 5 or 10 columns simultaneously. To accomplish this, a 1-liter, stainless steel, solvent reservoir is used; it has five outlets (at the bottom) for Teflon tubing, which is connected to each column with a straight union reducer.

TABLE II
ELUTION VOLUMES OF VITAMIN D METABOLITES FROM SEPHADEX LH-20

| Metabolite | Length of column, 0.8 cm diameter (ml) | |
|---|---|---|
| | 15 cm | 23 cm |
| 25-Hydroxyvitamin $D_3$ | 2–12 | 2–18 |
| 24,25-Dihydroxyvitamin $D_3$ | 12–20 | 24–34 |
| 1,25-Dihydroxyvitamin $D_3$ | 20–30 | 40–60 |

SAMPLE CHROMATOGRAPHY. Chloroform is evaporated in a 50° water bath under a stream of nitrogen, and the dried lipids are dissolved in 0.3 ml hexane : $CHCl_3$ (65 : 36, v/v). After the sample is applied to the top of the column, the reservoir is attached to the column via its straight union reducer. With a head of 1 meter, the solvent flow rate is approximately 0.5–1.0 ml/min.

The elution volumes of 25-hydroxyvitamin $D_3$, 24,25-dihydroxy-vitamin $D_3$, and 1,25-dihydroxyvitamin $D_3$ are shown in Table II. One can collect 2.0-ml fractions throughout the elution or collect the peaks as single fractions with several 1.0- or 2.0-ml fractions between each peak. The elution volumes in Table II should be verified each time a new set of columns is prepared using tritium-labeled markers for each of the vitamin D metabolites.

The solvent in each column fraction is evaporated under air, and radioactivity is determined by liquid scintillation counting. A convenient scintillation fluid is 5 g of phenylbiphenyl-1,3,4-oxadiazole per liter of toluene.

The percentage of the total radioactivity eluting from the column which occurs in each peak is determined, and the four time points from each incubation are used to calculate the rate of production of 1,25-dihydroxyvitamin $D_3$ or 24,25-dihydroxyvitamin $D_3$ unless the final point is clearly showing nonlinearity, in which case only the first three data points are used. Knowing the original substrate concentration allows the calculation of picomoles of product formed per minute per incubation, and results are normally expressed as picomoles per minute per milligram of protein.

Other methods have been described for the separation of the metabolites of 25-hydroxyvitamin $D_3$. Paper chromatography[7] has been introduced but is not in wide use. Longer Sephadex LH-20 columns result in greater separation of the metabolites but require more time and elaborate

[7] D. D. Bikle and H. Rasmussen, *Biochim. Biophys. Acta* **362,** 425 (1974).

fraction-collection equipment. High-pressure liquid chromatography[8] is probably the most effective at separating the steroids, but the equipment and personnel costs make this technique impractical for routine chromatography of many samples.

[8] G. Jones and H. F. DeLuca. *J. Lipid Res,* **16,** 448 (1975).

# [53] Purification, Characterization, and Quantitation of the Human Serum Binding Protein for Vitamin D and Its Metabolites

*By* John G. Haddad, Jr.

The introduction of radioactively labeled vitamin D of high specific activity led to elucidation of the metabolism of vitamin D. Whether derived from the cutaneous photoconversion of 7-dehydrocholesterol to cholecalciferol ($D_3$), or from the intestinal absorption of $D_3$ and ergocalciferol ($D_2$), the parent vitamin acquires a hydroxyl group at the C-25 position in the liver. The hepatic product, 25-hydroxycholecalciferol (25-OH-$D_3$) is further metabolized by the kidney to 1,25-dihydroxycholecalciferol [1,25-$(OH)_2D_3$], the most potent metabolite in promoting intestinal absorption of calcium and phosphorus and in effecting skeletal resorption.

The human blood transport of antiricketic factors was initially recognized to be a function of a protein with $\alpha$-mobility,[1] which had a density greater than the lipoprotein classes. In addition to bioassay and use of radioactive vitamin D sterols, the development of a competitive protein-binding radioassay for 25-OH-D provided another measure whereby the carrier protein could be identified by its ligand content or ligand-binding capacity during serum fractionation and purification procedures.[2] However, it was the use of radiolabeled sterols that led to the recent demonstration[3] that the group-specific component (Gc) serum protein and the serum binding protein for vitamin D and its metabolites (DBP) were the same protein. The purification of Gc protein had been published,[4] but

[1] W. C. Thomas, Jr., H. G. Morgan, T. B. Conner, L. Haddock, C. E. Bills, and J. E. Howard, *J. Clin. Invest.* **28,** 1078 (1959).
[2] J. G. Haddad, Jr. and K. J. Chyu, *J. Clin. Endocrinol. Metab.* **33,** 992 (1971).
[3] S. P. Daiger, M. S. Schanfield, and L. L. Cavilli-Starza, *Proc. Natl. Acad. Sci. U.S.A.* **72,** 2026 (1975).
[4] H. Cleve, J. H. Prunier, and A. G. Bearn, *J. Exp. Med.* **118,** 711 (1963).

three laboratories, without knowledge of its identity to DBP, worked to isolate human plasma DBP.[5-7] In fact, earlier studies had suggested the nonidentity of Gc and DBP.[8] Several studies prior to the "discovery" of DBP, which had previously been recognized as the genetic marker Gc protein, are of interest, for a considerable amount of clinical and epidemiological information had been gathered.[9] Whether the Gc variants might serve different functions in the conservation, transport, or metabolism of vitamin D or its metabolites has been the subject of recent speculation, but there are no presently available data to support these possibilities.

### Starting Source of Material for DBP Purification

Human plasma or serum can be utilized, but Cohn fraction IV serves as a more convenient starting material.[5] Furthermore, the ligand (25-OH-D) binding capacity per protein is enriched 2-fold in this fraction when compared to plasma.[5] Since the DBP content is increased in plasma from pregnant subjects and those who are receiving estrogens,[10] such plasma can also serve as a richer source.

Normally, a very small fraction of circulating DBP is saturated with vitamin D sterols,[5] and this fact was earlier suggested by the nearly identical D-sterol-binding capacity of normal versus D-deficient sera.[11] The concentration of DBP in plasma is high compared to the concentrations of vitamin and hormone-binding proteins. At nearly 10 $\mu M$ relatively small amounts of plasma can be used. Thus, 100 ml of plasma, purified to a recovery of only 10%, would yield 5 mg of DBP.

### Purification Methods

#### Column Purification Procedures

Four grams of human plasma Cohn fraction IV powder is solubilized in 200 ml of 10 m$M$ Tris, pH 8.3, and dialyzed against the same buffer for 24

[5] J. G. Haddad, Jr. and J. Walgate, *J. Biol. Chem.* **251**, 4803 (1976).

[6] M. Imawari, K. Kida, and D. S. Goodman, *J. Clin. Invest.* **58**, 514 (1976).

[7] R. Bouillon, H. Van Baelen, W. Rombauts, and P. De Moor, *Eur. J. Biochem.* **66**, 285 (1976).

[8] P. A. Peterson, *J. Biol. Chem.* **241**, 7748 (1971).

[9] H. Cleve, *Isr. J. Med. Sci.* **9**, 1133 (1973).

[10] J. G. Haddad, Jr., L. Hillman, and S. Rojanasathit, *J. Clin. Endocrinol. Metab.* **43**, 86 (1976).

[11] J. G. Haddad, Jr. and T. C. B. Stamp, *Am. J. Med.* **57**, 57 (1974).

hr at 4°. One buffer change is made, and the retentate supernatant (5000 rpm for 15 mins) is saved. The supernatant is then added to a flask in which 100 pmol of [26,27-$^3$H]25-OH-D$_3$ in ethanol had been N$_2$-dried. At 12.2 Ci/mmol (Amersham), this provides approximately 2.5 million dpm of the sterol. After 1–2 hr at 4°, the solution is applied to a 3 × 45 cm column of DEAE-cellulose (Cellex-D, Bio-Rad) equilibrated in the same buffer. After application of the protein solution, the column is washed in the same buffer until eluates do not contain material that absorbs UV light at 280 nm.

With 500 ml of starting buffer in the first reservoir at constant magnetic stirring, a second reservoir containing 10 m$M$ Tris and 0.3 $M$ NaCl, pH 8.3, is connected. Protein containing the tracer sterols is retained on the column until the salt gradient is well established. Aliquots of the eluates are analyzed for their optical density and radioactivity content. Those fractions with radioactivity are pooled, dialyzed against water at pH 7.4, and lyophilized.

The lyophilate is solubilized in 10 m$M$ Tris, pH 7.4, containing 0.15 $M$ NaCl and 0.02% NaN$_3$, and applied to a 10 × 90 cm column of Sephadex G-200 slurried in the same buffer. Since ligand-binding is favored in the cold and at neutral to alkaline pH,[12] alkaline buffers and coldroom purification steps are utilized. Gel filtration is carried out in ascending fashion, with 110-ml fractions collected at a flow rate of 150 ml/hr. Under these conditions, the material containing radioactivity is detected in the 4270–5240 ml elution volume. Those fractions containing radioactivity are pooled and concentrated in a filtration apparatus (UM-10 membrane, Amicon Corporation) under 40 psi N$_2$ at 4°. The concentrated pool is dialyzed against 0.1 $M$ Tris buffer, pH 8.3, for 48 hr and three buffer changes.

The dialysis retentate is then applied to a 3 × 45 cm column of DEAE-Sephadex A-50 equilibrated in the same buffer (Fig. 1). At this stage, essentially no protein is excluded from the column, but it elutes during a linear gradient of NaCl. The second buffer reservoir contains 0.4 $M$ NaCl in 500 ml of 10 m$M$ Tris, pH 8.3, and flows into the first reservoir of 500 ml 10 m$M$ Tris, pH 8.3, during continual magnetic stirring. Under these conditions, a dominant part of the major contaminating protein, serum albumin, is eluted prior to those fractions containing the DBP-radioactive sterol complex. By this step (see Table I), there is approximately a 50-fold purification of the binding protein.

Alternative procedures are available in the purification sequence. For example, Blue Sepharose 6B chromatography can be used to retain albumin and remove it from the DBP.[6] Similarly, preparative isoelectric focus-

---

[12] J. G. Haddad, Jr. and K. J. Chyu, *Biochim. Biophys. Acta* **248**, 471 (1971).

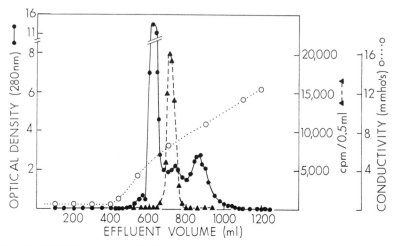

Fig. 1. DEAE-Sephadex A-50 chromatography of DBP-containing material from the Sephadex G-200 column. The sample (2.2 g of protein) was dialyzed in the equilibrating buffer of 10 m$M$ Tris, pH 8.3, and applied to a 3 × 45 cm column of DEAE-Sephadex. Fractions of 10 ml were collected at a flow rate of 25 ml/hr. After 40 fractions were collected, a second buffer reservoir containing 500 ml of 10 m$M$ Tris and 0.4 $M$ NaCl, pH 8.3, was connected to the initial buffer (500 ml) during continuous magnetic stirring of the buffer in the first reservoir. From J. G. Haddad, Jr. and J. Walgate, *J. Biol. Chem.* **251**, 4803 (1976).

ing (isoelectric point 4.8) on polyacrylamide gel can further help isolate the binding protein.[6] It is interesting to note that repeated intermediary steps of purification have been utilized by most groups.[5-7]

Another 2-fold purification is achieved by a repeat of the DEAE-Sephadex step, employing a more gradual salt gradient (0 to 0.225 $M$ NaCl in 10 m$M$ Tris, pH 8.3) and a repeat Sephadex G-200 filtration on a smaller column (2.5 × 45 cm). At this stage, 7% polyacrylamide gel elec-

TABLE I

PROCEDURES IN THE PURIFICATION OF HUMAN PLASMA DBP

| Purification step | Fold purification compared to serum |
|---|---|
| 1. Cohn IV | 2.1[a] |
| 2. DEAE-cellulose | 4 |
| 3. Sephadex G-200 | 9 |
| 4. DEAE-Sephadex | 49 |
| 5. DEAE-Sephadex | 86 |
| 6. Sephadex G-200 | 109 |
| 7. Preparative gel electrophoresis | 167 |

[a] As estimated by comparison of 25-OH-D$_3$ saturation analysis of normal human plasma and the Cohn fraction.

trophoresis (PAGE) revealed anodal (albumin) and cathodal contaminants to the DBP band that contained the 25-OH-D$_3$ radioactivity.

## Preparative Polyacrylamide Gel Electrophoresis

Preparative gel electrophoresis is performed with 7% polyacrylamide gels in a Polyprep (Buchler Instruments) apparatus (Fig. 2). Separating gel columns are 11 cm, with a 1-cm stacking gel, using Tris-glycine buffer at pH 8.3; 15 mg of the DBP preparation are applied, with bromophenol blue anodal tracking dye, and the major portion of each run is carried out with 10 mA current. Fractions of 3 ml are collected at a flow rate of 9 ml/hr. The complete procedure, at 4°, requires 12–16 hr. Occasionally, lower separating gel fragmentation occurs, and this interference can be avoided by banding a layer of gauze or porous material (tea bag paper) to the distal end of the glass column holding the separating gel.

Under these conditions, albumin elutes earlier than the DBP fractions, and a more cathodal peak is eluted later. The central fractions contain protein and radioactive sterol at a constant specific activity. In subsequent studies, we observe that the material eluted from the first DEAE-Sephadex column can be applied to the preparative gel step with identical results. Thus, steps 5 and 6 in Table I can be deleted from the purification

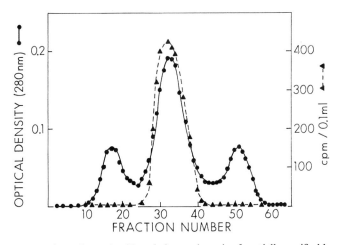

FIG. 2. Preparative polyacrylamide gel electrophoresis of partially purified human serum DBP. The DBP preparation (14 mg) was applied to a 1-cm stacking gel on top of a 7% polyacrylamide separating gel column 11.2 cm high (about 15 ml volume). The major portion of the procedure was carried out with 10 mA current at 4°, using Tris-glycine buffer at pH 8.3. Bromophenol blue was used as the anodal tracking dye, and fractions of 3 ml were collected at a rate of 9 ml/hr. From J. G. Haddad, Jr. and J. Walgate, *J. Biol. Chem.* **251,** 4803 (1976).

sequence. Under these conditions, however, relatively more albumin is presented for exclusion by the preparative gel electrophoresis step.

Properties of DBP

The purified DBP exhibited one stainable band when 25 $\mu$g were analyzed by 7% polyacrylamide gel electrophoresis, and migrated to the inter-alpha region of cellulose acetate strips in barbital buffer, pH 8.6 (Beckman microzone electrophoresis). Following overnight incubation in 1% sodium dodecyl sulfate (SDS) and 1% 2-mercaptoethanol, analysis by SDS-PAGE showed that reduced DBP migrated to an $R_f$ identical to that of DBP which had not been reduced. Under these conditions, DBP behaved as a single polypeptide chain. Molecular weight estimates by SDS-PAGE indicated a mobility consistent with that of a 58,000 molecular weight protein. Similarly, gel filtration on a Sephadex G-200 column (2 × 45 cm, 10 m$M$ Tris, pH 7.4, containing 0.15 $M$ NaCl) revealed a $V_e/V_o$ indicative of a globular protein of 60,000 molecular weight.

Pure DBP was used in sedimentation velocity analysis by observing the moving boundary in a Spinco Model E ultracentrifuge with absorption optics. For these studies, 0.55 mg of DBP in 0.7 ml of 10 m$M$ sodium phosphate buffer, containing 0.15 $M$ NaCl, pH 7.4, was centrifuged at 60,000 rpm at 20°. These studies revealed a single homogeneous protein with a sedimentation velocity of 3.46 S ($s_{20, w}$).

Amino acid composition of pure DBP was determined with 0.49 mg of protein that had been heated at 110° in 6 $N$ HCL. After removal of HCl *in vacuo* over KOH, analyses were done in duplicate on a Spinco Model 120B amino acid analyzer. The protein is rich in acidic residues, and others have found it to contain about 1% carbohydrate.[7] Recent amino acid analyses of DBP are in a reasonable accord with those reported for group-specific component.[4-7,13]

Spectral analyses were carried out with isotonic buffer solutions of DBP in a Beckman DB spectrophotometer. The extinction coefficient at 280 nm was determined by measuring the protein content of solutions of DBP analyzed in the spectrophotometer. An $E^{1\%}_{1cm}$ of 9.12 was observed, and a 280/260 nm absorption ratio of 1.21 was found.

Binding affinity and capacity studies were carried out by incubating pure DBP with increasing amounts of [³H]25-OH-D$_3$ (of known specific activity) which had been N$_2$-dried. After incubation at 4° for 24–96 hr, bound sterol was calculated from the amount of radioactivity observed in the 3–4 S region of 5 to 20% linear sucrose gradients after centrifugation at 40,000 rpm in a SW 50.1 rotor for 18 hr. Other trials with adsorbents

[13] N. Heimburger, K. Heide, H. Haupt, and H. E. Schultze, *Clin. Chim. Acta* **10**, 293 (1964).

TABLE II

BINDING AFFINITY AND CAPACITY OF PURIFIED HUMAN DBP FOR VITAMIN $D_3$ AND ITS METABOLITES

| Sterol | $K_d$ (M) | Mole of sterol bound per mole of DBP |
|---|---|---|
| $D_3$ | $4.3 \times 10^{-7}$m | 1.04 |
| 25-OH-$D_3$ | $6.4 \times 10^{-8}$m | 0.90 |
| 1,25-$(OH)_2D_3$ | $3.4 \times 10^{-7}$m | 1.07 |

(dextran-coated charcoal) and filters (Whatman DE-81 filter disks) proved these techniques to be more disruptive than the sucrose density gradient ultracentrifugation analyses. For each incubation tube, an aliquot for analysis of the recovery of material layered onto each sucrose gradient was performed. Similar analyses (after 3–4 days of incubation at 4°) can be performed with vitamin $D_3$ and 1,25-$(OH)_2D_3$. Results of such experiments are indicated in Table II. DBP binds antiricketic sterols mole per mole and has higher affinity for 25-OH-$D_3$ than for either 1,25-$(OH)_2D_3$ or vitamin $D_3$.

Under appropriate conditions, the microheterogeneity of DBP can be demonstrated by PAGE. For these purposes, 10-cm separating gels and specific discontinuous buffer systems are employed.[3,14]

### Antisera to DBP

One milligram of pure DBP in 50 m$M$ sodium phosphate buffer, pH 7.4, is emulsified with an equal volume of complete Freund's adjuvant and injected into multiple subcutaneous sites of a 3- to 4-month-old New Zealand rabbit. After 4 weeks, a booster injection of 0.5 mg of DBP emulsified in incomplete Freund's adjuvant is injected subcutaneously. Two to three weeks after the booster injection, the animal is bled from the central artery of the ear. Recently, a similar immunization schedule using a goat has also been successful.

Antiserum to DBP produces a single line of immunoprecipitation when allowed to diffuse against human serum or purified DBP in gels made of 0.6% agar in 10 m$M$ Tris buffer, pH 7.5. Immunoelectrophoresis in agarose also produces a single chevron of precipitate when anti-DBP diffuses against human serum. Further, all specific binding of antiricketic sterols by human serum is removed by prior incubation of serum with anti-DBP. Supernatants of this immunoprecipitate are devoid of specific

---

[14] M. Imawari and D. S. Goodman, *J. Clin. Invest.* **59**, 432 (1977).

sterol binding when analyzed by PAGE, sucrose gradient ultracentrifugation, or radiocompetitive assay utilizing adsorbent removal of unbound sterol.[5,6]

## DBP Immunoassays

### Radial Immunodiffusion

Since the concentration of DBP in serum is quite high, suitable quantitation of serum DBP can be performed by the radial immunodiffusion technique.[14,15] Agarose gels (1%) containing 1–4% rabbit anti-DBP antiserum in an alkaline buffer are used for this purpose. Wells 2–3 mm in diameter are punched out, and 5 $\mu$l of standard DBP solutions or serum samples are applied into the wells. The plates are kept in a humid chamber at room temperature for 48 hr, and the diameter of the immunoprecipitin rings is observed, often with the assistance of a magnifying comparator.

When the immunoprecipitin area or square of the diameter is plotted against the amount of DBP employed, a linear plot is described. Staining of the precipitin lines is not required, and intra-assay and inter-assay variation is quite low. Results from such an assay have correlated well with those obtained by radioimmunoassay of the same sera and the radial immunodiffisuion assay does not apparently distinguish between apo- and holo-DBP.[14]

### Radioimmunoassay

A highly sensitive radioimmunoassay for DBP can be developed by utilizing DBP made radioactive with $^{125}$I-chloramine-T[16] or lactoperoxidase.[14] In a glass tube, 3–6 $\mu$g of pure, lyophilized DBP are solubilized in 25 $\mu$l of 0.5 $M$ sodium phosphate buffer, pH 7.4. Two millicuries of carrier-free [$^{125}$I]Na in 25 $\mu$l of 50 m$M$ sodium phosphate buffer, pH 7.4, is added, and 50 $\mu$g of chloramine-T in 50 $\mu$l of the latter buffer is added. After 20 sec of mixing, 100 $\mu$g of sodium metabisulfite in 50 $\mu$l of 50 m$M$ sodium phosphate buffer, pH 7.4, is added, followed by 20 $\mu$g of KI in 50 $\mu$l of the same buffer. One milliliter of egg albumin, 0.025% in 10 m$M$ Tris · HCl buffer, pH 7.4, is added to the iodination tube to reduce adherence of the iodinated DBP to the glass tube. The contents are transferred to a 1 × 45 cm column of Sephadex G-200 equilibrated in the egg albumin buffer, containing 0.025% sodium azide. Ascending flow with the same

[15] R. Bouillon, H. Van Baelen, and P. De Moor, *J. Clin. Endocrinol. Metab.* **45**, 225 (1977).
[16] J. G. Haddad, Jr. and J. Walgate, *J. Clin. Invest.* **58**, 1217 (1976).

buffer is carried out in a 4° room at a flow rate of 15 ml/hr. Under these conditions, aggregate material appears at the $V_0$ (20 ml), [$^{125}$I]DBP elutes at 28–38 ml, and free $^{125}$I elutes at 50–60 ml.[16] Suitable aliquots of the pooled [$^{125}$I]DBP fractions are further diluted in the egg albumin buffer, which serves as the incubation buffer for the radioimmunoassay. The specific activity of the [$^{125}$I]DBP produced was usually 50–100 $\mu$Ci/$\mu$g, and 95% of the [$^{125}$I]DBP purified by G-200 gel filtration was bound by an excess of rabbit anti-DBP antiserum.

Incubation tubes containing 15,000 cpm of [$^{125}$I]DBP in 0.01 $\mu$l, suitable antiserum diltuion (to bind 40–50% of the [$^{125}$I]DBP) in 0.1 ml, standard DBP or serum dilution in 0.1 ml, and buffer to a total volume of 0.5 ml. After incubation at 24–48 hr at 4°, goat anti-rabbit $\gamma$-globulin antiserum is added, and the incubation is allowed to continue overnight at the same temperature. When high dilutions of rabbit anti-DBP are used, carrier nonimmune rabbit serum or IgG must be added to allow sufficient antigen for the second antibody to provide visible precipitates. The tubes are centrifuged at 2000 rpm for 20 min at 4°, and the supernatants are removed. The precipitates are washed with 0.5 ml of the Tris-egg albumin buffer and centrifuged again. The supernatants are removed, and the precipitates are assayed for radioactivity in a gamma spectrometer.

Each assay should contain tubes without antibody (index of nonspecific [$^{125}$I]DBP precipitation) replicate tubes with antibody and without competing DBP or serum, replicate tubes with varying concentrations of reference DBP or reference serum dilutions. Nonspecific [$^{125}$I]DBP precipitation is usually 8% or less of the radioactivity used. Radioactivity precipitated in the presence of standard DBP or serum is divided by that precipitated in the absence of unlabeled DBP to calculate "$B/B_0$."

As indicated in Fig. 3, the assay sensitivity and specificity is high, requiring considerable dilution of serum for its recognition in the useful part of the displacement curve. The accuracy and intra- and inter-assay variation are quite good if dilutions are carefully performed (Table III).

### Concentrations of DBP or Gc in Human Sera

Minor differences in serum DBP concentration have been reported from various laboratories,[14–16] but it is clear that DBP constitutes a major constituent (6%) of the serum $\alpha$-globulins. The concentration of DBP in normal human serum is $5 \times 10^{-6}$ $M$ to $10^{-5}$ $M$. Earlier quantitation of serum Gc by immunoprecipitation assay yielded somewhat higher values.[17] The importance of the liver in DBP synthesis is suggested by its

[17] F. D. Kitchin and A. G. Bearn, *Proc. Soc. Exp. Biol. Med.* **118**, 304 (1965).

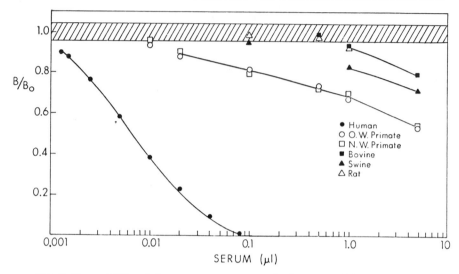

FIG. 3. Human DBP radioimmunoassay serum standard curve, and the specificity of the DBP radioimmunoassay for human serum. Also noncompetitive in displacing [$^{125}$I-]DBP from the antibody were up to 10 μg/tube of human albumin and γ-globulin. From J. G. Haddad, Jr. and J. Walgate, *J. Clin. Invest.* **58**, 1217 (1976).

lower concentration in liver disease, and increased levels of DBP have been observed during pregnancy and in women receiving estrogen-progestogen therapy.[15,16] Since DBP concentrations are similar during vitamin D deficiency and vitamin D excess, no apparent relationship be-

TABLE III
PRECISION AND ACCURACY OF HUMAN SERUM DBP RADIOIMMUNOASSAY

| Parameter | Number | DBP (μg/ml ± SD) | Coefficient of variation (%) |
|---|---|---|---|
| Intra-assay | 10 | 407 ± 23 | 5.7 |
| variation | 10 | 1,189 ± 38 | 3.2 |
| Inter-assay | | | |
| variation | 8 | 633 ± 58 | 9.2 |

| | Number | Added ng/tube | Measured ng/tube ±SEM | % Recovery, mean (range) |
|---|---|---|---|---|
| Accuracy | | | | |
| | 6 | 2.53 | 2.58 ± 0.08 | 102 (95–117) |
| | 6 | 4.03 | 4.10 ± 0.15 | 102 (87–112) |

tween DBP serum concentration and availability of vitamin D sterols is apparent.[16,18]

Recently available information suggests that the normal plasma levels of vitamin D, 25-OH-D, 24,25-$(OH)_2$D, and 1,25-$(OH)_2$D collectively represent a 1 to $2 \times 10^{-7} M$ concentration of antiricketic sterols. Since DBP is present at $5 \times 10^{-6} M$ to $1 \times 10^{-5} M$ concentrations, only 1–4% of the DBP is occupied with sterol. The dominant moiety of serum DBP in normal man is the apoprotein.

### Acknowledgment

This work was supported in part by Grants AM14570 and CRC RR00036 and Career Development Award AM70343 from the NIH. I am indebted to J. Walgate for technical assistance and to C. Melloh for typing the manuscript.

[18] S. Rojanasathit and J. G. Haddad, Jr., *Endocrinology* **100**, 642 (1977).

## [54] Competitive Binding Assay for 25-Hydroxyvitamin D Using Specific Binding Proteins

### By D. E. M. LAWSON

The importance of 25-hydroxyvitamin D is that it is the main chemical form in which the vitamin D activity of plasma occurs. Investigations on the metabolism of radioactive vitamin D in both normal and rachitic animals showed most of the radioactivity to be due to the 25-hydroxy metabolite,[1] and in addition chemical assays of the level of this latter substance in plasma were in good agreement with the values obtained for the total biologically assayed vitamin D activity of the same sample.[2] Consequently there is now acceptance that, in general, plasma 25-hydroxyvitamin D levels are a good indication of vitamin D status.[3] Early attempts to measure vitamin D levels involved either color reactions with antimony trichloride or ultraviolet spectroscopy, but neither method of detection of the vitamin D type of molecule has the necessary specificity or sensitivity to measure the levels found in animals. However, the reaction of vitamin D with antimony trichloride is used in the measurement of the vitamin

[1] D. E. M. Lawson, P. W. Wilson, and E. Kodicek, *Biochem J.* **115**, 269 (1969).
[2] S. Edelstein, M. Charman, D. E. M. Lawson, and E. Kodicek, *Clin. Sci. Mol. Med.* **46**, 231 (1974).
[3] M. A. Preece, S. Tomlinson, S. Ribot, C. A. Ribot, J. Pietrek, H. T. Korn, D. M. Davies, J. A. Ford, M. G. Dunningan, and J. L. H. O'Riordan, *Quart. J. Med.* **44**, 575 (1975).

level in fish oils, pharmaceutical products, and other high-potency products.[4] Gas–liquid chromatography, although it has the necessary sensitivity to detect low levels of vitamin D, has not been developed to achieve the necessary specificity. The discovery that vitamin D and its metabolites were present in plasma attached to specific proteins[5] meant that in theory it was possible to develop so-called competitive protein-binding assays for these substances using these proteins. In practice it has been possible to develop such assays only for 25-hydroxy-[2,6-14] and 24,25-dihydroxyvitamin D,[15-16] but not for the parent vitamin nor its hormonal form 1,25-dihydroxyvitamin D. The solubility of vitamin D in water is low, and all attempts to improve it apparently affects the interaction of the vitamin with the binding protein. As regards 1,25-dihydroxyvitamin D, the problem is the apparent low affinity of this substance for the binding-protein in plasma. Consequently, assays for this latter steroid involve tissue receptors that do have a sufficiently high affinity for it.[17]

Since 1977 there has been continuing interest in high pressure liquid chromatography as a means of measuring the levels of vitamin D and metabolites in biological samples. Several reports have appeared of the use of this technique to estimate 25-hydroxyvitamin D levels in plasma and tissues[17a] and it is the only method currently available for measuring plasma vitamin D levels[17b]. This technique is now in routine use in this laboratory for the estimation of both of these substances and similar assays for 24,25- and 25,26-dihydroxyvitamin D could be developed also.

[4] E. Kodicek and D. E. M. Lawson, in "The Vitamins" (P. György and W. N. Pearson, eds.), Vol. 6, p. 211. Academic Press, New York, 1967.

[5] H. Rikkers, R. Kletziens, and H. F. DeLuca, Proc. Soc. Exp. Biol. Med. 130, 1321 (1969).

[6] J. G. Haddad and K. J. Chyu, J. Clin. Endocrinol. 33, 992 (1971).

[7] R. E. Belsey, H. F. DeLuca, and J. T. Potts, J. Clin. Endocrinol. 33, 554 (1971).

[8] F. Bayard, P. Bec, and J. P. Louvet, Eur. J. Clin. Invest. 2, 195 (1972).

[9] G. Offerman and F. Dittmar, Horm. Metab. Res. 6, 534 (1974).

[10] M. A. Preece, J. L. H. O'Riordan, D. E. M. Lawson, and E. Kodicek, Clin. Chim. Acta 54, 235 (1974).

[11] R. Bouillon, P. Van Kerklove, and P. de Moore, Clin. Chem. 22, 364 (1976).

[12] J. F. Morris and M. Peacock, Clin. Chim. Acta 72, 383 (1976).

[13] B. Garcia-Pascual, A. Peytremann, B. Courvoisier, and D. E. M. Lawson, Clin. Chim. Acta 68, 99 (1976).

[14] T. Shimotsuji, Y. Seino, and H. Yabuuchi, Tohoku J. Exp. Med. 118, 233 (1976).

[15] C. M. Taylor, S. E. Hughes, and P. de Silva, Biochem. Biophys. Res. Commun. 70, 1243 (1976).

[16] J. G. Haddad, M. Chong, and J. W. Walgate, in "Vitamin D" (A. W. Norman et al., eds), p. 651. de Gruyter, Berlin, 1977.

[17] P. F. Brumbaugh, D. H. Haussler, K. M. Bursac, and M. R. Haussler, Biochemistry 13, 4091 (1974).

[17a] K. T. Koshy and A. L. VanderSilk, J. Agr. Food Chem. 25, 1246 (1977).

[17b] G. Jones, Clin. Chem. 24, 287 (1978).

General Considerations

*Choice of Binding Protein*

A number of binding-proteins have been used in the various 25-hydroxyvitamin D assays so far described, including plasma from rachitic[2,7,10,13,14] and normal[11] rats, and from osteomalacic[8,12] and normal[18] humans and also kidney cytoplasm from rachitic rats.[6,9,14] Direct comparisons have been made of several of these preparations, and similar binding, displacement and assay values were obtained.[19] The problem of the poor solubility of vitamin D in aqueous solutions has been overcome for 25-hydroxyvitamin D by including up to 8–10% ethanol in the system, but some investigators have also added $\beta$-lipoprotein[7,11] and others have included 1% albumin.[10,18] Direct comparisons of the methods using these three solubilizing techniques have shown that the incorporation of $\beta$-lipoprotein results in greater variability in assay results,[13,19] and the use of some albumin preparations can lead to additional problems.[20]

*Preparation of Sample*

25-Hydroxyvitamin D in plasma is stable for many months if kept frozen at $-20°$. With a suitably sensitive assay only 0.5 ml of plasma (or serum) is required for extraction and chromatography to yield sufficient steroid for the actual protein-binding assay. Several solvents have been used to extract the 25-hydroxyvitamin D, including ether,[6] chloroform/methanol,[2,10] ethanol or dichloromethane/methanol,[8,11] but whichever is chosen, recoveries should be greater than 90%.

A number of attempts have been made to develop an assay for 25-hydroxyvitamin D that does not involve chromatography,[9,12,13,21,22] but all careful comparisons of the two types of assays have shown that values obtained with chromatography are lower than those obtained without its use.[18,19] Although thin-layer chromatography has been used,[8] most procedures involve column chromatography with silica gel[6,7,10,18,19] or Sephadex LH-20.[2] Although the latter material is an increasingly popular support for all types of analysis of vitamin D metabolites, several accurate and precise methods have been described using silica-gel adsorption chromatography. The virtues of the latter process include speed and cheapness, both in terms of the cost of the support itself and volume of solvent used.

[18] R. K. Skinner and M. R. Wills, *Clin. Chim. Acta* **80**, 543 (1977).
[19] R. S. Mason and S. Posen, *Clin. Chem.* **23**, 806 (1977).
[20] J. M. Pettifor, F. P. Ross, and J. Wang, *Clin. Sci. Mol. Med.* **51**, 605 (1976).
[21] R. Belsey, H. F. DeLuca, and J. T. Potts, *Clin. Endocrinol. Metab.* **33**, 554 (1971).
[22] V. Justova, L. Starka, H. Wilczek, and V. Pacovsky, *Clin. Chim. Acta* **70**, 97 (1976).

Although extraction and chromatography of the plasma steroid is readily accomplished, great care must be undertaken to prevent losses and destruction of the 25-hydroxyvitamin D. It should be noted that recovery of the radioactive metabolite added to the plasma may be good, but the molecule itself may be damaged so as to interfere with its binding to the protein. Vitamin D steroids must be protected from direct sunlight, water, and oxygen, so that, when solvent is being evaporated, care must be taken not to allow water to condense and not to leave the steroid dry for any time, as it is at this stage that it is particularly sensitive to oxygen.

*Evaluation of 25-Hydroxyvitamin D Assays*

Although competitive protein-binding assays are simple in principle, many difficulties can arise in their execution (some having been referred to above). Consequently, one must ensure that the selected assay is sufficiently reliable, accurate, and precise, and it should be borne in mind that apparently conflicting claims in the literature may be due to different degrees of success that people achieve with different methods. The first two most important decisions facing the investigator are the choice of binding-protein and whether or not to include a chromatographic step. Availability will to some extent govern the choice of binding protein, but on theoretical grounds proteins from rachitic animals should give the most sensitive assay. A chromatographic step should be omitted only if the number of samples to be assayed makes the time required to carry out the assays prohibitively lengthy. Accuracy is generally found to lie between 90 and 110%, the intra-assay precision has a coefficient of variation of <10%, and the inter-assay coefficient of variation should be <20%. Inter-assay variation may be a major problem that can be improved by including with each batch of plasma samples for analysis an aliquot of a large quantity of a previously assayed plasma. The values obtained for the samples in each assay are then adjusted to a constant value for this plasma sample. Assays with a sensitivity to measure 50 pg/tube can be readily developed, and there are reports of assays measuring as little as 12 pg/tube, so that after correcting for recoveries and dilutions, etc., it is possible to measure 25-hydroxyvitamin D levels as low as 0.5 ng/per milliliter of plasma. Methods and plasma blanks are traditional sources of problems in competitive, protein-binding assays, but for a discussion of these and other aspects of this type of assay the reader is referred to a previous volume in this series.[23]

[23] D. Rodbard and H. A. Feldman, this series, Vol. 36, p. 3; C. A. Strott, this series, Vol. 36, p. 34.

Assay of Plasma 25-Hydroxyvitamin D

*Materials*

[26-$^3$H]25-Hydroxyvitamin D$_3$ of specific activity 11 Ci/mmol can be purchased from the usual suppliers. This material is stable if kept in benzene/ethanol (3 : 7) solution at $-20°$. From time to time the radioactive steroid should be analyzed by silica gel thin-layer chromatography with chloroform/methanol (9 : 1, v/v) as the solvent system; 90–95% of the applied radioactivity should cochromatograph with authentic 25-hydroxyvitamin D. A stock solution of 6000 dpm of [26-$^3$H]25-hydroxyvitamin D$_3$/40 $\mu$l of absolute ethanol is prepared.

25-Hydroxyvitamin D$_3$. Since its discovery this compound has been provided by several pharmaceutical companies, including Upjohn Co. Limited, Phillips-Duphar, Amsterdam, The Netherlands, and Roussel Laboratories, Paris, France. This steroid is stable in the crystalline state and can be kept in ethanolic solution under N$_2$ at $-20°$ for about 6 months. Several solutions should be prepared containing 25-hydroxyvitamin D at the following concentrations: zero, 0.05, 0.075, 0.12, 0.15, 0.24, and 0.30 ng/40 $\mu$l of ethanol, each solution to contain also 6000 dpm of [26-$^3$H]25-hydroxyvitamin D/40 $\mu$l.

Binding Protein. A quantity of plasma from rachitic rats is obtained, and divided into 0.5–1.0 ml aliquots, and kept at $-20°$. Samples have been stored in this manner for at least 6 months without loss of binding activity for 25-hydroxyvitamin D. Before use it is necessary to establish the dilution with Tris buffer required so that under the conditions of the assay about 40% of the undiluted [26-$^3$H]25-hydroxyvitamin D is bound to the protein. This dilution factor is almost constant during the life of the protein. Immediately before an assay is carried out, an aliquot of the plasma is diluted accordingly. This solution should be kept at 4°.

The cytosol from the kidneys of rachitic rats is also a good source of a suitable binding protein because, in my experience, assays using this material have an improved sensitivity over those using plasma binding protein. The cytosol is prepared by homogenizing the kidneys (6 pairs are a convenient number) in TKM buffer (50 m$M$ Tris-Cl, pH 7.4, 25 m$M$ KCl, 10 m$M$ MgCl$_2$) to give a 10% homogenate which is then centrifuged at 100,000 $g$ for 90 min. The supernatant is removed and stored at $-20°$ preferably in 0.2-ml aliquots. Again it is necessary to establish the dilution factor for each cytosol preparation before use in an assay. Whichever binding protein is used stability will be aided if the protein is not frozen and thawed continuously.

Solvents are analytical grade and are not purified further with the exception of light petroleum 40–60° (supplied in the United States as Skellysolve B). Some batches of this latter solvent contain a material that interferes with the binding of the steroid to the protein, and consequently it is advisable routinely to purify the light petroleum. A suitable procedure is to wash 4 liters of solvent 10 times with 100–200 ml of conc. $H_2SO_4$ followed by six washings with 200–300 ml of distilled water. The light petroleum is dried overnight with KOH pellets and distilled over KOH, collecting the fraction of light petroleum between 40° and 52°.

Dextran-coated charcoal is prepared from commercially available charcoal such as Norit GSX and can be used without further treatment. Dextran used to coat the charcoal is clinical grade with a molecular weight of 60,000–90,000. The charcoal (5 g) is dissolved in 200 ml of Tris buffer, pH 7.5, together with 0.5 g of dextran and left overnight. This solution will keep at 4° for 3–4 weeks before deteriorating. Each day that an assay is to be carried out, this solution should be diluted with buffer (1 : 6) before use. Because the charcoal is in suspension, and tends to settle, it must be thoroughly mixed before pipetting. The amount of dextran-coated charcoal added should be the minimum required to remove at least 95% of [26-$^3$H]25-hydroxyvitamin D held in solution in the Tris buffer by 8% ethanol. The amount of radioactivity in the blank tube (i.e., methods blank) will rise as the stock solution of dextran-coated charcoal deteriorates.

Sephadex LH-20 chromatography is carried out by allowing the support to swell for several hours in chloroform/methanol (45 : 55). The Sephadex LH-20 gel is then poured into a glass column (0.8 × 22 cm) plugged with glass wool, and washed thoroughly before use.

Scintillation fluid is prepared by mixing 0.5% 2,5-diphenyloxazole and 0.3% 1,4-bis(5-phenyloxazol-2-yl)benzene in a solution of Triton X-100 in toluene (1 : 2, v/v). Tris buffer consists of 10 m$M$ Tris · HCl, pH 7.5, 1.5 m$M$ EDTA, and 1 m$M$ mercaptoethanol.

### Method

*Extraction.* An aliquot of [26-$^3$H]25-hydroxyvitamin $D_3$ containing 4500 dpm is added to 0.5 ml of heparinized plasma in a stoppered tube followed by 5 ml of chloroform/methanol (2 : 1). The solutions are shaken, left for 30 min at 4°, and then 1 ml of water is added. The phases are separated by centrifugation at 1500 rpm for 5–10 min, and the bottom chloroform phase is removed and retained. The top aqueous layer is further extracted for 20 min with 5 ml of chloroform, and the phases are again separated by centrifugation. The bottom layer is added to the first chloroform extract, and the whole is evaporated to dryness under a stream of $N_2$. The residue is dissolved in 0.2 ml of chloroform/light petroleum (45 : 55).

*Chromatography.* The dissolved plasma lipid extract is applied to the Sephadex LH-20 column and the tube is washed out successively with 0.2 ml and 0.4 ml of chloroform/petrol (45 : 55). The washings are transferred to the column, which is developed with the same solvent. Each person must establish for himself the position of the 25-hydroxyvitamin D in the column eluate. Under the conditions used in this laboratory, the steroid is found in the 10 ml of solvent eluted after the first 8 ml have passed through the column. This fraction is evaporated to dryness under $N_2$, the residue is dissolved in 1.0 ml of ethanol, and a 0.2-ml aliquot is taken for measurement of tritium to assess recovery of added steroid.

*Protein-Binding Assay.* Forty microliters of the ethanolic solution containing 6000 dpm of [26-$^3$H]25-hydroxyvitamin $D_3$ and increasing amounts of 25-hydroxyvitamin $D_3$ in the range 0–0.3 ng (as indicated above) were dispensed in triplicate into suitable tubes. Forty microliters of ethanol are also added to each tube. For each plasma sample to be measured three tubes are prepared containing 40 $\mu$l of [26-$^3$H]25-hydroxyvitamin D and 40 $\mu$l of the plasma 25-hydroxyvitamin D extract. One milliliter of the plasma binding protein is then added to two tubes in each set, and 1 ml of the Tris buffer to the third tube. This latter tube is the methods blank. The solutions are stirred vigorously in a vortex-type mixer and then left in an ice-box for 1 hr, after which time 1 ml of the dextran-coated charcoal is added. Again the solutions are stirred vigorously. Ten minutes later the tubes are centrifuged at 2000 rpm for 10 min at 4°, and 1 ml of the supernatant in each tube is transferred to a counting vial containing 10 ml of the Triton/toluene scintillation fluid.

*Calculation.* The radioactivity in the methods blank is subtracted from the radioactivity in the supernatant to give the amount of true bound radioactivity. This is converted to dpm and expressed in percentage terms. The percentage of bound radioactivity in the supernatants at each concentration of 25-hydroxyvitamin D is plotted against nanograms of steroid per tube. The percentage of bound radioactivity in the supernatant of the plasma samples can, by reference to the standard curve, give the amount of 25-hydroxyvitamin D present. The final plasma concentration can be calculated after due allowance for plasma volume assayed, recovery after procedural losses, and variations in the level obtained for the internal standard. Some investigators find the logit transformation useful, since the standard curve now gives a straight line over a much wider range of steroid concentrations. In this case, log %B/(100-%B) is plotted against 25-hydroxyvitamin D concentration.

## [55] Competitive Protein Binding Assay for Plasma 25-Hydroxycholecalciferol

### By Tsunesuke Shimotsuji and Yoshiki Seino

## Assay Method

### Principle

Different assays for 25-hydroxycholecalciferol have been reported since 1971;[1-13] however, almost all assays were based on the radiocompetitive protein binding technique originally developed by Belsey et al.[1] The present method was developed in our laboratory, using a specific 25-hydroxycholecalciferol (25-OH-D$_3$) binding protein from the plasma or kidney cytosol of vitamin D-deficient rat. It has been shown that 25-hydroxycholecalciferol and 25-hydroxyergocalciferol react identically in this assay,[4] so that it can be used for measurement of 25-hydroxy derivatives of vitamin D$_2$ and vitamin D$_3$.

### Reagents

Buffer: 40 ml of 0.2 $M$ solution of monobasic sodium phosphate (27.8 g of NaH$_2$PO$_4$ in 1000 ml) is added to 210 ml of 0.2 $M$ solution of dibasic phosphate (53.65 g of Na$_2$HPO$_4 \cdot$ 7H$_2$O in 1000 ml), and the solution is adjusted with 1 $N$ sodium hydroxide to give a pH of 7.5 and diluted to a total of 1000 ml.

[1] R. Belsey, H. F. DeLuca, and J. T. Potts, Jr., *J. Clin. Endocrinol. Metab.* **33**, 554 (1971).

[2] J. G. Haddad and K. J. Chyu, *J. Clin. Endocrinol. Metab.* **33**, 992 (1971).

[3] F. Bayard, P. Bec, and J. P. Lonvet, *J. Clin. Invest.* **2**, 195 (1973).

[4] M. A. Preece, J. L. H. O'Riordan, D. E. M. Lawson, and E. Kodicek, *Clin. Chim. Acta* **54**, 235 (1974).

[5] S. Edelstein, M. Charman, D. E. M. Lawson, and E. Kodicek, *Clin. Sci. Mol. Med.* **46**, 231 (1974).

[6] G. Offermann, *Muenchen. Med. Wochenschr.* **116**, 1569 (1974).

[7] R. E. Belsey, H. F. DeLuca, and J. T. Potts, Jr., *J. Clin. Endcrinol. Metab.* **38**, 1046, (1974).

[8] G. Offermann and F. Ditter, *Horm. Metab. Res.* **6**, 534 (1974).

[9] T. Shimotsuji, Y. Seino, and H. Yabuuchi, *Tohoku J. Exp. Med.* **118**, 233 (1976).

[10] V. Justova, L. Starka, H. Wilczek, and V. Pacovsky, *Clin. Chim. Acta* **70**, 97 (1976).

[11] B. Garcia-Pascual, A. Peytremann, B. Courvoisier, and D. E. M. Lawson, *Clin. Chim. Acta,* **68**, 99 (1976).

[12] I. Björkhem, and I. Holmberg, *Clin. Chim. Acta* **68**, 215 (1976).

[13] J. F. Morris and M. Peacock, *Clin. Chim. Acta* **72**, 383 (1976).

METHODS IN ENZYMOLOGY, VOL. 67

Isotope: [26(27)-Me-$^3$H]25-OH-D$_3$ (spec. act., either 2.9 Ci/mmol or 6.9 Ci/mmol) can be purchased from the Radiochemical Centre, Amersham. It is stored with an activity of 25 $\mu$Ci. Before use in an assay, an aliquot is dried under a stream of nitrogen, redissolved in 500 $\mu$l of 95.0% ethanol, and diluted with an appropriate amount of ethanol.

Standard: Crystalline 25-OH-D$_3$ (supplied by Chugai Pharm. Co., Japan) is dissolved in ethanol to give a concentration of 280 $\mu$g/ml and stored at $-20°$. Before use in an assay, an aliquot is serially diluted with ethanol to give several standard solutions with a concentration range of 25-OH-D$_3$ from 10 ng/50 $\mu$l to 0.5 ng/50 $\mu$l.

Charcoal–dextran suspension: The suspension is prepared by mixing 3.0 g of Norit A charcoal (Sigma) and 0.3 g of Dextran 20 (Pharmacia) in 100 ml of 50 m$M$ phosphate buffer and stored at 4°.

The following compounds have been tested for their ability to displace 25-OH-D$_3$ from the binding protein: cholecalciferol (D$_3$), ergocalciferol (D$_2$), dehydrotachysterol, cortisone, progesterone, and cholesterol (Sigma London Chemical Co.), 1$\alpha$,25-(OH)$_2$D$_3$ and 1$\alpha$,-OH-D$_3$ (supplied by Chugai Pharm. Co., Japan).

## Preparation of Binding Protein

Young Wistar rats (1 or 2 weeks old) are fed the vitamin D- and calcium-deficient diet listed in Table I for 3 or 4 weeks and allowed free

TABLE I
COMPOSITION OF VITAMIN D AND CALCIUM-DEFICIENT DIET

| Component | Percentage |
|---|---|
| Vitamin-free casein | 25 |
| Cornstarch | 38 |
| $\alpha$-Starch | 10 |
| Granulated sugar | 5 |
| Cellulose powder | 8 |
| Linole oil | 6 |
| Minerals$^a$ | 6 |
| Vitamins$^b$ | 2 |
| | 100 |

$^a$ Minerals (/kg feed): NaCl, 2.808g; MgSO$_4$, 4.312 g; NaH$_2$PO$_4$, 5.628 g; K$_2$HPO$_4$, 5.730 g; KH$_2$PO$_4$, 17.000 g; Fe citrate, 1.912 g; KI, 3.3 mg; ZnCl$_2$, 62.55 mg; CuSO$_4$, 19.65 mg; MnSO$_4$, 76.90 mg.
$^b$ Vitamins (/kg feed): vitamin A, 1000 IU; E, 100 mg; B$_1$, 24 mg; B$_2$, 80 mg; B$_6$, 0.01 mg; C, 600 mg; K$_3$, 104 mg; biotin, 0.4 mg; folic acid, 4 mg; PABA, 100 mg; niacin, 120 mg; inositol, 120 mg; pantothenic-sodium, 97.8 mg; choline-chloride, 4 g.

access to deionized water. The weight gain of each rat is markedly decreased, compared with that of the control, after about 10 days on the vitamin D and calcium-deficient diet.

Each rat is then anesthetized and bled from the abdominal aorta; heparinized whole plasma is immediately frozen and stored at $-20°$ until use.

The kidneys are removed and homogenized in 50 m$M$ sodium phosphate buffer (pH 7.5) and spun at 1000 $g$ for 15 min at 4°, and the supernatant obtained is spun at 105,000 $g$ for 1 hr at 4°. The supernatant is also used as a binding protein.

The protein concentrations of the plasma and kidney cytosol are found to be about 60 mg/ml and 6.6 mg/ml, respectively, by Lowry's method. To 100 ml of the solution of binding protein is added 100 mg of human serum albumin (Sigma) at assay in order to prevent loss of labeled 25-OH-D$_3$ from the solution.

### Extraction and Fractionation of 25-OH-D$_3$

Before assay, samples of plasma must be extracted and the various metabolites of vitamin D should be separated by column chromatography. To 0.5 ml of the plasma are added 300–1000 pg (1000 dpm) of labeled 25-OH-D$_3$ in 20 $\mu$l of ethanol to assess the efficiency of the extraction and fractionation. After about 0.5 hr at room temperature, 0.5 ml of water, 1.25 ml of chloroform, and 2.5 ml of methanol are successively added, and the mixture is allowed to stand at room temperature. After a further 0.5 hr, another 1.0 ml of distilled water and 1.25 ml of chloroform are added and the tube is allowed to stand overnight at room temperature. The mixture is separated into two layers; a chloroform layer is at the bottom of the tube and an aqueous methanol layer at the top, with a layer of protein at the interphase. The lower chloroform layer is aspirated with a disposable syringe or pipette.

Reextraction may be omitted, as recovery from the reextracted chloroform layer is only about 4–6%. The aspirated chloroform layer is evaporated to dryness under a stream of nitrogen. The lipid extract is dissolved in approximately 0.5 ml of chloroform/$n$-hexane mixture (1:1, v/v) and applied to a 1 × 25 cm Sephadex LH-20 column and eluted by the same solvent at the rate of 0.4 ml/min and 2.0 ml/tube.

Figure 1 shows the chromatographic separation of [$^{14}$C]D$_3$ from [$^3$H]25-OH-D$_3$. Labeled 25-OH-D$_3$ was eluted in the 24- to 44-ml fractions. The more polar metabolites, [$^3$H]-24,25-(OH)$_2$D$_3$ and [$^3$H]1$\alpha$,25-(OH)$_2$D$_3$, were shown to be eluted after the fractions containing 25-OH-D$_3$. Reproducibility of the elution curve was found to be excellent.

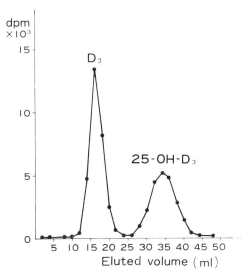

FIG. 1. Chromatographic separation of [³H]25-OH-D$_3$ from [¹⁴C]D$_3$. [4-¹⁴C, 1α-³H]D$_3$ (3.0 × 10⁴ dpm; and [26(27)-³H]25-OH-D$_3$ (2.2 × 10⁴ dpm) were applied to a 1 × 25 cm Sephadex LH-20 column and eluted in chloroform/$n$-hexane (1 : 1) at the rate of 0.4 ml/min and 0.2 ml/tube. From T. Shimotsuji, Y. Seino, and H. Yabuuchi, *Tohoku J. Exp. Med.* **118**, 233 (1976).

Each 2-ml fraction containing 25-OH-D$_3$ is collected. The solvent is evaporated under a stream of nitrogen, and the residue is redissolved in 3–5 ml of the solvent. An aliquot (1.0 ml) of the solvent solution is used for counting recovery, and an aliquot (1.0–2.0 ml) of the remaining extract is evaporated under a stream of nitrogen, the residue is redissolved in 50 μl of the 99.5% ethanol used for assay of 25-OH-D$_3$.

*Assay Procedure*

An aliquot of the preparation of 25-OH-D$_3$ (280 μg/ml of ethanol) is serially diluted with ethanol, and a total of six standards with concentrations ranging from 2.5 to 400 ng/ml is finally prepared. Then 50 μl of these standards are pipetted into triplicate 9.5 mm × 50 mm glass test tubes, together with another set of tubes containing 50 μl of ethanol only. As standards, 50-μl aliquots of the extract are pipetted in triplicate in the same manner. To each of the incubation tubes are added 0.3–1.0 ng (13,000 dpm) of labeled 25-OH-D$_3$ in 20 μl of the ethanol solution. Either 100–200 μg of D-deficient rat kidney cytosol or 30–50 μg of D-deficient rat plasma in 1.0 ml of cold sodium phosphate buffer (pH 7.5, 50 m$M$) is added to each tube. The contents are mixed thoroughly, and the tubes are allowed to stand at 4° for more than 30 min. In each assay is a set of tubes to which buffer without binding protein is added, so that the nonspecific binding of the labeled 25-OH-D$_3$ can be assessed.

After incubation, 0.2 ml of the suspension of Dextran 20-coated charcoal is added to each tube, and the suspension is mixed thoroughly. After standing at room temperature for 30 min, the tubes are centrifuged at 3000 rpm for 10 min at 4°; 500-$\mu$l aliquots of the supernatant, containing the bound 25-OH-D$_3$, are transferred to the counting vials, to which 10 ml of liquid scintillator are added. The scintillator consists of 6.0 g of PPO (Pharmacia) and 50 mg of POPOP (Pharmacia) in 1000 ml of xylene. A counting efficiency of 30–40% was achieved for tritium when vials were counted in the Nuclear Chicago liquid scintillation counter.

A standard curve is obtained by plotting the binding ratio (bound − blank)cpm/(total − blank)cpm. With various concentrations of D-deficient rat plasma used as binding protein, it was recognized that the addition of 13,000 dpm (2.9 Ci/mmol, 1.0 ng) of labeled 25-OH-D$_3$ and 1.0 ml of the 1200-fold diluted plasma to each standard was necessary to obtain a significant standard curve, as shown in Fig. 2.

Comparing the two kinds of binding protein (from plasma and from kidney cytosol), the plasma of D-deficient rat is more convenient for as-

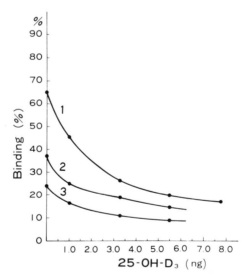

FIG. 2. Standard curves obtained by using various concentrations of the plasma protein of D-deficient rat. Different concentrations of the plasma protein of D-deficient rat (curve 1, 50 $\mu$g/ml; curve 2, 25 $\mu$g/ml; and curve 3, 12.5 $\mu$g/ml) and 13,000 dpm of [$^3$H]25-OH-D$_3$ were used for the assay. From T. Shimotsuji, Y. Seino, and H. Yabuuchi, *Tohoku J. Exp. Med.* **118**, 233 (1976).

say, although both binding proteins are useful. A range from 0.5 to 5.0 ng of 25-OH-D$_3$ has been found to be satisfactory for measurement.

## Properties

### Recovery and Precision

The recovery of the extract from plasma has been shown to be 86.1 ± 12.2% (mean ± SD, $n = 11$). In plasma samples, to which 10 and 20 ng of unlabeled 25-OH-D$_3$ in 50 $\mu$l of ethanol were added to give final concentrations of 10 and 20 ng/ml, the recovery of added 25-OH-D$_3$ has been shown to be 91.6 ± 9.6(SD)% ($n = 5$), being corrected with the recovery of labeled 25-OH-D$_3$.

The intra-assay variation was assessed by extracting five 0.5-ml aliquots of the same plasma sample from a healthy subject and assaying each concentrations of 25-OH-D$_3$ in one assay. A coefficient of intra-assay variation was 7.7%.

The inter-assay variation was estimated by extracting and assaying samples from a healthy subject on six different occasions. A coefficient of inter-assay variation was 11.8%.

### Specificity

The displacement of labeled 25-OH-D$_3$ with vitamin D and its analogs may take place, and cross-reaction of D$_2$ or D$_3$ was about 10% as potent as 25-OH-D$_3$, at the point of 50% inhibition. Although 1$\alpha$,25-(OH)$_2$D$_3$ could also displace labeled 25-OH-D$_3$, its cross-reaction was only about 1% as potent as that of 25-OH-D$_3$. A greater displacement has been shown to occur with 24,25-(OH)$_2$D$_3$ or 25,26-(OH)$_2$D$_3$ than with D$_2$, D$_3$ and 1$\alpha$,25-(OH)$_2$D$_3$.[13]

No displacement of labeled 25-OH-D$_3$ took place with 1$\alpha$-OH-D$_3$, dehydrotachysterol, cholesterol, cortisone, and progesterone in the amount of 10$^3$ ng (2 to 3 × 10$^{-6}$ $M$) per tube.

### Comments

Belsey et al.[7] first reported a direct-assay technique for 25-OH-D$_3$ without preparative chromatography. Twenty-seven plasma samples have been assayed with the method described here as well as by the direct assay without column chromatography. The results of both methods correlate well, as shown in Fig. 3, but the values obtained with the direct assay were slightly higher, about 30%, than those with the method de-

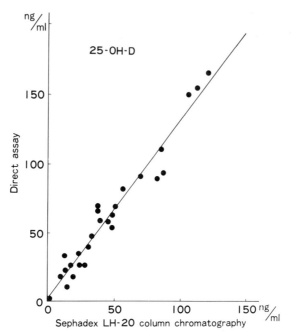

FIG. 3. Correlation of the apparent concentrations of the plasma 25-OH-D in 27 epileptic children assayed with and without chromatography. $y = a + bx$; $a = 1.390$; $b = 1.318$; $r = 0.973$; $p < 0.001$.

scribed. A direct assay might be useful to measure 25-OH-$D_3$ in a large number of plasma samples; however, the fact that the values obtained with a direct assay might include 24,25-$(OH)_2D$ and 25,26-$(OH)_2D$ as well as 25-OH-D must be considered.

It has been shown that the reactions of 25-OH-$D_2$ and 25-OH-$D_3$ were identical in this assay system,[4] and the separation of 25-OH-$D_3$ from 25-OH-$D_2$ was incomplete in Sephadex LH-20 preparative chromatography,[14] so that this assay can be used for measurement of 25-OH-D. With the method described, the plasma 25-OH-D level was found to be 21.6 ± 10.1 (SD) ng/ml ($n = 17$) in healthy infants and children from 1 to 15 years old, and 11.4 ± 8.6 (SD) ng/ml ($n = 17$) in mature neonates up to 2 days after delivery.[9]

[14] R. L. Horst, R. M. Shepard, N. A. Jorgensen, and H. F. DeLuca, *J. Lab. Clin. Med.* **93**, 277 (1979).

## [56] Steroid Competition Assay for Determination of 25-Hydroxyvitamin D and 24,25-Dihydroxyvitamin D

*By* ANTHONY W. NORMAN and PATRICIA A. ROBERTS

Vitamin D is now known to undergo obligatory metabolism prior to generation of its biological response.[1,2] The first step that occurs in the liver involves the introduction of a 25-hydroxyl group to give 25-hydroxyvitamin D [25-OH-D]. The second stage of metabolite transformation occurs in the kidney; here 25-OH-D may be converted into either $1\alpha,25$-dihydroxyvitamin D [1,25-(OH)$_2$D] or $24R,25$-dihydroxyvitamin D [24,25-(OH)$_2$D].

Currently there are no simple says to determine the blood levels of vitamin D, thus there is no facile assessment on animals of vitamin D nutritional status. However, in view of the fact that vitamin D is rapidly converted into 25-OH-D, and since this metabolite is the form of the vitamin that occurs in highest concentration in the blood, it has been proposed[3,4] that a measure of the plasma level of 25-OH-D is a useful index of vitamin D nutritional status. Several methods for the determination of the plasma concentration of 25-OH-D have been reported.[5-9]

The method presented here provides for a rapid and simple assay of 25-OH-D in biological samples. Also since 24,25-(OH)$_2$D$_3$ competes equally with 25-OH-D for their plasma binding protein, transcalciferan, it is possible to use this same assay with a modified sample purification scheme to effect a determination for this 24,25-dihydroxylated metabolite. With the use of a swinging-bucket rotor that can handle 48 tubes, it is possible to extract and assay as many as 16 unknown samples at two concentrations in 1 day.

[1] A. W. Norman and H. Henry, in *Recent Prog. Horm. Res.* **30**, 431 (1974).

[2] A. W. Norman, *in* "Vitamin D: The Calcium Homeostatic Steroid Hormone," pp. 1–490. Academic Press, New York, 1979.

[3] E. B. Mawer, G. A. Lumb, K. Schaefer, and S. W. Stanbury, *Clin. Sci.* **40**, 39 (1971).

[4] R. Neer, M. Clark, J. Friedman, R. Belsey, M. Sweeney, J. Buoncristiani, and J. T. Potts, Jr., *in* "Vitamin D: Biochemical, Chemical and Clinical Aspects Related to Calcium Metabolism" (A. W. Norman *et al.*, eds.), p. 595. de Gruyter, Berlin, 1977.

[5] R. E. Belsey, H. F. DeLuca, and J. T. Potts, *J. Clin. Endocrin. Metab.* **33**, 559 (1971).

[6] J. G. Haddad and K. J. Chyu, *J. Clin. Endocrinol.* **33**, 992 (1971).

[7] J. G. Haddad and S. J. Birge, *J. Biol. Chem.* **250**, 299 (1975).

[8] T. W. Osborn, Ph.D. Dissertation, Univ. of California, Riverside, 1976.

[9] R. Bouillon, P. Van Kerkhove, and P. DeMoor, *Clin. Chem.* **22**, 364 (1976).

Materials and Methods

*Animals.* The serum from normal male Sprague-Dawley rats is used as a source of 25-OH-D-binding protein (transcalciferan). Serum from 5–10 rats is pooled, centrifuged at 10,000 g for 30 min, lyophilized, and stored as a powder at $-20°$ until reconstitution with 50 m$M$ phosphate buffer shortly before use. The usual concentration for assay purposes is 1 mg/100 ml.

*Reagents, Solutions, etc.*

Sodium phosphate buffer, 50 m$M$, pH 7.4

Charcoal–dextran suspension: 2.5 g of activated charcoal and 0.25 g of Pharmacia Fine Chemical Dextran T-40 are stirred in 100 ml of the phosphate buffer at room temperature for 2 hr, then allowed to stand overnight at room temperature. Centrifuge 10 min at 5000 g and discard the supernatant. Resuspend the pellet in 100 ml 0.05 $M$ phosphate buffer. Store in a refrigerator until needed.

Radioactive ligand: A solution of [26,27-methyl-³H]25-hydroxy-cholecalciferol (either New England Nuclear or Amersham-Searle), specific activity approximately 11 Ci/mmol, containing approximately $3 \times 10^6$ dpm/ml absolute ethanol. Store in a freezer under argon.

Radioactive "spike": A radioactive "spike" of [³H]25-OH-D may be prepared by diluting the above radioactive ligand to approximately $1.6 \times 10^5$ dpm/ml with absolute ethanol. In the event an assay for 24,25-$(OH)_2$D is being conducted, a suitable "spike" of [³H]24,25-$(OH)_2D_3$ must be available. This can be prepared enzymatically from [26,27-methyl-³H]25-OH-$D_3$ according to the procedure of Taylor *et al.*[10] The specific activity of the "spike" should be high enough so that less than 0.1 ng of steroid is added to each sample. Use 25 $\mu$l per sample to determine recovery of 25-OH-D and/or 24,25-$(OH)_2$D during sample extraction and chromatographic purification.

Nonradioactive standards: Stock solutions of chemically synthesized 25-OH-$D_3$ (obtainable from Dr. John Babcock, The Upjohn Co.) and 24,25-$(OH)_2D_3$ (obtained from Dr. M. Uskokovic, Hoffmann-La Roche, Nutley, New Jersey) in absolute ethanol, ranging from 0.5 to 5.0 ng per 25 $\mu$l. The concentration of a primary stock solution must be determined accurately via UV absorbance measurements at 264 nm (the molar extinction coefficient of 25-OH-D and 24,25-$(OH)_2D_3$ at 264 nm is 18,300), and the subsequent dilutions must be made carefully. Store in the freezer under argon.

[10] C. M. Taylor, E. B. Mawer, and A. Reeve, *Clin. Sci. Mol. Med.* **49**, 391 (1975).

Solvents: Choroform, methanol, petroleum ether; all analytical reagents, all redistilled in glass. Absolute ethanol, redistilled.

Sephadex LH-20: From Pharmacia Fine Chemicals; equilibrate overnight in column solvent.

Sodium chloride, 1 $M$, reagent grade in glass-distilled water

*Equipment*

Centrifuge tubes, 50-ml polypropylene

Culture tubes, 18 × 150 and 10 × 75 mm, glass

Tank of nitrogen gas attached to manifold dryer

Centrifuge with swinging-bucket rotor

Refrigerated high speed centrifuge

Refrigerated shaker bath

Glass columns for chromatography 8 × 250 mm equipped with Teflon stopcocks, and a means of attaching the tops to a reservoir

Pipettes with disposable tips

Scintillation cocktail suitable for tritium in aqueous samples

Vortex mixer

Counting vials

*Sample Preparation*

*Extraction.* Place a 2-ml sample of plasma or other biological fluid in a 50-ml centrifuge tube. Add 25 $\mu$l of [$^3$H]25-OH-D$_3$ and/or [$^3$H]24,25-(OH)₂D$_3$ "spike" and mix with vortexing. Add 10.8 ml of a mixture of chloroform : methanol (1 : 1) and mix. Add 2.8 ml of 1 $M$ sodium chloride, vortex, and centrifuge 10 min at 10,000 $g$. Carefully remove the lower layer into a clean 18 × 150 mm tube. Wash the upper (water) layer with 2 ml of redistilled chloroform, vortexing and centrifuging as above. Combine the two chloroform extracts. Discard the water layer and protein. Dry the chloroform extract under nitrogen. The sample may be stored in the refrigerator in a small volume of CHCl$_3$ at this point.

*Chromatography.* A column of Sephadex LH-20, 0.8 × 25 cm is used for each sample. The dried sample is redissolved in a small volume of column solvent consisting of chloroform : petroleum ether (60 : 40) and applied with two washes to a column preequilibrated with this solvent. The first 10 ml of eluate, containing esters of vitamin D and free vitamin D, are discarded. The next 12 ml, containing the 25-OH-D, are collected and dried under nitrogen. In the event the sample 24,25-(OH)₂D levels are to be assayed, the next 8 ml of column eluate are discarded and the succeeding 30 ml of eluate containing the putative 24,25-(OH)₂D are collected and saved for further purification.

The 24,25-$(OH)_2D$ peak obtained from Sephadex LH-20 columns must be rechromatographed prior to steroid competition assay on a column of silicic acid via high-pressure liquid chromatography (HPLC) using an elution solvent of isopropanol : hexane (1 : 9) as described by Jones and DeLuca.[11] This additional HPLC purification of the 24,25-$(OH)_2D$ peak is essential to removal of interfering substances that lead to erroneous "high" values.

The 25-OH-D peak from the Sephadex LH-20 column and/or the 24,25-$(OH)_2D$ peak from the HPLC step are evaporated to dryness under nitrogen and resuspended in 1.0 or 0.50 ml, respectively, of absolute ethanol. The recovery of the added "spike" may be determined by removal of a 20% aliquot. For a series of 23 plasma samples, recovery of 24-OH-D averaged 78.2 ± 9.3 SD, and the recovery of 24,25-$(OH)_2D$ averaged 68.5 ± 7.8 SD.

*Steroid Competition Assay*

Place pairs of 12 × 75 mm culture tubes in an ice bath. To each tube add in the following order: 25 $\mu$l of nonradioactive standard or unknown sample; 10 $\mu$l of radioactive ligand; 1.0 ml of binding protein solution.

Vortex, and place in a refrigerated shaker bath at 6° for 30 min. After the incubation period, add 0.5 ml of thoroughly stirred charcoal–dextran suspension, vortex, and replace in the shaker bath for 10 min. Then place the tubes in a table-top centrifuge equipped with a swinging-bucket rotor and centrifuge at top speed for 5 min.

*Radioactivity Determination.* Remove a 1-ml aliquot of each supernatant into a scintillation counting vial. To each add 10 ml of suitable scintillation counting cocktail, mix, and count to 2% accuracy.

*Calculations.* The standard curve is plotted as $1/cpm \times 10^3$ vs nanograms of cold 25-OH-D or 24,25-$(OH)_2D$ per tube, which provides a linear relationship. Typical standard curves for both the 25-OH-D and 24,25-$(OH_2D$ assays are shown in Fig. 1. The amount of unknown per assay tube is then determined from this curve. This determination may be made either by hand or by computer. The amount of unknown per tube is then corrected for dilution and recovery, and the results are expressed in nanograms per milliliter of sample.

$$\frac{(\text{ng steroid/tube [unknown]}}{(\mu\text{l unknown sample/tube)}} \times \frac{(\text{total } \mu\text{l ethanol solution)}}{(\text{volume of original solution)}}$$
$$\div \text{ (percent recovery)} = \text{ng unknown steroid/ml original sample}$$

[11] G. Jones and H. F. DeLuca, *J. Lipid Res.* **16,** 448 (1975).

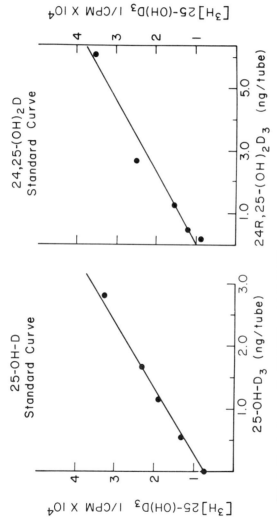

FIG. 1. Standard curve for 25-OH-D and 24,25-(OH)$_2$D competition assays. In each assay the indicated nonradioactive steroid competed with [$^3$H]25-OH-D$_3$ (1 pmol) in a preparation of lyophilized rat serum binding protein.

Discussion

Rat serum is used as a source of binding protein because neither it nor human serum protein discriminates between 25-OH-$D_2$ and 25-OH-$D_3$ or 24,25-$(OH)_2D_2$ and 24,25-$(OH)_2D_3$.[5,8,9] Further, since 25-OH-$D_3$ and 24,25-$(OH)_2D_3$ both compete equally with [$^3$H]25-OH-$D_3$ for binding to the rat serum transcalciferan, it is feasible to develop steroid competition assays for both vitamin D metabolites using only tritiated 25-OH-D as the marker ligand[8,12] (see Fig. 1). A large pool of rat serum is desirable, since its bound/free (B/F) ratio and useful range do not need to be determined frequently.

A B/F ratio and [$^3$H]25-OH-$D_3$ saturation curve must be determined for every new preparation of binding protein.

The protein concentration to be used in the assay may be determined from the linear portion of a [$^3$H]25-OH-$D_3$ binding curve—bound/free vs milligrams of protein per milliliter with a constant amount of [$^3$H]25-OH-D. If a protein concentration is chosen where the B/F is approximately 0.5, an appropriate standard curve can be constructed.

It is essential that a complete set of standards be run with each incubation to accommodate minor daily variations within the assays. The reproducibility of the standard curve can be monitored from day to day by evaluation of any changes in the slopes and intercepts of the linearized plots.

Significant competition of [$^3$H]25-OH-$D_3$ (>2 SD above the mean in the absence of unlabeled metabolite) occurs at 0.3 ng/tube (0.6 pmol/tube) for both the 25-OH-D and 24,25-$(OH)_2D$ assays; thus this represents the lower detection limit for both steroids. When aliquots of a human plasma pool were measured in 15 assays, the mean concentration of 25-OH-D was 38.5 ng/ml (coefficient of variation > 17%), while for 24,25-$(OH)_2D$ the concentration was 2.1 ng/ml. The interassay variation on the same pool of plasma run in different assays was 17% for 25-OH-D and 29% for 24,25-$(OH)_2D$.

This procedure has been utilized for determination of 25-OH-D levels in plasma obtained from experimental studies on vitamin D in rats, chicks, humans, and a variety of other animals.[8]

[12] A. E. Caldas, R. W. Gray, and J. Lemann, Jr., *J. Lab. Clin. Med.* **91**, 840 (1978).

## [57] Some Features of the Receptor Proteins for the Vitamin D Metabolites

*By* D. E. M. LAWSON

The discovery[1] that the kidney is the only site for the controlled conversion of 25-hydroxyvitamin $D_3$ (25-OH-$D_3$) to 1,25-dihydroxyvitamin $D_3$ [1,25-$(OH)_2D_3$] means that this latter substance should properly be classified as a steroid hormone. As expected of such a substance, the target tissues contain rceptor proteins that can be recognized by their high affinity for it. Such receptor proteins for 1,25-$(OH)_2D_3$ have now been found in the intestine[2-6] and bone[4,7] of rats and chicks, and in addition similar proteins have also been described for 25-OH-D.[3,8-10] These latter proteins are present in a much wider range of tissues than appears to be the case for the 1,25-$(OH)_2D_3$ receptor proteins. Although a function for the 1,25-$(OH)_2D_3$ receptor protein can be envisaged in the control of gene activity in the target tissue, the discovery of the 25-OH-D binding protein raises the question of its function, as no physiological or biochemical process is known which is controlled solely by 25-OH-D. The study of these proteins has invariably taken the approach of establishing various properties, including their affinity for the hormone, steroid specificity, tissue concentration, size of the protein, intracellular distribution, and stability.

### Techniques Used in the Study of Receptor Proteins

A previous volume in this series (Vol. 36) has an excellent coverage of the experimental conditions for the preparation of receptor proteins and of the techniques used to establish their properties and physical characteris-

[1] D. R. Fraser and E. Kodicek, *Nature (London)* **228**, 764 (1970).

[2] D. E. M. Lawson and P. W. Wilson, *Biochem. J.* **115**, 269 (1974).

[3] B. E. Kream, R. D. Reynolds, J. C. Knutson, J. A. Eisman, and H. F. DeLuca, *Arch. Biochem. Biophys.* **176**, 779 (1976).

[4] P. F. Brumbaugh and M. R. Haussler, *J. Biol. Chem.* **249**, 1251 (1974).

[5] D. A. Procsal, W. H. Okamura, and A. W. Norman, *J. Biol. Chem.* **250**, 8382 (1975).

[6] B. E. Kream, S. Yamada, H. K. Schnoes, and H. F. DeLuca, *J. Biol. Chem.* **252**, 4501 (1977).

[7] B. E. Kream, M. Jose, S. Yamada, and H. F. DeLuca, *Science* **197**, 1086 (1977).

[8] D. E. M. Lawson and J. S. Emtage, *in* "Calcium Regulating Hormones" (R. V. Talmage, M. Owen, and J. A. Parson, eds.) p. 330. Excerpta Med. Found., Amsterdam, 1975.

[9] D. E. M. Lawson, M. Charman, P. W. Wilson, and S. Edelstein, *Biochim. Biophys. Acta* **437**, 403 (1976).

[10] J. G. Haddad and S. J. Birge, *J. Biol. Chem.* **250**, 299 (1975).

METHODS IN ENZYMOLOGY, VOL. 67

tics. This chapter therefore will not include experimental details except inasmuch as they apply only to the binding proteins for the vitamin D metabolites; the reader should refer to Vol. 36 Chapters 14–19, for detailed discussion of the theoretical background and full descriptions of the working practices for the separation of the bound and free fractions of the steroid, estimation of sedimentation constants, chromatographic fractionation of binding proteins, etc.

The steroid-binding proteins are found in the cytoplasm and nuclei of the target tissues. The manner of preparation of the cytoplasmic protein produces a fraction of the cell usually referred to as the cytosol fraction, as it is not completely equivalent to the cytoplasm. Intestinal nuclei and cytosol fractions are prepared by scraping the mucosal cells from the underlying muscle layers, using a glass slide or a spatula.

The mucosal cells of either chicks or rats are then homogenized in 10 mM Tris · HCl buffer (pH 7.5) with 1.5 mM EDTA and 1 mM mercaptoethanol (TEM buffer). The homogenate is filtered, and the filtrate is centrifuged at 800 g for 10 min. The cytosolic fraction is obtained by centrifuging this supernatant at 100,000 g for 1 hr and the clear supernatant or cytosol is removed. The nuclear proteins are obtained from the pellet produced by the 800 g centrifugation. This pellet is washed twice with an equal volume of buffer and centrifuged each time at 800 g for 10 min. Proteins are extracted from the pellet by suspending it in TEM buffer containing 0.4 M KCl. After 1 hr at 4°, the suspension is centrifuged at 100,000 g for 1 hr, and the clear supernatant is removed. The protein content of this extract and the cytosolic fraction can be determined spectrophotometrically.

Sedimentation constants for the protein were obtained by centrifuging the proteins at 50,000 rpm for 20 hr through 5 to 20% (w/v) sucrose gradients in TEM buffer. The 25-OH-$D_3$-binding proteins can be analyzed by electrophoresis in 10% polyacrylamide gels, but the 1,25-$(OH)_2D_3$-binding proteins seem to dissociate in this system.

### 1,25-Dihydroxyvitamin $D_3$-Binding Protein

#### Nuclear Protein

*Studies in Vivo.* The nuclear protein can be obtained from chick intestinal nuclei preloaded with labeled hormone by treatment of this organelle with buffer solutions containing KCl (approximately 0.6 M). This treatment solubilizes about 80% of the nuclear protein (Fig. 1b). The protein can also be solubilized by treating nuclei with buffers, about 55% being solubilized by solutions at pH 9.0 (Fig. 1a). The protein nature of the

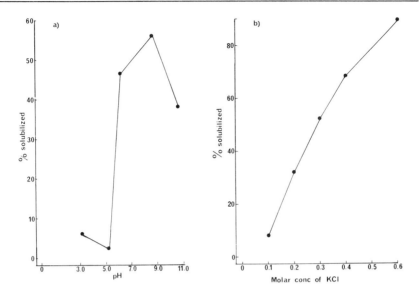

FIG. 1. Extraction of 1,25-$(OH)_2D_3$ receptor proteins from chick intestinal nuclei. Rachitic chicks were dosed with 125 ng of [26-$^3$H]25-OH-$D_3$, and pure intestinal nuclei were isolated. The nuclei were treated for 1 hr at 4° with (a) varying concentrations of KCl in TEM buffer or (b) buffers over the range pH 3–11 in the absence of KCl. The suspension was then centrifuged for 1 hr at 100,000 $g$, and the radioactivity in the supernatant was measured. This was expressed as percentage of the total radioactivity in the original nuclei.

nuclear receptor complex is indicated by the finding that proteolytic enzymes effect a significant release of the hormone from the receptor into the incubation solution. Further, the radioactivity in the extracts (>75%) is precipitated by half-saturation with $(NH_4)_2SO_4$ and can be eluted at the void volume from columns of Sephadex G-25. Stability of the physical features of the protein is improved by 3 m$M$ EDTA and 1 m$M$ mercaptoethanol, showing the importance of free SH groups in the protein for the binding activity. The sedimentation coefficient value of this protein in a sucrose gradient is about 3.2 S.

*Studies in Vitro.* Treatment of chick intestinal nuclei with 0.4 $M$ KCl solution in TEM buffer yields a protein fraction with 1,25-$(OH)_2D_3$-binding activity *in vitro*. The use of 0.4 $M$ KCl, rather than a solution containing 0.6 $M$ KCl, results in a lower extraction rate of inactive nuclear proteins and a product with a higher specific activity. The proportion of 1,25-$(OH)_2D_3$ bound *in vitro* to the nuclear extract is related to the protein concentration in the incubation mixture such that about 40–60% is bound by 0.5–1.0 mg of nuclear protein. With increasing concentrations of steroid, the amount bound to the hormone increases in a biphasic manner, but at any given protein concentration the binding sites are not saturable

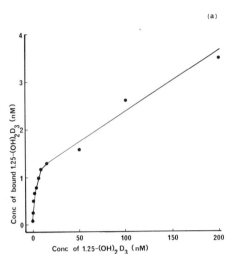

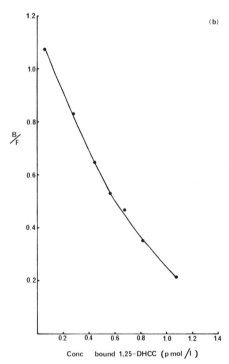

FIG. 2. (a) Uptake of radioactive 1,25-$(OH)_2D_3$ by chick intestinal nuclear extracts. A 0.4 $M$ KCl extract of chick intestinal nuclei was incubated for 1 hr at 4° with increasing amounts of [26-$^3$H]1,25-$(OH)_2D_3$. The mixture was treated with dextran-coated charcoal, and the bound radioactivity in the supernatant was removed. (b) The uptake of [26-$^3$H]1,25-$(OH)_2D_3$ by the chick intestinal nuclear protein, of the type shown in Fig. 2a, can be analyzed by

(Fig. 2a). The affinity of the nuclear protein in the KCl extract for the 1,25-$(OH)_2D_3$ is measured by recording the displacement of the bound radioactive steroid by increasing amounts of the nonlabeled hormone (Fig. 2b). Analysis of the results show that two classes of binding sites are present with association constants of $1.7 \times 10^9 \, M^{-1}$ and $3.1 \times 10^8 \, M^{-1}$ (see the table). The number of sites in the intestinal mucosa for these two classes is 2.4 pmol/g of tissue and 5.3 pmol/g of tissue. The high-affinity binding activity can be observed in KCl extracts of intestinal nuclei from both rachitic and normal chicks. The nuclear binding protein can also be analyzed by centrifugation through a linear 5 to 20% sucrose gradient. The position of the 1,25-$(OH)_2D_3$ in the gradient shows that it is bound to only one class of protein which gives a sedimentation coefficient of 3.5 S (see the table), showing that the hormone is most probably attached to the same protein *in vitro* as it is *in vivo*. KCl extracts of chick kidney nuclei can also bind radioactive 1,25-$(OH)_2D_3$ in a manner similar to binding of the chick intestinal nuclear proteins, and this bound radioactivity can be displaced by low concentrations of the unlabeled hormone. The association constant for the binding site in the kidney with the highest affinity for 1,25-$(OH)_2D_3$ is $5.8 \times 10^8 \, M^{-1}$. Proteins with an affinity for 1,25-$(OH)_2D_3$ have not so far been found in nuclei of chick liver and muscle. Liver nuclear extracts bind only a small proportion of radioactive hormone (approximately 10%), and the bound fraction cannot be displaced by amounts of the steroid that are effective in displacing bound radioactivity from intestinal and renal nuclear extracts. Competitive displacement experiments have shown the importance of the hydroxyl groups at C-1 and C-25 for the binding activity of the 1,25-$(OH)_2D_3$.[2,5]

Conditions for maintaining the stability of the chick intestinal 1,25-$(OH)_2D_3$-binding protein have not so far been described. The preservation of free SH groups on the protein, by the inclusion of mercaptoethanol, is necessary for the maintenance of binding activity. Storage of the protein at 4° results in complete loss of binding activity over 3–4 days, and until this loss can be prevented it will not be possible to purify the protein. Lowering the temperature to −20° or addition of 20% glycerol to the binding protein improves its stability, so that some activity is retained for up to 7 days under these conditions. Storage of the binding protein in liquid $N_2$ enables binding activity to be retained for several weeks.

---

plotting the ratio of the concentrations of bound (B) to free (F) steroid against the concentration of bound steroid. In the case of this protein, a curve was obtained that indicates that the total binding activity is due to the presence of more than one binding site in the nuclear proteins. Analysis of this curve [D. E. M. Lawson and P. W. Wilson, Biochem. J. **115**, 269 (1974)] showed that two classes of binding sites were present.

ASSOCIATION AND SEDIMENTATION CONSTANTS OF THE BINDING PROTEINS FOR 25-OH-D AND 1,25-(OH)$_2$D$_3$

| Constant | Chick | | | Rat | | |
|---|---|---|---|---|---|---|
| | Plasma<br>25-OH-D$_3$ | Intestine | | Plasma<br>25-OH-D$_3$ | Intestine | |
| | | 25-OH-D$_3$ | 1,25-(OH)$_2$D$_3$ | | 25-OH-D$_3$ | 1,25-(OH)$_2$D$_3$ |
| $K_a$ ($M^{-1}$) | $3 \times 10^8$ | $3.1 \times 10^{9a}$ | $1.7 \times 10^{9b}$ | $8 \times 10^9$ | $7.2 \times 10^{9a}$ | ND[c] |
| $s$ (S) | 4.0 | $5.0^a$ | $3.5^b$ | 4.1 | $5.8^a$ | $3.2^a$ |

[a] Cytosolic protein.
[b] Essentially the same value obtained for both nuclear and cytosolic proteins.
[c] ND, Not determined.

*Cytosol Protein*

A protein able to bind 1,25-$(OH)_2D_3$ can also be found in rat and chick intestinal cytosol fractions providing that sufficient care is taken to distinguish it from the 25-OH-D-binding protein.[6,10] As with the nuclear protein, that in the chick intestinal cytosol has a high affinity and a low capacity for the 1,25-$(OH)_2D_3$ with an association constant $>10^9 M^{-1}$. The sedimentation constant in a sucrose gradient is the same as for the nuclear protein and is not changed by variations in the salt concentration. The importance of the preservation of SH groups in the protein for binding activity has also been noted. These characteristics also apply to the rat intestinal cytosol protein ($s$ value is 3.2 S), although the $K_a$ does not yet appear to have been reported. A small amount of a 1,25-$(OH)_2D_3$-binding protein can be detected in the cytosol of rat testes but not in spleen, heart, liver, kidney, and thyroid.

The relative importance of the various features of 1,25-$(OH)_2D_3$ that affect extent of binding to these proteins has been recorded. The presence of the three double bonds in the molecule is undoubtedly one of the most important features, but additional hydroxyl groups are also necessary. Substances with the double-bond system altered or without the hydroxyl groups can bind only very weakly to the protein. With the double-bond structure intact, a substance with a hydroxyl group at C-1 binds more strongly than one with the hydroxyl at C-25. Changes in the side chain of the vitamin D molecule are least disruptive. Thus, although vitamin $D_2$ is much less active in chicks than vitamin $D_3$, it has been found that 1,25-$(OH)_2D_2$ binds to the protein as strongly as 1,25-$(OH)_2D_3$, and 1,24,25-$(OH)_3D_3$ is the next most potent derivative of 1,25-$(OH)_2D_3$ in this respect.

Properties of the 25-Hydroxyvitamin D-Binding Protein

25-OH-$D_3$-binding activity can be observed with cytosols of a wide range of tissues from both rachitic and vitamin D-replete animals, the only exception being red blood cells, extracts of which do not have any binding activity.[3,8–10] The time course of the uptake of 25-OH-$D_3$ by chick intestinal cytosol consists of an initial rapid uptake for 15–20 min, followed by a slower uptake until by 60 min the maximum amount of steroid is bound. The concentration of bound steroid then remains constant for the following 3 hr. Analysis of the interaction of the 25-OH-$D_3$ with the intestinal cytosol shows that saturation of the binding sites is achieved at relatively low concentrations (i.e., $<3$ n$M$). In some tissues, e.g., muscle, skin, and kidney, this interaction primarily involves high-affinity sites, whereas in others, such as intestine and liver, both high- and low-affinity sites are involved (Fig. 3). The association constant for the chick intestinal high

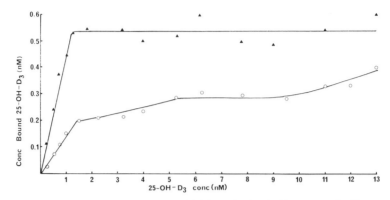

FIG. 3. Uptake of [26-³H]25-OH-D₃ by chick intestinal and kidney cytosols. The cytosols of the two tissues were incubated for 1 hr at 4° with increasing amounts of [26-³H]25-OH-D₃. The mixture was treated with dextran-coated charcoal, and the bound radioactivity in the supernatant was measured. The shape of the curve with kidney cytosol suggests that only one class of binding site is present, but with intestinal cytosol the shape suggests the presence of more than one class. Confirmation of this possibility requires analysis of the curves as in Fig. 2b; from such analysis, association constants and binding capacities of the cytosols can be calculated. ▲, kidney cytosol; ○, intestinal cytosol.

affinity sites is $3.1 \times 10^9 \, M^{-1}$, and for the other tissues the values range from $2.5 \times 10^9$ to $4.6 \times 10^9 \, M^{-1}$.

The binding activity of the cytosols for 25-OH-D₃ is greatly decreased by preincubation of the cytosols at 28° with 0.5 mg of chymotrypsin in 1 ml of TEM buffer. At the end of this period, binding activity for 25-OH-D₃ cannot be detected, showing the protein nature of the activity in the cytosols.

The sedimentation constant of the chick intestinal 25-OH-D₃-binding protein has been measured on a sucrose gradient, and a value of 5–6 S was found. This shows the active component to be distinct from the 3.5 S protein with 1,25-(OH)₂D₃-binding activity. Furthermore the tissue binding protein is distinct from the chick plasma 25-(OH)D₃-binding protein, as the sedimentation constant value for this protein is 3.5–4.0 S (see the table). The intestinal binding protein for 25-OH-D₃ can be separated from that in plasma by ion-exchange chromatography. The system consists of $1.5 \times 25$ cm columns of DEAE-Sephadex A-50 at 4° equilibrated with TEM buffer containing 0.1 $M$ NaCl. The proteins are eluted from the column with a gradient of 0.1 to 0.6 $M$ NaCl in TEM buffer. The plasma and intestinal cytosol proteins can also be resolved by gel electrophoresis. In this case a suitable gel system consists of 10% acrylamide, 0.5% bisacrylamide, 0.1% ammonium persulfate, 0.025% TEMED in a buffer of 1.5 $M$ Tris and 0.4 $M$ glycine, pH 8.8. The stacking gel consists of 4% acrylamide, 0.26% bisacrylamide, 0.1% ammonium persulfate, and

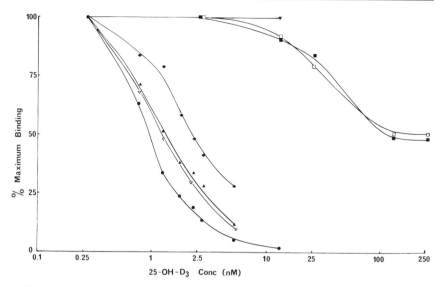

FIG. 4. Competitive binding curves using rachitic chick intestinal cytosol. A series of solutions was prepared containing a constant amount of $[26\text{-}^3\text{H}]25\text{-OH-D}_3$ and increasing amounts of the steroid shown. The cytosol was diluted with TEM buffer, so that 1 ml incubated for 1 hr at 4° with 40 $\mu$l of an ethanolic solution of $[26\text{-}^3\text{H}]25\text{-OH-D}_3$ ($10^4$ dpm) bound about 40% of the radioactivity. Using this dilution of cytosol, incubations were carried out with the steroids and the amount of radioactivity bound to the cytosolic protein was measured using dextran-coated charcoal. The percentage of the radioactive $25\text{-OH-D}_3$ bound at each of the concentrations of the individual steroids was plotted. $D_3$ has almost no ability to replace $25\text{-OH-D}_3$ from the binding sites on the cytosolic protein, whereas those steroids most similar to $25\text{-OH-D}_3$ have the highest replacement activity. ●, $25\text{-OH-D}_3$; ▽, $24,25\text{-(OH)}_2\text{D}_3$; ▲, $25,26\text{-(OH)}_2\text{D}_3$; ◆, $25\text{-OH-27-nor-D}_3$; □, $D_3$; ■, $1\text{-OH-D}_3$; ▼, $25\text{-OH-iso-T}_3$; $25\text{-OH-5,6-}trans\text{-D}_3$.

0.05% TEMED in the above buffer. The relative mobility of the 25-OH-D-binding protein in rat plasma and that in rat intestinal cytosol is 0.29 and 0.24, respectively.

Despite these readily demonstrable differences between the binding proteins from these two sites, it has been reported[11] that intestinal cytosol contains a protein that reacts immunologically with antisera prepared against the plasma protein. Van Baelen and colleagues have shown that the rat intestinal cytosolic protein can be formed from the plasma 25-OH-D-binding protein and a protein present in all nucleated rat tissues. This latter protein is heat-labile, sediments around 4 S, and is not able to bind 25-OH-D. It is suggested that the cytosolic 25-OH-D-binding protein arises from contamination of the tissues with the plasma.

[11] H. Van Baelen, R. Bouillon, and P. De Moor, *J. Biol. Chem.* **252**, 2515 (1977).

The nature of the groups on the 25-OH-D$_3$ molecule that the binding sites on the protein recognize has also been established. In this case the ability of increasing amounts of various isomers and derivatives of 25-OH-D$_3$ to compete for the binding sites with 25-OH-D$_3$ has been compared (Fig. 4). These studies showed the importance of the double-bond structure and of the 25-hydroxyl group. Thus the 25-hydroxy derivatives of 5,6-*trans*-D and of isotachysterol were unable to displace 25-OH-D$_3$ from the binding sites. Vitamin D$_3$ and 1-OH-D$_3$ compete effectively with 25-OH-D only at 100-fold higher concentrations. Alterations to the side chain, however, have very little effect on binding activity. Thus 25-hydroxy-27-norvitamin D$_3$ is about 2.5 times less effective than 25-OH-D$_3$ in binding to the protein, and 24,25-(OH)$_2$D$_3$ and 25,26-(OH)$_2$D$_3$ are only about 1.5 times less active. The intestinal binding protein is much more stable than the 1,25-(OH)$_2$D$_3$-binding protein, as it retains binding activity for up to 2 months in liquid nitrogen and for up to 3 weeks if left at $-20°$.

## [58] Measurement of Kinetic Rate Constants for the Binding of 1α,25-Dihydroxyvitamin D₃ to Its Chick Intestinal Mucosa Receptor Using a Hydroxyapatite Batch Assay

*By* WAYNE R. WECKSLER and ANTHONY W. NORMAN

1α,25-Dihydroxyvitamin D$_3$ [1α,25-(OH)$_2$D$_3$] is believed to act as a classical steroid hormone. Several workers have demonstrated that there are specific cytoplasmic[1-4] and chromatin-extractable[4-6] receptors for this steroid in chick intestine[1-6] and chick parathyroid glands.[4,7] In addition, cytoplasmic receptors for 1α,25-(OH)$_2$D$_3$ have recently been demonstrated in rat intestine[8] and chick and rat calvaria.[9] The lack of a conve-

[1] H. C. Tsai and A. W. Norman, *J. Biol. Chem.* **248**, 5967 (1973).

[2] P. F. Brumbaugh and M. R. Haussler, *J. Biol. Chem.* **249**, 1258 (1974).

[3] B. E. Kream, R. D. Reynolds, J. C. Knutson, J. A. Eisman, and H. F. DeLuca, *Arch. Biochem. Biophys.* **176**, 779 (1976).

[4] W. R. Wecksler, H. L. Henry, and A. W. Norman, *Arch. Biochem. Biophys.* **183**, 168 (1977).

[5] P. F. Brumbaugh and M. R. Haussler, *J. Biol. Chem.* **249**, 1251 (1974).

[6] D. E. M. Lawson and P. W. Wilson, *Biochem. J.* **144**, 573 (1974).

[7] P. F. Brumbaugh, M. R. Hughes, and M. R. Haussler, *Proc. Natl. Acad. Sci. U.S.A.* **72**, 4871 (1975).

[8] B. E. Kream, S. Yamada, H. K. Schnoes, and H. F. DeLuca, *J. Biol. Chem.* **252**, 4501 (1977).

[9] B. E. Kream, M. Jose, S. Yamada, and H. F. DeLuca, *Science* **197**, 1086 (1977).

nient assay system for these macromolecular complexes and their relative instability have hindered their detailed characterization. We have adapted a hydroxyapatite batch assay for the quantitation of $1\alpha,25$-$(OH)_2D_3$ receptor complex from chick intestinal mucosa cytosol. This has provided the first assay for this system that is facile enough to be used to study the kinetics of the binding processes. In this report we shall describe the assay procedure and its application to the measurement of association and dissociation rate constants for the chick intestinal mucosa $1\alpha,25$-$(OH)_2D_3$ receptor.

## Methods

*Animals.* White Leghorn cockerels are obtained on the day of hatching and raised for 4 weeks on a synthetic rachitogenic diet.[10] During this time, their vitamin D stores are depleted and the animals become rachitic.

*Tissue Preparation.* Chicks are killed by decapitation; intestines are removed, expressed of their contents, slit longitudinally, and placed in ice-cold saline. The mucosal layer is scraped from the underlying serosa using glass microscope slides on an ice-cold, inverted glass dish. The mucosa is minced and washed three times by centrifugation ($350\,g/5\,min$) in 3–5 volumes of ice-cold saline. The washed mucosa is then weighed, 4 volumes of $0.25\,M$ sucrose–$50\,mM$ Tris · HCl, $25\,mM$ KCl–$5\,mM$ MgCl$_2$, pH 7.5, are added, and the samples are homogenized in a glass–Teflon, motor-driven homogenizer. The homogenate is centrifuged at $4300\,g$ for 10 min, and the resulting supernatant fraction is centrifuged at $105,000\,g$ for 60 min to yield a cytosol preparation. Cytosol is either used fresh or lyophilized and stored at $-20°$ under argon. Lyophilized cytosol is reconstituted by addition of distilled water to give a cytosol protein concentration of 6–10 mg/ml. Since lyophilization does not appear to alter the binding activity, it is convenient to prepare a large batch (100–150 ml) of cytosol, which is then aliquoted into 5-ml samples and lyophilized. Thus, a homogeneous pool of cytosol is generated from which a portion can be reconstituted for assay as desired.

*Hydroxyapatite Preparation.* BioGel HTP (hydroxyapatite) is obtained from Bio-Rad Laboratories (Richmond, California). A 50% slurry of the resin is made as follows: 10 g of resin are added to 60 ml of 10 mM $K_xHPO_4$–$0.1\,M$ KCl, pH 7.5, in a 125-ml Erlenmeyer flask while swirling. The resin is allowed to settle for 10–20 min, the buffer is decanted off, and 30 ml of fresh buffer are added. The resin is resuspended and allowed to settle one more time, and 30 ml of fresh buffer are added. This mixture,

[10] A. W. Norman and R. G. Wong, *J. Nutr.* **102**, 1709 (1972).

when resuspended by swirling, gives a 50% slurry and must be resuspended prior to each pipetting of resin.

*Steroids.* Tritiated 25-hydroxy[25,26-methyl-$^3$H]vitamin $D_3$ (3–12 Ci/mmol) (obtained from Amersham/Searle) is converted *in vitro* into [$^3$H]$1\alpha$,25-$(OH)_2D_3$. Nonradioactive crystalline $1\alpha$,25-$(OH)_2D_3$ is synthesized as previously described.[11]

*Assay Procedure:* Tritiated $1\alpha$,25-$(OH)_2D_3$, in the presence or the absence of excess nonradioactive $1\alpha$,25-$(OH)_2D_3$ (to determine nonspecifically bound steroid), dissolved in 20 $\mu$l of ethanol, is incubated as described below in 1.7-ml microfuge tubes (Sarstedt) with 0.2 ml of cytosol. The reactions are stopped by removing the assay tubes to ice and immediately adding 0.4 ml of the hydroxyapatite slurry. The samples are vortexed and left on ice for 10 min with vortexing every 5 min. After this incubation, which allows the steroid-receptor complex to bind to the resin, the samples are centrifuged in a Beckman Microfuge B for 3 sec.[12] The supernatant is discarded, and the resin washed three times by vigorous vortexing and centrifuging as above with 0.8 ml of 10 m$M$ Tris · HCl–0.5% Triton X-100, pH 7.5. The final washed pellet is extracted twice with 0.8 ml of 2 : 1 methanol : chloroform. The organic solvent extracts are removed to scintillation vials and air-dried. Scintillation cocktail is added and the tritium content of the samples determined by liquid scintillation counting. The counts per minute (cpm) are converted to disintegrations per minute (dpm) by the external standard method.

*Incubations.* For determining the association rate constant, 4–6 time points are selected, based on the incubation temperature, and for each time point there are four incubations—two with tritiated steroid alone and two with tritiated steroid plus a 200-fold excess of nonradioactive steroid to measure nonspecific binding. The concentration of tritiated $1\alpha$,25-$(OH)_2D_3$ used in this study was 5.9 n$M$, but other concentrations can be used and lower concentrations may be useful at higher temperatures in order to slow down the reaction. At 0°, incubations are carried out for 0, 5, 10, 15, and 20 min, and at 24°, incubations are carried out for 0, 2, 4, 6, and 8 min. The length of the incubation should be sufficient to maintain binding linearity over several time points. Two conditions have been used to determine $k_{on}$. One is with $R_0 < S_0$ and the other is with $R_0 = S_0$. $S_0$ (initial steroid concentration) is determined from the specific activity of the tritiated ligand and is known. $R_0$ (initial receptor concentration) must be determined for each cytosol preparation. A Scatchard analysis is car-

---

[11] T. A. Narwid, J. I. Blount, J. A. Iacobelli, and M. R. Uskoković, *Helv. Chim. Acta* **57**, 781 (1974).

[12] This centrifuge holds 18 samples and spins at 15,000 rpm. It pellets the hydroxyapatite as fast as it takes to reach full speed and greatly facilitates these manipulations of the assay.

ried out on each cytosol preparation and $N_{max}$ (X intercept of the B/F vs B plot) gives the molar concentration of binding sites.[13] This is also why it is advisable to make a large batch of cytosol and lyophilize aliquots. One can periodically run Scatchard analyses on the same batch of cytosol to make sure that the binding site concentration has not changed upon storage. To calculate the association rate constant, the following plots are made:

for

$$R_o < S_o: 1/(S_o - R_o) \ln [S_o - [RS]_t/R_o - [RS]_t] \text{ vs } t$$

and for             $R_o = S_o:$ [free steroid]$^{-1}$ vs $t$

where $[RS]_t$ is the measured concentration of specifically bound [$^3$H]1α,25-(OH)$_2$D$_3$. The slope of each of these plots is equal to the association rate constant in units of $M^{-1}$ time$^{-1}$.

To measure the dissociation rate constant, one selects 5 or 6 (or more) time points, in duplicate, with tritiated ligand in the presence or the absence of excess nonradioactive ligand. First, steroids and cytosol are incubated together for several hours to allow the ligand to bind to the receptor. Then, the dissociation process is initiated by the addition to each incubation of a very large excess of nonradioactive 1α,25-(OH)$_2$D$_3$ (in 5 $\mu$l of ethanol). Time points are selected beginning with $t = 0$ as dictated by the temperature. At 4° the time points are selected at 24-hr intervals, while at 24°, at 15 to 20-min intervals. Since over a long time period there is loss of binding activity due to receptor degradation (of an unknown nature), a parallel series of incubations is carried out in which the off rate is initiated by the addition of ethanol (5 $\mu$l). This allows for a correction for the loss of binding activity that is not due to dissociation, but is due to degradation. The dissociation rate is a first-order process, and a plot of $\ln[RS]_t$ vs $t$ yields a straight line where the slope is equal to $-k_{off}$. The degradation rate is subtracted from the dissociation rate to yield a corrected dissociation rate.

## Results

*Association Rate Constant Determinations.* Figure 1 shows the results of a determination of the association rate constant for the 1α,25-(OH)$_2$D$_3$ receptor at 0° under conditions where $R_o = 1.1$ nM as determined by Scatchard analysis[13] and $S_o = 5.9$ nM. The rate constant was found to be $4.4 \times 10^6 M^{-1}$ min$^{-1}$. Figure 2 shows the results of another determination of $k_{on}$. This was carried out at 23° under conditions where $R_o = S_o = 1.1$

[13] G. Scatchard, *Ann. N. Y. Acad. Sci.* **51**, 660 (1949).

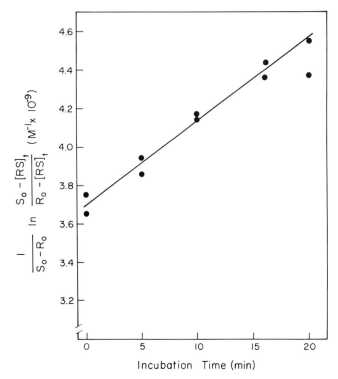

FIG. 1. Determination of the second-order association rate constant for $1\alpha$-25-$(OH)_2D_3$ with the chick intestinal mucosa cytosolic receptor at $0°$. Conditions were designed so that $R_0$ = 1.1 n$M$ and $S_0$ = 5.9 n$M$. Incubations were stopped at the indicated times by addition of ice-cold, hydroxyapatite slurry. Specifically bound $1\alpha,25$-$(OH)_2D_3$ was determined, in duplicate, for each time point ($[RS]_t$). Unweighted linear regression analysis ($r$ = 0.99) of data yielded an association rate constant (slope) of $4.4 \times 10^6 \ M^{-1} \ min^{-1}$.

TABLE I
SUMMARY OF ASSOCIATION RATE CONSTANTS

| Temperature (°C) | Associate rate constant[a] ($M^{-1}min^{-1} \times 10^{-7}$) | | | |
| | $R_0 < S_0$[b] | $n$ | $R_0 = S_0$[c] | $n$ |
|---|---|---|---|---|
| 0 | $0.55 \pm 0.10$ | 3 | $0.92 \pm 0.12$ | 3 |
| 6 | — | | $2.2 \pm 0.5$ | 2 |
| 13 | $2.5 \pm 0.7$ | 3 | $4.7 \pm 1.3$ | 3 |
| 18 | — | | 7.0 | 1 |
| 23 | — | | $9.7 \pm 2.1$ | 2 |
| 24 | $8.1 \pm 1.9$ | 3 | — | |
| 30 | — | | $14.5 \pm 0.7$ | 2 |

[a] Results are expressed as mean $\pm$ SD of $n$ determinations.
[b] $R_0$ = 1.1 n$M$, $S_0$ = 5.9 n$M$.
[c] $R_0 = S_0$ = 1.1 n$M$.

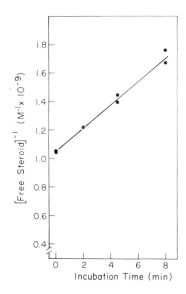

FIG. 2. Determination of the second-order association rate constant at 23° under conditions where $R_0 = S_0 = 1.1$ n$M$. Specifically bound $1\alpha,25$-$(OH)_2D_3$ was determined at the indicated times and used to calculate the free steroid concentration. The association rate constant was found to be $8.2 \times 10^7 \, M^{-1} \, min^{-1}$ ($r = 0.99$).

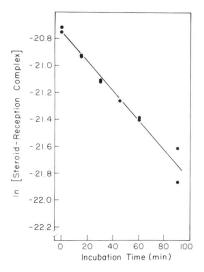

FIG. 3. Determination of the pseudo-first-order dissociation rate constant for the $1\alpha,25$-$(OH)_2D_3$-receptor complex at 24°. Cytosol ($R_0 = 1.1$ n$M$) was incubated with [$^3$H]$1\alpha,25$-$(OH)_2D_3$ (5.9 n$M$) for 4 hr. The dissociation was initiated by the addition of 0.5 $\mu M$ non-radioactive ligand in 5 $\mu$l of ethanol. Specifically bound $1\alpha,25$-$(OH)_2D_3$ was determined at the indicated times. The data were corrected for receptor degradation. The dissociation rate constant was found to the $1.1 \times 10^{-2} \, min^{-1}$, which yields a $t_{1/2}$ for the process of 63 min ($r = 0.99$.)

TABLE II
SUMMARY OF DISSOCIATION RATE CONSTANTS

| Temperature (°C) | $k_{off}$ (min$^{-1} \times 10^5$)$^a$ | $n$ | $t_{1/2}$(min)$^b$ |
|---|---|---|---|
| 4 | 4.1 ± 1.5 | 3 | 16,900 |
| 8 | 9.9 ± 0.4 | 2 | 7,000 |
| 13 | 14.6 ± 3.3 | 3 | 4,750 |
| 18 | 140 ± 100 | 3 | 495 |
| 24 | 530 ± 170 | 4 | 131 |
| 30 | 3000 ± 1500 | 3 | 23 |

$^a$ The results are expressed as mean ± SD of $n$ determinations and are corrected for receptor degradation.
$^b$ Half-lives of the steroid–receptor complex were calculated from the equation $t_{1/2} = \ln 2/k_{off}$.

n$M$. The association rate constant was found to be $8.2 \times 10^7 M^{-1}$ min$^{-1}$ in this assay. Table I summarizes the results of a series of determinations of $k_{on}$ at various temperatures.

*Dissociation Rate Constant Determinations.* Figure 3 shows the results of a determination of rate of dissociation of the $1\alpha,25$-(OH)$_2$D$_3$-receptor complex at 24°. The dissociation rate was found to be $1.1 \times 10^{-2}$ min$^{-1}$. Table II summarizes the results of a series of determinations of $k_{off}$ at various temperatures.

These results are in good agreement with those obtained by workers in other steroid hormone systems.[14-16]

[14] P. A. Bell and A. Munck, *Biochem. J.* **136**, 97 (1973).
[15] P. E. Hansen, A. Johnson, W. T. Schrader, and B. W. O'Malley, *J. Steroid Biochem.* **7**, 723 (1976).
[16] B. M. Sanborn, B. R. Rao, and S. G. Korenman, *Biochemistry* **10**, 4955 (1971).

# [59] Structural Aspects of the Binding of $1\alpha,25$-Dihydroxyvitamin D$_3$ to Its Receptor System in Chick Intestine

By WAYNE R. WECKSLER and ANTHONY W. NORMAN

$1\alpha,25$-Dihydroxyvitamin D$_3$ [$1\alpha,25$-(OH)$_2$D$_3$] is believed to act as a classical steroid hormone in its best-characterized target tissue, chick intestinal mucosa. In this organ, the steroid directs the de novo synthesis of a vitamin D-dependent calcium-binding protein.[1] These nuclear effects are mediated through a specific receptor system for $1\alpha,25$-$(OH)_2D_3$. The steroid enters the mucosal cell and binds stereospecifically to a cytoplasmic receptor protein[2,3]; this steroid-receptor complex subsequently becomes tightly associated with nuclear chromatin.[4]

The $1\alpha,25$-$(OH)_2D_3$ molecule possesses some structural features that are unique to classical steroid hormones. The hormone possesses the entire 8-carbon cholesterol side chain and is hydroxylated at C-25. It is also a secosteroid[5] in which the 9,10 carbon-carbon bond has undergone fission to yield an open B ring (top panel of Fig. 1). Not only does this provide a nonclassical A/B/C/D steroid ring system, but it also provides the A ring with a great deal of conformational freedom. Thus, the A ring undergoes a rapid chair-chair conformational inversion[6,7] in which the $1\alpha$- and $3\beta$-OH groups are continually flipping between an axial and equatorial conformation (bottom panel of Fig. 1). These unique structural features of the $1\alpha,25$-$(OH)_2D_3$ molecule combined with the fact that the target organ receptor system can discriminate between closely related and naturally occurring molecules (such as vitamin D itself and 25-hydroxyvitamin $D_3$ [25-$(OH)_2D_3$], the immediate metabolic precursor to the hormone) make this an interesting system to study.

The biological response of increased intestinal calcium transport induced by $1\alpha,25$-$(OH)_2D_3$ occurs over a fairly short time interval,[8] so that maintenance of a constant level of calcium uptake in the intestine is dependent upon the continuous metabolism of 25-OH-$D_3$ to the hormonally active form. Structure–function and structure-binding studies using chemically synthesized structural analogs of the $1\alpha,25$-$(OH)_2D_3$ molecule are directed at quantitatively elucidating the structural features of the $1\alpha,25$-$(OH)_2D_3$ molecule that contribute to ligand–receptor interaction.

In this chapter we describe the assay system used in our laboratory to study structure-binding relationships and the results obtained to date on the molecular topology of the $1\alpha,25$-$(OH)_2D_3$ molecule which results in its efficient interaction with its receptor system.

[1] J. S. Emtage, D. E. M. Lawson, and E. Kodicek, *Nature (London)* **246**, 100 (1973).

[2] P. F. Brumbaugh and M. R. Haussler, *Biochem. Biophys. Res. Commun.* **51**, 74 (1973).

[3] H. C. Tsai and A. W. Norman, *J. Biol. Chem.* **248**, 5967 (1973).

[4] M. R. Haussler, J. F. Myrtle, and A. W. Norman, *J. Biol. Chem.* **243**, 4055 (1968).

[5] The chemical name for vitamin $D_3$ (cholecalciferol) is 9,10-seco-5Z,7E,10(19)-cholesta-triene-3$\beta$-ol.

[6] G. D. Lamar and D. L. Budd, *J. Am. Chem. Soc.* **96**, 7317 (1974).

[7] R. M. Wing, W. H. Okamura, M. R. Pirio, S. M. Sine, and A. W. Norman, *Science* **186**, 939 (1974).

[8] J. F. Myrtle and A. W. Norman, *Science* **171**, 79 (1971).

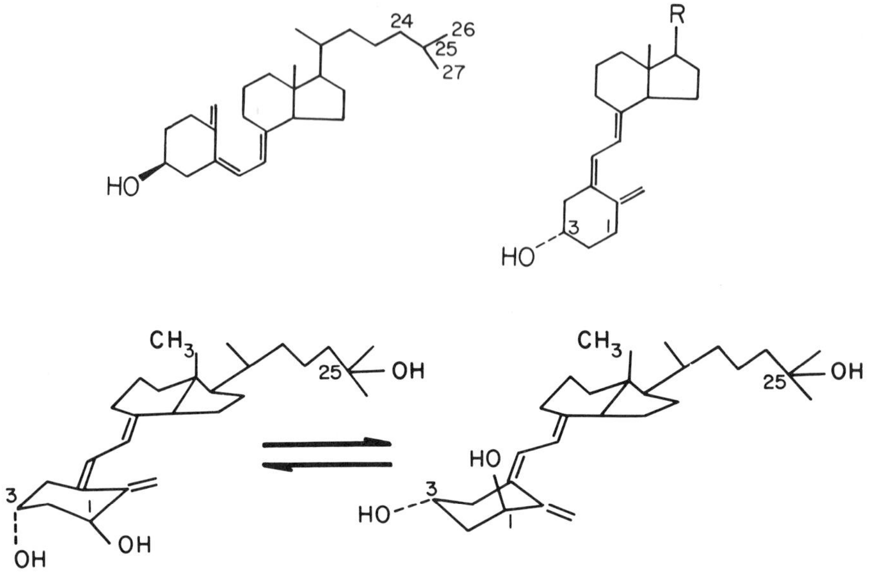

FIG. 1. Structural representations of vitamin $D_3$ and $1\alpha,25\text{-}(OH)_2D_3$. The structure on the left-hand side of the top panel shows vitamin $D_3$ in a manner that emphasizes its similarities with more classical steroids. The right-hand side of the top panel gives a representation of vitamin $D_3$ in its normal "opened up" configuration. The lower panel shows the chair–chair conformational equilibrium for $1\alpha,25\text{-}(OH)_2D_3$ that occurs in solution. Note that the $1\alpha$- and $3\beta$-hydroxyl groups oscillate between an equatorial and an axial position.

## Methods

*Animals.* While Leghorn cockerels are obtained on the day of hatching and placed on a synthetic rachitogenic diet.[9] The animals are sacrificed by decapitation in their fourth week, by which time they are depleted of vitamin D and have become rachitic.

*Tissue Preparation.* The intestine is excised, expressed of its contents, slit longitudinally, and removed to ice cold saline. The mucosa is then scraped from the underlying serosa using a glass microscope slide on an inverted glass dish embedded in ice. Scraped mucosa from 3–5 animals is weighed and homogenized in 1.5 volumes of 0.25 $M$ sucrose, 50 m$M$ Tris · HCl–25 m$M$ KCl–5 m$M$ MgCl$_2$, pH 7.5, with 10 strokes of a glass–Teflon homogenizer. The homogenate is centrifuged at 4300 $g$ for 10 min. The resulting supernatant fraction is centrifuged at 105,000 $g$ for 60 min to yield a cytosol preparation. The 4300 $g$ pellet is homogenized sequentially in 10 m$M$ Tris · HCl–0.5% Triton X-100, pH 8.5, and 50 m$M$ Tris · HCl, pH 7.5, with harvesting between homogenizations by centrifugation at

9 A. W. Norman and R. G. Wong, *J. Nutr.* **102**, 1709 (1972).

12,000 g for 15 min. This yields a crude chromatin fraction. One-fourth of the total chromatin fraction is recombined with all the cytosol by homogenization. This protocol results in approximately 15 mg of cytosol protein and 750 µg of DNA per milliliter of reconstituted chromatin and cytosol.

*Steroids.* Tritiated 25-hydroxy[26,27-methyl-$^3$H]vitamin D$_3$ (3–12 Ci/mmol) (obtained from Amersham/Searle) is converted *in vitro* into 1 α,25-dihydroxy[26,27-methyl-$^3$H] vitamin D$_3$. The steroids are purified on 1 × 60 cm Sephadex LH-20 columns in 35:65 hexane:chloroform. The specific activity of the tritiated 1α,25-(OH)$_2$D$_3$ can be diluted to a specific activity of approximately 500–2000 dpm/pmol with synthetic 1α,25-(OH)$_2$D$_3$.

*Competitive Binding Radioligand Assay.* Assays are carried out in 1.5 × 9.5 cm polypropylene centrifuge tubes. Five picomoles of [$^3$H]1α,25-(OH)$_2$D$_3$ are added to a series of incubation tubes. Increasing concentrations of competitor (over a 4- to 6-fold range) are also added to the series. The steroids are dissolved in 50 µl of absolute ethanol. To each tube, 0.50 ml of the reconstituted chromatin–cytosol preparation is added and the samples are vortexed. Incubations are carried out at room temperature for 45–60 min. The assay tubes are then centrifuged (Sorvall SM-24 rotor) at 10,000 g for 5 min; the supernatant fractions are discarded. The pellets are washed by homogenization in 3 ml of 10 mM Tris · HCl–0.5% Triton X-100, pH 8.5, and harvested by centrifugation as above. This Triton wash procedure is repeated two more times. The resulting washed chromatin pellets are then extracted for [$^3$H]1α,25-(OH)$_2$D$_3$ with 4 ml of 2:1 methanol:chloroform. The extracts are decanted into scintillation vials and air dried. Nine milliliters of butyl-PBD (Amersham/Searle) scintillation cocktail are added, and the tritium radioactivity is determined in a Beckman LS-233 liquid scintillation counter. Assays are carried out in duplicate or triplicate.

*Treatment of Competitive Data.* Maximum bound steroid is determined from incubations in which no competitor was added. The percentage of maximum binding for each concentration of competitor is calculated on the basis of this value. The ability of a competitor to decrease the amount of chromatin-bound tritiated steroid can be described by the following equation.[10]

$$\text{Percent maximum bound} = \frac{[^3\text{H-ligand}]}{[^3\text{H-ligand}] + \alpha[\text{competitor}]}$$

In this equation, α is the competitive index for the analog. For 1α,25-(OH)$_2$D$_3$, α equals 1.0. Taking the reciprocal of this equation leads to a

---

[10] S. Liao and T. Liang, this series, Vol. 36, p. 313.

linear relationship between Percent Max$^{-1}$ and [competitor]/[$^3$H-ligand] where the slope of the line equals $\alpha$ and the intercept equals 1.0.

Each time that the assay is run to obtain an $\alpha$ for a competitor, a $1\alpha,25$-(OH)$_2$D$_3$ standard curve, using nonradiolabeled steroid, is also determined. To normalize the slight day-to-day variations in the assay [5% as determined by the $\alpha$ obtained for $1\alpha,25$-(OH)$_2$D$_3$], the results of the competition data are normalized to the value of the competitive index obtained for $1\alpha,25$-(OH)$_2$D$_3$ on that day, yielding a relative competitive index (RCI).

$$\text{RCI} = \frac{\alpha_{\text{competitor}}}{\alpha_{1\alpha,25-\text{dihydroxyvitamin D}_3}} \times 100$$

The RCI for $1\alpha,25$-(OH)$_2$D$_3$ is, therefore, defined as 100.

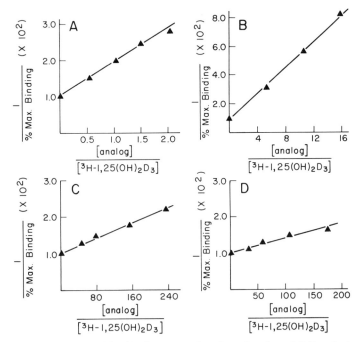

FIG. 2. Results of competitive binding assays for selected analogs. (A) Standard curve for $1\alpha,25$-(OH)$_2$D$_3(\alpha = 1.0)$. (B) Binding curve for 24-nor-$1\alpha,25$-(OH)$_2$D$_3(\alpha = 0.44)$. (C) Binding curve for 25-hydroxyvitamin D$_3$ ($\alpha = 0.005$). (D) Binding curve for $1\alpha$-OH-D$_3$ ($\alpha = 0.004$). Assays were carried out as described in the text and expressed in linearized form. Competitive indices ($\alpha$) are equal to the slope of the line and were determined by unweighted linear regression analysis. Each data point represents a mean of duplicate determinations.

SUMMARY OF COMPETITION BINDING RESULTS

| Analog | RCI |
|---|---|
| Side chain analogs of $1\alpha$,25-dihydroxyvitamin D$_3$ | |
| $1\alpha$,25-Dihydroxyvitamin D$_3$[a] | 100 |
| 24-Nor-$1\alpha$,25-dihydroxyvitamin D$_3$[e] | 46 |
| $1\alpha$,24$R$,25-Trihydroxyvitamin D$_3$[a] | 41 |
| $1\alpha$,24$S$,25-Trihydroxyvitamin D$_3$[a] | 33 |
| $1\alpha$-Hydroxyvitamin D$_3$[e] | 0.4 |
| Side chain analogs of 25-hydroxyvitamin D$_3$ | |
| 25-Hydroxyvitamin D$_3$[c] | 0.5 |
| 24$R$,25-Dihydroxyvitamin D$_3$[a] | 0.03 |
| 24$S$,25-Dihydroxyvitamin D$_3$[a] | 0.03 |
| 24-Nor-25-hydroxyvitamin D$_3$[e] | 0.012 |
| 24-Homo-25-hydroxyvitamin D$_3$[e] | 0.012 |
| 23,24-Dinor-25-Hydroxyvitamin D$_3$[e] | 0 |
| A-Ring analogs of $1\alpha$,25-dihydroxyvitamin D$_3$ | |
| $1\alpha$,25-Dihydroxy-5,6-*trans*-vitamin D$_3$[a] | 12.4 |
| 3-Deoxy-$1\alpha$,25-dihydroxyvitamin D$_3$[e] | 5.7 |
| 25-Hydroxyvitamin D$_3$[c] | 0.5 |
| 25-Hydroxydihydrotachysterol$_3$[d] | 0.64 |
| 25-Hydroxy-5,6-*trans*-vitamin D$_3$[a] | 0.51 |
| 19,25-Dihydroxy-10$S$(19)dihydroxyvitamin D$_3$[e] | 0.087 |
| 19,25-Dihydroxy-10$R$(19)dihydroxyvitamin D$_3$[e] | 0.087 |
| Side chain and A-ring analogs of $1\alpha$,25-dihydroxyvitamin D$_3$ | |
| 3-Deoxy-$1\alpha$-hydroxyvitamin D$_3$[e] | 0.03 |
| $1\alpha$-Hydroxy-3-epivitamin D$_3$[e] | 0.01 |
| 24-Nor-25-hydroxy-5,6-*trans*-vitamin D$_3$[e] | 0.10 |
| 24-Homo-25-hydroxy-5,6-*trans*-vitamin D$_3$[e] | 0.06 |
| 3-Deoxy-$3\alpha$-methyl-$1\alpha$-hydroxyvitamin D$_3$[e] | 0 |
| 5,6-*trans*-Vitamin D$_3$[e] | 0 |
| Dihydroxytachysterol$_3$[b] | 0 |
| 19-Hydroxy-10$S$(19)-dihydroxyvitamin D$_3$[e] | 0 |
| 19-Hydroxy-10$R$(19)-dihydroxyvitamin D$_3$[e] | 0 |
| Vitamin D$_3$[b] | 0 |

[a] Provided by Dr. M. Uskoković, Hoffmann–La Roche, Nutley, New Jersey.
[b] Obtained from Philips-Duphar, Weesp, The Netherlands.
[c] Provided by Dr. J. Babcock, Upjohn.
[d] Provided by Dr. P. Bell, Cardiff, Wales.
[e] Provided by Dr. W. H. Okamura, University of California, Riverside.

Assay Results

Figure 2 shows the type of data obtained from these competitive binding assays. Determinations are made, when possible, on different synthetic batches of analog, and the results of 4–10 determinations are used to assign an RCI value.

The analogs that we have examined to date fall into the following four major categories: (A) side-chain analogs of $1\alpha,25$-$(OH)_2D_3$, (B) side-chain analogs of 25-OH-$D_3$, (C) A-ring analogs of $1\alpha,25$-$(OH)_2D_3$, and (D) side-chain and A-ring analogs of $1\alpha,25$-$(OH)_2D_3$. The table summarizes the results of the analog studies that we have completed to date.[11]

These results have led to the following conclusions regarding the topological features of the $1\alpha,25$-$(OH)_2D_3$ molecule that are required for productive binding to the receptor system.

1. The $1\alpha$- and 25-hydroxyl groups are critical to the interaction.
2. The $3\beta$-hydroxyl is about one-tenth as important as the $1\alpha$- and 25-OH.
3. There is a great deal of tolerance for side-chain modification at C-24 provided that both the $1\alpha$- and 25-hydroxyl groups are present.
4. Rotation of the A ring about the 5,6-double bond to produce the 5,6-*trans* series of compounds is less favorable than the naturally occurring *5,6-cis* series.
5. Reduction of the C10(19)-double bond to a C-10 methyl or hydroxymethyl group hinders binding.
6. The $3\beta$-OH cannot be replaced by a $3\alpha$-hydroxyl or $3\alpha$-methyl group.

In conclusion, it appears that there are combinations of interactions between the ligand and the receptor at carbon atoms 1, 3, 4, 10, 19, 24, and 25 that all contribute to the effective binding of the ligand to its receptor.

[11] These studies have been made possible in large part through the determined efforts of Professor W. H. Okamura and his associates from the Department of Chemistry, University of California, Riverside, who synthesized and made available to us for assay many of the non-naturally occurring analogs.

# [60] Radioimmunoassay for Chick Intestinal Calcium-Binding Protein

*By* SYLVIA CHRISTAKOS and ANTHONY W. NORMAN

Vitamin D through its hormonally active steroid metabolite, 1,25-dihydroxyvitamin $D_3$ [1,25-$(OH)_2D_3$] is known to induce the de novo synthesis in its target intestine of a calcium-binding protein (CaBP).[1]

[1] E. J. Friedlander, H. Henry, and A. W. Norman, *J. Biol. Chem.* **252**, 8677 (1977).

The most widely used methods for quantitating chick CaBP have been the relatively insensitive Chelex binding assay[2] and the radial immunodiffusion assay, which uses antisera produced in the rabbit against highly purified chick intestinal CaBP.[3] However, a more sensitive and specific method for quantitation of mammalian CaBP levels is a radioimmunoassay that was first developed for pig intestinal CaBP.[4] More recently a radioimmunoassay has also been developed for rat intestinal CaBP.[5] However, because both the pig and the rat are relatively difficult to make vitamin D deficient, a study of CaBP production and distribution in the chick system has been of fundamental importance to our current understanding of the mechanism of action of vitamin D. Therefore, a radioimmunoassay for chick intestinal CaBP has been developed.[6]

## Assay Method

*Principle.* The assay system is based on the quantitative measurement of the displacement of radioactive antigen CaBP from its binding antibodies by the unlabeled antigen; i.e., there is competitive binding of radioactive and unlabeled CaBP to the specific antiserum. Since there is a fixed, limited amount of labeled CaBP and antibody and varying amounts of unlabeled CaBP, the percentage of label that is bound is inversely related to the concentration of unlabeled CaBP.

*Reagents*

Reagents used in [125]I-iodination procedure
  Sodium phosphate buffer, 0.5 $M$, pH 7.4
  Sodium phosphate buffer, 0.1 $M$, pH 7.4
  Chloramine-T solution: 1 mg/ml chloramine-T (Calbiochem) in 0.1 $M$ sodium phosphate buffer, pH 7.4, freshly prepared
  Sodium metabisulfite solution: 10 m$M$ sodium metabisulfite in 0.1 $M$ phosphate buffer, pH 7.4, freshly prepared
Reagents used in the formation of the antigen–antibody complex
  Sodium phosphate, 10 m$M$, 0.15 $M$ NaCl, 0.1% sodium azide at pH 7.4 (phosphosaline)

[2] R. H. Wasserman, R. A. Corradino, and A. N. Taylor, *J. Biol. Chem.* **243**, 3978 (1968).
[3] A. N. Taylor, *Arch. Biochem. Biophys.* **161**, 100 (1974).
[4] T. M. Murray, B. M. Arnold, W. H. Tam, A. J. W. Hitchman, and J. E. Harrison, *Metabolism* **23**, 829 (1974).
[5] P. Marche, P. Pradelles, C. Gros and M. T. Thomasset, *Biochem. Biophys. Res. Commun.* **76**, 1020 (1977).
[6] S. Christakos, E. Friedlander, B. Frandsen, and A. W. Norman, *Endocrinology* **104**, 1495 (1979).

Bovine serum albumin, 0.1%, in phosphosaline
Normal rabbit serum, 3%, in 50 mM EDTA, 10 mM sodium phosphate, 0.15 M NaCl, sodium azide, 0.1%, at pH 7.4

*Preparation of Antisera.* Purified chick CaBP (prepared by the method
of Friedlander and Norman,[7] which was a modification of the method of
Wasserman *et al.*[2] is emulsified with an equal volume of Freund's complete adjuvant. Each rabbit is injected with 1 mg of pure CaBP at 6–8 sites
on the back. Eighty days after the initial injection, all rabbits are chal-
lenged by an injection of 1 mg pure CaBP (emulsified with an equal
volume of Freund's incomplete adjuvant) near the lymph node in the back
leg region. The animals are bled by central ear artery puncture 14 days
later. Booster injections of 1 mg of CaBP are administered at monthly
intervals for 6 months, and animals are subsequently bled as described.

### Procedure

*Iodination of CaBP.* Purified CaBP is labeled with carrier-free $Na^{125}I$
(17 Ci/mg; ICN) using a modification of the chloramine-T procedure of
Greenwood *et al.*[8] To 10 μg of CaBP in 10 μl of 0.1 M sodium phosphate
buffer, pH 7.4, is added in sequence 20 μl of 0.5 M sodium phosphate pH
7.4, 10 μl of $Na^{125}I$ (1 mCi), and 20 μl of freshly prepared chloramine-T
solution (1 mg/ml chloramine-T in 0.1 M sodium phosphate). The reaction
is terminated after 30 sec by the addition of an excess of sodium metabisulfite (50 μl of a 10 mM solution in 0.1 M sodium phosphate). The free
iodine is then separated from the radioactive CaBP by filtration through a
1.5 × 15 cm column of Sephadex G-75 previously equilibrated with 0.10
M sodium phosphate, pH 7.4, and coated with 1 ml of 1% BSA in 0.1 M
$PO_4$, pH 7.4. Fractions of 0.5 ml are collected. The first peak is the
radioactive CaBP, and the second peak is the free iodine. The specific
activity and recovery of labeled CaBP is calculated as 40–60 μCi/μg. After
iodination, the radioactive CaBP (the 3 or 4 fractions containing the most
radioactivity in the first peak) is stored frozen until use. Prior to radioimmunoassay, the labeled CaBP is repassed through the G-75 Sephadex
column if 48 hours have passed since the original iodination in order to
eliminate damaged peptide and unreacted [125]I.

*Immunoassay Conditions.* The chick intestinal CaBP radioimmunoassay utilizes the double-antibody technique, which reduces background to
a minimum and maximizes sensitivity. The assay involves the addition of

[7] E. J. Friedlander and A. W. Norman, *in* "Vitamin D: Biochemical, Chemical and Clinical
Aspects Related to Calcium Metabolism" (A. W. Norman *et al.,*) p. 241. de Gruyter,
Berlin, 1977.
[8] F. C. Greenwood, W. M. Hunter, and J. S. Glover, *Biochem. J.* **89,** 114 (1963).

sample or standard (highly purified chick CaBP) in 0.4 ml of 0.1% bovine serum albumin (BSA), phosphosaline, pH 7.4 The antisera is diluted in 1% normal rabbit serum, 50 m$M$ EDTA phosphosaline, pH 7.4, so that 200 $\mu$l of the antiserum solution precipitates 50% of the total counts ([$^{125}$I]CaBP) added. In our assay the antisera are diluted 1 : 55,000. Diluted antibody in 0.2 ml is added to each tube except the tubes used to determine nonspecific binding. Iodinated CaBP is diluted in 0.1% BSA phosphosaline to give 20,000 cpm/0.2 ml. To each tube is added 0.2 ml of this diluted $^{125}$I-labeled CaBP. Each tube is then vortexed and incubated at 4° for 48 hr. On the third day, 50 $\mu$l of goat anti-rabbit $\gamma$-globulin (Calbiochem) is added. After an additional 24-hr incubation at 4°, the samples are centrifuged for 30 min at 3000 rpm. The supernatant solution is decanted, and the precipitate is counted to a standard error of 5% or less in a Beckman Biogamma II counter. The unknown samples are compared to the percentage of total counts precipitated with the CaBP reference preparation. The radioactivity precipitated in all other tubes (B) is expressed as a percentage of the activity precipiated in the 100% or maximum binding tubes (i.e., tubes containing only iodinated CaBP and antibody) ($B_0$) by the formula $B/B_0$ or

Percent activity bound =

$$\frac{\text{(cpm of standards or samples)} - \text{(nonspecific binding)}}{\text{(cpm of maximum binding tube)} - \text{(nonspecific binding)}} \times 100$$

Nonspecific binding refers to the counts precipitated when no rabbit antibody is added. Relative potency estimates are based on a logit percent bound vs log dose relationship.

## Sensitivity, Precision, and Accuracy of the Radioimmunoassay for Chick Intestinal CaBP

The mean 50% intercept of the standard curve (plotted as logit percent bound vs log dose CaBP) is 9.2 ng ± 0.58 (SE). The mean slope is −0.949 ± 0.044 (SE). The sensitivity of the assay (i.e., the minimum concentration of CaBP needed to reduce the maximum iodinated CaBP by 2 standard deviations) is 1 ng. The interassay variability determined after measuring the same extract in five successive radioimmunoassays, is 16.3% ± 5.5 (SE). The intraassay variability is 4.0% ± 1.1 (SE).

## [61] Purification of Chick Intestinal Calcium-Binding Protein

*By* ERNEST J. FRIEDLANDER and ANTHONY W. NORMAN

Intestinal calcium-binding proteins (CaBP) have been identified and purified from several species of animals.[1-5] The method described enables isolation of chick intestinal CaBP in an electrophoretically homogeneous state. The procedure employed is a modification of the method of Wasserman *et al.*;[1] it utilizes an extra $(NH_4)_2SO_4$ fractionation step and substitutes two anion-exchange steps for the originally reported preparative acrylamide disc gel electrophoresis procedure.

*Step 1. Preparation of Crude Extract.* Duodenum obtained from vitamin D-repleted chickens was immediately cooled to 4°, slit open, and rinsed with cold $0.12 M$ NaCl. The mucosal tissue was scraped from the underlying muscle layers with a glass slide and then homogenized in a chilled buffer mixture (20% w/v) with a Potter–Elvehjem homogenizer utilizing a Teflon pestle. The composition of the buffer utilized was 13.7 mM Tris, $0.12 M$ NaCl with 0.01% $\beta$-mercaptoethanol; this solution was adjusted to pH 7.4 with HCl. The homogenate was centrifuged at 38,000 $g$ in a refrigerated centrifuge at 4° for 20 min, and the supernatant was recovered. All subsequent steps in the purification were carried out at 4°. Calcium-binding activity in the crude supernatant and from fractions in all subsequent steps was assayed by the Chelex-100 ion-exchange procedure as described by Wasserman *et al.*[1]

*Step 2. $(NH_4)_2SO_4$ Precipitation.* The crude supernatant fluid was transferred to an Erlenmeyer flask immersed in an ice bath. The fluid was stirred with a magnetic mixer while $(NH_4)_2SO_4$ was slowly added until 75% saturation (0.48 g/ml). Three hours after $(NH_4)_2SO_4$ addition, the supernatant was centrifuged at 12,000 $g$ for 30 min. The supernatant was again transferred to an Erlenmeyer flask as before and brought to 99% saturation with $(NH_4)_2SO_4$. This was allowed to stand overnight with gentle stirring. After centrifugation at 100,000 $g$ for 1 hr at 4°, the supernatant was discarded and the pellets were resuspended in 8–10 ml of the Tris buffer. The resuspended protein solution was then placed in Visking dialysis tubing and dialyzed against a liter of 13.7 mM Tris buffer, pH 7.4, for 3 hr. If the volume after dialysis was greater than 12 ml, the protein

[1] R. H. Wasserman, R. A. Corradino, and A. N. Taylor, *J. Biol. Chem.* **243**, 3987 (1968).
[2] A. J. Hitchman, M. K. Ken, and J. E. Harrison, *Arch. Biochem. Biophys.* **155**, 221 (1973).
[3] M. Harmeyer and H. F. DeLuca, *Arch. Biochem. Biophys.* **133**, 247 (1969).
[4] A. J. Hitchman and J. E. Harrison, *Can. J. Biochem.* **50**, 758 (1972).
[5] C. S. Fullmer and R. H. Wasserman, *Biochim. Biophys. Acta* **317**, 172 (1973).

solution was concentrated in an Amicon pressurized molecular sieve unit with a PM-10 membrane.

*Step 3. Gel Filtration.* The dialyzed and concentrated protein was applied to a 2.5 × 85 cm Sephadex G-100 column suitably swelled in the Tris buffer previously described. Flow was in a downward direction by gravity, and the rate was adjusted to 1 ml/min by adjusting the height of the buffer reservoir. Calcium-binding activity appears in the effluent between 224 and 290 ml and is coincident with one of the major protein peaks eluted from the column.

*Step 4. Anion Exchange I.* Protein pooled from the calcium-binding activity peak of the Sephadex G-100 column was dialyzed against 5 liters of the following buffer mixture: 13.7 m$M$, Tris, 70 m$M$ NaCl, 1 m$M$ EDTA, and 0.01% $\beta$-mercaptoethanol at pH 7.4. The protein was then applied to a 1.5 × 20 cm DEAE (A50)-Sephadex column previously equilibrated in the above buffer. Protein elution was controlled by gravity as previously described. Total protein was continuously monitored at $A_{280}$. Elution with above buffer was continued until $A_{280}$ dropped to base line, after which a linear gradient from 0.07 $M$ to 0.40 $M$ NaCl in the same buffer was applied over 500 ml. Calcium-binding activity is associated with the last major elution protein peak, which occurs at a NaCl gradient concentration of 0.22 $M$.

*Step 5. Anion Exchange II.* The peak containing calcium-binding activity from step 4 was dialyzed for 10–12 hr against 3 liters of the following buffer: 13.7 m$M$ Tris, 68 m$M$ NaCl, 1 m$M$ CaCl$_2$, and 0.01% $\beta$-mercaptoethanol, pH 7.4. The protein was then applied to a 1 × 20 cm DEAE (A50)-Sephadex column previously equilibrated in the same buffer. Protein was monitored and elution controlled as before. Elution was continued with the column buffer for the first 120 ml, after which a linear gradient from 68 m$M$ NaCl to 0.40 $M$ NaCl, in the same buffer, was applied over 500 ml. Only one protein peak is observed to elute from the column, and calcium-binding activity is exactly coincident with this material. This peak now occurs at a NaCl gradient concentration of 0.14 $M$. Electrophoresis on Tris–glycine system at pH 8.9 and the Tris–borate–EDTA system at pH 8.2 demonstrates only one protein band.

The final stages of the purification procedure reported herein takes advantage of the change in anionic charge upon binding calcium. Figures 1A-C demonstrate the changes in anionic properties of CaBP that occur when calcium is bound. When pooled CaBP from the Sephadex G-100 column is chromatographed on DEAE (A50)-Sephadex utilizing a buffer system with neither EDTA nor CaCl$_2$ two calcium binding peaks are observed (Fig. 1A). By contrast, when the pooled CaBP from the Sephadex G-100 column is dialyzed with buffer containing 1 m$M$ EDTA and

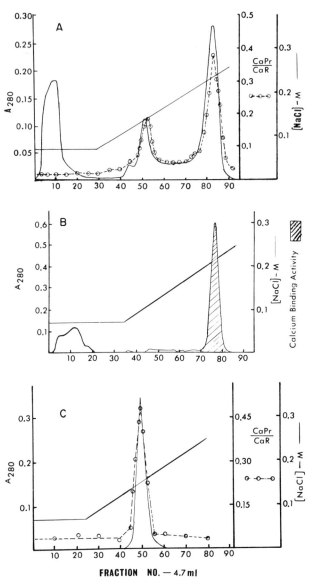

Fig. 1. (A) Separation of the proteins from the pooled calcium-binding peak of a Sephadex G-100 column on DEAE (A50)-Sephadex. Conditions were similar to those described in the text except that neither $CaCl_2$ nor EDTA were included in the buffers. (B) Separation of the proteins from the pooled calcium-binding peak of a Sephadex G-100 column on DEAE (A50)-Sephadex with EDTA. The tubes demonstrating a substantial $A_{280}$ were pooled and extensively dialyzed in order to reduce EDTA concentrations. An aliquot from the entire pooled peak was then assayed for CaBP by the Chelex-100 procedure. (C) Chromatography of CaBP from (B) on DEAE (A50)-Sephadex with $CaCl_2$. Protein on all columns was monitored at $A_{280}$, and CaBP activity was measured by the Chelex-100 procedure.

Purification Chart Demonstrating Specific Activities, Amounts of Protein, and Yields at Each Step in the Purification Procedure for Intestinal CaBP

| Procedure | Total protein | Volume (ml) | Units[a]/ ml | Total units | Protein (mg/ml) | Units[a] per mg protein | Yield (%) | Purification (fold) |
|---|---|---|---|---|---|---|---|---|
| Crude | 3770 | 248 | 2.17 | 538 | 15.2 | 0.14 | 100 | 1 |
| (NH₄)₂SO₄, 75–99% | 150 | 20 | 14.3 | 286 | 7.5 | 1.91 | 53 | 13.6 |
| Sephadex G-100 | 75 | 60 | 4.27 | 256 | 1.25 | 3.42 | 48 | 24.4 |
| DEAE (A50)-Sephadex with EDTA | 30.7 | 53 | 2.43 | 129 | 0.58 | 4.19 | 24 | 29.9 |
| DEAE (A50)-Sephadex with CaCl₂ | 22.6 | 48 | 2.02 | 97 | 0.47 | 4.30 | 18 | 30.7 |

[a] CaPr/CaR.

chromatographed on a similar DEAE (A50)-Sephadex column previously equilibrated with buffer containing 1 m$M$ EDTA, only one peak containing calcium-binding activity is observed (Fig. 1B). This peak chromatographs in the same part of the gradient as the late-eluting calcium-binding peak in Fig. 1A. Upon repeat chromatography of this peak in the system containing CaCl$_2$, the CaBP eluted in the earlier part of the gradient where the first calcium binding peak of Fig. 1A elutes.

A summary of the purification procedure is given in the table. The 36-fold purification necessary to attain homogeneity is an indication of the extisting high concentration of this protein.

## [62] Characteristics and Purification of the Intestinal Receptor for 1,25-Dihydroxyvitamin D

By J. WESLEY PIKE and MARK R. HAUSSLER

Vitamin D is known to be an active principle and general physiologic regulator of calcium and phosphate metabolism. However, only in recent years have aspects of the vitamin's metabolism and fundamental mode of action been defined. It is now clear that the primary metabolic pathway of vitamin D involves the hepatic conversion of the parent compound to 25-hydroxyvitamin D,[1] and then subsequently an all important hormonally and ionically regulated hydroxylation of 25-OH-D in the kidney to 1,25-dihydroxyvitamin D [1,25-(OH)$_2$D].[2] Current research indicates that this form of the vitamin represents the active metabolite, and that it acts in a manner suggestive of steroid hormones, such as estrogen and progesterone. This novel sterol hormone first binds to a specific, high-affinity component within the cytoplasm[3,4] and then rapidly accumulates in the nuclear compartment[5,6] as an apparent functionally active complex capable of stimulating RNA synthesis.[7,8] Subsequent ex-

[1] H. F. DeLuca and H. K. Schnoes, *Annu. Rev. Biochem.* **45,** 631 (1976).

[2] M. R. Haussler and T. A. McCain, *N. Engl. J. Med.* **297,** 974 (1977).

[3] H. C. Tsai and A. W. Norman, *J. Biol. Chem.* **248,** 5967 (1973).

[4] P. F. Brumbaugh and M. R. Haussler, *J. Biol. Chem.* **249,** 1251 (1974).

[5] P. F. Brumbaugh and M. R. Haussler, *J. Biol. Chem.* **250,** 1588 (1975).

[6] M. Zile, E. C. Bunge, L. Barsness, S. Yamada, H. R. Schnoes, and H. F. DeLuca, *Arch. Biochem. Biophys.* **186,** 15 (1978).

[7] H. C. Tsai, R. J. Midgett, and A. W. Norman, *Biochem. Biophys. Res. Commun.* **54,** 622 (1973).

[8] J. E. Zerwekh, T. J. Lindell, and M. R. Haussler, *J. Biol. Chem.* **251,** 2388 (1976).

METHODS IN ENZYMOLOGY, VOL. 67

traction of the chromatin from isolated intestinal nuclei[5] demonstrates the presence and presumed ultimate site of action of the hormone-charged receptor. Since the receptor for 1,25-(OH)$_2$D has only recently been purified to near homogeneity, it will be of extreme interest, in fact essential, to compare the properties of the crude cytosolic receptor with those obtained after purification. These studies will be of highest priority in elucidating the function of the 1,25-(OH)$_2$D–receptor complex.

Tissue Source

Most research has focused on receptors for 1,25-(OH)$_2$D in the intestines of rachitic (D-deficient) chicks 4–6 weeks old,[2] although a similar cytoplasmic receptor has been identified in other tissues, such as parathyroid glands,[9] pancreas,[10,11] bone,[12] and kidney.[10,11] Preparations derived from rachitic chicks benefit from the primary advantage that titers of endogenous 1,25-(OH)$_2$D are virtually nonexistent. As a result, the concentration of receptors with available sterol binding sites (nascent receptors) is maximal, providing unique advantages in the overall characterization, quantification, and purification of the protein. The use of intestinal tissue in most studies is due predominantly to the extent of tissue available (3–5 g per chick), the ease of extraction and subsequent manipulation, and, most important, the fact that titers of receptor appear to be 10- to 100-fold greater than in other tissue examined. D-deficient rat tissues have also been used to study 1,25-(OH)$_2$D receptors. However, the difficulty of raising and maintaining these animals far outweighs the advantage of predominantly hormone-free receptor populations. Further, titers of nascent receptors do not appear to be as great as titers found in chicks. It must be remembered, however, that despite the relative abundance of the 1,25-(OH)$_2$D receptor in rachitic chick intestinal tissue, it still remains an extremely rare cytosolic protein.

The 1,25-(OH)$_2$D receptor can also be extracted from the tissue of normal chicks, rats, and pigs. The extent of nascent vs endogenous hormone-bound receptor populations is not presently known, nor are the potentially extensive factors that might govern these populations.

Receptor extractions are most successfully made from freshly excised tissue, since preparations made from frozen whole tissue, although possible, suffer from relatively lower yields. In our laboratory, all tissue de-

[9] M. R. Hughes, and M. R. Haussler, *J. Biol. Chem.* **253**, 1065 (1978).

[10] S. Christakos, W. Wecksler, and A. W. Norman, *Fed. Proc.* **38**, 385 (1979).

[11] J. S. Chandler, J. W. Pike, and M. R. Haussler, *Fed. Proc.* **38**, 385 (1979).

[12] B. E. Kream, M. J. Jose, S. Yamada, and H. F. DeLuca, *Science* **197**, 1086 (1977).

stined to be used for receptor purification are obtained from chicks within minutes of sacrifice.

## Extraction and Properties of the Cytosolic 1,25-(OH)$_2$D Receptor

### Buffers

Buffers used are as follows

STKM: 0.25 M sucrose, 50 mM Tris · HCl, pH 7.4, 25 mM KCl, 5 mM MgCl$_2$, 12 mM thioglycerol, and 1 mM EDTA

KETT-buffer: 10 mM Tris · HCl, pH 7.4, 1 mM EDTA, 12 mM thioglycerol, and various concentrations of KCl (KETT-0 = no KCl; KETT-0.1 = 0.1 M KCl, etc.)

HAP-buffers: 50 mM Tris · HCl, pH 7.4, 1 mM EDTA, 12 mM thioglycerol, 0.1 M KCl, and various concentrations of KH$_2$PO$_4$ (HAP-0 = no KH$_2$PO$_4$; HAP-0.3 = 0.3 M KH$_2$PO$_4$)

STKM is used for homogenization of tissues and contains isotonic sucrose, thioglycerol to stabilize the receptor preparation, and EDTA to bind calcium and inactivate proteases. KETT buffers are used for resolubilization of receptor precipitates, receptor assays, and the equilibration and chromatography of all receptor preparations on ion-exchange and pseudo-affinity resins. HAP buffers are used for chromatography of receptors on hydroxylapatite. Receptors have been prepared in phosphate buffer and barbitol buffers containing glycerol with equivalent success. However, the homogenization of tissues in KETT-0 results in almost complete loss of receptor activity.

### Tissue Homogenization

Rachitic chicks (1–50) are sacrificed by decapitation; their intestines quickly removed, rinsed extensively in either ice cold STKM or 0.25 M sucrose, and placed in cold STKM. The guts are blotted, the mucosa is scraped free of the underlying muscular serosa and then homogenized in 2.5 volumes (w/v) of ice-cold STKM with six passes in a Potter–Elvehjem homogenizer (with intermittent cooling on ice). The homogenate is then centrifuged in a Sorvall HB4 rotor at 16,000 g for 15 min. The resultant supernatant is decanted through several layers of cheesecloth, and then used for the preparation of high speed cytosols. It should be emphasized that all operations must be rigorously performed at 0–4° or on ice. Slight warming of the tissue or receptor preparation at any stage will result in major decreases in receptor specific activity.

Large-scale preparation of receptors entails the use of the whole intestine (vs mucosa) by first rinsing and then homogenizing with either a Waring blender or Polytron (PT-10, Brinkmann). Either procedure results in receptor specific activity that is almost equivalent to treatment with the Potter–Elvehjem. The resulting homogenate is then centrifuged 16,000 $g$ for 15 min with the use of a GSA rotor, and the supernatant is treated as described above.

### Preparation of Cytosol and Ammonium Sulfate Precipitates of Cytosol

The 16,000 $g$ supernatant is centrifuged routinely in a Beckman 50.2 rotor at 220,000 $g$ for 45 min at 2° to prepare the soluble cytoplasmic fraction. Although cytosol can be assayed for receptor activity, it has been found beneficial to precipitate the receptor by the addition of preground ammonium sulfate crystals to 40% of saturation. After stirring for 30 min, the precipitate is sedimented at 16,000 $g$ for 15 min. The pellet can be resolubilized in one-fourth to one-fifth the original cytosol volume, with subsequent yield of receptors of approximately 70%. The resultant routine nascent-receptor preparation (2-fold purified) can then be labeled with hormone and subsequently used for all analyses of receptor characteristics.

### Receptor Concentration and Labeling with $1,25$-$(OH)_2[^3H]D_3$

The $1,25$-$(OH)_2D$ receptor exists in cytosol as less than 0.001% of the total protein, i.e., approximately 200 $\mu g$ per kilogram of intestinal tissue. Ammonium sulfate precipitation leads to a loss of receptor, leaving approximately 120–160 $\mu g$ of nascent receptors per kilogram available for analysis or purification. However, the advantages of salt precipitation or other initial purification procedures far outweigh losses of receptor incurred. Further, high salt does not appear to affect the receptor, since no salt-dissociable dimer has yet been observed.

25-Hydroxyvitamin $D_3$ (110 Ci/mmol, Amersham) is generated biologically to $1,25$-$(OH)_2[^3H]D_3$ (110 Ci/mmol) using rachitic chick kidney homogenates and purified as described.[4,13] The radiolabeled sterol is stored at $-20°$ in 50% ethanol/toluene as a stock solution, and then nitrogen dried, redissolved in ethanol, and diluted by the addition of nonradioactive $1,25$-$(OH)_2D_3$ to a working specific activity (usually approximately 7 Ci/mmol).

[13] D. E. M. Lawson, D. R. Fraser, E. Kodicek, H. R. Morris, and D. H. Williams, *Nature* (*London*) **230**, 228 (1971).

Receptor preparations are labeled to saturation with 1,25-$(OH)_2[^3H]D_3$, added in ethanol to 5% (v/v), and then incubated with stirring at 0° for 2–12 hr (depending upon experiment). Alternatively, the receptor preparation can be incubated at 25° for 15–30 min and then returned to 0–4° with similar results. The latter reaction, however, must be rigorously tested, owing to the lability of receptors at temperatures above 0–4° for extended periods of time.

### Stability of the 1,25-$(OH)_2D$ Receptor

Nascent receptors in cytosol demonstrate an approximate half-life in the absence of sterol of 4–6 hr at 0°. Elevated temperatures serve to cause an extremely rapid irreversible deterioration of receptor activity. These observations also hold true for crude receptor preparations prepared by ammonium sulfate or other precipitation techniques. However, if nascent receptors are labeled in the presence of a saturating concentration of 1,25-$(OH)_2D_3$, binding activity can be demonstrated to be stable for several days to weeks at 0° and at least several hours at 24°. This temperature sensitivity and the unusual lability of nascent receptors dictates that most receptor preparations be labeled to saturation and all work be performed at 0–4°.

### Sedimentation Property of the 1,25-$(OH)_2D$ Receptor

The receptor for 1,25-$(OH)_2D$ sediments in 5 to 20% sucrose gradients prepared in KETT-0.3 as a 3.3 S macromolecule. This pattern is seen in cytosol (Fig. 1A), in ammonium sulfate precipitates (Fig. 1B), and in other forms of precipitation to which crude cytosol is subjected. As yet, there is no definitive evidence for a larger form of the receptor in crude cytosol as is typical of several other steroid hormone receptors.

Receptor Binding to 1,25-Dihydroxyvitamin D

### Assays for 1,25-$(OH)_2D$-Receptor Activity

DEAE-Filter Technique. The receptor for 1,25-$(OH)_2D$ can be assayed by a modification[4] of the DEAE filter technique described by Santi et al.[14] Aliquots of nascent receptor solutions (10–500 μl) are incubated with a saturating concentration of labeled 1,25-$(OH)_2[^3H]D_3$ or incubated in

[14] D. V. Santi, C. H. Sibley, E. R. Perriard, G. M. Tompkins, and J. D. Baxter, *Biochemistry* 12, 2412 (1973).

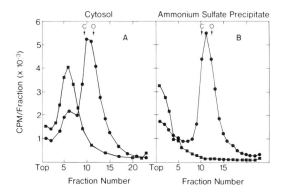

FIG. 1. Sucrose gradient analysis of the 1,25-dihydroxyvitamin D-receptor. Aliquots of receptor material (400 $\mu$l) were incubated with either 4 n$M$ 1,25-$(OH)_2$[$^3$H]$D_3$ (110 Ci/mmol) for 1 hr at 0° (●——●) or similarly with the inclusion of 0.4 $\mu M$ unlabeled 1,25-$(OH)_2D_3$ (■——■). The aliquots were layered onto a 5 to 20% sucrose gradient prepared in KETT-0.3, and centrifuged at 220,000 $g$ for 22.5 hr. Gradients were then fractionated into 5-drop fractions and counted for $^3$H. (A) Analysis of receptor from crude cytosol. (B) Analysis of receptor from resolubilized ammonium sulfate precipitates. Standard proteins: C, chymotrypsinogen (2.5 S); O, ovalbumin (3.7 S).

parallel with both labeled 1,25-$(OH)_2$[$^3$H]$D_3$ and a 100-fold excess of non-radioactive hormone. After incubation, either for 2 hr at 0–4° or 15 min at 25°, the samples are diluted with 1–3 ml of 10 m$M$ Tris · HCl, pH 7.4, 1% Triton X-100 (so final salt concentration is less than 0.1 $M$ KCl), and filtered at 1 ml/min through two presoaked DE-81 (Whatman) filters. Filters are then extracted in scintillation vials with acetone for 5 min, extracts and filters dried, and counted for tritium by liquid scintillation techniques (33% efficiency). Receptor preparations that have already been labeled can also be assayed as above by omitting the sterol incubation step. However, once the preparations are labeled, estimates of nonspecific binding, generally 5–10% of the total receptor binding activity, cannot be measured. The simplicity, rapidity, and reproducibility of assay make this technique preferable for receptor measurements, despite an estimated 70–80% yield when assaying receptors in cytosol. As purification proceeds, however, estimates of receptor activity in samples approaches 100%.

*Hydroxylapatite Assay*. Receptors can be assayed by hydroxylapatite essentially as described by Williams and Gorski.[15] Incubations of receptor containing samples are carried out as previously described. After incubation, two volumes of a 50% slurry of hydroxylapatite (BioGel HTP, Bio-Rad, equilibrated in HAP-0.01) are added and mixed. The resin is washed three times with HAP-0.01, collected by centrifugation, and then ex-

[15] D. Williams and J. Gorski, *Biochemistry* **13**, 5537 (1974).

tracted in 1 ml of ethanol. The supernatant is removed after centrifugation, dried, and counted by liquid scintillation.[16]

*DNA-Cellulose.* Receptors can also be assayed with DNA-cellulose. After incubation of receptor with sterol as above (salt concentrations less than 0.1 $M$), the incubate is applied to a 1-ml microcolumn of DNA-cellulose and washed with 5 volumes of KETT-0.1; then the receptor is eluted with 4 volumes of KETT-0.5. An aliquot of the KETT-0.5 eluate is counted by liquid scintillation techniques. Further, slurry techniques with DNA-cellulose can also be used. Since independent assessment of purification of the 1,25-(OH)$_2$D-receptor demonstrates that DNA-cellulose is far superior to either DEAE-cellulose or hydroxyapatite, this is the method of choice for accurate determinations of 1,25-(OH)$_2$D-receptor binding activity.

*Sucrose Gradient Analysis.* Standard sucrose gradient analysis of the receptor in KETT-0.3 can be performed. This technique is used primarily for identification of the receptor in a sample since extensive receptor–hormone dissociation occurs owing to nonequilibrium conditions created during the lengthy period of resolution (16–24 hr). Further, few samples (generally six) can be processed during a day.

### Determination of $K_d$

The affinity of the cytoplasmic receptor for 1,25-(OH)$_2$D is determined by incubating aliquots of a receptor preparation with increasing concentrations of 1,25-(OH)$_2$[$^3$H]D$_3$ (total binding) and parallel incubations with 100-fold excess unlabeled hormone (nonspecific binding). Separation of bound from free sterol is carried out as previously described for receptor sterol-binding activity, and Scatchard analysis of the saturation curve generated from the data is then performed. The 1,25-(OH)$_2$D receptor demonstrates a dissociation constant of approximately $2.2 \times 10^{-10}M$, when incubations are carried out to equilibrium (30–60 min at 25° or 4–24 hr at 0–4°). Earlier measurements appear to have been determined prior to reaching equilibrium.[4]

### Sterol Specificity

The specificity of the receptor for vitamin D compounds can be assessed in intestinal tissue by competing increasing concentrations of unlabeled vitamin D or its metabolites with a saturating concentration of

[16] J. H. Clark and E. J. Peck, *in* "Laboratory Methods Manual for Hormone Action and Molecular Endocrinology" (W. T. Schrader and B. W. O'Malley, eds.), p. 2-1. Endocrine Society, Bethesda, Maryland, 1977.

PURIFICATION OF THE CHICK INTESTINAL RECEPTOR FOR 1,25-(OH)$_2$D[a]

| Step | Protein (mg) | Receptor (cpm $\times$ 10$^{-6}$) | Specific activity (cpm $\times$ 10$^{-3}$/ mg protein) | Yield | Purification (fold) |
|---|---|---|---|---|---|
| Cytosol | 22,400.0 | — | 0.58 | — | 1 |
| Polymin P Precipitate | 1,170.0 | 9.10 | 7.69 | 100 | 13 |
| DNA-cellulose, 0.28 $M$ | 19.8 | 9.00 | 454.50 | 99 | 780 |
| Sephacryl | — | 6.35 | — | 70 | — |
| Blue Dextran-Sepharose, 0.4 $M$ | 0.85 | 4.50 | 5,290.00 | 49 | 9,100 |
| DNA-cellulose, 0.28 $M$ | 0.216 | 2.75 | 12,700.00 | 30 | 22,000 |
| Heparin-Sepharose, 0.20 $M$ | 0.026 | 1.30 | 50,000.00 | 14 | 86,000 |

[a] Specific activity of the 1,25-(OH)$_2$D receptor was determined on an aliquot of cytosol. Receptors from the remaining cytosol (800 g-equivalents of intestinal mucosa) were then precipitated with Polymin P, resolubilized, and labeled with 35 n$M$ 1,25-(OH)$_2$[$^3$H]D$_3$ (7 Ci/mmol) for 12 hr at 0°, and then sequentially chromatographed on the resins listed. Initial DNA-cellulose: 180 ml/hr flow rate, 600-ml linear gradient of 0.1 to 0.8 $M$ KCl, 15-ml fractions; Blue Dextran-Sepharose: 60 ml/hr flow rate, 240 ml linear gradient from 0.2 to 0.6 $M$ KCl, 5-ml fractions; DNA-cellulose: 60 ml/hr flow rate, 120-ml linear gradient from 0.1 to 0.6 $M$ KCl, 3-ml fractions; heparin-Sepharose: 50 ml/hr flow rate, 100-ml linear gradient from 0.05 to 0.4 $M$ KCl, 3-ml fractions. Gel filtration on Sephacryl was achieved in KETT-0.3 at an ascending flow rate of 60 ml/hr, 3-ml fractions. Ammonium sulfate precipitation was used to concentrate the receptor peaks at early stages, and chromatography on an hydroxylapatite column (0.8 $\times$ 0.5 cm) was employed to concentrate the final heparin-Sepharose purified receptor peak. One percent of each fraction was counted for tritium.

labeled 1,25-(OH)$_2$D$_3$ (taken as a value of 1) for the receptor sterol binding site. A comparison of receptor binding is then made between the concentration of competing sterol required to reduce the binding of 1,25-(OH)$_2$D$_3$ by 50%. The 1,25-(OH)$_2$D receptor demonstrates a decided selectivity for 1,25-(OH)$_2$D$_3$: 25-OH-D$_3$, 1/500; 1$\alpha$-OH-D$_3$, 1/800; vitamin D$_3$, 1/100,000; 24$R$,25-(OH)$_2$D$_3$, 1/800.[17]

Purification of the 1,25-(OH)$_2$D-Receptor

The receptor for 1,25-(OH)$_2$D requires a theoretically calculated 200,000-fold enhancement of specific activity over cytosol to achieve homogeneity. Thus the purification procedures as outlined in the table are capable of isolating the receptor from rachitic chick intestine to approximately 50% of homogeneity (86,000-fold).

[17] M. R. Haussler and P. F. Brumbaugh, in "Hormone–Receptor Interaction: Molecular Aspects" (G. S. Levey, ed.), p. 301. Dekker, New York, 1976.

*Polymin P Precipitation*

The receptors from crude cytosol can be purified approximately 12- to 15-fold in 60% yield by the use of a selective precipitation procedure employing Polymin P.[18] Cytosol (from 30% homogenates of intestine) is titrated with Polymin P at 0–4° by slow addition of a 10% solution (made 10% with distilled water, pH 7.9 with HCl) to a concentration of 0.04% (v/v). After brief centrifugation (16,000 g for 15 min), the pellet is discarded, and the supernatant is made to 0.08% (v/v) by the repeated slow addition of an equal volume of 10% Polymin P, as above. After centrifugation, the supernatant is discarded, and the 1,25-$(OH)_2$D receptor is found in the precipitated pellet. Resuspension of the sediment in KETT-0.5 will resolubilize the receptor. After additional centrifugation to eliminate insoluble material, the supernatant (containing receptors) is made 40% saturated with ammonium sulfate and then pelleted again at 16,000 g for 10 min, rendering the sedimented receptor free of residual Polymin P and in a highly concentrated form. The final pellet is frozen in liquid nitrogen, and can be stored at −80° without loss of receptor activity for at least 6 months. Thus, this step not only provides an important initial purification, but is a convenient procedure permitting the storage and accumulation of receptor material.

*DNA-Cellulose*

Polymin P purified receptors can be chromatographed on a DNA affinity column using cellulose as inert support and calf thymus DNA as ligand, noncovalently coupled by procedures described by Alberts and Herrick.[19] The DNA-cellulose is washed extensively with KETT-1 and then equilibrated in KETT-0.05. Receptors from Polymin P precipitates (or other crude sources) are resolubilized in KETT-0 for 15 min, the insoluble material is removed by centrifugation, and the receptors are labeled as previously described. They are then applied slowly to the resin in total salt concentration not exceeding 0.15 M KCl. After application, the column is washed with 3–5 volumes of KETT-0.15, which results in a large fall-through of protein and nonspecifically bound hormone. However, quantitative retention of the receptor is achieved as subsequent elution of this protein can be demonstrated as a single peak of macromolecular bound tritium during a linear salt gradient at 0.28 M KCl (Fig. 2A). Further washing of the DNA-cellulose with KETT-1 fails to elute any

[18] J. J. Jendrisak and R. R. Burgess, *Biochemistry* **14**, 4639 (1975).
[19] B. Alberts and G. Herrick, this series, Vol. 21, p. 198.

additional receptor-bound radioactivity. However, extraction of the resin with an organic solvent results in an abundance of radioactivity that represents free $1,25\text{-}(OH)_2[^3H]D$ that bound to the resin.

This procedure can purify the receptor approximately 50- to 100-fold over Polymin P precipitates and thus nearly 1000-fold over beginning cytosol. A kilogram of intestine, after homogenization and selective precipitation, can be chromatographed on a 250-ml bed of DNA-cellulose and reduced to 10–20 mg of total protein. Further, the 6 S cytosol D binding protein does not bind to DNA, and is thus eliminated from further consideration, as is the serum D binding protein.

DNA-cellulose has been extensively utilized in this laboratory as a technique for prepurifying nascent receptor for subsequent kinetic studies. Since the unlabeled receptor also binds to DNA-cellulose, crude receptor preparations (cytosol or ammonium sulfate precipitates) are chromatographed, receptor activity is located by DEAE filter assay (free of the 6 S component), the peak is pooled, and kinetic and sterol specificity studies are carried out. Further, it has been used as a means of identifying $1,25\text{-}(OH)_2D$ receptors in other issue, when receptor titers require the prepurification of large amounts of tissue that can be subsequently concentrated and analyzed on sucrose density gradients.

*Gel Filtration on Sephacryl*

The receptor can be chromatographed on Sephacryl S-200 (Pharmacia) in KETT-0.3 (Fig. 2B), with resultant purifications of 10- to 12-fold achievable over crude receptors from salt precipitates of cytosol and 2- to 3-fold over DNA-cellulose purified receptors. The chromatography results in the receptor's comigration with BSA at a molecular weight of 68,000 and leads to a purification primarily through its exclusion limits. Owing to high flow rates achievable with this rigid gel, receptor-hormone dissociation is minimized, and yields are high (60–80%). However resolution, and thus purification, under these conditions is minimal. Generally, the receptors from previous DNA-cellulose purification are ammonium sulfate precipitated, redissolved in 1 ml of KETT-0.3, and then chromatographed. The pooled receptor peak from this step is then used directly without salt dilution in the next step.

Receptors can also be chromatographed from crude sources (cytosol, salt precipitates) on agarose (BioGel A-0.5, Bio-Rad) in KETT-0.3. Although yield is poor, owing in part to the dissociation of receptor–hormone complexes, this technique has provided more precise information about molecular weight and other physical parameters. However, as a purification procedure it has not proved to be satisfactory.

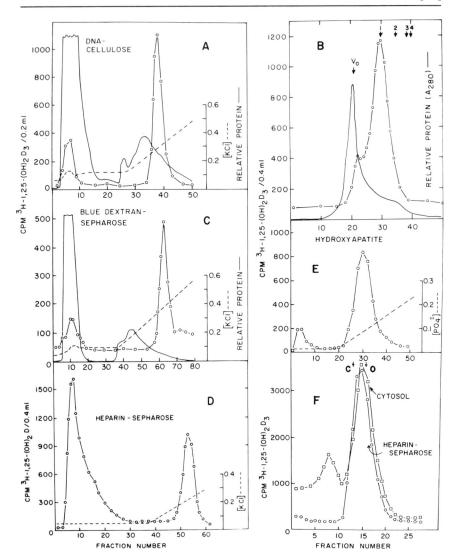

FIG. 2. Chromatography of the 1,25-dihydroxyvitamin D receptor from rachitic chick intestine. Individual preparation of cytosol from 4–8 g of mucosa were ammonium sulfate precipitated, resolubilized in KETT-buffer, labeled with a saturating concentration of 1,25-$(OH)_2[^3H]D_3$ (7 Ci/mmol added in ethanol, 5% v/v) for 1 hr at 0°, and then applied to the individual resins. After chromatography, aliquots of the collected fractions were counted for tritium by standard liquid scintillation techniques (efficiency, 33%). KCl gradients were measured by conductivity, using a Radiometer conductivity meter (----). Relative protein was measured by absorbance at 280 nm (——). All procedures were performed rigorously at 0–4°.

(A) DNA-cellulose chromatography of the 1,25-$(OH)_2$D-receptor. Receptors were

*Blue Dextran–Sepharose*

Blue Dextran can be linked to Sepharose as described by Ryan,[20] washed extensively with KETT-1, equilibrated in KETT-0.1, and successfully used to purify the 1,25-(OH)$_2$D receptor. The receptor can be chromatographed from crude sources (cytosol, etc.) or applied directly as the pooled peak from previous Sephacryl chromatography. After extensive washing with KETT-0.2, which removes noninteracting protein species, the quantitatively retained receptor can be eluted during a linear salt gradient as a single peak at 0.40 $M$ KCl (Fig. 2C). Again, a KETT-1 wash of the resin does not result in additional eluted radioactivity. This unique resin is used to purify both dinucleotide and DNA binding protein species,[20,21] and can purify the receptor approximately 100-fold from salt precipitates of cytosol, and 2- to 3-fold over the previous steps described in the table. As an independent purification procedure, it also does not bind

[20] L. D. Ryan and C. S. Vestling, *Arch. Biochem. Biophys.* **160**, 279 (1974).
[21] T. D. Lindell, B. P. Nichols, J. E. Donelson, and E. Stellwagen, *Biochim. Biophys. Acta* **562**, 231 (1979).

---

applied in KETT-0.05 to a 2 × 3 cm DNA-cellulose column equilibrated in KETT-0.05. The column was washed with 5 volumes of KETT-0.10, and then receptors eluted (○——○) during a linear 0.1 to 0.6 $M$ KCl gradient. Fractions were 3 ml.

(B) Chromatography of the 1,25-(OH)$_2$D-receptor on Sephacryl S-200. Receptors were resolubilized in 1 ml of KETT-0.3 and, after labeling, chromatographed on a 1.6 × 60 cm Sephacryl column in KETT-0.3 (○——○) with ascending flow. $V_0$ is determined with Blue Dextran 2000 (Pharmacia), and protein standards are: 1, BSA; 2, ovalbumin; 3, chymotrypsinogen; 4, ribonuclease; 3 ml fractions were collected.

(C) Blue Dextran–Sepharose chromatography of the 1,25-(OH)$_2$D-receptor. Receptors were applied in KETT-0.10 to a 2.5 × 3 cm Blue Dextran–Sepharose column equilibrated in KETT-0.1 and then receptors (○——○) were eluted during a linear 0.1 to 0.6 $M$ KCl gradient; 3-ml fractions were collected.

(D) Heparin-Sepharose chromatography of the 1,25-(OH)$_2$D-receptor. Receptors were applied in KETT-0.05 to a 2 × 2.5 cm heparin–Sepharose column equilibrated in KETT-0.05. The column was washed with 4 volumes of KETT-0.05, and receptors (○——○) were eluted during a linear 0.05 to 0.4 $M$ KCl gradient; 3-ml fractions were collected.

(E) Hydroxylapatite chromatography of the 1,25-(OH)$_2$D-receptor. Receptors were applied in HAP-0 to 2 × 3 cm hydroxylapatite column (Bio-Rad, BioGel HTP) pre-equilibrated with degassed HAP-O. After extensive washing with HAP-O, the receptor was eluted during a linear 0.01 to 0.3 $M$ KH$_2$PO$_4$ gradient. Fractions were 3 ml.

(F) Sucrose gradient analysis of the 1,25-(OH)$_2$D-receptor. Labeled cytosol (400 $\mu$l) was layered onto a 5 to 20% sucrose gradient prepared in KETT-0.3, and centrifuged at 220,000 $g$ for 22.5 hr (□——□). Five-drop fractions were collected and counted for tritium. Purified receptor (after heparin–Sepharose chromatography, see the table) (400 $\mu$l) was layered onto a 5 to 20% sucrose gradient prepared in KETT-0.3, and centrifuged at 220,000 $g$ for 22.5 hr (○——○). Five-drop fractions were collected and counted for tritium. Standard proteins are: C, chymotrypsinogen (2.5 S); O, ovalbumin (3.7 S).

the 6 S vitamin D binding protein, and thus it has utility in preparing prepurified receptors for subsequent kinetic studies.

### Heparin–Sepharose

Heparin has been successfully linked to Sepharose[22] and used as an immobilized affinity ligand to purify other steroid hormone receptors.[22,23] Likewise, this polysulfate-containing mucopolysaccharide will quantitatively bind the 1,25-$(OH)_2D$ receptor from crude cytosol or from material at various stages of purification (see the table). The resin is equilibrated after KETT-1 washes in KETT-0.05 and is charged with receptor applied in salt less than 0.10 $M$ KCl. After washing with equilibration buffer, receptors are eluted from the resin at 0.20 $M$ during a linear KCl gradient (Fig. 2D). The resin is capable of a 20- to 40-fold purification of receptor from cruce sources and approximately 4-fold purifications when used in the scheme listed in the table. Generally, as an in-tandem purification step, the DNA-cellulose rechromatographed receptors are applied directly, after salt dilution with KETT-0, to heparin–Sepharose, and the pooled peak subsequently concentrated for analysis.

### Hydroxylapatite

Hydroxylapatite (BioGel HTP, Bio-Rad) is washed by gentle suspension in HAP-0.3 and then extensively washed in HAP-0 prior to its equilibration in column form in HAP-0. Receptors derived from cytosol, or purified material, can be applied and will adsorb under conditions of very low phosphate. Other salts, particularly KCl, appear to have no effect on receptor binding. As a purification technique, the receptor is eluted from the gel during a $KH_2PO_4$ gradient at approximately 80 m$M$ (Fig. 2E), although its capacity to purify the receptor over crude material is minor. Of primary utility, however, is its use in concentrating and desalting receptor preparations, since small amounts of receptor in a large volume of high-salt eluate can be applied to a microcolumn of equilibrated hydroxylapatite, washed with HAP-0, and then eluted stepwise with HAP-0.3 in a much reduced volume. In practice, receptors that have been purified as in the table are concentrated by this technique to 1–2 ml prior to analysis.

[22] A. M. Molinari, N. Medici, B. Moncharmont, and G. A. Puca, *Proc. Natl. Acad. Sci. U.S.A.* **74**, 4886 (1977).
[23] V. Sica and F. Bresciani, *Biochemistry* **18**, 2369 (1979).

*Ion-Exchange and Other Resins*

Receptors can be chromatographed on an extensive number of exchange resins and commercially available group-selective affinity resins. The former include phosphocellulose, DEAE-cellulose, and DEAE-Sephadex.[24] The latter include the various Affi-Gels (Bio-Rad), and Matrex Red (Amicon). Further, immobilized ligands, such as phosvitin, casein, and a variety of histones, also bind receptor, although most likely by ionic interactions. However, most of these resins fail to achieve significant independent purifications of the receptor, and when several are placed in sequence, they are incapable of approaching the level of purification described in the table.

*Assessment and Analysis of Purification*

Receptors from rachitic chick intestine have only recently been purified to the extent outlined. As a result, the physical, kinetic, and sterol binding properties of the purified material remain to be studied and compared to that seen in the crude state. Further, the receptor remains to be purified to homogeneity and identified unequivocally through its hormonal ligand. Nevertheless, crude and purified receptors have been analyzed as described in the next three subsections.

*SDS–Polyacrylamide Gel Electrophoresis.* The sodium dodecyl sulfate–polyacrylamide gel electrophoresis system described by Laemmli[25] and Weber and Osborn[26] can be utilized to resolve protein species derived from crude material or from various stages of receptor purification. The Laemmli technique is preferred for large samples (0.5 ml) that contain some salt and provides excellent resolution of crude or purified samples. However, the Weber and Osborn technique is considered to give more reliable estimates of molecular weight. Analysis of protein species by the Laemmli system, from receptor containing samples derived from the final heparin–Sepharose step (see the table) demonstrates one major and three minor bands in the molecular weight range of 50,000–65,000, which represent 95% of the total protein stained. Although the receptor is not identified, theoretical considerations would suggest it to be one of the predominant species present.

*Nondenaturing Polyacrylamide Gel Electrophoresis.* The native gel system described by Davis[27] can be used to resolve receptors from cytosol

[24] T. A. McCain, M. R. Haussler, D. Okrent, and M. R. Hughes, *FEBS Lett.* **86**, 65, (1978).
[25] U. K. Laemmli, *Nature (London)* **227**, 680 (1970).
[26] K. Weber and M. Osborn, *J. Biol. Chem.* **255**, 4406 (1969).
[27] B. J. Davis, *Ann. N. Y. Acad. Sci.* **121**, 404 (1964).

and early purification steps. Receptor–hormone complexes are electrophoresed under rigorously cold conditions using a polyethylene glycol water bath. The gels are removed from the tubes, frozen, sliced (2 mm), extracted with acetone for several hours, and then counted for tritium. The results of these experiments suggest that the tritiated hormone is taken into the gel bound to a specific macromolecule, although yield is approximately 10–25%. Comigration of a tritiated peak (sliced gel) with a protein band (stained gel) has not been accomplished because of the need for more purified material than has heretofore been available. However, the technique will be utilized for the unequivocal identification of the receptor.

*Sucrose Gradient Ultracentrifugation.* The techniques for sucrose gradient analysis have been previously described and will not be discussed here. Nevertheless, purified material does demonstrate a 3.3 S macromolecule that retains tritiated hormone in a 5 to 20% sucrose gradient containing KETT-0.3 (Fig. 2F). Most important, this macromolecule is seen to retain the same sedimentation property as that observed for the receptor in crude cytosol.

# [63] A Sensitive Radioreceptor Assay for 1α,25-Dihydroxyvitamin D in Biological Fluids

By JOHN S. CHANDLER, J. WESLEY PIKE, LAURA A. HAGAN, and MARK R. HAUSSLER

Vitamin D has long been recognized as a principal regulator of calcium and phosphate homeostasis in a variety of animal species, including humans. More recently, 1α,25-dihydroxyvitamin D [1α,25-(OH)$_2$D] was described as the hormonal metabolite responsible for mediating intestinal transport of these minerals[1-3] and hence necessitated development of a precise method for quantitating this sterol in biological fluids, particularly serum. Brumbaugh and Haussler demonstrated that chick intestinal mucosa contained a cytosolic protein that rapidly and specifically bound 1α,25-(OH)$_2$D$_3$ both *in vivo* and *in vitro* with high affinity. Furthermore, the

[1] M. R. Haussler, D. W. Boyce, E. T. Littledike, and H. Rasmussen, *Proc. Natl. Acad. Sci. U.S.A.* **68**, 177 (1971).
[2] D. E. M. Lawson, D. R. Fraser, E. Kodicek, H. R. Morris, and D. H. Williams, *Nature (London)* **230**, 228 (1971).
[3] M. F. Holick, H. K. Schnoes, H. F. DeLuca, T. Suda, and R. J. Cousins, *Biochemistry* **10**, 2799 (1971).

METHODS IN ENZYMOLOGY, VOL. 67

sterol–protein complex was rapidly translocated to the intestinal cell nucleus, suggesting a classic steroid hormone action.[4-6]

The ligand affinity and specificity of this cytosolic receptor has been exploited to develop a sensitive competitive protein binding assay for 1α,25-(OH)$_2$D. This radioreceptor assay has been instrumental in the phosphate-related disease states, as well as in characterizing the physiological regulation of 1α,25-(OH)$_2$D.

## Materials

### Extraction and Purification of 1α,25-(OH)$_2$D

Organic solvents: acetone, diethyl ether, hexane, isopropanol, ethanol (100%), chloroform, and ethyl acetate. All solvents are glass-distilled except diethyl ether.

Chromatography resins: silicic acid (Bio-Sil HA -325 mesh); Sephadex LH-20 (Sigma); Celite (Johns-Manville), as an alternative to HPLC; Dupont Zorbax-SIL (25 cm × 4.6 mm) column for high-performance liquid chromatography (HPLC) and a suitable system for HPLC. (Ten-milliliter glass pipettes fitted with glass-wool plugs are used as columns.)

### Radioreceptor Assay

Rachitic chicks: 3- to 4-week old chicks raised on a rachitogenic diet as described by McNutt and Haussler[7]

Nonradioactive 1α,25-(OH)$_2$D$_3$; obtained commercially (Hoffmann-La Roche) and stored at −20° in ethanol (100%) at a concentration of 0.2 ng/ml

Radioactive 1α,25-(OH)$_2$[³H]D$_3$, 110 Ci/mmol, obtained from the Amersham Corp. The sterol is stored in toluene : ethanol (1 : 1) at −20° at a concentration of 1 ng/ml.

Millipore filter manifold (30-place) and vacuum source

Gelman Type A/E glass fiber filters

[4] P. F. Brumbaugh and M. R. Haussler, *Life Sci.* **16**, 353 (1975).
[5] P. F. Brumbaugh and M. R. Haussler, *J. Biol. Chem.* **249**, 1251 (1974).
[6] P. F. Brumbaugh and M. R. Haussler, *J. Biol. Chem.* **249**, 1258 (1974).
[7] K. W. McNutt and M. R. Haussler, *J. Nutr.* **103**, 681 (1973).

*General*

In addition to the specific components described, common laboratory equipment required includes the following items: refrigerated centrifuge, preparative ultracentrifuge, Potter–Elvehjem homogenizer, shaking water bath (25°), glass-distilling apparatus, liquid scintillation counting equipment, and nitrogen gas ($N_2$) drying apparatus. All glassware is routinely prerinsed with acetone, $N_2$ dried, and prerinsed with working solvent.

Procedures

*Hormone Extraction*

Each sample (2 ml) is "spiked" with 1000 cpm of $1\alpha,25\text{-(OH)}_2[^3H]D_3$ (110 Ci/mmol) and extracted with 5 volumes of acetone. Insoluble material is sedimented by centrifugation, and the acetone (supernatant) is evaporated under nitrogen. The residual aqueous portion of the supernatant is reextracted with 15 ml of diethyl ether, and the water phase is discarded. The ether fraction is evaporated to dryness under nitrogen. The tritium spike is used to follow the endogenous hormone during extraction and purification and also to calculate final yields of hormone. Glass scintillation vials are convenient for sample storage, collection, and transfer.

*$1\alpha,25\text{-(OH)}_2D$ Purification*

Extensive purification of the hormone is essential, since excess lipids will adversely affect the radioreceptor assay. Individual sample extracts ($N_2$ dried) are subjected to the following chromatographic procedures.

*Silicic Acid.* Activated silicic acid (120° for 24 hr under vacuum) is slurried in 25% diethyl ether in hexane, and 3.0-ml columns (1.5 g of resin) are packed under light $N_2$ pressure (2–4 psi). Samples are solubilized in small volumes (0.5 ml) of ether/hexane and applied to columns. The columns are washed with 8 ml of 25% diethyl ether/hexane (total volume including sample application) under slight $N_2$ pressure to achieve flow rates of 2–3 ml/min. The hormone is stripped from the columns with 8 ml of acetone. This fraction is $N_2$ dried for further purification.

*Sephadex LH-20.* Six-milliliter columns (2.3 g of resin) are prepared (under gravity) from equilibrated resin (65% chloroform in hexane, slurried for 10 min). The post-silicic acid acetone fractions ($N_2$ dried) are solubilized in 65% chloroform in hexane (0.4 ml) and applied to the columns. The columns are eluted with 65% chloroform in hexane, and the hormone emerges between 20 and 40 ml elution volume. The 20–40-ml fraction is dried under nitrogen for final purification on HPLC or Celite.

*Celite*. As mentioned, Celite liquid–liquid partition chromatography is a suitable alternative to HPLC for final hormone purification. Celite (Johns-Manville) is acid and solvent washed.[8] The solvent system employed is 10% ethyl acetate in hexane (mobile phase) and 45% water in ethanol (stationary phase). The two solvents are cross-equilibrated for 10 min (5 volumes of 10% ethyl acetate in hexane to 1 volume of 45% water in ethanol), and the phases are split. The lower (stationary) phase is used to equilibrate the Celite, and the upper (mobile) phase is used for column elution.

To prepare Celite columns, the resin is thoroughly mixed with stationary phase (1 ml/g) and excess mobile phase is added to obtain a fluffy suspension. The suspension is stirred for 10 min, and 4-ml columns (1.5 g of Celite, dry weight) are tightly packed in 10-ml glass pipettes fitted with glass-wool plugs.

Post LH-20 (20–40-ml fraction) samples are $N_2$ dried, solubilized in mobile phase (0.4 ml), and applied to Celite columns. The samples are eluted with mobile phase under $N_2$ pressure (flow rates of 1.0 ml/min), and the hormone emerges between 7 and 22 ml of solvent. This fraction is $N_2$ dried, resolubilized in 0.4 ml of ethanol, and used for assay and yield determinations.

*High-Performance Liquid Chromatography (HPLC)*. Several HPLC systems have been described that resolve 1α,25-$(OH)_2$D from contaminating lipids and metabolites. Our system consists of a Dupont 830 liquid chromatograph fitted with a DuPont Zorbax-SIL-850 column (25 cm × 4.6 mm). Samples are injected with a Rheodyne Model 70-10 sample injection valve and Model 70-11 Loop-filler port with a 200-$\mu$l sample loop. For hormone resolution the system is equilibrated with 15% isopropanol in hexane at a flow rate of 1 ml/min.

Post LH-20 samples (20–40 ml fraction) are $N_2$ dried and resolubilized in 15% isopropanol in hexane (0.1 ml) for HPLC purification. The hormone elutes between 14 and 17 min (14–17 ml) from time of injection. This fraction is $N_2$ dried and resolubilized in 0.4 ml of ethanol for assay and yield determinations. Yields of 50–75% are obtained, since these procedures do not separate 1α,25-$(OH)_2D_2$ from 1α,25-$(OH)_2D_3$, a total concentration of 1α,25-$(OH)_2$D is assayed. If desired, the two sterols can be resolved by modified HPLC.[9]

[8] L. L. Engle, C. B. Cameron, A. Stoffyn, J. A. Alexander, D. Klein, and N. D. Trofinnow, *Anal. Biochem.* **2**, 114 (1961).

[9] G. Jones and H. F. DeLuca, *J. Lipid Res.* **16**, 445 (1975).

*Preparation and Reconstitution of Cytosol-Chromatin*

The preparation and reconstitution of the intestinal mucosal chromatin fraction and the receptor-rich cytosol fraction is a critical procedure owing to the lability of the unliganded receptor. All procedures during preparation of this system are performed rapidly at 0–4°. The procedure can easily be scaled up to provide ample quantities that are stable for several months when properly prepared and stored.

The intestine(s) from one (or more) rachitic chicks (3–4 weeks old) are removed, rinsed in ice-cold sucrose (0.3 $M$) and slit lengthwise to expose the mucosal surface. The mucosa is scraped from underlying musculature and homogenized (25 ml/2 g mucosa) in STKM buffer (0.25 $M$ sucrose, 50 m$M$ Tris-HCl, pH 7.4, 25 m$M$ KCl, 5 m$M$ MgCl$_2$, 1 m$M$ EDTA, and 12 m$M$ thioglycerol).

Crude nuclei are sedimented from the initial homogenate by centrifugation at 1000 $g$ for 10 min. The supernatant is centrifuged at 100,000 $g$ for 1 hr to prepare the cytosol fraction. Chromatin is prepared from the crude nuclear pellet by sequential homogenization in one 25-ml portion of 8 m$M$ EDTA, 25 m$M$ NaCl (pH 8); one 25-ml portion of 1% Triton X-100, 10 m$M$ Tris · HCl (pH 7.5); and one 25-ml portion of 10 m$M$ Tris · HCl (pH 7.5). The chromatin is harvested by sedimentation at 30,000 $g$ for 10 min after each of the washes.

The entire chromatin pellet from the final wash is reconstituted with half the cytosol fraction by gentle homogenization to create the cytosol–chromatin receptor system for the radioreceptor assay. This homogenate is forced through a No. 22 needle, and 1-ml aliquots are quickly frozen in liquid N$_2$ and stored at −80°.

*Radioreceptor Assay Procedure*

New glass culture tubes (13 × 100 mm) are used for individual assay incubations. In a typical 30-tube assay, 15 tubes are used to construct a standard competition curve (5 points in triplicate) and the remaining tubes for sample assays. To each tube, 20 $\mu$l (20 pg) of the radioactive 1$\alpha$,25-(OH)$_2$D$_3$ solution is added. The standard curve is prepared by adding various concentrations of nonradioactive 1$\alpha$,25-(OH)$_2$D$_3$ to the appropriate tubes, i.e., (0. pg, 2 pg, 4 pg, 6 pg, 10 pg). For sample assay, 100 $\mu$l per assay incubation is used (this volume may be reduced if the competition is greater than the highest point in the standard curve). The tubes containing radioactive sterol (20 $\mu$l) and unlabeled sterol or sample are dried under nitrogen and resolubilized in 10 $\mu$l of ethanol.

The reconstituted cytosol–chromatin (1 ml portion) is quickly thawed and diluted with 7 ml of STKM buffer. The solution is again sheared by forcing it through a No. 22 needle and then kept stirring on ice, while aliquoting. To each tube, 100 μl of the cytosol–chromatin solution is added (10–13 μg of DNA, approximately 50 μg of protein). The assay components are then incubated for 60 min at 25° in a shaking water bath. The reactions are stopped by adding 1 ml of cold Triton X-100 (1%) in 10 mM Tris · HCl (pH 7.5) to each tube. Bound hormone is separated from free by filtering on prerinsed Gelman A/E glass fiber filters. Filtration rates are adjusted to 1 ml/min with low vacuum pressure. Following filtration and washing (5 ml Triton X-100 solution) the filters are removed from the manifold, placed in glass scintillation vials and extracted with 5 ml of acetone. After 5 min the acetone and filters are thoroughly dried and the sterols are solubilized in a suitable scintillation cocktail for tritium counting. Efficiency for tritium should approach 50% and is identical for all samples.

Sample levels of 1α,25-(OH)$_2$D are determined by interpolating competition values (cpm/filter) from the standard curve and correcting this

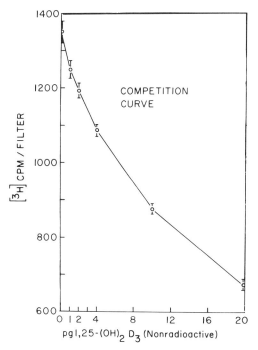

Fig. 1. Standard competition curve for the measurement of 1α,25-(OH)$_2$D. The curve was generated by competing increasing concentrations of nonradioactive 1α,25-(OH)$_2$D$_3$ with 20 pg of 1α,25-(OH)$_2$[³H]D$_3$ for cytosol-chromatin binding. The incubations contained 10 μl of ethanol for sterol solubilization. Incubation was for 60 min at 25°.

1α,25-(OH)$_2$D Values for Various Plasma Samples Determined by Radioreceptor Assay

| Sample (plasma)[a] | 1α,25-(OH)$_2$D (pg/ml) ± SEM ($N$) |
|---|---|
| Normal human | 33 ± 6 (100) |
| Anephric human | Undetectable (8) |
| Rachitic chick (4 weeks) | Undetectable (5) |
| Normal (+D) chick (4 weeks) | 66 ± 18 (8) |
| Normal rats | 100 ± 7 (7) |
| Pregnant rats (21 days) | 179 ± 9 (7) |
| Lactating rats (17–18 days) | 160 ± 10 (14) |
| Trout (male) | 170 ± 22 (10) |
| Trout (spawning females) | 75 ± 8 (10) |

[a] Two-milliliter samples were assayed according to the procedure described herein.

value for hormone yield (as determined by counting a sample aliquot for remaining tritium spike). Values are generally expressed in picograms of 1α,25-(OH)$_2$D per milliliter of sample. A typical standard curve is shown in Fig. 1 and demonstrates the sensitivity of the assay in the picogram range.

When the cytosol-chromatin system is prepared as described, 20 pg of 1α,25-(OH)$_2$[$^3$H]D$_3$ is sufficient to saturate the cytosolic receptor component in the presence of excess chromatin (used to trap the hormone–receptor complex). The dissociation constant of this interaction has been determined to be on the order of $2.0 \times 10^{-10}$ $M$ sterol. Total binding efficiency is about 25% as determined by recovery of 1α,25-(OH)$_2$[$^3$H]D$_3$ on filters in the absence of competing nonradioactive sterol. Interassay variation is 10–15% except at very low 1α,25-(OH)$_2$D sample levels, i.e., <5 pg/ml. Samples are routinely assayed in triplicate to ensure accuracy. The table gives the 1α,25-(OH)$_2$D levels of several samples that have been assayed in this laboratory. The anephric patient and the rachitic chick plasma samples are good examples of negative controls. The 1α,25-(OH)$_2$D hormone is clearly measurable in a number of species and circulates at 33–179 pg/ml (80–430 p$M$). Modulation of human levels occur during several physiologic perturbations of mineral metabolism.

## [64] Use of Chick Kidney to Enzymatically Generate Radiolabeled 1,25-Dihydroxyvitamin D and Other Vitamin D Metabolites

*By* MARK R. HAUSSLER, JOHN S. CHANDLER, LAURA A. HAGAN, and J. WESLEY PIKE

### Introduction

With the realization that vitamin D is metabolized to hydroxylated derivatives prior to performing its biologic function of calcium and phosphate regulation,[1,2] it has become necessary to obtain these metabolites for biochemical, physiologic, and clinical studies. Of fundamental importance is the availability of radiolabeled active metabolites for use in competitive binding assays, receptor analyses, and in further metabolism studies. Radioactive 25-hydroxyvitamin D$_3$ (25-(OH)[$^3$H]D$_3$), the proximal metabolite, has been chemically synthesized and has been on the market for a number of years, but its more active dihydroxylated derivatives proved much more difficult to synthesize. Of particular significance is 1,25-dihydroxyvitamin D$_3$ (1,25-(OH)$_2$D$_3$), which is now accepted as the hormonal form of vitamin D which carries out most, if not all, of the functions of the parent vitamin.[1,2] Also of current interest is 24,25-dihydroxyvitamin D$_3$ (24,25-(OH)$_2$D$_3$) which is proposed as a "bone mineralization" form of vitamin D by some workers in the field.[3] Since 25-(OH)[$^3$H]D$_3$ is now available in very high specific activities ($\sim$100 Ci/mmol) from the Amersham Corporation, we have undertaken to develop a reliable procedure for biosynthetically preparing 1,25-(OH)$_2$[$^3$H]D$_3$ and 24,25-(OH)$_2$[$^3$H]D$_3$ from this common precursor molecule. This technique involves the use of the renal 1$\alpha$-hydroxylase (1$\alpha$-OHase) or 24-hydroxylase (24-OHase) enzyme systems obtained from chickens and the isolation and purification of the desired product by high performance liquid chromatography (HPLC). The present chapter will describe this procedure after first detailing the properties of the renal 1$\alpha$-OHase enzyme of vitamin D-deficient chicks. This enzyme is of course pivotal in the biosynthesis, *in vitro*, of the most active 1$\alpha$-hydroxylated D-vitamins. We will also outline the protocol for inducing the 24-OHase enzyme in

[1] H. F. DeLuca and H. K. Schnoes, *Ann. Rev. Biochem.* **45**, 631 (1976).

[2] M. R. Haussler and T. A. McCain, *New Engl. J. Med.* **297**, 974, 1041 (1977).

[3] J. A. Kanis, T. Cundy, M. Barlett, R. Smith, G. Heynen, G. Warner, and R. G. G. Russell, *Brit. Med. J.* **1**, 1382 (1978).

METHODS IN ENZYMOLOGY, VOL. 67

chicks and its utilization in the *in vitro* generation of 24-hydroxylated metabolites of vitamin D.

## Materials

### Rachitic Chickens

Male chicks are raised from hatching in standard brooders in a room excluding sunlight and ultraviolet or fluorescent fixtures. They are fed *ad libitum* a vitamin D-deficient diet[4] composed of the following: (%) ground wheat (36.0), ground corn (35.37), soybean meal-solvent extracted (16.0), alcohol-extracted casein (8.0), NaCl (1.0), Wesson oil (1.0), $CaHPO_4 \cdot 2H_2O$ (2.0), charcoal (0.5), water-soluble vitamins (0.1), $MnSO_4 \cdot H_2O$ (0.02), and choline chloride (0.01). The composition of the water-soluble vitamin mixture is: (%) ascorbic acid (10), thiamin (0.5), riboflavin (0.5), pyridoxine · HCl (0.5), calcium pantothenate (2.8), nicotinamide (2.0), *i*-inositol (10), folic acid (0.02), vitamin $B_{12}$ (0.002), biotin (0.01), and crude dextrose filler (73.668). Also added to the diet in each feed trough twice weekly with the aid of a "squirt" bottle is approximately 5 ml of a solution of vitamins A, E, and K in Wesson oil. This supplement consists of 400 mg $\beta$-carotene, 5 g $\alpha$-tocopherol, and 600 mg of menadione per liter of Wesson oil. After about 4 weeks, the chicken's weight should level off at 100–150 g and rickets should be obvious. Serum calcium is approximately 5–7 mg% at this point (N = 10–11 mg%). Although lack of growth and hypocalcemia are good indications of rickets, this can be confirmed by radiographs of tibia or via bone histology. When kidneys are excised from such rachitic chickens, they possess an enhanced 25-(OH)D-1$\alpha$-OHase enzyme activity (at least three times that of normal chicks[5]) and little or no 24-OHase activity, *in vitro*.[5]

### 24-OHase Induced Chicks

To suppress renal 1$\alpha$-OHase activity and stimulate the 24-OHase enzyme, chicks are first raised for 2 weeks on the rachitogenic diet described above. They are then changed to an identical diet supplemented with 2.5% (additional) calcium, added as $CaCO_3$; the total calcium content of the diet is therefore 3% (w/w). From this point, each chick is also dosed orally every other day with 320 ng of nonradioactive 1,25-$(OH)_2D_3$ (Dr. M. Uskoković, Hoffmann La Roche). The 1,25-$(OH)_2D_3$ is initially dissolved in

---

[4] K. W. McNutt and M. R. Haussler, *J. Nutr.* **103**, 681 (1973).
[5] G. Tucker, R. E. Gagnon, and M. R. Haussler, *Arch. Biochem. Biophys.* **155**, 47 (1973).

diethyl ether (ca. 200 $\mu$l) and then 1,2-propanediol is mixed in to give a final concentration of 320 ng of 1,25-(OH)$_2$D$_3$ per 0.2 ml of 1,2-propanediol. Diethyl ether is removed by gently bubbling nitrogen through the solution for several hours and the dosing solution then is stored at 0–4°. After 2–3 weeks on this combined high-calcium 1,25-(OH)$_2$D$_3$ regimen, the chicks' renal 24-OHase is maximally induced. Doses (320 ng) of 1,25-(OH)$_2$D$_3$ are given for the 2 days immediately prior to sacrifice, with the last dose being 16 hr before the kidneys are removed for *in vitro* incubation with 25-(OH)[$^3$H]D$_3$.

### 25-(OH)[$^3$H]D$_3$ Precursor

25-Hydroxy[23,24(n)$^3$H]vitamin D$_3$ (60–110 Ci/mmol) is obtained from the Amersham Corporation, Arlington Heights, IL. This radioactive substrate is stored at $-80°$ in a solution of toluene–ethanol (1 : 1, v/v). To minimize radiolytic or chemical decomposition, concentration should not exceed 1 mCi/ml. Routine monitoring of radiochemical purity must be carried out using HPLC,[6] especially if the substrate is stored for several months before use. *Immediately* prior to incubation with kidney enzymes, the toluene–ethanol is removed with a stream of N$_2$ gas and the 25-(OH)[$^3$H]D$_3$ substrate reconstituted in (pure) distilled ethanol. The concentration should be 35 nmol/2 ml ethanol or, at a specific activity of 110 Ci/mmol, 3.85 mCi of substrate should be dissolved per 2 ml of ethanol.

### Buffer plus Additives

*Phosphate buffer.* Weigh 32.6 g KH$_2$PO$_4$ and 0.96 MgCl$_2$ · 6H$_2$O, dissolve in deionized water (ca. 950 ml), pH to 7.4 with 10 $N$ NaOH and then dilute to 1 liter with H$_2$O.

*NADPH regenerating system.* Weigh the following ingredients (dry) and keep separate until just before mixing the composite incubation buffer: 0.39 g L-malic acid, 0.39 g D-glucose-6-phosphate, and 0.09 g NADP$^+$ (Sigma Chemical Co., St. Louis, MO); 500 $\mu$l of glucose-6-phosphate dehydrogenase (crystalline suspension in 3.2 $M$ (NH$_4$)$_2$SO$_4$; Sigma Cat. No. G-7877) will also be required.

### Properties of Chick Renal 1$\alpha$-OHase

Since the *in vitro* generation of 1,25-(OH)$_2$[$^3$H]D$_3$ was judged to be of primary importance, the parameters governing this enzymatic reaction were first studied in detail. Then, employing the optimal conditions de-

---

[6] G. Jones and H. F. DeLuca, *J. Lipid Res.* **16**, 445 (1975).

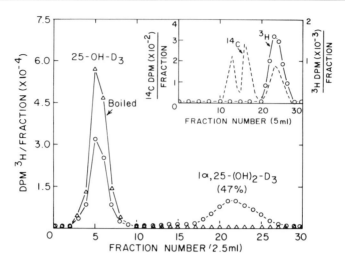

FIG. 1. Chromatographic resolution of enzymatically produced $1,25\text{-}(OH)_2[^3H]D_3$. The reaction was run as described by Tucker *et al.*[5] A 9% homogenate of rachitic chick kidney in 0.3 $M$ sucrose was prepared and 0.4 ml added to 9.6 ml of a phosphate buffer, pH 7.4, containing magnesium and a NADPH-regenerating system.[5] The reaction was initiated by adding 65 pmol of $25\text{-}(OH)[^3H]D_3$ (final concentration $6.5 \times 10^{-9}\ M$) in 50 $\mu$l of ethanol and was allowed to proceed under air for 20 min at 37°. The lipid extract of the incubation mixture was chromatographed on a $1 \times 13$ cm Sephadex LH-20 column; elution was with 65% $CHCl_3$ in hexane. ○, Normal reaction mixture; △, results when the kidney homogenate is boiled first. The inset illustrates rechromatography of the suspected $1,25\text{-}(OH)_2[^3H]D_3$ peak (○) from Sephadex LH-20 on Celite by the procedure of Haussler and Rasmussen.[9] Included with the tritiated sample is a $[^{14}C]$dihydroxyvitamin $D_3$ fraction (----) generated by dosing rachitic chicks with 1 nmol of $[^{14}C]$vitamin $D_3$ and isolating the aggregate 24,25-, 25,26-, and $1,25\text{-}(OH)_2D_3$ fraction from blood (15 hr after injection) via silicic acid chromatography.[9]

fined for the $1\alpha$-OHase reaction, quantities of $1,25\text{-}(OH)_2[^3H]D_3$ could be biosynthesized on a larger scale. Also, the *in vitro* reaction and subsequent purification of the product could serve as a model for the less well understood 24-OHase catalyzed conversion.

## Demonstration of $1\alpha$-OHase Enzymatic Activity

Using conditions similar to those described by Fraser and Kodicek[7] and by Norman *et al.*,[8] we first incubated radioactive $25\text{-}(OH)D_3$ with rachitic chick kidney homogenate for 20 min and then analyzed the extracted sterols on a $1 \times 13$ cm Sephadex LH-20 column. Figure 1 illustrates

[7] D. R. Fraser and E. Kodicek, *Nature (London)* **228**, 764 (1970).

[8] A. W. Norman, R. J. Midgett, J. R. Myrtle, and H. G. Nowicki, *Biochem. Biophys. Res. Commun.* **42**, 1082 (1971).

that 47% of the radioactivity recovered from a typical column migrates in the area of 1,25-(OH)$_2$D$_3$. The fact that this conversion is an enzymatic one is proven by the observation that boiling the homogenate for 10 min prior to the incubation with substrate abolishes·the appearance of product. Furthermore, rechromatography of putative 1,25-(OH)$_2$[³H]D$_3$ from an unboiled reaction (Fig. 1) on Celite[9] indicates that all of the tritium elutes (fract. No. 23) with 1,25-(OH)$_2$[¹⁴C]D$_3$ generated in chick blood by injection of [¹⁴C]vitamin D$_3$ (see inset to Fig. 1). On such a Celite column, the other two known dihydroxylated D-vitamins—24,25-(OH)$_2$[¹⁴C]D$_3$ and 25,26-(OH)$_2$[¹⁴C]D$_3$, elute in fractions No. 13 and 16, respectively.[9] Therefore, it is clear that homogenates of rachitic chick kidney convert 25-(OH)D$_3$ exclusively to 1,25-(OH)$_2$D$_3$, and that the reaction is approximately 50% complete in 20 min.

Properties of the 1α-OHase Enzyme

Next we studied various parameters of the enzymatic generation of 1,25-(OH)$_2$D$_3$, using the procedure described in the legend to Fig. 1 as the basic protocol. Amount of product 1,25-(OH)$_2$[³H]D$_3$ formed was found to be linear with respect to added kidney homogenate up to 3 mg kidney protein per 10 ml final incubation volume. This rate (200 fmol/mg protein/min) was at least three times that of reactions run with homogenates of normal chick kidney.[6,10] Under these conditions, rate of 1,25-(OH)$_2$D$_3$ formation using D-deficient kidneys was linear for 20 min, although about 50% more product could be formed by extending the incubation to 2 hr. The pH optimum for the reaction was determined to be 7.4 and maximum conversion of 25-(OH)[³H]D$_3$ substrate occurred at a temperature of 37° (data not shown). Finally, Fig. 2 pictures the determination of the $K_m$ (Fig. 2A) as first reported by our group in 1972.[11] The value for $K_m$ ($1.5 \times 10^{-7} M$) is about an order of magnitude higher than the prevailing circulating concentration of 25-(OH)D$_3$ in normal chicks.[12] Importantly, the $K_m$ determination indicates that concentrations of 25-(OH)D$_3$ substrate, *in vitro,* should be in the neighborhood of $10^{-7} M$ to achieve efficient conversion to 1,25-(OH)$_2$D$_3$ hormone. Figure 2B illustrates inhibition of the enzyme by metyrapone, a well-known blocker of P$_{450}$-mediated hydroxylases such as the 11β-OHase of the adrenal cortex. Such an inhibition of the 1α-OHase

[9] M. R. Haussler and H. Rasmussen, *J. Biol. Chem.* **247,** 2328 (1972).
[10] D. J. Cork, M. R. Haussler, M. J. Pitt, E. Rizzardo, R. H. Hesse, and M. M. Pechet, *Endocrinology* **94,** 1337 (1974).
[11] M. R. Haussler, *Fed. Proc.* **31,** 639 (1972).
[12] M. R. Hughes, D. J. Baylink, W. A. Gonnerman, S. U. Toverud, W. K. Ramp, and M. R. Haussler, *Endocrinology* **100,** 799 (1977).

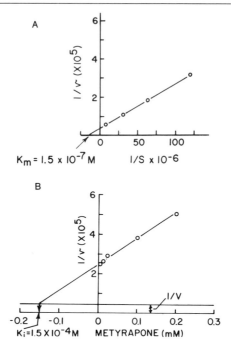

FIG. 2. Determination of $K_m$ for 25-(OH)D$_3$ substrate (A) and $K_i$ for metyrapone (B)—two properties of the renal 1α-OHase. Reaction conditions were as indicated in Fig. 1, except that substrate 25-(OH)D$_3$ or metyrapone concentrations were varied accordingly. (A) $K_m$ as calculated from a Lineweaver–Burk plot; (B) a Dixon plot to arrive at the $K_i$ for metyrapone.

was first reported by Gray *et al.*,[13] and is represented for our data in the form of a Dixon plot (Fig. 2B). The $K_i$ for metyrapone (recrystallized) was found to be $1.5 \times 10^{-4}$ $M$. Thus, all of the properties of the 1α-OHase outlined above are consistent with the findings of other groups,[1,7,14] indicating that this enzyme is a P$_{450}$-containing hydroxylase requiring molecular oxygen and NADPH. Characterization of the reaction provides a rational basis for the generation of 1,25-(OH)$_2$[$^3$H]D$_3$ on a larger scale.

## Procedure for Generating 1,25-(OH)$_2$[$^3$H]D$_3$

### Incubation

Ten to twenty rachitic chicks are killed by decapitation and the kidneys immediately excised, blotted, and placed in ice cold 0.3 $M$ sucrose (unbuf-

[13] R. W. Gray, J. L. Omdahl, J. G. Ghazarian, and H. F. DeLuca, *J. Biol. Chem.* **247**, 7528 (1972).
[14] H. L. Henry, R. J. Midgett, and A. W. Norman, *J. Biol. Chem.* **249**, 7584 (1974).

fered). The kidneys are rinsed 3–4 times in cold 0.3 $M$ sucrose and weighed (12 g should be used). The kidneys are gently homogenized in 100 ml of 0.3 $M$ sucrose with a Potter-Elvehjem homogenizer equipped with a Teflon pestle; do 6–8 passes on a very low motor setting (No. 2). The resulting homogenate (110 ml) is poured into a beaker and stored on ice for no more than 2–3 min before initiating the reaction.

In the meantime, the incubation buffer should be formulated as follows: dissolve the three dry chemicals listed above (malic acid, glucose-6-phosphate, and NADP⁺) in 15 ml of deionized water each. Add each sequentially to 255 ml of phosphate buffer (see Phosphate Buffer) at room temperature. Mix and add 500 μl of glucose-6-phosphate dehydrogenase suspension (162 units). The incubation buffer plus additives is now complete and should be divided into five 60-ml portions and placed in 250-ml Erlenmeyer flasks. Quickly, 22 ml of kidney homogenate should be mixed into each of the five flasks. Substrate 25-(OH)[³H]D₃ (400 μl) in ethanol should then be added to each flask with swirling to achieve a final substrate concentration of $0.85 \times 10^{-7} M$. All five flasks are then placed in a shaking incubator at 37° and incubated for 90 min open to the air with gentle rotary shaking.

### Extraction

The reaction is stopped by adding the contents of the five incubation flasks to a total of 1400 ml of methanol–chloroform (2 : 1). Note that all organic solvents should be glass distilled. The solution, which should consist of one phase, is then stirred for 30 min. After stirring, the denatured protein is allowed to settle out and the supernatant is carefully decanted into a large separatory funnel. The last few milliliters of liquid are harvested by centrifugation. Discard pellets and take total supernatant (ca. 1800 ml) and add 360 ml of CHCl₃ and 190 ml deionized H₂O, shake, and allow phases to separate. Harvest lower CHCl₃ phase and save. Wash upper methanol–H₂O phase once with 750 ml CHCl₃, shake, and obtain a second lower phase to be combined with the first CHCl₃ phase. The combined CHCl₃ layers contain unreacted 25-(OH)[³H]D₃ and the generated 1,25-(OH)₂[³H]D₃ product; the methanol–H₂O layer is discarded.

Roto-evaporate the CHCl₃ solution, taking care not to heat above 50° while drawing the solvent off with a vacuum. Flash evaporate to dryness and solubilize the lipid residue in diethyl ether using several washes of the flask to yield a final volume of 20–30 ml. Clarify this solution via centrifugation at 0° for 10 min at 10,000 $g$. Decant off the supernatant and proceed to the purification.

Purification

*Silicic Acid Column.* Activate 25 g of Bio-Rad Bio-Sil HA minus 325 mesh silicic acid for 24 hr at 120° under a vacuum. Cool and slurry-up the silicic acid in diethyl ether and pour into a 1.8 × 18-cm column. The column is packed and run with approximately 2 lb of nitrogen pressure applied. The lipid sample from the extraction is dried completely under a stream of nitrogen and resolubilized in 10 ml of diethyl ether for application to the column. After two 5-ml ether washes of the vial are also applied to the column, elution is carried out with 5% acetone in diethyl ether. One hundred-milliliter fractions should be collected and 5 $\mu$l counted to locate peaks; unconverted 25-(OH)[$^3$H]D$_3$ will elute in fraction No. 2 and 1,25-(OH)$_2$[$^3$H]D$_3$ product in fraction No. 6. Very infrequently, the 1,25-(OH)$_2$[$^3$H]D$_3$ peak will appear later, perhaps at fraction No. 10 or beyond. A more common problem is improperly activated silicic acid or water in the applied sample, causing the 1,25-(OH)$_2$D$_3$ not to be resolved from 25-(OH)D$_3$, with all tritium emerging in fraction No. 2–4. An alternate silicic acid system which avoids this problem and yields more pure unconverted 25-(OH)[$^3$H]D$_3$ and 1,25-(OH)$_2$[$^3$H]D$_3$ is the following: apply the sample in 50% diethyl ether in hexane and elute with that solvent. Collecting 100-ml fractions, 25-(OH)D$_3$ will emerge very "clean" in fraction No. 8–10 and then at the end of fraction No. 17, a solvent change to 10% acetone in diethyl ether is made. 1,25-(OH)$_2$[$^3$H]D$_3$ subsequently elutes in approximately fraction No. 19–20. Whichever system is used, the 1,25-(OH)$_2$[$^3$H]D$_3$ peak is harvested, roto-evaporated, and again solubilized in diethyl ether. If the product is to be stored overnight at this point, the ether should be dried off and 2 ml of toluene–ethanol (1:1) should be added; storage is at −20°.

*Sephadex LH-20 Column.* Five grams of Sephadex LH-20 is slurried up in 65% CHCl$_3$ in hexane (use glass-distilled solvents for this procedure and all future steps; allow solvents to come in contact with only clean glass or Teflon). Using gravity flow, the slurry is formed into a 1 × 15-cm column and equilibrated about 15 min. The sample is taken to dryness with N$_2$ and resolubilized in 200 $\mu$l of 65% CHCl$_3$ in hexane. It is then applied to the column, run on, and followed with several 200–400 $\mu$l washes of the vial. During the actual elution, two fractions are collected, the first being 40 ml and the second 75 ml. Ten-microliter aliquots are counted to assure that the product is in fraction II.

*MicroCelite Column.* Liquid–liquid partition chromatography on Celite (Johns Manville, Co.) is performed using Celite[9] that is first washed extensively with HCl[15] and then with copious quantities of water, distilled

[15] L. L. Engle, C. B. Cameron, A. Stoffyn, J. A. Alexander, D. Klein, and N. D. Troffinow, *Anal. Biochem.* **2**, 114 (1961).

methanol, and distilled hexane. The solvent system[9] for Celite is as follows: 50 ml of ethyl acetate is diluted to 500 ml with hexane (upper-mobile phase) and 45 ml of water is diluted to 100 ml with ethanol (lower-stationary phase). The two phases are cross-equilibrated by shaking in a separatory funnel and letting stand for at least 30 min. Five grams of acid-washed Celite is weighed out (this is in excess of that required for the column) in a 250-ml beaker and 5 ml of lower phase thoroughly mixed in with a glass stirring rod. Excess mobile (upper) phase is next added until a suspension is obtained via stirring. This suspension is further equilibrated with vigorous mixing on a magnetic stirrer for 15 min until the Celite appears "fluffed." A portion of the Celite is then wet-packed into a 10-ml graduated pipet cut off at the top and fitted with a glass wool plug at the bottom. The column should be packed very tightly with a glass rod, to a final volume of 4.0 ml. The sample is taken to dryness under N$_2$, solubilized in 200 $\mu$l of mobile phase, and applied to the column under 2–4 lb N$_2$ pressure. After several washes of the sample vial are applied, the column is run under nitrogen pressure to achieve a flow rate between 0.5 and 1.0 ml/min of mobile phase. The following fractions are collected: I = 6 ml, II = 15 ml, III = 6 ml; 1–2 $\mu$l aliquots are counted to locate 1,25-(OH)$_2$[$^3$H]D$_3$ which should be in fraction II.[9] This fraction should be dried down and prepared for HPLC. It should be noted that the microCelite step can be eliminated from this procedure, but the product may then require two runs on HPLC to achieve the desired radiochemical purity.

*HPLC.* Several HPLC systems have been described which resolve 1,25-(OH)$_2$D$_3$ from contaminating lipids and other vitamin D metabolites.[6] Our system consists of a DuPont 830 Liquid Chromatograph equipped with a 4.6 mm × 25 cm Zorbax-Sil 850 column. Samples are injected with a Rheodyne Model 70-10 sample injection valve and Model 70-11 Loop-filler port with a 200 $\mu$l sample loop. The column is run with 15% isopropanol in hexane at a flow rate of 0.7 to 1 ml/min. The 1,25-(OH)$_2$[$^3$H]D$_3$ sample from microCelite is solubilized in 100 $\mu$l of 15% isopropanol in hexane and injected along with two 40 $\mu$l rinses of the vial. 1,25-(OH)$_2$D$_3$ elutes in approximately 16 min from the time of injection at a flow rate of 1 ml/min. The purified product (monitored both by tritium counts on 1-$\mu$l aliquots of 1 min fractions and via UV absorbance at 254 nm) is pooled, dried with N$_2$, and solubilized in 10 ml of toluene–ethanol (1:1). Stock solutions should be stored at $-80°$, but a working solution of the radiochemical can be stored at $-20°$. The compound is stable for at least 6 months when properly stored.

### Summary of 1,25-(OH)$_2$[$^3$H]$_3$ Generation

Figure 3 depicts a typical result of the purification of product 1,25-(OH)$_2$[$^3$H]D$_3$ when 3.85 mCi of 25-(OH)[$^3$H]D$_3$ is incubated with kidney

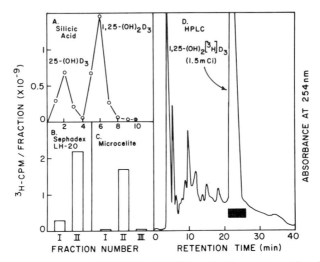

FIG. 3. Purification of 1,25-(OH)$_2$[$^3$H]D$_3$ (110 Ci/mmol). Chromatographic columns were developed sequentially on a single sample as detailed in the text. The silicic acid column (A) was eluted with 5% acetone in diethyl ether. The flow rate of the HPLC column (D) was 0.7 ml/min of 15% isopropanol in hexane and the dark shaded area represents the final harvested 1,25-(OH)$_2$[$^3$H]D$_3$.

homogenate from rachitic chicks as described above. Resolution of product from substrate on silicic acid (Fig. 3A) indicates about a 70% conversion to 1,25-(OH)$_2$[$^3$H]D$_3$. Figure 3B and C illustrate the radiochemical (and chemical) purification of the product on Sephadex LH-20 and microCelite, with the product being about 90–95% radiochemically pure after microCelite chromatography. Final purification is achieved via HPLC (Fig. 3D), where the product elutes between 21 and 25 min and is the dominant UV absorbing species present. 1,25-(OH)$_2$[$^3$H]D$_3$ (1.5 mCi) (110 Ci/mmol) was obtained, providing a 39% yield with respect to starting material. The radiochemical purity of the product was >96% by rechromatography on HPLC and it was virtually free of all residue or chemical contamination.

## Biosynthesis of 24-Hydroxylated Sterols

### General

As indicated under Materials, renal 24-OHase enzyme can be induced by chronic treatment of chicks with 1,25-(OH)$_2$D$_3$ and high calcium diet.[16]

[16] J. G. Haddad, C. Min, M. Mendelsohn, E. Slatopolsky, and T. J. Hahn, *Arch. Biochem. Biophys.* **182**, 390 (1977).

Using renal homogenates from such chicks, we tested both 25-(OH)[³H]D₃ and 1,25-(OH)₂[³H]D₃ as substrates for the 24-OHase enzyme. A comparison was made with renal homogenates from rachitic chicks and respective substrates and products were analyzed via Sephadex LH-20 chromatography. The data from this experiment, in which 200 μCi of substrate was employed, are pictured in Fig. 4. The top panel shows a standard profile of 1,25-(OH)₂[³H]D₃ production (75%) from 25-(OH)[³H]D₃ using vitamin D-deficient chicks and the protocol detailed under Procedure for Generating 1,25-(OH)₂[³H]D₃. Clearly, the 1α-OHase enzyme is enhanced in this renal homogenate. In sharp contrast, when 200 μCi of 25-(OH)[³H]D₃ is incubated with renal homogenates from 24-OHase-induced chicks, there is virtually no 1,25-(OH)₂[³H]D₃ produced and 36% conversion to 24,25-(OH)₂[³H]D₃ occurs (Fig. 4, center panel). Similarly (Fig. 4, bottom panel), use of 200 μCi of 1,25-(OH)₂[³H]D₃ as a substrate elicits metabolism to 55% 1,24,25-(OH)₃[³H]D₃. The fact that a greater conversion of 1,25-(OH)₂D₃ to 1,24,25-(OH)₃D₃ as compared to 24-hydroxylation of 25-(OH)D₃ occurs under identical conditions in this

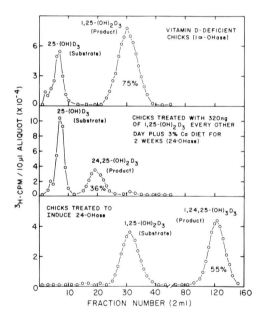

Fig. 4. Comparison of 1α-OHase and 24-OHase activities in chick kidney homogenates by resolution of sterols on Sephadex LH-20. Each panel represents a 1 × 15 cm Sephadex LH-20 profile obtained by elution with 65% CHCl₃ in hexane. The reactions were "scaled down" (200 μCi substrate in each case) versions of the generation procedures detailed for 1,25-(OH)₂[³H]D₃, 24,25-(OH)₂[³H]D₃, and 1,24,25-(OH)₃[³H]D₃ (see text). Conditions were identical to those outlined in the text, but extracts were chromatographed directly on Sephadex LH-20, with no preliminary silicic acid column. The identity of all three products was verified by exact comigration with authentic nonradioactive sterols on HPLC.

preliminary experiment suggests that $1,25\text{-}(OH)_2D_3$ may indeed be the natural substrate for the renal 24-OHase. In any event, such kidney preparations provide a valuable tool in generating radioactive 24-hydroxylated sterols.

### Generation and Purification of $24,25\text{-}(OH)_2[^3H]D_3$

*Materials and Incubation.* Buffers and additives are basically the same as those used for the production of $1,25\text{-}(OH)_2[^3H]D_3$, except that the concentration of the NADPH regenerating system is higher. Chicks are treated to enhance 24-OHase as indicated above. In a typical experiment, 1.4 mCi of $25\text{-}(OH)[^3H]D_3$ (110 Ci/mmol) is dissolved in 800 $\mu l$ of distilled ethanol to serve as the substrate. Seven chicks are killed to obtain about 12 g of kidney (wet weight). The kidney is homogenized in 5 volumes of 0.3 $M$ sucrose analogously to the method outlined under Incubation, except that the homogenate is more concentrated to compensate for the lower activity of the 24-OHase compared to the $1\alpha$-OHase. Meanwhile, the incubation buffer is made as follows: Dissolve 0.24 g L-malic acid, 0.24 g glucose-6-phosphate, and 0.14 g $NADP^+$ each in 1 ml of $H_2O$. Add each of these to 112 ml of phosphate buffer; mix in 700 $\mu l$ of glucose-6-phosphate dehydrogenase suspension (227 units). The incubation buffer is divided into two flasks (58 ml each) and 35 ml of freshly prepared kidney homogenate added to each flask. Quickly, 400 $\mu l$ of ethanol containing the $25\text{-}(OH)[^3H]D_3$ substrate is pipeted into each flask and then the reaction is run at 37° with gentle shaking for 90 min open to the air. Note that the substrate concentration in this case is $0.68 \times 10^{-7} M$.

*Extraction and Purification.* The extraction is carried out exactly as described for $1,25\text{-}(OH)_2[^3H]D_3$ under Extraction, except that all volumes are scaled down by a factor of 186/410 to accommodate the smaller total volume being extracted in this case. Purification is carried out by the following three columns:

SILICIC ACID. The second system under Silicic Acid Column is used, with elution being performed first with 50% diethyl ether in hexane and then changing to 10% acetone in ether at the end of fraction No. 17. $24,25\text{-}(OH)_2[^3H]D_3$ product will emerge in fraction No. 19–20, similarly to $1,25\text{-}(OH)_2D_3$.

SEPHADEX LH-20. The $24,25\text{-}(OH)_2D_3$ peak from silicic acid is next run on the $1 \times 15$-cm Sephadex LH-20 column described above under Sephadex LH-20 Column. In this case, 2 ml fractions are collected and 10-$\mu l$ aliquots counted to locate $24,25\text{-}(OH)_2[^3H]D_3$; the product should be present in fraction No. 15–25 during elution with 65% $CHCl_3$ in hexane.

HPLC. The peak from Sephadex LH-20 can be directly introduced to HPLC (see HPLC above) and elution carried out with 10% isopropanol in hexane. A typical profile from such a run in which 1.4 mCi of 25-(OH)[³H]D₃ was used as an initial substrate is pictured in Fig. 5. The 24,25-(OH)₂[³H]D₃ product is the major UV-absorbing species and elutes nicely between 16 and 18 min retention time. In this particular case, 144 μCi of product or a 10% radiochemical yield compared to starting material was achieved. However, since the substrate 25-(OH)[³H]D₃ was labeled with tritium in the number 23 and 24 carbons, it can be assumed that there is about a 25% loss of tritium during 24-hydroxylation. Thus, the *chemical* yield would be higher than the radiochemical yield and is estimated to be 14%. Also, this means that the specific activity of the product 24,25-(OH)₂[³H]D₃ is reduced from 110 Ci/mmol to 82 Ci/mmol.

### Production of 1,24,25-(OH)₃[³H]D₃

Although the generation of large amounts of 1,24,25-(OH)₃[³H]D₃ will not be detailed here, preliminary experiments indicate that this is feasible.

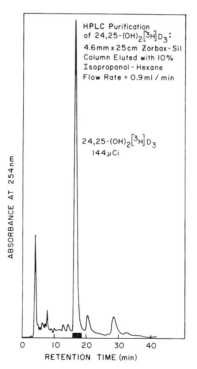

FIG. 5. HPLC purification of enzymatically generated 24,25-(OH)₂[³H]D₃. 24,25-(OH)₂[³H]D₃ was biosynthesized from 1.4 mCi of 25-(OH)[³H]D₃ as indicated in the text. The dark shaded area represents the harvested 24,25-(OH)₂[³H]D₃ product.

First one must generate $1,25\text{-(OH)}_2[^3H]D_3$ as discussed under Procedure for Generating $1,25\text{-(OH)}_2[^3H]D_3$. This $1,25\text{-(OH)}_2D_3$ is then used as a substrate for the 24-OHase as outlined under Generation and Purification of $24,25\text{-(OH)}_2[^3H]D_3$, with 1.4 mCi of $1,25\text{-(OH)}_2[^3H]D_3$ replacing the $25\text{-(OH)}[^3H]D_3$ employed previously. If the reaction is run identically, following extraction, the $1,24,25\text{-(OH)}_3[^3H]D_3$ can be purified by directly running the lipid extract on Sephadex LH-20 as shown in the lower panel of Fig. 4. The product will elute in a very broad, but purified, peak centered around 250 ml. This $1,24,25\text{-(OH)}_3[^3H]D_3$ fraction is then applied to HPLC with elution being carried out with 20% isopropanol in hexane. At a flow rate of 1 ml/min, the pure product emerges between 20 and 30 min retention time. We have prepared $1,24,25\text{-(OH)}_3[^3H]D_3$ in this manner, although like the $24,25\text{-(OH)}_2[^3H]D_3$, its specific activity is compromised to approximately 82 Ci/mmol. At the present time, the function of $1,24,25\text{-(OH)}_3D_3$ is not known. It may be physiologically active, since it binds efficiently to receptors for the $1,25\text{-(OH)}_2D_3$ hormone,[2] but it is more likely that it represents the initial metabolite in the functional degradation of $1,25\text{-(OH)}_2D_3$.[2,17]

## Conclusion

It is clear that the biosynthesis of radioactive D vitamins described here will be important to the elucidation of their biologic functions via receptor and autoradiographic studies, to the delineation of the pathway of vitamin D metabolism by *in vivo* and *in vitro* investigations, and finally to clinically relevant competitive binding and radioimmunoassays for these active sterols.

### Acknowledgments

We are grateful to Drs. Martin Bye, Donald Berry, and Anthony Evans of the Amersham Corporation for their consultations and for supplying the $25\text{-(OH)}[^3H]D_3$ substrate used in these experiments. This work was supported in part by NIH grant AM15781.

[17] H. F. DeLuca, *Arch. Int. Med.* **138**, 836 (1978).

# Section VII

# Miscellaneous Vitamins and Coenzymes

## [65] Coenzyme M: Preparation and Assay

By JAMES A. ROMESSER and WILLIAM E. BALCH

Although the biochemistry of methane production is poorly under-stood, recent research has led to the discovery of a new cofactor, coen-zyme M. This coenzyme has been identified as 2-mercaptoethanesulfonic acid and has been shown to function as a methyl transfer cofactor in the terminal reduction step of methanogenesis.[1]

Preparation of Coenzyme M and Derivatives

*Ammonium 2-Mercaptoethanesulfonate (HS-CoM).* A modified proce-dure of Schramm *et al.*[2] may be used to synthesize HS-CoM. A 500-mg amount of sodium 2-bromoethanesulfonate monohydrate (2.18 mmol) and 167 mg of thiourea (2.18 mmol) are added to 2 ml of 95% ethanol. The mixture is refluxed for 5 hr, then cooled to $-20°$, and the precipitate is collected by filtration. Concentrated ammonium hydroxide (0.7 ml) is added to the precipitate, and the solution is refluxed for 2 hr. The solution is then flash evaporated to dryness. The white residue is taken up in water and applied to a Sephadex SP-C25 (ammonium form) column ($1.2 \times 18$ cm) that has been previously equilibrated with water. HS-CoM is eluted with water. To locate HS-CoM, a sample of each fraction is spotted with nitrous acid. HS-CoM reacts with nitrous acid (prepared by mixing equal volumes of $1 N$ $NaNO_2$ and $1.5 N$ HCl) to form a red S-nitroso derivative. The appropriate fractions are pooled and flash evaporated to dryness. The residue is taken up in 3 ml of water, and acetone is added to cause precipi-tation. The precipitate is removed by filtration and discarded. The filtrate is flash evaporated to dryness, and 154 mg of HS-CoM (44.4% yield) is crystallized from the methanol-diethyl ether. This procedure may be used to synthesize $H^{35}S$-CoM.

Taylor and Wolfe[1] have developed an alternative procedure to prepare HS-CoM by use of 2-bromoethanesulfonate and hydrogen sulfide as start-ing materials. 2-Mercaptoethanesulfonic acid as the sodium salt may be purchased from Pierce Chemical Company (Rockford, Illinois) and E. Merck Co. (West Germany). Results of our analysis of the preparations by the Ellman assay[3] indicate that these preparations contain approximately 79% and 90% HS-CoM, and 12% and 5% of the oxidized form, $(S\text{-CoM})_2$,

[1] C. D. Taylor and R. S. Wolfe, *J. Biol. Chem.* **249**, 4879 (1974).

[2] C. H. Schramm, H. Lemaire, and R. H. Karlson, *J. Am. Chem. Soc.* **77**, 6231 (1955).

[3] G. Ellman, *Arch. Biochem. Biophys.* **82**, 70 (1959).

METHODS IN ENZYMOLOGY, VOL. 67

in the Pierce and Merck preparations, respectively. The remaining contaminating materials in each preparation were not characterized.

*Ammonium 2-(Methylthio)ethanesulfonate (CH₃-S-CoM)*. In an ice bath a 5-ml amount of concentrated $NH_4OH$ in a stoppered flask is made anaerobic by sparging with nitrogen for 20 min (10 cc/min). A second stoppered flask containing 200 mg of HS-CoM (1.25 mmol) and a stirring bar also is gassed with nitrogen. The $NH_4OH$ is then anaerobically transferred to the second stoppered flask, and the sparging is continued for 5 min. The gassing probes are removed, and 78.4 $\mu$l of $CH_3I$ (1.26 mmol) is added. The flask is covered with aluminum foil, and the reaction mixture is stirred at 4° overnight. The mixture is then flash evaporated to dryness, and the residue is taken up in a minimal amount of water. A 107-mg amount of $CH_3$-S-CoM (49.0% yield) is precipitated by the addition of acetone. This general procedure may be used for the synthesis of other alkyl derivatives of HS-CoM; it is of special value in the synthesis of [14]C-labeled alkyl derivatives. Taylor and Wolfe[1] offer an alternative synthesis of $CH_3$-S-CoM by use of 2-bromoethanesulfonate and methylmercaptan as starting materials.

*Ammonium 2,2-Dithiodiethanesulfonate (S-CoM)₂*. The oxidized coenzyme is prepared as described by Taylor and Wolfe.[1] A 0.5-g amount of HS-CoM is dissolved in 30 ml of 30% aqueous ammonium hydroxide, and the solution is bubbled with oxygen until HS-CoM cannot be detected with nitrous acid. $(S-CoM)_2$ crystallizes upon the addition of acetone and is recrystallized 3 times from aqueous acetone with an overall yield of 20%.

## Assays

### Bioassay

*Growth of Methane-Forming Bacteria in Tubes*. All species of methane bacteria now in pure culture are able to grow by the oxidation of hydrogen and reduction of $CO_2$:

$$4 H_2 + CO_2 \rightarrow CH_4 + 2 H_2O \qquad (-31 \text{ kcal})$$

A recent modification of the Hungate technique[4,5] has been devised for growing methanogens efficiently and reproducibly in small-volume cultures.[6] The technique uses 18 × 150 mm septum-lip serum tubes (Bellco

[4] R. E. Hungate, *in* ''Methods in Microbiology (J. R. Norris and D. W. Ribbons, ed.), Vol. 3B, p. 117. Academic Press, New York, 1969.

[5] M. P. Bryant, *Am. J. Clin. Nutr.* **25,** 1323 (1972).

[6] W. E. Balch and R. S. Wolfe, *Appl. Environ. Microbiol.* **32,** 781 (1976).

Glass Inc., No. 2048-00180) or standard serum bottles (mouth OD, 20 mm) which are pressurized to 3 atm. The general medium contains the following constituents in grams per liter and is prepared in a 500-ml round-bottom flask under a gas phase of 80% $N_2$ and 20% $CO_2$ by use of the Hungate technique[4,5]: $KH_2PO_4$, 0.45; $K_2HPO_4$, 0.45; $(NH_4)_2SO_4$, 0.45; NaCl, 0.9; $MgSO_4 \cdot 7H_2O$, 0.18; $CaCl_2 \cdot 2H_2O$, 0.012; $FeSO_4 \cdot 7H_2O$, 0.002; resazurin, 0.001; sodium formate, 2.0; sodium acetate, 2.5; $NaHCO_3$, 5.0; L-cysteine $\cdot$ HCl, 0.5; $Na_2S \cdot 9H_2O$, 0.5; yeast extract (Difco), 2.0; trypticase (BBL), 2.0; trace mineral solution,[7] 10.0 ml; vitamin solution,[7] 10.0 ml. The final pH should be 7.0 to 7.2.

Reduced medium (5 ml per tube) is transferred to the serum tubes in a Freter-type anaerobic hood[8] or is transferred to the tubes anaerobically by use of the Hungate technique. The use of an anaerobic hood for transfers is the method of choice. Each tube with medium is sealed by inserting a No. 0 rubber stopper (durometer hardness, 45) to a depth of 1 cm. The stopper is then trimmed so that it protrudes 3 mm above the glass lip of the tube. Alternatively, manufactured stoppers with a septum lip are available commercially (Bellco Glass Inc., No. 2048-11800). The stopper is crimped into place with an aluminum seal crimper (Pierce Co., No. 13212) and a one piece aluminum seal (Wheaton Glass Inc., No. 224183).

The gas in the sealed tube is exchanged for substrate, a commercial mixture of 80% hydrogen and 20% carbon dioxide, by means of a gassing manifold (Fig. 1), consisting of a 3-way valve (C) connected to a vacuum source, a gas source at 3 atm pressure (A), and a series of needle-tipped gassing channels (H). To add substrate, a gassing channel is inserted into a tube, and the gas is exchanged through three successive vacuum and pressurization cycles, providing a 99.9% effective exchange. The pressurized tube with medium can be autoclaved without additional equipment. All transfers to the tubed medium utilize glass-disposable syringes (Becton-Dickenson, Inc., No. 7045), which are flushed free of oxygen just prior to use.

During growth, when substrate has been exhausted, fresh substrate can be added to the tube aseptically by passage of the gas through a sterile, cotton-filled glass syringe. Repressurization to 3 atm requires only a few seconds. This procedure offers numerous advantages over the classical Hungate procedure for growing methanogens, minimizing the risk of oxidation or contamination of the medium and avoiding the necessity for frequent gas additions.

*Growth Factor Assay.* A sensitive and specific bioassay for picomole amounts of HS-CoM (and certain of its derivatives) utilizes *Methanobac-*

[7] E. A. Wolin, M. J. Wolin, and R. S. Wolfe, *J. Biol. Chem.* **238**, 2882 (1963).

[8] A. Aranki and R. Freter, *Am. J. Clin. Nutr.* **25**, 1329 (1972).

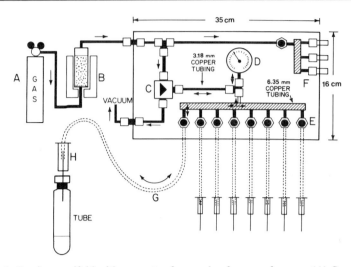

FIG. 1. Gassing manifold with apparatus for supply of oxygen-free gas. (A) Gas mixture tank; (B) reduced copper column (oxygen scrubber) with heater; (C) 3-way valve with Swagelok fittings; (D) vacuum-pressure gauge; (E) Nupro fine-metering valves; (F) alternate tubing connectors for gassing probes, which may be used in the normal Hungate procedure; (G) thin-bore polyethylene tubing; (H) Vacutainer-holder needle. Squares (□) indicate Swagelok brass fittings; 6.35-mm and 3.18-mm copper tubing are indicated by hatched and filled regions, respectively. Reprinted, with permission, from W. E. Balch and R. S. Wolfe, *Appl. Environ. Microbiol.* **32,** 781 (1976).

*terium ruminantium* strain M1; this organism requires HS-CoM as a growth factor. Strain M1 is cultivated in the general medium described above with the following modifications: the amount of $NaHCO_3$ is reduced to 2.5 g/liter; the trace mineral and vitamin solutions are omitted; isobutyric, methylbutyric, isovaleric, and valeric acids are each added at a final concentration of 0.05% (v/v); Tween-80 is added at a concentration of 0.002% (v/v). Each assay should contain in triplicate a series of tubes for an HS-CoM standard curve as follows (values in picomoles per milliliter): 3, 6, 12, 25, and 50; a control tube contains a saturating amount, 500. The biological activity of the cofactor is unaffected by normal sterilization procedures.

After sterilization, the culture tubes are inoculated with strain M1 which is maintained on a medium supplemented with 500 n$M$ HS-CoM. At an absorbance (660 nm) of 0.3 to 0.5 (48–72 hr), a 100-$\mu$l aliquot of the stock culture is transferred to 10 ml of CoM-free medium. Incubation is continued until an absorbance (660 nm) of 0.03 to 0.05 is reached (36–48 hr). To initiate an experiment, a 100-$\mu$l aliquot is transferred aseptically from the coenzyme-free medium to each serum tube in the assay. Inoculated serum tubes are maintained in a stationary position for 24 hr at 38°

followed by incubation on a reciprocating shaker (100 strokes per minute) at an angle of 10°. Growth is followed at 660 nm until maximal growth is observed.

With the bioassay, half-maximal growth-factor activity is observed at 25 n$M$ HS-CoM. A linear relationship is observed between maximal cell density (0.15 to 1.8 absorbance units at 660 nm) and a coenzyme concentration of 3 n$M$ to 50 n$M$. Use of commercially prepared HS-CoM yields identical results to our own preparations.

Results of extensive studies clearly show that the bioassay is specific for HS-CoM. Possible biosynthetic precursors and structural analogs to HS-CoM fail to replace activity. However, the HS-CoM derivatives $CH_3$-S-CoM, $CH_3CH_2$-S-CoM, $CH_3CO$-S-CoM, and (S-CoM)$_2$ show full coenzyme activity in the bioassay. Halogenated derivatives such as 2-bromoethanesulfonic acid, a precursor for the chemical synthesis of HS-CoM, strongly inhibit growth. It is essential that preparations be free of this compound for expression of activity.

*Enzymic Assay*

*Mass Culture of Cells.* A tube culture of a methanogen, such as *Methanobacterium* strain M.o.H., is transferred to a growth flask,[9] which is incubated with shaking under an atmosphere of 80% $H_2$ : 20% $CO_2$ until an optical density at 660 nm of 0.6–1.0 is attained. The contents of the flask are then added to a 12-liter fermentor of anaerobic medium previously equilibrated with the 80 : 20 gas mixture.[9] After 7 days of incubation, the fermentor is harvested.

*Preparation of Cell-Free Extracts.* The cells that have been harvested by centrifugation are suspended in one part (w/v) of 50 m$M$ N-tris(hydroxymethyl)methyl-2-aminoethanesulfonic acid (TES) buffer, pH 7.0. The slurry is then flushed with a stream of oxygen-free hydrogen until the cells are reduced; reduction is indicated by the disappearance of the fluorescence of coenzyme $F_{420}$,[10] a cofactor found in all methanogenic bacteria so far examined. When oxidized, coenzyme $F_{420}$ fluoresces blue-green under longwave ultraviolet light, but no fluorescence is observed when the coenzyme is reduced. The cells are broken by passage through a cold French pressure-cell at 16,000 psi, the effluent being collected in a stainless steel centrifuge tube under a stream of hydrogen. The centrifuge tube is then capped to maintain anaerobic conditions, and the cell debris is removed by centrifugation at 30,000 $g$ for 30 min. The supernatant solution is decanted into serum vials and gassed

---

[9] M. P. Bryant, B. C. McBride, and R. S. Wolfe, *J. Bacteriol.* **95**, 1118 (1968).
[10] P. Cheeseman, A. Toms-Wood, and R. S. Wolfe, *J. Bacteriol.* **112**, 527 (1972).

with hydrogen until reduced. Whole-cell suspensions or cell-free extracts may be stored under hydrogen at $-20°$ for several months without extensive loss of activity.

Cell-free extracts that are free of coenzyme M may be obtained by dialysis of the extracts against 50 m$M$ potassium phosphate buffer pH 7.0 at $4°$. To keep the extracts reduced, the buffer is continuously sparged with hydrogen. A final dilution factor of 1 : 10,000 in the dialyzate is recommended to be sure that background coenzyme activity is sufficiently low.

*Methyltransferase.* A methylcobalamin-coenzyme M methyltransferase in *Methanobacterium* strain M.o.H. may be used in the assay of coenzyme M.[11] It is important to realize, however, that there is no evidence that this methyl-transfer reaction is a part of the $CO_2$ reduction pathway in methanogens.

$$CH_3\text{-}B_{12} + HS\text{-}CoM \xrightarrow[\text{(P}_i)_3, \text{ATP}]{\text{anaerobic conditions, methyltransferase}} B_{12r} + CH_3\text{-}S\text{-}CoM$$

The enzymic reaction is followed in a 15 × 55 mm glass tube. Each reaction mixture contains 12.5 $\mu$mol of potassium phosphate buffer (pH 7.1); 2.5 $\mu$mol of sodium tripolyphosphate to inhibit enzymic demethylation of $CH_3$-S-CoM; 1.25 $\mu$mol of ATP; 120 nmol of HS-CoM; cell-free extract (10–400 $\mu$g); 120 nmol [methyl-$^{14}$C]methylcobalamin (specific activity, 1000 cpm/nmol) (prepared according to Wood *et al.*[12]); and $H_2O$ to a volume of 0.25 ml. All constituents except [methyl-$^{14}$C]methylcobalamin and cell-free extract are added to the reaction tube, and the tube is closed with a rubber serum stopper. The tube is flushed with oxygen-free argon (flow rate, 10 cc/min) for 10 min. The cell-free extract is then anaerobically added to each vial, and the reaction mixture is preincubated for 1 min at $40°$. The reaction is initiated by the addition of an anaerobic solution of [methyl-$^{14}$C]methylcobalamin. At appropriate times a 50-$\mu$l aliquot is anaerobically removed with a syringe and injected into a 3-ml aqueous slurry of 50% by volume Bio-Rad AG 50 W-X4 cation exchange resin, $H^+$ form. The excess [methyl-$^{14}$C]methylcobalamin binds tightly to the cation exchange resin; $^{14}CH_3$-S-CoM remains in solution. The beads are allowed to settle, then a 0.5-ml aliquot of the supernatant solution is counted in Bray's solution by liquid scintillation counting. The nanomoles of $CH_3$-S-CoM formed are calculated from the specific activity of the [methyl-$^{14}$C]methylcobalamin.

$$\frac{\text{nmoles CH}_3\text{-S-CoM formed}}{50 \ \mu\text{l aliquot}} = \frac{\text{cpm} \times 3.0 \times W}{0.5 \times \text{sp. act. [methyl-}^{14}\text{C]methylcobalamin}}$$

[11] C. D. Taylor and R. S. Wolfe, *J. Biol. Chem.* **249,** 4886 (1974).

[12] J. M. Wood, A. M. Allam, W. J. Brill, and R. S. Wolfe, *J. Biol. Chem.* **240,** 4564 (1965).

A correction factor $(W)$ must be included in the calculations in order to account for the deviations of the actual dilution volume from 3.0 ml due to the presence of the resin matrix in the slurry. This correction factor must be determined for each batch of resin and is obtained by comparing the difference in dilution of counts of $^{14}CH_3$-S-CoM when the dilution is made with $H_2O$ or with the resin bead slurry. A typical value of $W = 0.67$ has been obtained. Taylor and Wolfe find that, at a 100-fold stage of purity, ATP, $Mg^{2+}$, and $(P_i)_3$ are no longer required, and the enzyme is stable in air.[11]

The methylcobalamin-coenzyme M methyltransferase may be used as a crude assay for the detection and quantitation of HS-CoM. Reaction conditions identical to those noted above are used except that 2.2 $\mu$mol of [methyl-$^{14}C$]methylcobalamin and 600 nmol of $NaBH_4$ (to reduce chemically any oxidized coenzyme M) are added. The sample to be tested for HS-CoM should not contain more than 300 nmol of HS-CoM or 150 nmol of $(S$-$CoM)_2$. The reaction is allowed to proceed to completion (30 min). The nanomoles of $^{14}CH_3$-S-CoM formed are then determined. It is not known whether this methyltransferase is specific for HS-CoM; other sulfhydryls present in tissues may act as the methyl acceptor in the reaction.

*Methylreductase.* $CH_3$-S-CoM is reductively demethylated to yield methane by cell-free extracts of *Methanobacterium* strain M.o.H.[1]

$$CH_3\text{-S-CoM} \xrightarrow[\text{methylreductase}]{\text{anaerobic conditions, ATP, } Mg^{2+}, H_2} CH_4 + HS\text{-CoM}$$

The reaction is followed in a $15 \times 55$ mm glass tube. Each reaction mixture contains 12.5 $\mu$mol of potassium phosphate buffer pH 7.1, 1.2 $\mu$mol of magnesium sulfate, 1.3 $\mu$mol of ATP, 3.0 $\mu$mol of $CH_3$-S-CoM, cell-free extract (2–4 mg), and $H_2O$ to 0.25 ml. All constituents except the enzyme are added to the tube; the tube is sealed with a rubber serum stopper and flushed for 10 min with oxygen-free hydrogen. The reaction mixture is preincubated at 40° for 1 min; then the reaction is initiated by the injection of the anaerobic enzyme preparation. At appropriate intervals 50-$\mu$l samples of the gas phase are withdrawn from each tube and analyzed for methane by gas chromatography.[13] Rates of 222 nmol of $CH_4$ formed per hour per milligram of protein have been reported for dialyzed cell-free extracts of *Methanobacterium* strain M.o.H.[1] Gunsalus *et al.*[14] have examined several alkyl derivatives and analogs of $CH_3$-S-CoM for their activity in the methylreductase assay. Only 2-(ethylthio)ethanesulfonate ($CH_3CH_2$-S-CoM) was dealkylated at a measurable rate (20% the rate of

[13] B. C. McBride and R. S. Wolfe, *Biochemistry* **10**, 2317 (1971).
[14] R. Gunsalus, D. Eirich, J. Romesser, W. Balch, S. Shapiro, and R. S. Wolfe, *in* "Microbial Production and Utilization of Gases ($H_2$, $CH_4$, CO)" (H. G. Schlegel, G. Gottschalk, and N. Pfennig, eds.), p. 191. Erich Goltze K. G., Göttingen, 1976.

dealkylation of $CH_3$-S-CoM). Additional work by these authors indicated 2-bromoethanesulfonic acid and 2-chloroethanesulfonic acid to be potent inhibitors of the methyl reductase. Fifty percent inhibition was observed at concentrations of $10^{-6}$ $M$ and $10^{-5}$ $M$, respectively. Since 2-bromoethanesulfonate is one of the starting reagents for the synthesis of HS-CoM, particular care should be taken to remove all traces of 2-bromoethanesulfonate from the final product.

## [66] Simultaneous Determination of Vitamins B₁ and B₆ by Nuclear Magnetic Resonance Spectroscopy

### By SAAD S. M. HASSAN

The methods previously published on the analysis of the B-group vitamins in their pharmaceutical mixtures have often involved preliminary time-consuming separation procedures involving the use of thin-layer,[1] column,[2] and paper chromatography[3] followed by application of specific spectrophotometric finish. These methods, besides being tedious and cumbersome, give results within $\pm 5\%$ of the expected values.

The present investigation describes a new simple, rapid and accurate method for the determination of vitamins B₁ and B₆ either singly or simultaneously in their mixtures in the presence of vitamin B₂ without prior separation using nuclear magnetic resonance (NMR) spectrometry.[4]

Assay Method

*Reagents*

Maleic acid
Deuterium oxide, 99.7% minimum

*Apparatus*

60 MHz Perkin-Elmer (Model R 12A) and Varian T-60 NMR spectrometers

*Procedure.* Weigh accurately into the NMR tube about 20–200 mg of carefully dried powder of the vitamin sample. Using a micropipette, in-

---

[1] M. Frodyma and V. Lieu, *Anal. Chem.* **39**, 814 (1967).
[2] H. Nerlo, S. Pawlak, and W. Czarnecki, *Acta Pol. Pharm.* 26, 173 (1969).
[3] K. Paczek and H. Wardynska, *Acta Pol. Pharm.* 27, 573 (1970).
[4] S. S. M. Hassan, *J. Assoc. Off. Anal. Chem.* **61**, 111 (1978).

troduce exactly 0.5 ml of deuterium oxide solution containing 25–50 mg of maleic acid. Cap the sample tube and shake. Place the tube in a water bath at 50° for 5 min to affect complete solution of the sample. Vitamin $B_2$ (yellow color) and some other excipients in pharmaceutical tablets, if present, will not dissolve and can be left in the tube. The tube is then placed in the spectrometer, and the NMR spectrum is recorded. Adjust the phase control for a perfectly symmetrical peak shape. Switch to integrate, adjust the sweep rate, radiofrequency level, and sensitivity to give a convenient integral presentation. The integral of the peaks at δ 2.55 ppm (vitamin $B_1$), at δ 2.65 ppm (vitamin $B_6$ and/or vitamin $B_1$) and at δ 6.6 ppm (maleic acid) are recorded five times at the same sensitivity, and the average of each is taken.

For the analysis of pharmaceutical tablets, 10 tablets are weighed and powdered. An aliquot of the powder equivalents to 100 mg is accurately weighed in the NMR sample tube and the above procedure is followed.

Calculate the vitamin concentrations according to Eqs. (1)–(4).

$$\text{Vitamin } B_1 \text{ (mg)} = 1.4 A_1 C/A_3 \text{ (in the presence or absence of } B_6) \quad (1)$$
$$\text{Vitamin } B_1 \text{ (mg)} = 0.77(A_1 + A_2)C/A_3 \text{ (in the absence of } B_6) \quad (2)$$
$$\text{Vitamin } B_6 \text{ (mg)} = A_2 C/A_3 \text{ (in the absence of } B_1) \quad (3)$$
$$\text{Vitamin } B_6 \text{ (mg)} = (A_2 - 0.82 A_1)C/A_3 \text{ (in the presence of } B_1) \quad (4)$$

where $A_1$, $A_2$, and $A_3$ are the integral values at δ 2.55, 2.65, and 6.6 ppm, respectively, under the same instrumental conditions and C is the concentration of maleic acid used in each run in milligrams.

## Discussion

The NMR spectrum of vitamin $B_1$ is characterized by two singlets at δ 2.55 and 2.65 ppm, representing the protons of the two methyl groups. These two sharp signals were chosen as analytical peaks for the analysis of vitamin $B_1$ singly or in mixtures with vitamin $B_6$. For vitamin $B_6$, the three-proton singlet at δ 2.65 ppm is suitable as an analytical peak, although one of the methyl peaks of vitamin $B_1$ resonates at this position (Fig. 1). With mixtures of vitamins $B_1$ and $B_6$, calculation of the intensity of this signal can be easily made by subtracting the integral value of the peak at δ 2.55 ppm, equivalent to vitamin $B_1$ only, from the integral value of the combination peak at δ 2.65 ppm, which is equivalent to both vitamins $B_1$ and $B_6$.

The integral values of these signals are proportional to the number of protons present and the vitamin concentrations. However, with vitamin $B_1$, the intensities of the three-proton signals at δ 2.55 and 2.65 ppm are not equal. They are in the ratio of 1 : 0.82 with a relative standard devia-

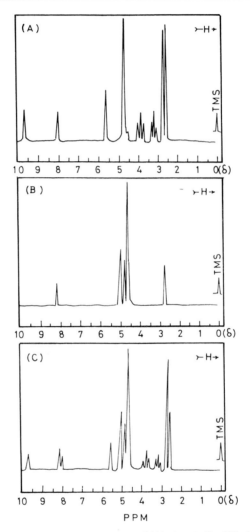

Fig. 1. Nuclear magnetic resonance spectra of (A) vitamin $B_1$, (B) vitamin $B_6$, and (C) a 1 : 1 (w/w) mixture of vitamins $B_1$ and $B_6$ in deuterium oxide as a solvent.

tion of about 1%. This is probably due to some overlapping of the integrals of both peaks and/or a variation in the relaxation time of the two environmentally different methyl groups. Careful adjustment of the phase control of the phase sensitive detector, changing the sweep rate and the intensity of the driving radiofrequency field, do not affect the aforementioned consistent and constant ratio. Similar results were obtained under sweep widths of 250, 100, 50, and 25 Hz. However, linear relationships between

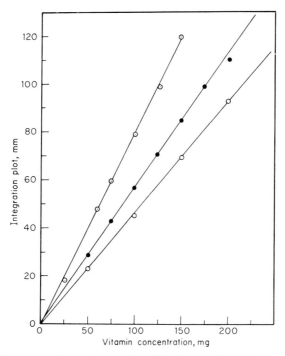

FIG. 2. Calibration graphs of vitamins $B_1$ and $B_6$. Integration of vitamin $B_6$ peak at 2.65 ppm (◎), vitamin $B_1$ peak at 2.55 ppm (●), and vitamin $B_1$ peak at 2.65 ppm (○).

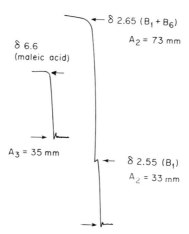

FIG. 3. Typical integration plot of a 1 : 1 (w/w) mixture of vitamins $B_1$ and $B_6$ with maleic acid as internal reference standard. Weight taken of vitamin $B_1$ = 60 mg, $B_6$ = 60 mg, and maleic acid = 45 mg. Weight found [according to Eqs. (1)–(4)] of vitamin $B_1$ = 59.4 mg and $B_6$ = 59 mg.

the vitamin concentrations and the integral values of these signals are obtained.

The relationship of the peak areas of the signals at $\delta$ 2.55 ppm (vitamin $B_1$) and at $\delta$ 2.65 ppm (vitamin $B_1$ + $B_6$) to that at $\delta$ 6.6 ppm (maleic acid) were derived, whereby the composition of the mixtures can be readily quantitated. The integration values of the signals at $\delta$ 6.6 ppm produced by 100 mg maleic acid as internal standard, at $\delta$ 2.65 ppm produced by 100 mg of vitamin $B_6$ or 175 mg of vitamin $B_1$, and at $\delta$ 2.55 ppm produced by 140 mg of vitamin $B_1$ are all equal (Fig. 2). Based on these findings, simple equations for these relationships were derived [Eqs. (1)–(4)] (Fig. 3).

The results obtained for the analysis of either vitamin $B_1$ or $B_6$ show average recoveries of 98.2% and 97.9% for vitamin concentrations in the range of 30–100 mg. Analysis of binary vitamin mixtures in the range of composition usually found in pharmaceutical preparations shows results of the same accuracy, and no effect was noticed due to vitamin $B_2$, if present. This procedure was also applied to the analysis of vitamins $B_1$ and $B_6$ in some pharmaceutical preparations. Results accurate to $\pm 2\%$ were obtained, and such active ingredients as folic acid and vitamin $B_{12}$, when present in amounts less than 1%, do not interfere. Insoluble excipients can be left in the sample tube without prior filtration.

# Author Index

Numbers in parentheses are reference numbers and indicate that an author's work is referred to although the name is not cited in the text.

## N

# Subject Index